WITHDRAWN
University of
Illinois Library
at Urbana-Champaign

The Measure of the MOON

Arizona Meteorite Crater. (Courtesy U.S. Air Force.)

The Measure of the MOON

RALPH B. BALDWIN

The University of Chicago Press

Library of Congress Catalog Card Number: 62-20025
The University of Chicago Press, Chicago & London
The University of Toronto Press, Toronto 5, Canada
© 1963 by The University of Chicago. All rights reserved. Published 1963
Composed and printed by THE UNIVERSITY OF CHICAGO PRESS
Chicago, Illinois, U.S.A.

TO MY WIFE, LOIS

WITH GRATEFUL THANKS FOR HER HELP AND PATIENCE

ACKNOWLEDGMENTS

Several valuable conferences were held with John J. Gilvarry and J. E. Hill concerning scaling and cratering problems. C. D. Shane furnished the beautiful Lick Observatory contact positive plates of the moon from which the new contour map was derived. Dean Simeon Leland of Northwestern University graciously allowed me to use the Dearborn Observatory Parallax Measuring Engine in my home for a year. Harry Rymer was of great help in the reductions of the measures by doing a portion of the work on the Dearborn Computing Center IBM 650.

C. B. Watts was of tremendous help. He computed the libration constants for the five Lick moon plates and gave freely of his time and encouragement during the measuring and reduction period prior to the completion of the contour map.

G. Schrutka-Rechtenstamm in Austria, P. Hédervári in Hungary, and G. Fielder in England corresponded frequently on various aspects of the work.

R. S. Dietz furnished several photographs and was most encouraging.

Mrs. Grace Savage and later Mrs. Marjorie S. Clopine, librarians of the United States Naval Observatory, dug out many obscure references for me.

G. P. Kuiper furnished numerous photographs and gave substantial help.

Conferences and interchanges of letters with H. C. Urey have contributed in large measure to the scope of this volume.

H. B. S. Cooke arranged for aerial photographs to be taken of the Pretoria Salt Pan.

Republican Congressman Gerald R. Ford of Michigan was able to obtain aerial photographs of the African Meteoritic Craters.

H. H. Nininger shared his knowledge of meteoritics and, in particular, the

Arizona Meteorite Crater. My discussions with him in and around this crater will long be remembered.

A visit to Ottawa, Canada, and C. S. Beals was most illuminating, and his correspondence concerning the ancient Canadian craters is deeply appreciated.

H. B. Hart showed me many courtesies on a visit to the Decaturville impact structure.

The correspondence with G. J. F. MacDonald on volcanic matters and on the heat balance of the moon and earth was helpful in formalizing my own ideas of the thermal history of the moon.

The manuscript has been read and important suggestions made by John J. Gilvarry and Robert S. Dietz.

It has been typed at least twice by Phoebe Jordan.

My debts are many. I can only say—Thank you!

R. B. B.

TABLE OF CONTENTS

LIST OF ILLUSTRATIONS xi

LIST OF TABLES xv

INTRODUCTION xvii

1. OUTLINE OF THE PROBLEM 1

2. MODERN TERRESTRIAL METEORITIC CRATERS 6

3. PROBABLE AND POSSIBLE TERRESTRIAL METEORITIC CRATERS . 52

4. ANCIENT METEORITIC CRATERS AND CRYPTOVOLCANIC STRUCTURES . 66

5. CRATER FREQUENCY ON EARTH AND MOON 106

6. CHARACTERISTICS OF EXPLOSION CRATERS, TERRESTRIAL METEORITIC CRATERS, AND LUNAR CRATERS 113

7. RELATIONSHIPS BETWEEN CRATER PARAMETERS 128

8. DETERMINATION OF ENERGIES NEEDED TO PRODUCE METEORITIC CRATERS 153

SUMMARY INSTRUCTIONS FOR USE OF EQUATIONS WHICH REASONABLY REPRESENT TERRESTRIAL METEORITIC CRATERS AND LUNAR CRATERS OF CLASS 1 186

9. VARIATIONS IN LUNAR CRATERS AS FUNCTIONS OF THEIR AGES . 188

10.	The Problem of the Moon's Motion and Shape	197
11.	The Shape of the Moon and the New Contour Map	212
12.	Nature of the Lunar-Surface Materials as Determined by Reflected Light	248
13.	Nature of the Lunar-Surface Materials as Determined by Heat Measures at Infrared and Radio Frequencies	268
14.	Tektites	287
15.	Analyses of Earlier Theories of the Moon's History	293
16.	The Circular Maria and Related Structures	314
17.	The Lava Flows	333
18.	The Atmosphere of the Moon	341
19.	The Lunar Rays	350
20.	The Central Peaks of Lunar Craters	361
21.	Rilles, Wrinkles, and Faults	371
22.	The Lunar Grid System	385
23.	Domes	390
24.	The Heat Balance of the Moon	395
25.	Magnetic Field of the Moon	412
26.	Recent Changes on the Moon	415
27.	Summary and Conclusions	420

APPENDIXES

1.	Derivation of the Relationship between the Distance of the Moon and Geologic Time	433
2.	Tables	435
3.	The Lunar Tidal Bulge as a Function of the Moon's Distance	473

Author Index	475
Subject Index	481
Lunar Map	In pocket at end of book

LIST OF ILLUSTRATIONS

PLATES

Arizona Meteorite Crater . . . . *Frontispiece*

FACING PAGE

I. New Quebec Crater	26
II. Wolf Creek Crater	26
III. Aouelloul Crater	26
IV. Crater of Talemzane	26
V. Pretoria Salt Pan	58
VI. Lake Bosumtwi in Ashanti Crater	58
VII. Cratère de Tennoumer	58
VIII. Cratère de Temimichat Gallaman	58
IX. Shatter Cones	74
X. Steinheim Basin	74
XI. The Rieskessel	74
XII. Holleford Crater	74
XIII. Brent Crater	74
XIV. Deep Bay Crater	74
XV. Wells Creek Basin	74
XVI. Sierra Madera Dome	74
XVII. Decaturville Structure	74
XVIII. Crooked Creek Structure	74
XIX. Burton-on-Trent Crater	122
XX. Mare Nectaris and Altai Scarp	314
XXI. Mare Crisium	314
XXII. Mare Humorum	314
XXIII. Mare Serenitatis and Haemus Range . . .	314
XXIV. Mare Imbrium	314

xi

xii *List of Illustrations*

FACING PAGE

XXV. PTOLEMAEUS REGION 314
XXVI. RICOCHET GROOVE 314
XXVII. ANCIENT RILLE 382
XXVIII. MUTED AREA BETWEEN UKERT AND ERATOSTHENES 382

FIGURES

1. DISTRIBUTION OF METEORITIC MATERIAL AROUND ARIZONA CRATER . . . 13
2. ODESSA METEORITIC CRATER—CROSS-SECTION 20
3. OESEL CRATER—KAALI JÄRV 32
4. OESEL CRATERS—NOS. 1, 2, 3, AND 4 AND KAALI JÄRV 33
5. THEORETICAL STRUCTURE BENEATH A TYPICAL METEORITE CRATER . . 72
6. GRAVITY MAP, HOLLEFORD CRATER 81
7. CROSS-SECTION, HOLLEFORD CRATER 82
8. CROSS-SECTION, BRENT CRATER 84
9. GRAVITY MAP, BRENT CRATER 85
10. GRAVITY MAP, DEEP BAY CRATER 87
11. GEOLOGIC MAP, WELLS CREEK BASIN 90
12. RADIO MAP OF KENTLAND STRUCTURE 97
13. SECTION ACROSS CRATER FORMED BY 40-GRAIN DYNAMITE CHARGE . . 120
14. SCHEMATIC RELATIONSHIPS BETWEEN DIMENSIONS AND DEPTHS OF EXPLOSIONS FOR TERRESTRIAL CRATERS 129
15. SCALED RADIUS AND SCALED DEPTH FOR APPARENT CRATER VERSUS SCALED DEPTH OF BURST 134
16. SCALED RIM HEIGHT FOR APPARENT CRATER VERSUS SCALED DEPTH OF BURST 135
17. R/S VERSUS $H/W^{1/3}$ 136
18. LOGARITHMIC DIAMETER (D) VERSUS LOGARITHMIC DEPTH (d) FOR APPARENT CRATER FOR FOUR SCALED DEPTHS OF BURST 138
19. LOGARITHMIC DIAMETER (D) VERSUS LOGARITHMIC DEPTH (d) IN TERRESTRIAL METEORITIC CRATER RANGE 139
20. LOGARITHMIC DIAMETER (D) VERSUS LOGARITHMIC DEPTH (d) IN LUNAR CRATER RANGE 141
21. LOGARITHMIC RIM HEIGHT (R_H) VERSUS LOGARITHMIC APPARENT CRATER DIAMETER (D) 144
22. LOGARITHMIC APPARENT RIM WIDTH (R_W) VERSUS LOGARITHMIC APPARENT CRATER DIAMETER (D) 144
23. RATIO OF TRUE CRATER DIAMETER TO APPARENT CRATER DIAMETER . . 147
24. RELATIONSHIPS BETWEEN DEPTH OF BRECCIATION FROM GROUND LEVEL AND TRUE CRATER DEPTH AND APPARENT CRATER DIAMETER . . . 149
25. IDEALIZED CROSS-SECTIONS THROUGH IMPACT CRATERS, SHOWING DISTORTIONS OF ROCK LAYERS AND ZONE OF BRECCIATION 150
26. CRATER-FORMING ENERGIES 159

List of Illustrations xiii

27. Observed Logarithmic Energy (E) versus Logarithmic Apparent Crater Diameter (D) for a Scaled Depth of Burst, $H/W^{1/3} = 0.10$ . . 161
28. Observed Logarithmic Energy (E) versus Logarithmic Apparent Crater Depth (d) for a Scaled Depth of Burst, $H/W^{1/3} = 0.10$ . . . 161
29. Illustration of Relationship between Meteoritic Size, Penetration into Ground, and Crater Size 166
30. Determination of Scaled Depth of Burst for Two Terrestrial Meteoritic Craters 181
31. Relationships between Logarithmic Apparent Crater Dimensions for Lunar Craters 191
32. Cross-Section of Lunar Crater Theophilus 192
33. Contour Map of the Moon According to Franz 217
34. Contour Map of the Moon According to Ritter 218
35. Contour Map of the Moon According to Schrutka and Hopmann . 219
36. No Title 226
37. Correlations of Individual Height Measures, Relative to a Sphere, of Lunar Features, Baldwin's Plate Pairs 2, 3 versus 3, 4 . . . 230
38. Correlations of Individual Height Measures, Relative to a Sphere, of Lunar Features, between Plate Pairs 2, 3 and 3, 4 231
39. Contour Map of the Moon 236
40. Limb-Height Measures by Watts 238
41. Hypsographic Curve of the Moon 241
42. Hypsometric Curve of the Moon 242
43. Radar Brightness of the Moon's Disk 279
44. Relationship between Diameter of Ray Pattern and Diameter of Central Crater 356
45. Correlations between Crater Diameter and Central Peak Statistics 366
46. Sketch of Ariadaeus-Hyginus Region 373
47. Major Families of Rilles 382
48. Major Families of Wrinkle Ridges 382
49. Temperature Distribution in the Moon 403
50. Development of Temperature with Time at Various Depths for a Cold Moon 404
51. Development of Temperature with Time at Various Depths for a Hot Moon 404
52. Melting Point of Diopside and Iron as a Function of Depth within the Moon 405
53. Change of Radius of a Moon Initially at 0° C, 4.5 × 10⁹ Years Ago 408
54. Change of Radius of a Moon Initially at 600° C, 4.5 × 10⁹ Years Ago 408

LIST OF TABLES

1. Critical Temperatures 34
2. Radiant and Blast Effects from Nuclear Bombs 42
3. Impact Structures of the Cryptovolcanic Type 107
4. Chemical and Nuclear Explosive Craters 436
5. Heat of Explosion for Chemical Explosives 441
6. Terrestrial Meteorite Craters 442
7. Lunar Craters 443
8. Scaled Crater Dimensions versus Scaled Depth of Burst . . . 447
9. Penetration of Centers of Meteorites for Specific Scaled Depths of Burst 140
10. Relationship between True and Apparent Crater Diameters for Explosion Pits 146
11. Investigation of Schröter's Rule 148
12. Limit of Brecciation of Rock beneath Explosion Craters . . 149
13. Crater-forming Energies (Terrestrial Craters) 157
14. Sizes of Nickel-Iron Meteorites Needed To Produce Various Sizes of Craters, Point-Source Model 165
15. Effects of Curvature of Moon's Surface on Crater Depths . . 169
16. Relationships between Crater Diameters and Meteoritic Parameters for Certain Lunar and Terrestrial Meteoritic Craters on the Assumption of a Point-Source Explosion Model 170
17. Time Needed To Stop 10-m.p.s. Meteorites Which Penetrate 2 Diameters into the Ground 175
18. Relationships between Crater Diameters and Meteoritic Parameters for Certain Lunar Craters Larger than 10 Miles in Diameter on the Assumption of a Surface-Source Explosion Model . . . 177
19. Distance of the Moon during Geologic Time 201
20. Determinations of f 203

List of Tables

21. Data for Lick Moon Plates	221
22. Fundamental Plate and Reduction Constants for Lick Moon Plates	225
23. Co-ordinates of Selected Lunar Features and Measured Heights Relative to a Sphere	450
24. The Albedos of Surfaces of Various Materials	251
25. The Albedos of Certain Lunar Formations	255
26. Comparison of Lunar Regions with Terrestrial Rocks Studied	259
27. Size-Frequency Distribution of Primary Impact Craters on Lunar Maria	296
28. Postdry Circular Maria, Prelava Craters	307
29. Cumulative Crater Counts	308
30. Possible Sizes of Planetoids Which Produced the Circular Maria	315
31. Data on Valleys Radial to Mare Imbrium	327
32. Ray Craters	354
33. Central Peak Data for Class 1 Craters	362
34. Central Peak Data for Class 2 Craters	363
35. Central Peak Data for Class 3 Craters	363
36. Central Peak Data for Class 4 Craters	364
37. Central Peak Data for Class 5 Craters	364
38. Eccentric Central Peak Data for Classes 1–5 Craters	365
39. Average Content of Uranium and Thorium in Igneous Rocks	400
40. Uranium Content of Dunites	400
41. Heat Production by Igneous Rocks	401
42. Results of Investigations of Various Models of Homogeneous Moons	405
43. Original Measures of Positions of Selected Lunar Formations on Lick Observatory Photographs	464

INTRODUCTION

This work is not simply a revision of *The Face of the Moon*. It forms a sequel, carrying the work on beyond the point that was possible twelve years ago. The earlier volume was published in 1949. Its prime contributions were the dating of many of the formations on the moon as products of the terminal phases of the building of the moon and the establishment of general correlations which demonstrated to the satisfaction of most students of the subject (1, 2)[1] that the craters and certain other formations were produced by impacts of bodies from space. Crude attempts were there made to draw quantitative, rather than just qualitative, conclusions. Usually these gave order-of-magnitude solutions.

Through 1949 there were only occasional articles written which could aid in the solution of lunar problems. Most of the new data published in that general period were by amateurs and consisted of detailed drawings or notes on specific lunar objects.

Since 1949 the trickle of valuable contributions has become a flood. At present it is almost impossible to keep up with the volume of new and thought-provoking articles and books in many languages. Harold Urey, Gerard Kuiper, and Gilbert Fielder have contributed mightily to this increase in interest in lunar problems.

Another factor has been the program to break the earth's gravitational bonds and land men and instruments on the moon. This has stimulated great public interest and has brought other scientists, such as Eugene Shoemaker, into the picture. Hopefully, this last program has added a new discipline to the students of the moon, for in no previous period have astronomers known that their work might soon be checked by on-the-spot observations.

The magnificent achievement of the Russians in photographing the back side

---

[1] Numbers in parentheses refer to the References at the end of each chapter.

of the moon will not be discussed. For our purposes, the photographs are not of sufficient resolving power to add to, or detract from, the arguments used here.

Only two things stand out as possible differences from the front face. One is the paucity of maria, but if Mare Imbrium were removed from the front, the two sides would appear much more alike. The second is the long, light marking which has been named the Soviet Range. If confirmed, the structure is a new type, but it does not seem probable that it is a mountain range.

The publications and data of the last 13 years have invalidated none of the basic conclusions reached in *The Face of the Moon*. Instead, they have been strengthened. Minor misinterpretations and errors have been pointed out by others. They will be discussed in the text.

Certain ideas are here accepted as correct. One of these is that the great majority of the craters on the moon are of meteoritic origin. The large circular maria are included in this category. There will be no attempt to prove or disprove this idea. Great efforts will be made to define and clarify the things which happen when a meteorite strikes and to apply the resulting empirical equations to terrestrial and lunar problems.

Urey's postulate that the earth and moon were formed from the coalescence of cold planetesimals is also accepted, and conclusions are drawn from this idea.

The age of the terminal portion of the moon's growth is considered to be 4,500,000,000 years.

With these three premises, plus the wealth of other observations, I have attempted to establish a logical sequence of happenings on an unambiguous time scale which will yield a history of the moon, the earth, and the earth-moon system that is consistent with most of the lunar markings and certain terrestrial formations.

Clearly, other forces than meteoritic impacts have operated on the moon. There are many evidences of heat, of cracking of the crust, of erosional forces, of out-gassing, of uplift and downthrow, of isostatic changes. There are observations which point toward varying ages for lunar features. The conclusions reached here must be in harmony with all these myriad facts.

It is beyond hope that we shall ever have a complete and definitive answer to all lunar problems. Geologists and others have been studying the earth at close range for generations, but there are many problems for which opposing solutions are still offered and others on which general agreement is not yet possible. In a real sense they have only scratched the surface. How much more difficult it is, then, to solve all the problems concerning a distant satellite. Landing on the moon and analyzing its materials will help greatly but will raise more problems than are solved.

The broad historical development of the moon and its surface features, as presented here, differs substantially from the conclusions reached by other

Introduction xix

students from less extensive data. It is believed that the many known facts about the moon have been synthesized into a consistent theory. It is hoped that this volume will encourage others to study even more thoroughly the mysteries of the nearest, strangest, and most beautiful of the heavenly bodies.

REFERENCES

1. UREY, H. C. "The Origin of the Moon's Surface Features," *Sky and Telescope,* **15,** 2, 1956.
2. GOLD, T. "The Lunar Surface," *M.N.,* **115,** 586, 1955.

NOTE

Maps.—The lunar map furnished with this book is reproduced by special permission of *Sky and Telescope* magazine, Harvard College Observatory. The original underlying map is by Karel Andel; the grid and keying were added by the Sky Publishing Corporation. The map shows most, but not all, of the features described in the text. Several other good maps have been published and are in most libraries. These include an excellent one by Goodacre in *Splendour of the Heavens,* edited by Phillips and Steavenson (London: Hutchinson & Co., 1924); a large, detailed map in *The Moon* by Wilkins and Moore (London: Faber & Faber, 1955); the I.A.U. map by Blagg (1935) is recommended.

Ideally, of course, the student should have available *The Photographic Lunar Atlas,* edited by G. P. Kuiper and published by the University of Chicago Press.

Units.—Throughout this book, the English system of units is used, because it is the one normally employed in the United States. However, scientists in general and people of certain countries are more familiar with the metric system. Therefore, all major equations have been given in both English and metric forms. Whenever another investigator is quoted or referred to in the text, the units with which he worked are the ones reported.

1
OUTLINE OF THE PROBLEM

Every investigation of the moon raises more problems than it solves. It is hanging up in space within easy naked-eye and telescope range. Uncounted numbers of observations of many types have been made, and yet even today there is no universal agreement concerning the moon's shape, its surface contours, its surface materials, the processes which have operated upon it, the nature of the various markings and their ages, its history, the nature of its inner layers, its temperature variations, and many other factors. This volume constitutes an attempt to solve certain of these problems.

With the naked eye the moon is seen to be essentially round. The moving terminator adds a third dimension, and it shows that the moon approximates a sphere. Irregularities in the terminator illustrate the local irregularities in the surface.

The naked eye clearly distinguishes two major brightness regions. These bright and dark areas remain fixed on the surface, but, even with the naked eye, one can see that the face of the moon presented to the earth is always the same one and that there is an oscillation which permits us to see a little farther around the edges at different times, so that about 59 per cent of the surface is visible.

This effect has been known for ages, but only recently has the explanation been given. The moon is not spherical, but there is a bulge aligned with the earth, and a gravitational couple acting on this bulge will force the moon to remain aligned with the earth. The height of the bulge may be determined dynamically, and it turns out to be roughly 3,000 feet.

If the moon were of uniform density and the surface were smooth, the geometrical bulge would agree with the dynamical bulge. Very rough determinations by Franz, Saunder, Schrutka-Rechtenstamm, Hopmann, Ritter, Hayn, and others all indicate that the bulges do not agree in height. This raises the possibility that all determinations of the geometry of the moon's shape are in error. This is entirely possible, inasmuch as the measurements and their reductions are exceedingly long and tedious and subject to error, because the quantities to be measured are almost at the limit of measurement.

If the dynamical determinations of the height and shape of the bulge actually do differ from the geometrical lunar bulge, we are left with two alternatives or a combination of them. There may be density variations in the lunar materials, or there may be surface irregularities which will systematically distort the surface. The dynamically determined bulge height is actually real and positive and represents a blurred average determination of the moments of inertia around each of the three axes.

The determination of the shape of the moon is of paramount importance. If the bulge proves to be a fossil tidal bulge, it will tell us that the moon was once capable of adjusting its shape to the tidal pull of the earth. It was once close to the earth; it then receded from the earth as tidal friction in the shallow seas of the earth or as bodily friction slowed the earth's rotation, and angular momentum was transferred from the earth to the moon, forcing the moon farther from the earth. Relatively early in this recession, the moon solidified to the point where it could no longer adjust its shape to the decreased tidal deforming action of the earth; and, hence, throughout all subsequent history, the moon possessed a fossil tidal bulge.

If, on the contrary, the moon does not possess a fossil tidal bulge, the observation of the existence of a dynamical bulge can be due to accidental, systematic variations of the surface, or there may be variations in density of the outer layers. If the moon had been formed cold and had always had a cool, strong, thick crust, it could never have been very close to the earth. It must have been formed at a considerable distance from its primary, or possibly it is a captured planet. Efforts will be made to clarify certain of these points.

In order to know more concerning the early history of the moon, we must derive certain approximations to its present temperature distribution. Studies of the heat balance of the moon and the thermal history have been made by several scientists, but usually from very arbitrary assumptions. It is now possible to pinpoint with a greater degree of accuracy the evolution of the moon's internal temperature from its effect on surface structure. This will lead to a closer approximation to the present temperature distribution.

The naked eye clearly distinguishes the bright and dark regions on the face. The usually accepted theory is that the bright areas are somewhat acidic rock and that the dark areas are lava, which is more basic in nature. Some authorities suggest that the lavas came from the body of the moon and were released by the great impacts which formed the circular maria. Another hypothesis is that the circular maria were produced by the impacts of great projectiles which themselves were melted by the energies of the low-velocity collisions. It has also been advocated that the lavas came from the body of the moon long after the impacts and were not directly associated with the collisions. A late-comer is the idea that the maria are not lava at all but are vast fields of dust which had been eroded from high areas and translated by certain hypothetical mechanisms to the low areas. These cannot all be true.

Coincidental with the foregoing investigations is the need to ascertain whether or not the moon ever adjusted its shape isostatically to correct for the variations in load on its surface. This brings up the matter of the strength of the lunar rock layers. We must also try to determine whether or not the moon ever became differentiated or whether the composition is uniform throughout.

Associated with the dark maria are hundreds of rilles. These are cracklike formations which are from practically zero to several miles wide. The bottoms are depressed but not excessively so. They are mostly grouped into four types: (1) those which roughly outline the dark areas; (2) those which parallel the edges of the circular maria; and (3) those which are perpendicular to the second group. Some of the rilles show *en echelon* effects. As a fourth type, many of the great cracks are marked by crater-like formations which clearly are not of meteoritic origin. These appear to differ from the usual rille, and sometimes they are found on the bright uplands. What is their significance, and where do they fit into the evolutionary picture?

There is a general agreement that the rilles represent a modified form of crack or fault; but the reason for their being, the reasons for their being where they are, and their mode of origin are still controversial. One school leans toward compressional and shear causes, another invokes tension.

At several places on the moon we find great faults, where one side has moved vertically relative to the other. The prime example is the Straight Wall. Horizontal motion does not seem to have been present.

Distributed widely on the maria are numerous great domes which resemble shield volcanoes. Usually, but not always, they have central crater pits, which differ in appearance from normal lunar craters. What is the origin of these features? How do they fit into the normal sequence of events? Are they affiliated with other more irregular forms, both on the maria and on the uplands?

Radio-wave reflection and radiation from the moon have yielded many new facts concerning the moon's outer surface. How do these facts tie into the overall picture? Can the measurements taken at different radio wave lengths be placed in a consistent setting? Similar measures have been made at other wave lengths—infrared, visual, and ultraviolet. Much can be learned from them.

Does the moon have an atmosphere? If so, what are its composition and its density? What effect does its presence or absence have on the lunar surface?

The most obvious features on the moon are the myriads of craters, large and small and old and new in appearance. While it is generally assumed here that the meteoritic nature of most of these craters has been established, we still know relatively little concerning the processes which occur when a meteorite strikes, the energy expended, the efficiency of the process, and the relationship between depth of burst and crater dimensions. Accurate relationships must be established between energy expended and crater dimensions. Why do some craters show central peak craterlets, while others do not? Are crater dimensions functions of impact energy or impact momentum? Did the lunar crater-forming meteorites come from one or more sources?

The moon presents a hodgepodge of craters of all ages and sizes, but only the visible surface can be studied. On the earth we have numerous terrestrial meteoritic craters available—some new, some old. Mostly they are smaller than their lunar counterparts. Tremendous amounts of work have been done in studying them, both on the surface and below. The terrestrial meteoritic craters are the Rosetta stone for the moon.

Correlations can be made which will aid in deciphering the coded message left on the moon. In this book the correlations will be purely empirical. In most cases it is still too early for theoretical explanations, which must await accurate statements of observed relationships.

These are the fields within which this book will be written. Its aim is to synthesize a consistent history of the moon from the time when it reached approximately its present size to the present. In order to do this, it has been necessary to collect a forest of facts, both from the literature and originally determined, on those portions of the problem just outlined briefly. Within these limited areas, an attempt has been made to be as complete as feasible, so that the conclusions drawn can rest on as firm a foundation as possible. Someone else attacking this general historical problem might well have placed the emphasis differently, but most or all of the data proved to be pertinent and necessary.

Inevitably, when the scope of an investigation is as broad as this one and the observations have been made by different people and to differing degrees of precision, some of the observations will prove to be in error to a greater or lesser extent. Some of the conclusions may be similarly erroneous, but the major thesis, as outlined in the final chapter, is in full agreement with the observations as we now know them, yet permits no inconsistencies in time scale and leads to a moon in its present condition. It differs in various ways from

Outline of the Problem 5

any theories of the history of the moon that have been published before and flatly contradicts some of these theories. It is in general accord with the ideas presented in *The Face of the Moon* in 1949 but goes into greater detail than the earlier study, goes beyond it in many ways, and in numerous fields gives quantitative results rather than primarily qualitative.

To study the moon, we shall start on the earth and remain on the earth for a considerable time. The increasing number of terrestrial meteoritic craters will first be analyzed and compared. We can study, for example, the Arizona Meteorite Crater far more thoroughly than one of the same size on the moon. The latter would scarcely be visible.

2

MODERN TERRESTRIAL METEORITIC CRATERS

When an irresistible force meets an immovable object, the only thing that can happen is a meteorite crater. The magnitudes of the energies released in collisions of objects at planetary velocities beggar the imagination. The usual laws of mechanics which suffice in the low-energy world in which we normally live show unexpected extensions. Rock-hard substances suddenly become compressed to unusual densities. Matter acts as though it were liquid, or at least extremely plastic. It may even behave as a degenerate gas (1) in parts of a large-scale explosion. Ordinarily, rock layers will fracture if suddenly bent, although they will slowly warp over millennia. In a great meteoritic impact, the rock layers of the earth writhe and roll in a vain effort to escape. Tremendous shock waves race outward from the focus, producing ring synclines and anticlines, which soon become permanent parts of the landscape. Ordinary inert matter is far more violently explosive than TNT when stopped from such speeds. Compression effects will make rock rebound like rubber. The velocity of a shock wave at the impact point is roughly proportional to the velocity with which the meteorite strikes. It often can be substantially higher than in an uncompressed medium.

An observer, on a clear and moonless night, will see about ten meteors per hour (2) under average conditions. Actually, he can see only a very small portion of the atmosphere of the earth, but it can be calculated that this small number represents more than 70,000,000 meteors bright enough to be seen by the naked eye per day over the entire earth.

Each day something of the order of 100 tons of meteoritic material strike the earth's atmosphere. One per cent of this comes from meteorites which crash through the air and land on the surface, while the rest gradually sifts down in finely divided form. The latter is in two main forms—the residual matter from tiny burned-out meteorites and micrometeorites captured by the earth. This rate of accretion sounds large but would have yielded a layer only 1 centimeter thick over the entire earth in the last 40,000,000 years.

Data gathered by the United States rockets (3) have shown that the concentration of micrometeorites near the earth is many orders of magnitude greater than that farther out. All studies indicate that there is rapid decrease in the numbers of meteorites as their masses increase. It is feasible to determine the distribution of masses up to a few pounds, but it would be unwise to put too much weight on an extrapolation that could tell how many meteorites of extreme mass would strike the earth in any given time interval. We simply do not have enough facts as yet.

Almost all the small meteoritic masses which fall to the ground daily are the remnants of larger bodies which have been so slowed up by the resistance of the air that they have reached terminal velocity. Their velocity is controlled by a balance between the acceleration due to gravity and the slowing-up effects of the air resistance. They do not possess a high velocity or a high kinetic energy. They do not explode on impact.

Under favorable conditions, rather large masses can be so reduced in velocity by the air that they reach the earth at very low speeds and come to rest on, or close to, the surface. Prime examples are the four great nickel-iron meteorites found by Perry on the Greenland ice (4). The largest, Ahnighito ("the tent"), weighs 33 tons. It and two smaller masses are on exhibition at the Hayden Planetarium in New York City, while a fourth is in Copenhagen. The largest known meteorite is called "Hoba West" and weighs about 60 tons. It lies in limestone, where it was found near Grootfontein, South-West Africa. If allowance is made for the mass of nickel-iron lost through rusting and now found as a layer of laminated iron shale around the metal, the original mass was nearly 80 tons. There are no craters at these falls.

The frequency of collisions between the earth and meteorites massive enough to penetrate the atmosphere and strike the surface with a high velocity is exceedingly low. In all recorded history, only three such impacts have been observed—two in Siberia in 1908 and 1947 and one probable case in the Pamir some centuries earlier.

But time is long. We are only now beginning to realize how pock-marked the earth is. A rather large number of small modern craters has been positively identified. Others are probable or possible meteoritic craters. Various ancient structures show that the earth has been the target of the celestial bombs throughout many geologic periods.

Geologists, in particular, have been somewhat tardy in their acceptance of meteoritic impacts as a significant factor in the history of the earth. This, of course, is only natural, as geology, geophysics, and geochemistry are tremendously complex fields which still have many unknown parts. J. D. Boon, and C. C. Albritton, Jr. (107–10), in the 1930's and R. S. Dietz (111, 112) in 1946 are exceptions to this statement. Their early pioneering should be remembered.

The subject of meteoritic study itself is very new. There are ancient records of stones falling to earth, as witness the biblical story in which God slew the enemies of Joshua by casting down great stones from heaven, but these were not accepted generally as authentic. Even as recently as the time of Thomas Jefferson, that esteemed gentleman, who was one of the most advanced thinkers of his time, stated that he would rather believe that two Yankee professors would lie than to believe that stones could fall from heaven. Inevitably, the proofs of the non-terrestrial nature of these bodies accumulated until even the general public came to accept meteors and meteorites for what they are.

When these facts became evident, searches were made for meteorites and records of their falls. Progress was slow, and it was not until the turn of the century that the Barringers (5–8) positively identified the Coon Butte crater of Arizona as having been caused by the impact and subsequent explosion of a large nickel-iron meteorite. Slowly at first but then more rapidly, an imposing number of terrestrial meteoritic craters have been discovered and identified.

To identify a crater as meteoritic, certain rules are needed. Although the list of tests has been expanded, the basic principles and criteria with which the Barringers worked are still accepted as standard today. All the more familiar forms of crater genesis—volcanic action, steam blowouts, sinks, maars, etc.—must be eliminated. Meteoritic material or derivatives from it must be discovered associated with the crater. Where such materials are lacking, modifications in the rocky layers of the earth which can be produced by extremely high pressures, such as the coesite form of SiO_2, the stishovite form of silica, or the shatter cones pointed upward which are formed as a result of the violent shock waves, or masses of rock flour, are strong evidences that a meteoritic impact formed the crater. The original shape of the crater—the contour before erosion modified it—must conform to the relationships shown in equations (7-2), (7-6), and (7-9) in Chapter 7. The presence of symmetrical or bilaterally symmetrical ring synclines and anticlines is noted at all well-observed craters of this type except, perhaps, the very smallest. A lens-shaped mass of angular breccia is often found below the floor of a meteoritic crater. A significant number of these tests must be passed without any evidence in opposition before a crater can be accepted as having been produced by meteoritic impact.

When a small meteorite hits at a high velocity, the resultant crater is purely one of explosion. Other processes, including major shock-wave phenomena, enter into the production of the tremendous pits like those we see on the moon. In all but the very smallest bodies, meteorites reaching the surface of the earth will possess far more energy than an equal mass of any known chemical explosive.

When meteoritic impacts were first considered as being able to cause craters, it was thought that the simple process of splashing of material away from a high-velocity punch was the method of forming these craters. Öpik (9) and Gifford (10, 11) were among the first to point out that such craters must be formed by explosions. They demonstrated the great energies developed in collisions of the moon with meteorites and the corollary that the size of the meteorite was quite small relative to the crater it could produce. An object moving nearly 2 miles per second possesses the same kinetic energy as is released by an explosion of an equal mass of TNT. The kinetic energy of a meteorite changes with the square of the velocity and hence goes up rapidly with increasing velocity.

Gifford drew the correct inference that even oblique impacts would lead to circular craters. Statistically, it is far more probable that the angle of impact of a meteorite with earth or moon is low rather than high. Formerly it was thought that these low-angle impacts would produce elliptical craters or grooves. These were conspicuous by their absence on earth and most parts of the moon, and this was used as an argument against the impact theory. Gifford's first paper did not stop this argument from being used, but it should have done so. Öpik's (9) 1916 article, written in Esthonian, reached similar conclusions, but its merit was not recognized.

The ground offers a tremendous resistance to penetration by shells and bombs, regardless of striking velocity. They usually hit at approximately the speed of sound, or about 1,100 feet per second. They penetrate only a very few feet and so are brought to rest in perhaps 0.01 second. The higher the velocity of impact, the greater is the rate of deceleration, and the more rapid is the rate of release of energy of motion. Even the high-velocity meteoritic masses moving more rapidly than the velocity of shock waves in the earth's crust must be brought to rest within a very small fraction of a second. Hence the three conditions necessary for the formation of an explosion crater are fulfilled. The meteorites bring with them large quantities of energy, and the energy is released rapidly, close to the surface.

In the remainder of this chapter are brief descriptions of the relatively numerous terrestrial craters which are known to be meteoritic.

The North American Meteoritic Craters

THE ARIZONA CRATER

There really is not the slightest question but that the Arizona Meteorite Crater is meteoritic in origin. Of course, every so often someone livens things up by claiming otherwise (12-14), but this merely adds a little spice to a subject which has stopped being controversial and now requires the keenest type of thorough scientific investigation. This crater is an ideal subject for study. It is large. It is accessible. It is recent enough that the original conditions can be reasonably well determined.

It would be well for the nation and well for the future research programs on the mechanism of meteoritic impacts if the whole region were made into a national monument, with control vested in some organization dedicated to protecting the crater and yet allowing it to be available to public view. "Unlike the national parks, they [the monuments] are not established specifically for scenery . . ." (15) but to preserve some historical or scientific treasure.

It is often said that this crater resembles a butte from a distance, and it is true that the rim rises to a considerable height above the plain. It was called Canyon Diablo from a nearby dry creek, or Coon Butte, but resembles no ordinary butte, as it has no capstone.

Like all true meteoritic craters, the outer rim slopes gradually up from the plain. The angle here is about 13°. The outer limits of the rim are indeterminate, but the rim proper is only about 800 feet wide. The present height at the crest is from 130 to over 200 feet and averages perhaps 180 feet.

The upper layers of the rim are loose detritus ejected from the pit. This material ranges in size from the finest dust to boulders larger than houses. It is a mixture of shattered and metamorphosed landscape and meteorite. The ejectamenta in the rim were deposited in reverse order. The layers above the explosion were peeled back and overturned, much like the opening of a flower. These upper layers are rather thin. Much of the height of the rim comes from rock compressed and distorted by the violence of the explosion and pushed outward and upward.

From the rim itself, the crater outline seems tolerably circular, but aerial photographs show clearly that the crest deviates slightly from a circle and gives an impression of squareness. Nininger (16) points out that this resemblance to a square is more apparent than real and that the crater never departs more than about 300 feet from the circle and is in most places within 100 feet. Shoemaker (17) shows that the sides of the "square" are roughly parallel to the two main sets of faults in the neighborhood. The form of the crater may have been affected by structural weaknesses in the rock.

The inner walls are steep and rocky, but several paths to the floor exist. The original crater was approximately parabolic in cross-section, but the floor has been flattened by fine material blown in from the outside and by erosion from the walls. The crater has had a central lake on at least one occasion when the water table was higher than at present. There is no visible central peak, and, because of the infill, the nature of the original crater bottom cannot be seen.

Numerous drill holes have been thrust deeply into the crater bottom and nearby areas (18). In general, they showed that there was a buried surface of solid rock, deeper in the center than near the rim and about as far below the crater bottom as the bottom is below the rim. A typical core record from hole No. 12 is given in the accompanying table (19). This drill hole was about 350 feet east-southeast of center.

	Feet
1. Surface soil, blown sand, etc.	0– 30
2. Lake-bed deposits.	30– 90
3. Sand (rock flour), sandstone, in part metamorphosed.	90–630
4. Rock, at first soft and shattered but becoming gradually harder as greater depths were reached.	630–830

Shoemaker (17), of the U.S. Geological Survey, recently completed a detailed analysis of the Arizona Meteorite Crater and gives a chart showing the contour to undisturbed rock. He found a maximum depth to the undisturbed rock of 1,100 feet from the original surface.

There is a considerable divergence of opinion as to the exact size of the crater. Merrill (18) lists original measures as giving a diameter of 3,808 feet along an east-west line and 3,654 feet along a north-south line, but he quotes Lombard (20) as finding the major diameter as 3,950 feet and the lesser as 3,850 feet. Barringer (5) stated that the crater rim was, "roughly speaking, about 4,000 feet or something over three-quarters of a mile in diameter," with an average depth of about 570 feet. Fisher (22) indicated that this hole was about 4,200 feet from rim to rim, but Nininger later modified this to 4,000 feet (23). Shoemaker (17) lists the original diameter as 3,850 feet, and, from measures on his cross-sectional chart, the present apparent diameter is shown as 3,925 feet. Probably a present diameter of 4,000 feet and an original diameter of very slightly less are sufficiently accurate for all practical purposes.

Shoemaker shows an original depth of 730 feet, with a rim height of 250 feet and a true crater depth of 480 feet. The present crater is about 570 feet deep, and the rim varies from perhaps 130 to just over 200 feet in height. Because the detritus ejected from the crater still lies on the rim, although some has certainly been eroded away, I prefer a value of 180 feet for the original average rim height and 680 feet for the original apparent crater depth. This leads to an original true crater depth of 500 feet.

The crater exhibits a definitely bilaterally symmetrical shape. This is interpreted to mean that the impact velocity was low and hence the mass was relatively large. The effects of the transfer of momentum from the meteorite to the rock layers is clearly visible in addition to the greater effect of the explosive energy. The meteorite came from slightly west of north and struck at an angle of perhaps 45° or more from the vertical at a velocity estimated to be 10 miles per second or less.

Rinehart (24) suggests that the object struck from a direction slightly south of west. He bases this on the lack of symmetry of the meteoritic debris—the metallic spheroids—found in his tests and because there are two patches of boulders on the rim, one slightly north of east and the other opposite to it. There is a better explanation of the asymmetrical distribution of the spheroids, and the two boulder fields are symmetrical with the usually assumed approach from slightly west of north.

Shoemaker (17) also questions the approach from the north theory and suggests that the zenith angle of impact was small; but the small zenith angle of approach does not explain the observed bilateral symmetry of the crater wall rock layers, the bilateral symmetry of the distribution of spalled-off meteoritic fragments and stripped-off meteoritic fragments as described below, or the bilateral symmetry in distribution of ejected rim materials as shown in Shoemaker's Figure 3. All these are closely symmetrical around an axis tilted very slightly from a north-south line. The present state of knowledge suggests strongly that the meteorite struck at a low velocity at a moderate zenith angle and that it came from slightly west of north.

When the object struck, it penetrated below ground and then exploded. At the instant of impact, a shock wave was sent charging through the meteorite. When such a shock wave reached the rear surface, there was a reflection of a tensional or rarefaction wave from the surface, which should have acted, in principle, like cracking a whip. Material could well have been spalled off, and, since these fragments would have come from the rear, they would have been scattered at lower than impact velocity. Such spalled-off fragments should not show great heating effects.

Figure 1 is redrawn from Nininger (23). The small dark circle is the crater. Within a radius of 6 miles, many thousands of meteorites have been recovered. The regions within which the meteorites of different types are found are outlined on the chart.

Region A is symmetrical relative to the supposed meteoritic path and to the bilateral symmetry of the crater. Within this area, thousands of irons, ranging from a few ounces to more than 1,000 pounds, have been recovered. None of these showed heat alterations, except those in Area C. Much oxidized meteoritic material of all sizes up to more than 100 pounds is present in Area A.

Region B is also bilaterally symmetrical along the same axis. It contains thousands of irons of a few grams in weight, most of which show heat alterations. Nininger suggests that they were stripped from the main mass as it penetrated the ground.

Region C is the area of heat-altered irons of 1 ounce to about 10 pounds.

Region D is the area of metallic spheroids.

Areas C and D show a strong preference for the northeast area near the crater.

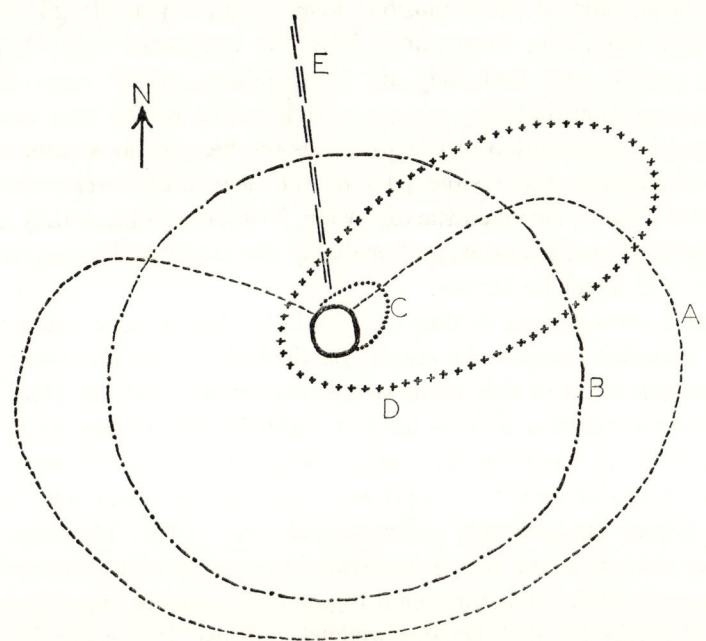

FIG. 1.—Distribution of meteoritic material around Arizona Crater. *A:* area within which thousands of irons of a few ounces to more than 1,000 pounds were recovered, *none of which show heat alteration,* except those within Area C. Also, meteoritic oxide is present in fragments of all sizes up to more than 100 pounds. *B:* area within which thousands of irons of a few grams weight were recovered. Mostly showing heat alterations and believed to be rock-penetration strippings. *C:* area of heat-altered irons of 1 ounce to about 10 pounds. *D:* area of metallic spheroids. *E:* supposed course of colliding mass.

The fragments in Region A are those which did not go through the explosion. It has been suggested (26), to account for the variations in the meteoritic structures, that two dissimilar objects struck in the same hole at the same time. This is still a possibility, but it is not a necessity. The meteorite was large enough that it need not have been homogeneous.

The spalled-off fragments would be expected to have the sizes, structures, and distribution of the observed objects, if the meteorite came from a relatively low angle from the approximate north. The shock wave, when it reached the back side of the meteorite, i.e., the side opposite the impact point, would spray off fragments. These lesser objects would possess a considerable range of velocities and directions of motion. They would also be affected by the horizontal component of the striking velocity of the main body. Those fragments that were projected oppositely to the motion of the original body had to overcome its motion. Clearly, the spalling velocity was low, for the unaltered irons are rare or absent north of the crater.

The small fragments in Region B, most of which show effects of heat, did go through at least part of the explosion. Their slight asymmetrical distribution indicates that they, too, were affected by the horizontal component of the meteorite's motion and that they are not products of the main detonation. Nininger's suggestion that they are pieces stripped off by the rock layers as the missile began its penetration of the ground is the best so far advanced.

Region C contains the visible part of the heat-altered fragments, which, through some chance, survived the explosion. This zone is intimately associated with the ejecta from the crater, and probably the true distribution of this type of fragment will never be known.

Barringer's surveys and drillings up to 1928 showed that there was some meteoritic material beneath the crater, particularly under the south rim. The amount and condition of this material are not known, but the 1931 report of International Geophysics written by J. J. Jakosky and presented to the New York meeting of the Institute of Mining and Metallurgical Engineers suggested that similar material under the southwest floor of the crater was "not in the form of a sphere but probably a fragmental zone" (27). The original churn drill hole on the south rim, which penetrated to a depth of 1,376 feet before its abandonment in 1923, went through a region of increasing amounts of meteoritic material before the drill became wedged and lost. It is generally conceded that no large solid mass exists below the crater or rim. This conclusion bears out a prediction made by Moulton in his amazingly modern study of the crater issued in 1929 (28). There may be some smaller masses mixed with the breccia under the crater floor. These objects would have been spalled off the rear of the meteorite on impact at such low velocity that they could not escape the crater.

Region D is in many ways the most interesting. It marks the area within which tiny metallic spheroids are found. Rinehart (24) has made an excellent and very thorough study of the distribution of these objects, and a detailed map is given in his article. He covered systematically an area of 8 by 11 miles centered on the crater. His results show that Nininger's earlier map is approximately correct.

Nininger's interpretation of the spheroids, slightly modified, is as follows. The meteorite struck and buried itself before exploding. The object exploded with essentially spherical symmetry. This effort dug the crater, pushed its walls

Modern Terrestrial Meteoritic Craters

up and out, and scattered much debris. Shock waves produced rock flour from the sandstone layers, ejected much of it, and left some in the crater.

The main mass of the meteorite was vaporized, and the tremendous amount of released hot gases ascended violently into the sky, forming a great mushroom cloud. The oxygen of the air was banished from the inside of this column. The gaseous metals condensed into spheroids, which then dropped out of the cloud. The larger the spheroid, the more quickly it dropped, and the closer to the crater it landed. At the time of the explosion, there was a wind from the southwest. The fallout is thus found mainly in an oval region extending miles to the northeast.

Nininger found that the percentages of nickel and cobalt in the spheroids are substantially greater than in the irons spalled off at collision. He stated that, in an area of only 2 square miles adjacent to the crater in the northeast, there were spheroids of the order of 2,000–3,000 tons in sizes of 0.1–0.2 millimeters in diameter. More than half this weight is in the form of bright, unoxidized metal (within a covering of oxide). These 2,000–3,000 tons contain an amount of nickel equal to 5,000–7,000 tons of the original meteoritic material and enough cobalt to equal 10,000–15,000 tons of the original meteorite. In addition to these tonnages, there is at least an equal amount of a less concentrated deposit spread over an area within a 2-mile radius of the crater's center. Rinehart's sampling indicated a total amount of meteoritic debris around the crater of 12,000 short tons.

It must not be thought that the droplets represent a major part of the original body. Far from it. When the meteorite was vaporized, the metallic cloud shot upward. Only the central regions were protected from oxygen, and so the metallic spheroids were formed in that region only. The outer layers of the cloud were mixed, and the mixture of heat and vaporized metal and oxygen formed an ideal combination. A large part of the metal was oxidized and, in the form of finely divided oxide, was blown great distances by the wind. Nininger found many little spherules of oxide in an outer zone beyond where the metallic spheroids were plentiful. Nowhere have they been found in such abundance as the metallic spheroids.

The total amount of meteoritic material observed and inferred is larger than the minimum masses deduced for the meteorite, around 12,000 tons. It is very much less than the highest estimates of several million tons. It is less than, but reasonably consistent with, the intermediate sizes. A value of 288,000 tons at a striking velocity of 10 miles per second will be derived in Chapter 8.

The concentrations of the metallic spheroids and the oxides near the crater preclude their having been formed as strippings from the object as it passed through the air. Such materials stripped from a nickel-iron meteorite should be completely oxidized before reaching the ground. Such iron oxide spherules are known from widely separated locations.

Several other associated materials have been found near or in the crater. Small amounts of lechatelierite were found by Barringer and Tilghman in the pit. It is a spongelike material, almost pure white and containing no nickel-iron spheroids. It has been identified as having been produced by great heat in connection with steam from the Coconino sandstone. It is surprising that so little has been found.

Nininger has also identified large numbers of bombs and bomblets of silica glass, now called "impactite." They are from the melted sandstone and limestone layers. The sandstone came from both the Moenkopi and the Coconino formations. Many of the bombs were spherical, others irregular and highly vesicular. Essentially, all contained metallic bits which gave a positive test for nickel. Many contained fragments of local rocks. They were of all sizes, with the smallest objects being most numerous. They represent the ground materials in the fringe zone of the explosion, where the materials were melted, but not vaporized, before ejection. Similar fusion of materials is noted at atomic bomb explosion sites at Yucca Flats, Nevada.

Nininger has identified "sluglets," which apparently are irregular masses comparable to the metallic spheroids but fewer in number. They seem to be the tiniest of the observable heat-altered, but not vaporized, fragments of the meteorite. Their chemical composition is normal and not enriched in nickel and cobalt.

Diamonds in the meteorites from this crater have long been known. Koenig (29) identified them in 1891. Nininger studied them in wholesale lots. He found that the carbonado inclusions may be expected in these meteorites with a frequency of about 1 in 32 square inches of sectioned surface. They range from about 1/20 millimeter in diameter to 2 millimeters. This is equivalent to finding six diamonds in every pound of meteorite. Smaller than 1/20 millimeter, they are likely to be dragged out of their matrix by the polishing process without being noticed. The key point is that, in spite of a search, Nininger states that "in our experience, no diamonds have been found in any irons that did not show evidence of heat above 760° C" (30). It must be remembered that the larger meteoritic fragments which showed no evidence of heat were spalled off from the body before the major shock of impact and explosion could deform them. Nininger drew the logical inference that the diamonds did not exist before the impact and were formed in those fragments which were badly shocked and heated but not vaporized.

Recently, Anders, of the Enrico Fermi Institute for Nuclear Studies, University of Chicago, has reached the same conclusions (31). The pressure needed to form such diamonds is over 30,000 atmospheres, or about 500,000 pounds per square inch. This is far less than that resulting from the impact of a large meteorite. Kennedy (104) reports that it requires this pressure and a minimum temperature of 1400° C.

Urey (32) has felt that these diamonds were formed in bodies as large as the moon, but it does not seem necessary to call upon this mechanism here. Diamonds may well be formed by the process that Urey suggests, but the absence of diamonds in the unaltered Canyon Diablo meteorites is strong evidence against that body's coming from a parent object as large as the moon. In 1961 DeCarli and Jamieson (33) produced small diamonds in samples of graphite exposed to an explosive shock of 300,000-atmospheres estimated intensity. Diamonds have, however, been found in several other meteorites which did not come from crater-forming impacts unless the impacts occurred between such bodies in outer space.

A powerful new tool has recently been placed in the hands of those searching for evidences of meteoritic impact. In 1953 a new mineral—coesite—was produced in the laboratory (34). It is a high-pressure polymorph of SiO_2, quartz. This material is about 16 per cent denser than quartz. It has been searched for, unsuccessfully, as a naturally occurring mineral since its identification.

Nininger predicted in 1956 that it would be found in the Arizona Meteorite Crater converted from the Coconino sandstone (35). Recently, scientists of the U.S. Geological Survey have found it to be abundant in the sheared sandstone of this crater (36, 37). Identification was made from its X-ray powder diffraction pattern.

The relatively undamaged Coconino sandstone in the walls of the crater is a white, fine-grained saccharoidal cross-bedded quartzose sandstone. Coesite occurs chiefly in compressed and sheared Coconino sandstone, which constitutes a major part of the layer of mixed debris under the crater floor and is dispersed in the underlying breccia lens. Coesite-bearing sandstone fragments are also a major constituent of drill cuttings from near the base of the breccia lens, 600 to 650 feet beneath the crater floor. Some coesite-bearing fragments of sandstone are also found in Pleistocene and Recent alluvium on the rim of the crater, mainly in association with sintered rocks. Coesite also is a subordinate constituent of sandstone that has largely been converted to glass (lechatelierite). The glassy fragments form large frothy chunks in the base of the Pleistocene lake beds in the crater floor, are also found as lapilli and bombs incorporated in the alluvium on the crater rim and are dispersed as finer fragments in the mixed debris and breccia under the crater. In some samples from this crater, fine-grained coesite had previously been thought to be glass or partially devitrified glass by Merrill [38], Rogers [39], and Shoemaker [17]. Coesite occurs in the fine-grained, nearly isotropic, matrix in which the subrounded fractured quartz grains are imbedded [37].

The occurrence of coesite at Meteor Crater has significant implications for the fields of both geology and physics. First, it demonstrates that the polymorphic transformation from quartz to coesite may occur under shocks generated by meteorite impact. It is too early at this stage to say what the pressure and temperature conditions were when coesite was formed by impact at the Meteor Crater. The presence of coesite indicates pressures in excess of 20 kilobars. The additional presence of silica glass may indicate temperatures, at least locally, of about 1000° C or higher. DeCarli and Jamieson [40] failed to find coesite in single quartz crystals shock-loaded to pressures up to 800 kilobars, and one of us (E.M.S.) has failed to find coesite in quartose media shocked to similar high pressures by experimental hypervelocity impact and by nuclear explosion. These results suggest that the transformation is too sluggish to take place in shock waves of very short duration, and that the sluggish quartz-coesite transformation may occur some distance behind the shock front in a shock wave of much longer duration, such as was probably produced by impact at Meteor Crater [37].

The occurrence of coesite at Meteor Crater suggests that the presence of coesite may afford a criterion for the recognition of other impact craters on the earth and perhaps ultimately on the moon and other planets. According to the data of Boyd and England (41), coesite probably cannot form at pressures of less than about 20 kilobars—a pressure not likely to be reached near the surface of a planet for a long enough period of time for coesite formation except by the mechanism of impact. However, coesite must persist in the low-pressure regime for a significant period of geologic time if it is to be a useful tool in the recognition of ancient geologic structures. The discovery of coesite traces (105) at the Holleford Crater shows that coesite is stable over long periods of time.

Recently Chao, of the U.S. Geological Survey, found a small shatter-cone fragment of sandstone in the fallout debris on the southern rim (42). This is an important confirmation of this test for a meteoritic crater.

More recently a team of Russian geochemists headed by S. M. Stishov has identified in the laboratory a new form of silica, now called "stishovite." It is another high-pressure polymorph of SiO_2, 46 per cent denser than coesite. Stishovite requires pressures of 160,000 atmospheres for its creation, more than twice that required to manufacture diamonds. Chao has identified stishovite in the shocked sandstone of the Arizona Crater.

THE ODESSA CRATERS

Many thousands of years ago, a nickel-iron meteorite plunged through the earth's atmosphere and struck, at a speed of possibly 10 miles per second, what would some day be southwestern Texas near the town of Odessa. It weighed 315 tons.[1] Accompanying the main body were at least four smaller companions. They also struck, exploded, or partially exploded and formed lesser craters. This all occurred well back in the Pleistocene period, when the American con-

---

[1] These approximate values result from calculations given in Chap. 8.

tinents were devoid of man. The C^{14} content of the meteorites indicates an age of the fall of at least 11,000 years (21).

The rains came, and the winds blew. The craters and their rims were gradually filled up and worn down. While the main crater was still a prominent feature, a primitive horse entered the crater. Possibly it was old, or perhaps it was trapped there by same predator. In any event, the horse died. Its bones were soon covered by silt and sand and were discovered only when the Bureau of Economic Geology of the University of Texas undertook the excavation of the crater (43, 44). This amazing discovery tells us that the crater is not recent but that it was formed many thousands of years ago, before this type of horse became extinct.

The main crater was discovered in 1921. It is in a region of horizontal, light-gray limestone strata. Monnig and Brown (45) produced a contour map of the crater and found a diameter of 550 feet, with a greatest depth of 14 feet. They suggest that erosion is the cause of the present irregular form. The rim averages 7 feet high now, occasionally rising to 12 feet, and is marked by numerous rock fragments. The rock strata, where they are exposed in outcrops on the inner slope, dip 20°–30° radially outward. They are buff-colored, for they contain numerous minute particles of iron oxide.

Nininger has gone over the crater and the surrounding area with a magnetic plow and collected 1,500 metallic fragments ranging up to 8 pounds. Magnetic measurements by Sellards (43) indicated a rather large mass under the crater, 164 feet from the surface; but when it was reached by shaft, it was found to be a very hard, firmly cemented sandstone, quartzitic in nature.

Sellards and Evans (44) report that the crater bottom originally was 81.5 feet below the surrounding plain and that the base of the rock flour formed in the pit lay 24.5 feet farther down. It is clear that the explosion did not clean out the pit; i.e., much fragmental and pulverized material stayed in or fell back into the crater. This phenomenon is well known in shell craters, on the one hand, and in the Arizona Crater, on the other. The rim must have risen about 25 feet.

The vertical cross-section of the Odessa Crater, as given in Figure 2, shows that the tangential thrust of the explosion compressed the rock layers until they buckled, producing a ring anticline surrounding the pit under the upraised rim. It exhibits bilateral symmetry.

The dip in the rock strata flattens rapidly outward from the crater, as shown in test wells located on the north and south rims of the crater. In the north crater rim, the limestone stratum flattens to almost its normal position within the short horizontal distance of 110 feet. In the south rim of the crater, the strata flatten somewhat less rapidly.

Trenches which have been dug through the postmeteor fill and into the rock walls of the crater reveal complex structural features in the Cretaceous strata. In some places the explosion of the meteorite so shattered, compressed, and distorted the beds that it is now impossible to determine accurately their original thickness or to interpret the structural attitude of the beds. In other places, particularly in trenches on the south and east sides of the crater, the individual beds can be distinguished and their structure interpreted. In a trench at the south side, nearly all the beds shown in the section of this region have been exposed. The lower portion of this section, including the resistant fossiliferous limestone, is folded into an asymmetrical anticline. The upper part of the anticline is somewhat overturned away from the crater and is wedged against and into the steeply dipping higher beds. In addition to being folded, these beds are

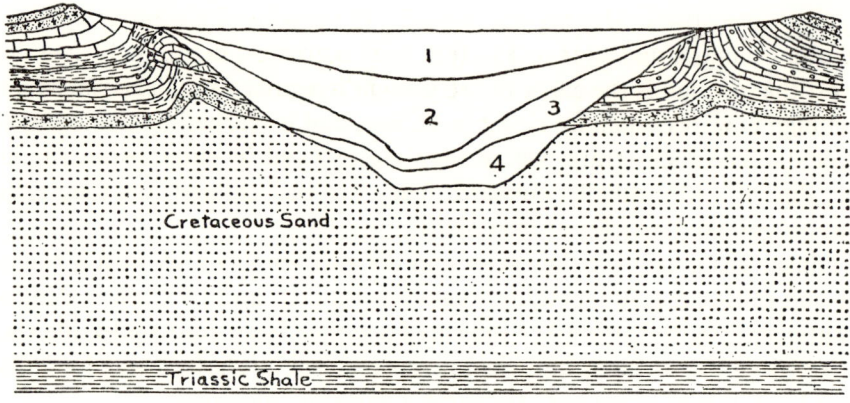

FIG. 2.—Section across Odessa meteorite crater (redrawn from Ref. 43). *1:* latest silt and sand; *2:* older silt, sand, caliche, and pebbles; *3:* fragmental rock; *4:* rock flour.

thrust-faulted. One small thrust originates in the broken, overturned portion of the anticline and extends outward through the overlying beds. The other thrust fault, which cuts the same beds, originates at the base of the anticline. The strata where cut by the faults stand nearly vertical, while the plane of the faults is nearly horizontal. The beds making up the craterward flank of the anticline have a relatively low but constant angle of inclination for about 25 feet from the crest of the anticline, where they terminate abruptly.

Crater No. 2 lies nearby. It has been rather completely excavated and was found to have been originally 70 feet in diameter and 17 feet deep. The pit does not go down to bedrock but is entirely surrounded by incoherent material. A maximum of 6 feet of fragmental rock fell back into this crater, probably from No. 1.

The sectioning of this pit has proved to be of prime importance. Approximately 1,000 small meteorites have been recovered from the layers of material immediately underneath the crater. Many more still remain buried. Estimates of a total weight of 6 tons of meteoritic material have been made, but this seems rather high. It has been suggested (43) that No. 2 was formed by the combined energies of a swarm of small masses; but this seems improbable because ballistics of small masses are far different from those of larger size. If the meteorites which produced the five known craters differed greatly in mass, they would not have landed so closely together. It seems more probable that crater No. 2 was formed from the impact of a single body, which then exploded, driving numerous smaller fragment masses from a few inches to 3 feet or more into the bottom and sides of the crater. The pit thus represents a transition type in which much of the meteoritic material remains in the pit.

Three other craters, much like No. 2 but smaller, have also been identified. Magnetometer surveys show other magnetic highs, so there may be other small craters near by.

Crater No. 3 has been partially excavated. It is smaller than No. 2 but also was found to contain numerous meteoritic fragments. At Station H-8, 30–35 meteorites were removed from a circular area about 10 feet in diameter. They were 5–9 feet below the present ground level. The total weight was 600 pounds.

The Odessa group was produced by the collision between the earth and a small cluster of meteoritic masses. As will be seen, many of the other recently discovered meteoritic craters occur in bunches. Hence it may confidently be inferred that numerous small swarms of metallic bodies held together by mutual gravitational forces travel together through space. (It is a crying shame that the county of Ector or the state of Texas has not done something to preserve this ancient group of craters. They are now overrun by an oilfield and are being virtually destroyed by dumping and digging.)

These small masses travel in their orbits around the earth or sun like miniatures of a planet with its satellites. Usually there is one rather large crater and numerous smaller pits. Let us see what kind of orbits such bodies would possess. It is assumed here that a period of revolution of 1 month for the small body around the large meteorite would be near to the maximum. The meteoritic mass has been arbitrarily chosen to be small.

Kepler's third law may be written:

$$\frac{P}{P_1} = \left(\frac{a}{a_1}\right)^{3/2} \sqrt{\frac{M_1}{M_0}},$$

where

$P = 1$ year,
$P_1 = 1$ month = the period of the meteorites around each other,
$a = $ the semimajor axis of the earth's orbit, or 1.495×10^{13} cm,
$a_1 = $ the semimajor axis of the meteoritic orbit, the quantity to be found,
$M_1 = $ the meteoritic combined mass $= 21.9$ tons $= 1.987 \times 10^7$ gm,
$M_0 = $ the sun's mass $= 1.987 \times 10^{33}$ gm.

The semimajor axis of the meteoritic orbit is found to be 6,145 cm, or 202 feet, when the period is 1 month. If the semimajor axis is assumed to be 1.495×10^5 cm, or about 1 mile, the period is 10 years.

If the period is 1 month and the combined masses are one hundred times as large, 1.987×10^9 gm, or 2,190 tons, the semimajor axis of the orbit will be 2.852×10^4 cm, or 936 feet.

It does not seem possible that meteoritic masses of the order of a few tons, or even a few hundred tons, can be separated by more than a few hundred feet if the orbits are to remain stable. These distances are approximately those found separating multiple meteoritic craters.

The magnitude of the gravitational forces on meteoritic pairs due to the sun at a distance of 1 A.U. is tremendously greater than the attraction between the meteorites, but the disruptive force is much smaller. A satellite orbit is stable, regardless of the absolute magnitude of the sun's pull, until such a point is reached that the difference in the sun's attraction for the satellite, when the satellite is at the near and far sides of its orbit, is greater than the attraction between the satellite and its primary. In the case just described, this solar disruptive attraction is many orders of magnitude too small at 1 A.U. to cause the meteoritic system to disintegrate. Unless the multiple meteoritic system passes very close to the sun or a major planet, the satellite orbits will be stable.

THE NEW QUEBEC CRATER
(Longitude 73°40′ W., Latitude 61°17′ N.)

In northern Quebec lies a tremendous crater generally considered, but not absolutely proved, to be meteoritic in origin. Its form is correct, but, so far, no meteoritic material has been recovered. Suggestions have been made that this object was formed by the impact of a great non-metallic meteorite. Leonard (46) and LaPaz (47) report that, at the insistence of the Quebec government, the crater will have the name "New Quebec Crater," inasmuch as the Ungava Peninsula has been renamed "New Quebec." This name replaces the names "Chubb" and "Ungava," which have been in common usage. Leonard and LaPaz alone seriously question the meteoritic origin of this crater.

The Royal Canadian Air Force in 1946 took a photograph which showed the crater clearly. In February, 1950, F. W. Chubb called the attention of V. B. Meen, director of the Royal Ontario Museum of Geology and Mineralogy, to the formation. Later, Meen found that a U.S. Air Force plane had first photographed it during the early part of World War II while on a routine weather flight. Meen felt that the meteoritic origin of the crater was the most likely, but Chubb hoped for a volcanic origin, remembering the Kimberly diamond mines in the extinct volcano in South Africa.

The combined interests of these two men led to a privately underwritten (by anonymous donors) expedition in July, 1950. They reported (48) that the great broad rim rose out of a central lake at an angle of 45°, except for a lesser slope at a small part of the east side. The outer slope varied considerably but averaged about 25° for the first quarter of a mile. Farther out, the slope is more gradual and, in general, is a talus slope made up of fragments thrown from the crater. Millman (49) later reduced these values to 31° for the inner walls and 8° average for 1,400 feet—then roughly 0°.7.

Meen states: "At first approach, the rim appeared to be a jumble of broken blocks of granite dropped at random into their present positions. On more thorough examination, I became convinced that the rim represents a mass of granite bed-rock which has been fractured by a tremendous explosion and has been lifted bodily to its present position" (50). The rock layers dip radially outward at about 15°. Numerous rifts are evident, partially cutting through the walls.

The water in the lake is about 50 feet higher in level than that in Museum Lake, about 2 miles away. This additional head of water affords evidence that the discrete nature of the top layers of the crater rim does not extend all the way down to the lake level. The lower part of the rim is certainly solid rock, although undoubtedly it was twisted and tortured in the violent explosion that produced this crater.

At roughly ⅗ mile from the crest, or 1.7 miles from the center, and again at 1 mile from the north rim, or about 2.1 miles from the center, there are ridges of jointed bedrock which rise out of the general level of fragments. These ridges are about 35–50 feet high and show joints pointed toward the crater and at right angles to this direction. The third joint plane is steeper than that shown on the crater rim. Similar ridges could be seen outside the crater at many points. Meen on his first trip did not establish that these ridges were continuous around the crater but stated that their positions seemed to indicate a series of wrinkles or ripples, either continuous or overlapping, which surrounded the crater. Figure 1 in Meen's article shows snow patches which seem to define these rings.

Harrison (51) reported that the ripples which were found on the north did not go all the way around the crater. Those on the west, if present, remain as isolated lumps which are probably glacial remnants.

Meen (52) then revisited the crater and confirmed the existence of the ridges of granite bedrock circling the crater. They are not continuous but are concentric and are particularly noticeable on the north flank. The outer ring shows less strong jointing.

Something of the nature of one or more concentric rings or partial rings of raised rock seems to be present surrounding the crater. It is tempting to think of these as frozen shock waves, possibly similar to, but much smaller than, the

Altai Mountain Range concentric with Mare Nectaris on the moon. This ring is also incomplete.

The crater was originally reported by Meen as 10,000 feet in diameter, but Millman (49) feels that 11,290 feet wide, 1,185 feet deep, and a lip height of 330 feet best represent the structure at present. The crater is very slightly out of round.

Portions of the rim now extend well above 500 feet. The glaciers have certainly gouged some valleys through the rim and probably have rounded off the crest, which is of far smoother appearance than other terrestrial meteoritic craters. A figure of 500 feet probably is reasonably close to the original average rim height. This would make the original depth at least 1,355 feet.

Meen's preliminary report was that the rifts in the crater wall were radial and that the crater was postglacial. Harrison showed that the rifts exhibited a preferential direction parallel to the known ice movement in the area and that the crater had once been under glacial ice. Its age is thus greater than the 3,000–15,000 years assigned by Meen. Meen's later revaluation of this problem (52) agrees with Harrison, but he believes that the rifts and tongues of material aligned with the rifts date from the explosion and have been modified only by the ice.

LaPaz and Leonard (46, 47) believe that an ice lake filled the crater during the ice age and that the glacier slid over the full and frozen pond without seriously distorting the walls. Meen and Harrison agree with this. Shoemaker visited the New Quebec Crater in the summer of 1961 and mapped it. Although little new material was gathered, he agrees that it certainly is meteoritic (106).

THE HAVILAND CRATER

A small buffalo wallow near Haviland, Kansas, was excavated in 1933 by Nininger (53–55), that tireless searcher after craters, meteorites, and meteoritic information. It turned out to be a real meteoritic crater, 10 feet deep and of oval shape, 56 feet long, and 36 feet wide. The long dimension lies west-northwest to east-southeast. The fact that the crater is oval and not round is evidence that the impact velocity was low and the angle of fall low. The meteorite fragmented, driving flinders into the crater walls and bottom. Nininger found within the crater several meteorites weighing up to 125 pounds, along with hundreds of small, partly oxidized meteorites. This object is a structure comparable with the Odessa No. 2 Crater. Nininger also found nickeliferous pellets ranging from 1 to 2 millimeters in diameter, much like those found at the Arizona Crater. A model of this crater was a feature of the unusual and interesting Meteorite Museum formerly maintained at Sedona, Arizona, by Dr. Nininger.

The Argentine Craters

At Campo del Cielo in the Gran Chaco of Argentine is a large aggregation of small meteoritic craters. It is an unusual group, and Watson (56) predicts that it may prove to be the world's largest. The craters so far recognized range in diameter from 20 to 254 feet. They are so old that they have been filled in with debris until they appear as round, shallow depressions, with the rim of the largest having been eroded so that it rises only 4 feet above the surrounding pampa. There is a lake in the pit, Laguna Negra.

On the surface are many pieces of large and small meteoritic iron, the largest weighing about 1,400 pounds. The irons have been known since the discovery of the craters in 1576, although the meteoritic nature of the craters was not recognized until much later. Rubin de Celis in 1783 saw a meteoritic iron variously estimated to weigh $13\frac{1}{2}$–45 metric tons.

One of the craters, 175 feet in diameter and now 16 feet deep, has been partially excavated. Beneath it were found "white ash" and "transparent glass" and small bits of nickel-iron. These, of course, are identical with the rock flour and silica glass which are standard materials at other meteoritic craters.

Spencer notes: "There are other suggestive features worthy of investigation in this district. Many small lakes and pozos are scattered around; and in particular a chain of small lakes extends southward from the spot where the large masses of meteoritic iron have been found for a distance of nearly one hundred miles into the province of Santa Fe" (57).

Several conclusions may be drawn. Multiple craters are known elsewhere; hence a swarm or cluster of meteorites traveling through space may be indicated. The craters are small, so the impact energies were low. Beneath one of the largest craters are found rock flour and silica glass, and therefore a strong shock wave and an explosion did occur; but the presence of meteoritic fragments in the crater shows that the explosion was incomplete. Consequently, the striking velocity was 4–5 miles per second. The presence of large meteoritic masses on the surface, assuming that they came from the same fall, indicates a very low striking velocity. Unlike other meteoritic crater groups, such as Odessa and Henbury, the craters are scattered over a considerable distance, possibly over 100 miles.

When all these facts are considered together, it seems probable that the Campo del Cielo craters were not formed by the separate members of a cluster but that a single object entered the earth's atmosphere at a low angle along a north-south line and, while still at a substantial height above the ground, exploded. This mechanism accounts for the large number of craters, their wide distribution, and the range of impact velocities shown. It thus may be a larger version of the Sikhote-Alin occurrence in Siberia in 1947.

The Australian Craters

THE CRATER GROUP AT HENBURY

The fall of the Karoonda meteorite on November 25, 1930, in central Australia led to the organization of an expedition under Professor Kerr Grant. This expedition brought back word that natives had found fragments of meteoritic iron surrounding several crater-like depressions near the Henbury Cattle Station. The number of craters was variously estimated as three and five. When this information became available, the South Australian Museum commissioned A. R. Alderman to survey the area.

Alderman visited the site and gathered as much pertinent data as possible. From his report it is evident that the view of the craters is decidedly unimpressive. They lie in an arid region, and from a distance the only indication of the craters is the presence of green mulga trees. In spite of the paucity of rainfall—only about 6 inches per year—the inwash of rain has partially filled them up. The finer sediments formed a hard, almost impervious layer which holds water longer than normally. As a result, the mulga trees, which are usually confined to the watercourses, grow in several of the craters, aiding in their identification.

Brief descriptions of each of the thirteen recognized craters follow. They are numbered according to Alderman (58), who predicts that there are additional pits still to be found. All the craters lie in a square, $\frac{1}{2}$ mile on the side.

The Henbury series, modified as it is by erosion and infilling, beautifully demonstrates the transition from a small crater formed by simple splashing or gouging of the ground to the true explosion crater in which the meteorite is broken up into small bits, most of which are backfired out of the crater.

Crater No. 1 has been completely removed by erosion. All that remains is a clay-pan floor and numerous surrounding fragments of nickel-iron. It was probably circular and about 75 feet in diameter.

Crater No. 2 was similar to No. 1 but slightly larger. It may have been 90 feet in diameter.

Crater No. 3 is larger, being 135 feet in diameter, and still shows a slightly raised rim. The pit has been largely filled in but is still 10–18 feet deep. Within the crater a 13-pound meteorite was found, and nearby lay 160 fragments, many of which were small. Eighty per cent of these masses were found to the west of the depression.

Radiating outward from the crater walls into the plain for about 90 feet can be seen five or six low ridges of sandstone, apparently identical with that found anywhere in the neighborhood (Ordovician age). They are a few inches high and consist of small blocks whose surfaces are blackened by weathering; hence they may be easily distinguished from the prevailing reddish color of the surroundings. Traces of similar ridges may be seen around other craters, particularly No. 4, but in no case are they as well defined as at No. 3.

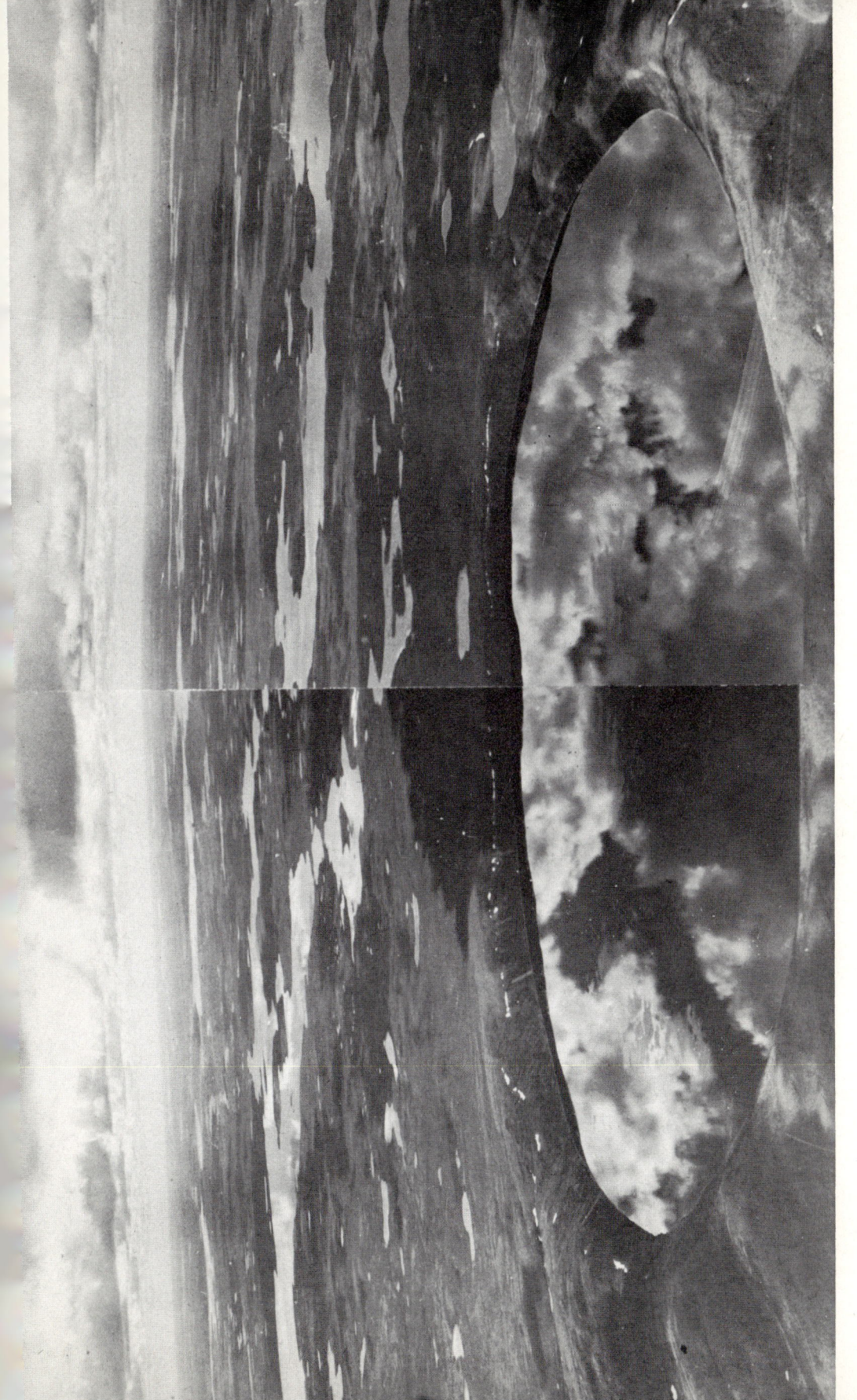

PLATE I.—New Quebec Crater. (Courtesy Dominion Observatory, Ottawa, Canada.)

PLATE II.—Wolf Creek Crater. (Courtesy Robert S. Dietz.)

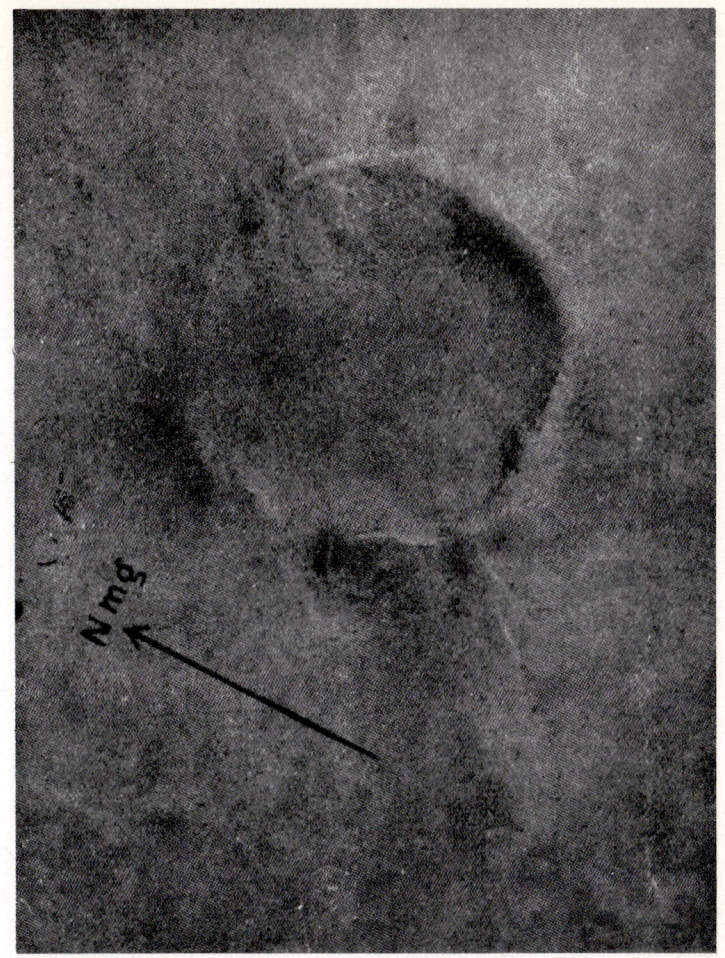

PLATE III.—Aouelloul Crater. (Photograph Air A.O.F. EOM 81, No. 64.)

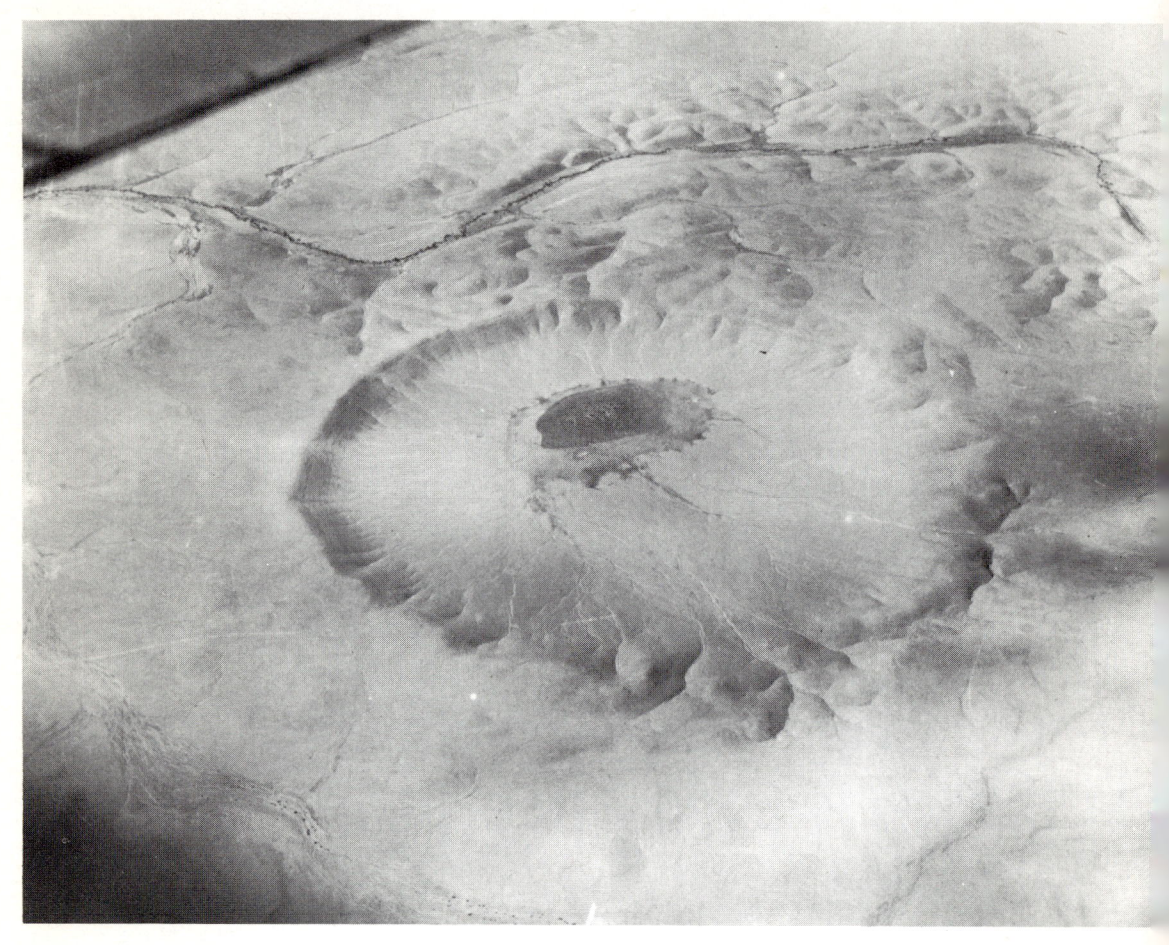

PLATE IV.—Crater of Talemzane. (Courtesy M. Baussart, Secrétaire Général Adjoint de l'U.G.G.I.)

The existence of these radiating ridges of raised blocks indicates a set of radial faults produced by the impact and explosion. They may be analogous to the radial markings around such lunar craters as Aristillus; and they could conceivably be related to the great system of radial valleys, including the Rheita Valley, which were produced when Mare Nectaris was formed. The subject would bear a good deal more study.

Crater No. 4 is identical with No. 3. Of about 500 fragments of nickel-iron associated with this crater, 400 lay to the west. One hundred of these were in an area 6 feet square. This suggests the breaking-up of a larger body.

Crater No. 5 is the same size as No. 1, which it much resembles, except that low walls, 4 feet high, still remain. A shaft sunk in this crater passed through 8 feet of fine soil before coarse rock fragments, apparently washed in from the walls, prevented further digging.

Crater No. 6, known as the "water crater," is roughly circular, 240 feet in diameter and from 12 to 25 feet deep, and is the larger of two adjacent to the main crater, No. 7. A watercourse, normally dry, has broken through its walls, and after each rain the crater is flooded, the moisture being retained for some weeks. Mulgas and acacias, larger than usual, grow in the pit. Some of the latter are 45 feet high and 21 inches in diameter. The walls of the "water crater" are highest on the side abutting the main crater.

Crater No. 7, the main crater, probably should be considered as being two or even three craters formed by the nearly simultaneous impact of the two or three largest meteorites, for it is oval, 660 by 360 feet at the surface. The depth averages 40–50 feet and, in places, is 60 feet, but quite obviously was once much greater. The walls are of the usual form, steep near the summit and with talus slopes at the foot. They are formed mainly of shattered and crushed fragments of sandstone and slaty rock, varying from the finest rock flour to large blocks several cubic feet in volume.

Black, glassy material was found near the main crater. Alderman described it as being like the glass of fulgurites (lightning tubes), vesicular in some cases, containing rock fragments in others. It is certainly heat-fused rock—the silica glass so well known at other meteoritic craters. On the west side of the crater, large cellular masses and pieces of partly fused sandstone are found close to the rim, while on the east side small tear-shaped drops and threads with a smooth glossy surface are found along a narrow strip of ground extending eastward a mile from the crater.

Relatively few metallic fragments were found on the surface near the large craters. A small number were near the silica glass from the main crater, but probably most of the pieces were covered up by erosion from the rim. Most of those found were in the watercourses.

Crater No. 8 is large and well defined, although rather shallow. It is about

175 feet in diameter but is only from 3 to 15 feet deep. Like No. 6, the walls are highest on the side next to the main crater.

Crater No. 9 is ill defined and doubtful. The topography suggests a small crater.

Crater No. 10, like Nos. 11, 12, and 13, is situated on a low ridge of sandstone to the south of the larger craters. It is circular, about 60 feet in diameter, with low walls.

Crater No. 11 is small and circular, about 45 feet in diameter. It has been partially excavated without the finding of any iron.

Crater No. 12 is very well defined. Although it is only 60 feet in diameter, its walls still reach 12 feet on the highest side, the one toward the ridge.

Crater No. 13 is small and rather indefinite. It is about 30 feet wide and only about 3 feet deep. Bedford, in 1932, found four masses of 292, 120, 24, and 5 pounds in this pit, buried 7 feet below the bottom under broken blocks of sandstone. This is the first instance of a considerable mass of meteoritic iron found buried in a crater. It clearly represents a case in which the mechanical impact, rather than the subsequent explosion, produced the crater. Bedford showed that all four pieces came from one mass. The energy needed to produce No. 11 is about four times that needed to produce No. 13. As no meteoritic iron was found in No. 11, this defines the dividing line between the mechanical and the explosive modes of crater formation.

About 2,000 pieces of nickel-iron have been collected over the entire region by various men. The 292-pound mass is the largest. Most of these fragments are curiously twisted. Etched sections of these show that the bands of the lamellar structure are bent and crumpled. The pieces were undoubtedly torn from larger masses by the force of the explosions which made the craters. Larger pieces with the normal pittings of meteoritic irons are seen in etched sections each to consist of the usual large crystals. The normal lamellar octohedral structure (Widmanstätten figures) extends up to the very edge of the mass without any granulation of the kamacite. This clearly proves that these masses are merely the weathered remains of still larger masses, the cores of which had not been raised to a temperature of 760° C by the conduction of heat from the outside. It shows that meteoritic collisions happen so rapidly and are so violent that most of the matter in the explosion region is either vaporized or else not seriously heated. Very little liquefaction occurs.

No one knows how old these craters are. Trees hundreds of years old are found in the pits. Some of the meteorites have been completely oxidized, which implies a great age, as the climate there is extremely arid. Bedford (59) suggests that the craters are not ancient. Alderman (58) believes that they are thousands of years old, and yet the natives seem to have a tradition that they were produced in a fiery explosion, for they call the place "Chindu chinna waru chingi yabu," which means "Sun walk fire devil rock." Old blacks would not camp within a couple of miles of the craters. How long such a legend would be handed down is a moot point, but there does not seem to be any way for the natives to associate light and fire with these depressions in the ground unless their ancestors had actually witnessed the landings and resultant blasts. On the basis of the C^{14} content of the Henbury meteorites, Goel and Kohman (21) suggest that the craters are at least 7,000 years old.

THE BOXHOLE CRATER

In 1937 another large but ancient meteoritic crater (60) was found in Australia about 200 miles northeast of Henbury. Apparently it is similar to the main crater at Odessa, being about 600 feet in diameter. The pit has been filled in until it is now only 50 feet deep, i.e., probably about half its original depth.

Shale balls, meteoritic fragments, and a 180-pound meteorite have been identified nearby; otherwise, practically nothing is known of this fall.

THE DALGARANGA CRATER

A smaller crater was discovered in 1923 by the manager of the Dalgaranga Sheep Station. It was first reported by Simpson (61) in 1938, based on the earlier description. Simpson did not see the crater, and his listing of the crater dimensions have proved to be seriously in error. He suggested a diameter of 225 feet and a depth of 15 feet. Recently, Nininger (62) visited the spot and found that the crater is actually much smaller. It is only 70 feet wide and $10\frac{1}{2}$ feet deep. Around the rim, particularly on the northwest side, the rock layers have been shattered and tilted up. A considerable number of meteorites were found associated with this crater, many of which had been twisted and warped in a manner reminiscent of those at the Henbury fall.

THE WOLF CREEK CRATER

A large and interesting crater has been found recently in Western Australia. It is easily the least accessible of the known craters on that continent, for it lies about 70 "surface miles" south of Hall's Creek, on the edge of the essentially unexplored Western Desert. Cassidy (63), who visited the crater in 1953, reports that it is practicable to reach it only during the relatively drier season of the Australian winter, i.e., from June to September. A dirt track from Hall's Creek to Billiluna Cattle Station can be followed for the first part of the trip to the crater. This track has been unused for many years and eventually becomes impossible to follow, being in places completely erased by a masking growth of

"Spinifex" (porcupine grass), a spiny type of grass common to most Australian desert areas. A vehicle with a four-wheel drive is necessary to reach the crater, since the ground is very sandy in places and the dry bed of Wolf Creek, which must be crossed, is treacherous. The nearest water to the crater is Beaudesert Well, a somewhat saline bore 18 miles to the north, toward Hall's Creek.

The aborigines in the area know the crater under the name of "Kandimalal," a word which has no meaning in their language. No legend exists among these legend-loving people to give a hint of its origin. It is very probable, therefore, that the crater was formed before the arrival in the region of the present Djaru tribes, who replaced an earlier folk many generations ago.

The Wolf Creek Crater was formed from pre-Cambrian quartzite. The rock layers, where visible in the inner walls, dip steeply out from the center. The dip is about 20° on the east and 50°–60° elsewhere. In the southwest and northwest, there is some inward dipping. Many great blocks of stone have been hurled out onto the rim and even farther. The crater is asymmetrical. There is much more material in the southwest rim than in other parts. This indicates that the meteorite came at a low velocity from the northeast.

The age of the crater has permitted erosion and infilling to modify its appearance. The crater, which is about 2,700 feet wide, has a rim which rises now about 80–100 feet above the plain and perhaps 200 feet maximum above the floor of the crater.

The vegetation inside the crater is much more luxuriant than outside because of the rain-catching propensities of the great bowl. The general area is covered with loose sand, with an occasional low dune.

Guppy and Matheson (64) believe the crater's age to be Pleistocene or Recent.

Cassidy recovered more than 1,400 pounds of more or less completely oxidized meteoritic material, some of which proved to contain streaks and granules of unaltered metal in their interiors. These latter came from a small area outside the unusually extensive southwest rim below a slope where the rocky talus of the rim appeared much more irregular and hummocky than elsewhere. Included in this group were three unusually large specimens weighing 352, 336, and 324 pounds. These large masses were found in line with the probable path of the main body and seem to have been spalled off the rear face of the meteorite at impact.

The Eurasian Craters

THE ESTHONIAN CRATERS

On the large Baltic island of Oesel is another of the groups of multiple craters (65–68). The largest of the seven pits, the Kaali Järv,[2] appears as a wooded

---

[2] "Kaali Järv" actually means "Kaali Lake."

knoll, 20–25 feet high. Beyond the tree-covered rim, the crater itself becomes visible. It is nearly circular, with diameters ranging from 300 to 360 feet, averaging 319 feet. Most of the variation in diameter seems to be due to erosion. Much of the crater is filled with a lake 200 feet across. The crater is 53.4 feet deep, rim to lake bottom including 4 feet of muck, and it is known that the underlying rock layers are shattered. A layer of pulverized rock containing rock fragments lies above the shattered layers. The limit of brecciation seems to be about 57 feet below the original ground level. (See Figs. 3 and 4.)

The Kaali Järv has been known since J. von Luce described it in 1827, and many bad guesses have been made as to its origin. Some of the earlier thoughts were that it was a man-made battlement, a karst weathering of limestone, or even that it was produced by the solution of salt in a salt dome. These wild guesses completely ignored the fact that the region showed all the characteristics of a crater born in a violent explosion. It has the upraised rim with steep inner and gradual outer slopes. There are numerous blocks of rock and other shattered ejectamenta, and the local rock strata have been upraised until they dip radially outward at angles up to 30° or 40°.

This is the only known terrestrial meteoritic crater that shows a central peak. Reinvaldt's cross-sections of the crater show a distinct peak rising about 6 feet above the present lake bottom and hence about 10 feet above the original crater bottom. It is located very close to the center of the crater.

The normal geologic structure of the island is that of horizontally bedded Silurian dolomite with a covering of glacial deposits about a yard thick. Intermingled rock fragments from the craters and glacial debris show that the craters are postglacial. The fact that the meteorites landed in this material may have hindered the recognition of the true nature of the craters. No silica glass has ever been found, owing, of course, to the absence of quartz in the dolomite. J. Kalkun compared the craters with the Arizona Crater in 1922.

For many years, no meteoritic materials were found, but in 1937, after 10 years of effort, Reinvaldt completely established the meteoritic nature of the craters by finding 28 small fragments of nickel-iron, totaling 110 grams. As the island has been inhabited for many centuries and extensively cultivated, it is not strange that all surface metals have long since disappeared.

Nearby, in an area of less than half a square mile, lie six other smaller craters and at least three other depressions which have been filled in with stones from farmers' fields. Four of the craters are circular, 120, 100, 65, and 35 feet in diameter. The other is oval, 175 by 120 feet, and is probably a double crater much like the main crater at Henbury. Several of the smaller craters have been excavated. Two tiny meteorites were found in the smallest crater, the others in the oval one. Beneath the 65-foot crater was a funnel-shaped hole, 4 feet wide and 2 feet deep, blasted into the limestone. The adjacent rocks are cracked and have a burned appearance.

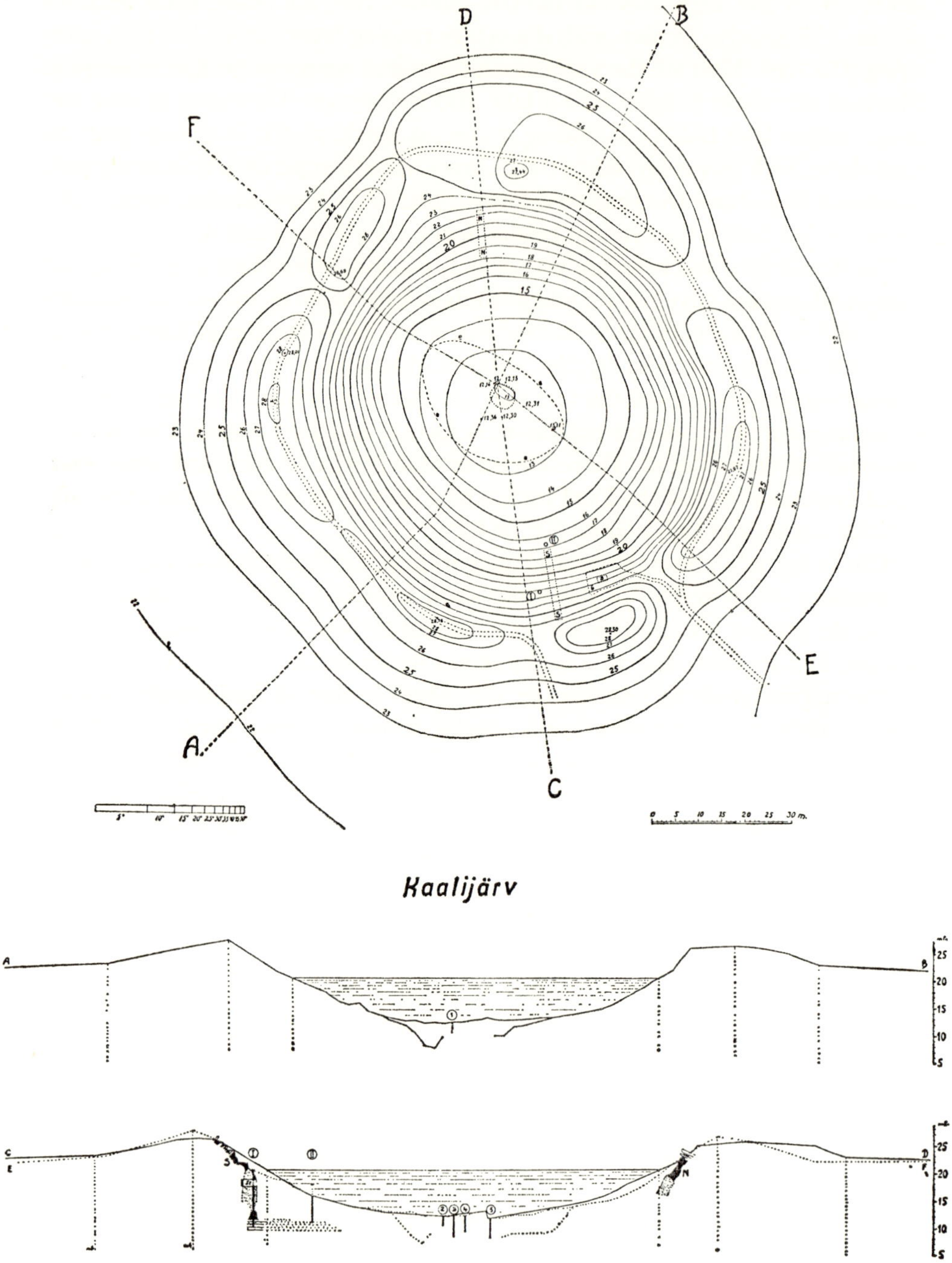

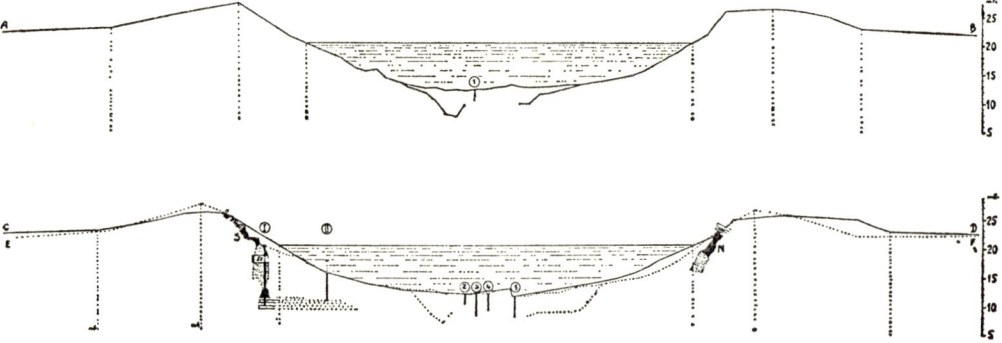

Kaalijärv

Fig. 3.—Oesel Crater, Kaali Järv. (Redrawn from Reinvaldt, Ref. 66.)

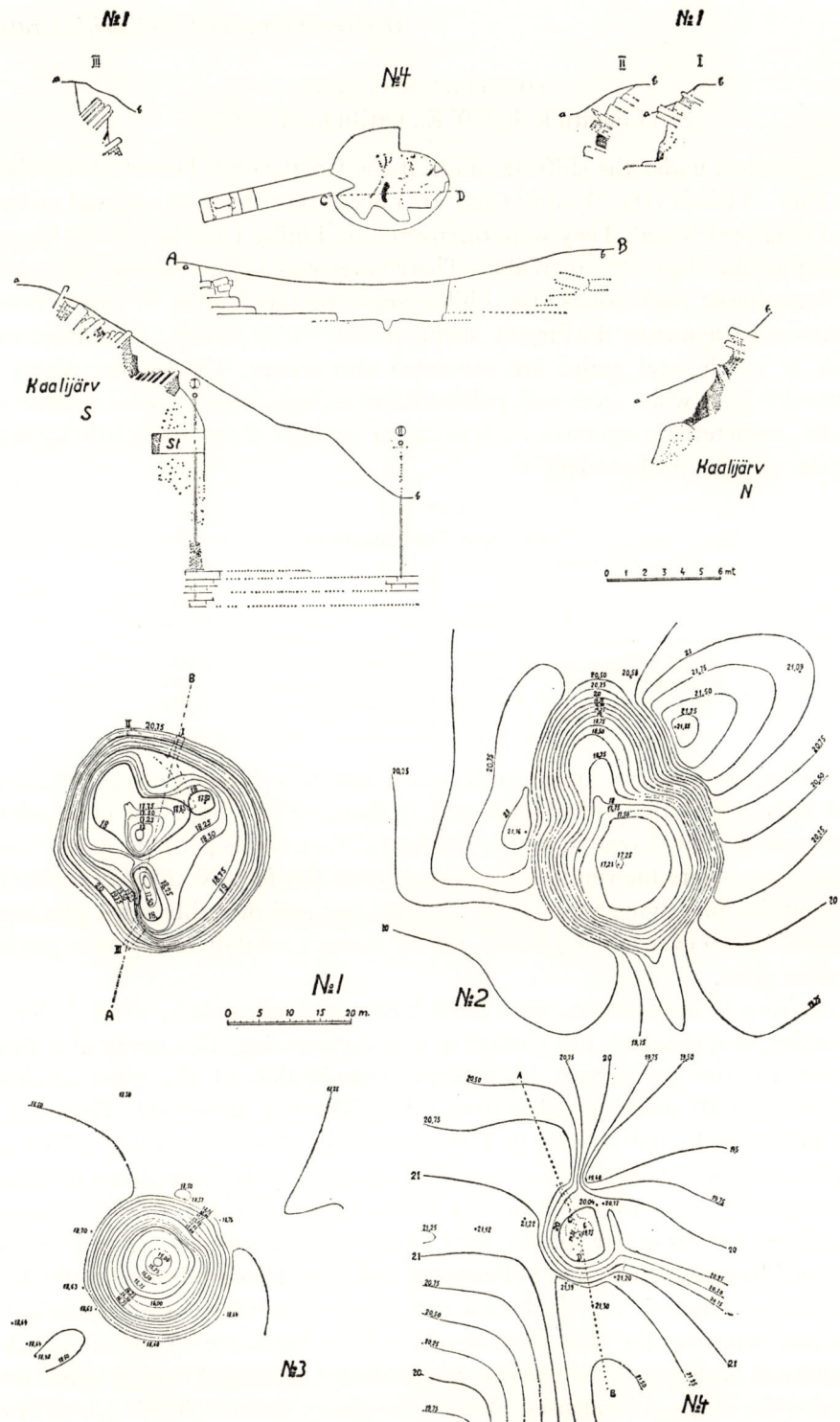

FIG. 4.—Oesel Craters—Nos. 1, 2, 3, and 4 and Kaali Järv. (Redrawn from Reinvaldt, Ref. 66.)

THE ARABIAN CRATERS
(Longitude 50°40′ E., Latitude 21°29′ N.)

Almost lost under the shifting sands of the Great South Desert of Arabia in the Rub' al Khali (the "Empty Quarter") are two of the most unusual meteoritic craters yet found. They were discovered by Philby (69) in 1932 in his epic search for the "lost city" of Wabar. There never was any question of the nature of these queer depressions, for Philby reported the finding of many rusted nickel-iron fragments, the largest having a mass of 25 pounds. Laminated iron shale in which sand grains are cemented also occurs. The smaller pieces of meteoritic iron, when sectioned, polished, and etched, show a partial destruction of the characteristic structure such as can be brought about artificially by heating the material to about 760° C.

TABLE 1
CRITICAL TEMPERATURES

	Degrees Centigrade		Degrees Centigrade
Nickel melts	1,452	Nickel boils	2,900
Iron melts	1,535	Iron boils	3,200
Silica melts	1,710	Silica boils	3,500

The larger of the two known craters at Wabar is perhaps 319 feet in diameter and, although partially filled with sand, is about 40 feet deep. The other, about 250 yards distant, is oval in shape, 180 by 130 feet, and is perhaps 30 feet deep. Silica glass covers the rim to a distance of over 100 feet. Nothing is known regarding the underlying rock layers. At least two and probably other pits exist, covered by the ever moving dunes. Philby shows a total of 12 separate patches of silica glass.

The rims of both craters seem to be built up of silica glass, which is highly vesicular, in appearance like cinders or iron-furnace slag. This material is clearly not volcanic in nature. Microscopic examination of the glass discloses numerous small specks ranging from 0.14 millimeter downward. They give a metallic luster by reflected light. There are usually from one to two million per cubic centimeter. Analysis indicates that they are actually little globes of nickel-iron. Spencer (57), formerly Keeper of the Minerals in the British Museum, concluded that these tiny spheres were formed in an atmosphere from which the oxygen had momentarily been eliminated and then drizzled down into the boiling silica. As there is very little moisture in the clean Wabar Desert sands, the vesicular nature of the glass must also have been due to the boiling of the silica. Additional evidence of the extreme temperatures generated here is found from the dewlike drops of condensed silica on the glassy surfaces. Table 1 is of interest in this connection. At the high pressures which probably existed, the corresponding figures were possibly somewhat higher.

At no other meteoritic crater is there any comparable evidence of liquefaction on such a large scale. It may also be noted that in no other known case did a meteorite land in deep, loose sand. The incoherent nature of the sand may be the factor which led to the generation of so much local heat.

Coesite was discovered at the Wabar Craters in 1961 (42, 70). Only about 1 per cent of the specimens of sandstone and white siliceous material proved to be coesite, but the identification is definite. This is the smallest meteoritic crater that has exhibited this impact-produced mineral.

Wabar is a legendary city mentioned in semiclassical Arabian writings as having been "destroyed by fire from heaven." It seems entirely probable that the "ruins" are the craters and the "cinders" the abundant silica glass. Perhaps here, too, is ancient race memory of the meteoritic fall.

THE SIBERIAN FALLS

THE 1908 FALL OF THE TUNGUSKA METEORITE
(Longitude 100°57′ E., Latitude 60°55′ N.)

In the early morning of June 30, 1908, at 0017 hours, 11 seconds G.M.T., the inhabitants of the north and east portions of distant Siberia were privileged to witness the greatest fall within historic times of a large meteorite. Strangely, in spite of the magnitude of the fall, no large expeditions were organized to investigate the region until 1927. It must be admitted, however, that World War I and the Russian Revolution occupied a part of the interval.

In that year the Russian scientist, Kulik, studied the area at the head of a small expedition. In 1928, 1929–30, and 1939 Kulik led three further expeditions under the sponsorship of the U.S.S.R. Academy of Sciences. E. L. Krinov (72, 73), who was with Kulik on the 1939 expedition, says:

The Tunguska Meteorite belongs to the class of gigantic meteorites whose fall results in their explosion and, in creation of meteorite craters. Unfortunately, study of location and of circumstances surrounding the fall of this meteorite first began nineteen years after its occurrence. This was the principal cause of the difficulties which investigators encountered in their work; and, indicates why investigation of the Tunguska Meteorite fall has been so long drawn out. In addition, scientific literature concerning this meteorite, even that appearing in the Soviet Union, as well as that appearing abroad, contains due to sundry circumstances . . . a series of inaccurate and actually false statements concerning the phenomena attendant upon the fall, exact location, and characteristic peculiarities of this meteorite. Consequently, many investigators do not have an entirely accurate picture of this significant meteorite.

It should be added that although no new work has been carried out since the conclusion of the last L. A. Kulik expedition (1939) to the present day (i.e., a lapse of two decades), our knowledge of circumstances surrounding the Tunguska Meteorite fall, has been expanded during the past few years by substantial new information.

The Tunguska Meteorite fall was accompanied by unusually violent visual, acoustical, and mechanical phenomena which were observed throughout a large area of central Siberia . . . about 1,500 kilometers in diameter. Through a cloudless sky, eyewitnesses saw the flight of a blindingly bright bolide visible for several seconds in its course from southeast to northwest. In its trajectory, the bolide left a great trail of smoke, which appeared, from points situated along the length of the projection of the bolide's path, as a gigantic pillar. From nearer localities a flame and a smoke cloud were observed over the site where the meteorite fell. Ear-splitting explosions (heard at 1,000 kilometers radius) thundered after the bolide disappeared. In many places earthquakes were observed; houses shook, window panes broke, standing objects slid, house furnishings fell, and hanging objects began to swing. In the Vanovara trading post about 60 kilometers south of the meteorite fall location . . . an inhabitant sitting outside a house was thrown several meters by the air wave; shortly before, he had experienced the effect of the "heat." Another inhabitant, standing by the inside wall of a house, had a feeling suddenly that his ears were burning. Several hundred kilometers radius from the fall site, men and animals were knocked down. However, in earlier accounts of the Tunguska Meteorite, the report that the meteorite had caused the death of thousands of reindeer, proves to be inaccurate. At that time there did not exist any such number of reindeer [71] either near the site where the meteorite fell, or in the far larger territory of the present day Evenkysk National District.

G. P. Merrill, of the United States National Museum, reviewed Kulik's original report (74), which was printed in Russian. A portion of this translation is given here:

The vibrations produced by the fall of the meteorite were detected and registered by the seismographs of the Physical Observatory at Irkutsk, where Mr. A. V. Vesnesenski[3] [75], who was in charge of the Observatory, calculated the epicenter of the "earthquake" to be located in the upper part of the Podkamennaya Tunguska.[4]

The phenomenon produced considerable panic, especially among the natives living in the basins of the Yenissei and the various Tunguska rivers and adjacent parts of the Lena River Basin.

Several attempts made in 1908 to find the body of the meteorite were fruitless as for some reason all parties were searching near the city of Kansk and not in the locality determined by A. V. Vesnesenski, whose observations unfortunately remained unpublished. Gradually interest in the new meteorite was almost forgotten, except as a tale among the natives.

In 1928 [sic], Mr. L. Kulik attempted to find the exact location of the meteorite and led an expedition to the Tungusk region. Owing to the lack of funds and the extreme

---

[3] Vosnessenski.

[4] Almost exactly at the correct position.

difficulties of transportation in the wilderness of taiga and tundra, the expedition was not altogether successful. However, Mr. Kulik was able to reach the area where the taiga bore distinct traces of the passage of the meteorite. An area struck by the meteorite is a water table between the upper part of the Podkamennaya Tunguska and its right tributary, the river Chuni. The area is largely covered with tundra in the process of formation, intersected by hills, small lakes, swamps and typical tundra. The immediate area is surrounded by high naked hills, deforested by the falling meteorite. All the trees are still on the ground, their tops spread out in fan-like fashion away from the central zone of the fall. Exceptions are noted only in the ravines or in the gorges and deep perpendicular valleys and also in a zone which can be considered as the interference zone. And even in these places the trees, in most cases, are scorched and dead.

The zone where the heat effect of the meteorite is evident is considered by L. Kulik to be thirty kilometers in diameter and the area of the air wave breaking the trees is fifty kilometers in diameter.

The center part of the "fire zone" is covered by shallow, funnel-shaped craters, reaching in some instances many tens of meters in diameter and not greater than four to five meters in depth. The bottoms of the craters are covered with swampy growth.

Unfortunately, Mr. Kulik was not able to find the body of the meteorite or determine the depth to which it had sunk.

He believes that the meteorite of 1908 was an aggregate of meteors, moving with a rate approaching seventy-two kilometers a second. Some of the aggregates undoubtedly exceeded one hundred thirty tons in weight. Hot gases (above one thousand degrees C) surrounded the meteorite and started fires before the meteorite had reached the ground and sunk into it, forming craters, uprooting trees, and burning everything that can burn in the center of its fall [76].

Based on the similarity between the uprooting damage to forests, and of some other phenomena accompanying the meteorite fall, to effects produced by hurricanes, volcanic eruptions, Astapovich [77] calculated the explosion energy of the Tunguska Meteorite to be 10^{21} ergs [72].

LaPaz (78) has suggested the possibility that it might be as high as 10^{24} ergs, but Whipple's determination of the energy in the air waves (79) would not seem to permit this interpretation.

At the Kew Observatory in England at 5:30 P.M. on the day of the collision, a series of unusual pressure waves in the atmosphere was found on the records of the microbarograph. Whipple calculated that the rate of travel of sound is such as properly to account for the difference in time within a quarter of a minute. European seismographs recorded a strong ground wave. Both sound and shock waves were recorded at many Siberian stations. It appears that about five thousand times as much energy went to make air waves as went to make ground waves. Krinov (72) states:

The Tunguska Meteorite fall was accompanied also by optical phenomena of further interest: the first night after the meteorite fall was unusually light everywhere in Europe and in western Siberia. Even in southern Russia, the Caucasus for instance, it was so light in the middle of the night that one could read easily without artificial light; also, following nights were abnormally light. The illumination, lessening slowly, did not disappear until the end of August [80, 81].

Schönrock's work seems to indicate that the effect existed days before the meteorite appeared (80).

The effect was explained first by Whipple (79) in 1930 and Astapovich (77) in 1939 as an indication that the Tunguska Meteorite was the nucleus of a small comet whose dusty tail, during moments of contact with the upper atmosphere, lay opposite the sun, or in a northwesterly direction. The dust, scattered in the atmosphere, produced the illumination phenomena. Krinov (72) states further:

In 1949, V. G. Fesenkov [82] followed with a detailed study, research made by Abbott [83] (1908) in California on transparency of the atmosphere. In the course of this study Fesenkov established a very interesting fact: From nearly the middle of July to the last half of August 1908, it appears that a marked drop in the coefficient of atmospheric transparency was noticed. Fesenkov's final conclusion stated that a violent atmospheric turbidity resulted from the sudden reduction to dust of a huge mass of substance of the Tunguska Meteorite. This took place during flight of the meteorite at cosmic speed through the atmosphere. From this conclusion it follows that the total mass of powdered material must have amounted to a million tons. The mass of dust suspended in the air absorbed solar radiation, darkening the atmosphere by day; and, conversely, reflected solar radiation during the night, illuminating the night sky.

Based on study of the various material gathered by Kulik's 1927 expedition, Krinov (72) says that

... it could be established later that the uprooted-forest region was ellipse-shaped, and, that the point at which direction lines of fallen trees intersected, lay in the northern part of this region; i.e., at a focus of the ellipse. This indicates to us the following conclusion: that in part, particularly towards the southeast, uprooting resulted from the shock wave accompanying the meteor during its flight through the atmosphere. In fact, as was later proved, the angle of the meteor trajectory was only 7° (from the horizontal) as it passed from southeast to northwest over the earth.

Kulik found that the area was covered by numerous round pits of a few to several dozen meters in diameter. The pits were filled with water and Sphagnum moss. He assumed that these were the meteoritic craters and that the fall was of a shower of meteorites rather than a single body. Actually, at least several hundred similar holes could be pointed out. It was later proved that these pits were purely terrestrial in origin. This is linked to the fact that this region is one of permanently frozen ground, often covered with standing water and mud.

While it was still thought that the pits were meteoritic craters, one of them,

30 meters across, called "Suslov Crater," was drained and sectioned. It was devoid of meteoritic material and proved not to be an impact structure.

Earlier reports of the finding of meteoritic material in a crater were due to the use of an old, inaccurate peat-magnetometer (84):

The probable fall (explosion site of the meteorite) is now thought to be in the southern part of the lowland, i.e., the so-called "Southern Swamp" [72].

The initial speed of the meteorite remains unclear. Until recently the opinion was held that the Tunguska Meteorite came in from a direction opposite to that in which the earth was moving, and consequently, must have had an extremely high initial velocity. According to the calculations of I. S. Astopovich [77], its velocity must have amounted to 50–60 km per sec. His conclusion was based on the proximity of position of the meteorite radiant to the apex of the earth. In 1952, however, B. Yu. Levin demonstrated that it did not necessarily follow that proximity of the meteorite radiant to the apex of the earth indicated the meteorite to have come from the opposite direction [85], paper by Levin, on speed and trajectory 1953. On the same position of the radiant the Tunguska Meteorite could have been moving either in the same, or opposite direction with respect to movement of the earth; i.e., it could have had either very large or proportionately minimal initial velocity. When one considers the extraordinarily great after-effects of the Tunguska Meteorite fall, however, one may well conclude that the meteorite must have had gigantic initial velocity and, consequently, must have been incoming. Because it possessed initially very great mass as well as extremely high velocity, the Tunguska Meteorite flew through the air at cosmic velocity and was moving, probably, at several km per sec when it struck the earth [72].

Let us examine these conclusions briefly. Krinov and Fesenkov suggest that the total mass of the original object was tremendous and that it lost a million tons of dust while moving through its long, low path. The explosive energy released was measured at about 10^{21} ergs. Equation (8-3) of Chapter 8 suggests that this amount of energy would yield a crater 739 feet across if the object had burst at a scaled depth of $H/W^{1/3} = 0.10$, normal for a meteoritic impact. This amount of energy would be carried by a sphere of nickel-iron 26 feet in diameter, weighing 2,000 metric tons, moving 10 km/sec. The original preatmospheric velocity seems to have been about 50–60 km/sec; therefore, the explosion energy would have to be multiplied by about 30, i.e., each unit mass of the meteorite lost 29/30 of its energy in passing through the atmosphere. In addition, the object lost 997,800 tons out of 1,000,000 in the fiery passage. The multiplication factor here is about 450. On these assumptions the preatmospheric energy was over 10^{25} ergs.

These conclusions and proportions do not seem to be reasonable. The seismograph and barograph measures indicate that about five thousand times as much energy went into the air wave as into the ground wave. The trees were felled, not in a circular area, but in an elliptical area. The impact point was closest to the northwest limit and farthest, about 30 kilometers, from the southeast limit. The meteorite came from the southeast. Krinov suggests that the air wave moving with the meteorite was responsible for the eccentric nature of the zone of fallen trees, while the explosion proper accounted for the larger part of the effect.

At 30 kilometers from the impact point the object would still have been less than 5 kilometers high. In other words, when the body was 5–8 kilometers high, it killed and knocked over very few trees, but when it reached a height of about 5 kilometers, it abruptly started to knock over and kill trees below it and to do the same to trees which were nearly 15 kilometers on either side of the trajectory.

From this peculiarity and from the observed fact that the trees which were felled point radially away from the impact point, it is concluded that the killing and breaking of the trees and the burning of much of the area were done more by the explosion than by the effects along the trajectory. If this is so and the energy of the explosion is 10^{21} ergs, the estimated mass of dust along the path is considerably too high.

The description of the amount of dust resembles the effects observed when a great stony meteorite arrives rather than a nickel-iron object. At a nickel-iron impact, there should have been some evidences of metallic remains, if searched for by a qualified team of scientists such as Kulik's groups. The observations seem to indicate a stony meteorite rather than a nickel-iron object. If so, the diameter of the body at the end of its path must have been about 35 feet if it exploded at 10 km/sec, on the assumption that the energy released was 10^{21} ergs.

Mr. Krinov was at the site for a long time, and his opinions should carry great weight, but his conclusion that the crater was formed and then disappeared within a year is certainly in need of verification. It would seem that some major traces of such a large crater would still remain in the area when we consider the amount of material which must have been ejected onto the rim, that the area is permanently frozen immediately under the surface, that there are hills and sufficient soil to hold tall trees, and that only part of the area is swamp. Even in the Suslov Crater the ground was solid enough to permit a rather large tree to grow. When the pit was drained, the men could walk in the crater in the warm summer.

This line of reasoning leads to the conclusion that there is not and never was a single great crater. It is more probable that the object exploded violently in the air just before striking. In this manner the excessive amount of energy in the air waves relative to the seismic waves may be explained. In a chemical or nuclear explosion of this magnitude with the burst center slightly below ground level, the ratio of energies is near to 100 or 200 rather than 5,000.

This conclusion may be subjected to a quantitative analysis. Let us see whether the observed effects at the Tunguska fall and the calculated energy released are consistent with the results observed in fission and fusion bombs as tested by the United States. The explosion of a nuclear bomb may not duplicate all the conditions of a large meteoritic explosion, but certainly the types of explosions are similar enough to warrant intercomparison.

Astapovich (77) says that the Tunguska Meteorite released 10^{21} ergs on exploding. One metric ton of TNT releases 4.18×10^{16} ergs. A 20-kiloton bomb would release 8.37×10^{20} ergs. This is a Hiroshima-size bomb. An explosion of 10^{21} ergs, then, is equal to a 23.9-KT bomb. Glasstone (86, 87) states:

> The maximum temperature attained in a fission bomb is probably several million degrees.... Due to the great heat produced by the nuclear explosion, all the materials are converted into the gaseous form.
>
> Within a few millionths of a second of the detonation of the bomb, the intensely hot gases at extremely high pressure formed in this manner appear as a roughly spherical, highly luminous mass. This is the ball of fire (or fireball).... Although the brightness decreases with time, after about seven-tenths (0.7) of a millisecond, the fireball from a 1-megaton nuclear bomb would appear to an observer 60 miles away to be more than 30 times as brilliant as the sun at noon.
>
> As a general rule, the luminosity does not vary greatly with the energy (or power) of the bomb. The surface temperatures attained, upon which the brightness depends, are thus not very different, in spite of differences in the total amounts of energy released.

A nuclear bomb explodes effectively from a single point. It releases a flood of energy, one-third of which (88) will be in the form of radiant energy. A meteoritic explosion starts the instant the body touches the ground and continues throughout an expanding volume of space. It is not a point-source model, but the energy density is so high that at moderate distances from the center the effects become essentially as though it were a point explosion. It is not known what percentage of the total energy is released as radiant energy, but calculations (89) indicate that extremely high temperatures are attained, and hence the radiant energy effects must be high.

For a surface burst, the effects of the radiant energy on regions outside the crater are considerably reduced by the geometry of the situation and by the tremendous obscuration from dust and debris raised by the explosion. Subsurface bursts will permit less and less radiant energy to escape as the focus is moved deeper.

Similar effects will be noted in the shock waves sent racing outward from the explosion. As the transition is made from an air burst to a surface burst to a subsurface burst, the energies which go to produce the crater become an increasing percentage of the total energy and the attenuation of the shock waves in the air becomes marked.

The maximum blast effects of a 20-KT bomb are greatest for a height of burst of about 1,850 feet (90).

Table 2 lists the observed data on the limits of the kindling of flash fires by radiant energy from two different sizes of bombs and the limits at which about 30 per cent of the trees will be blown down. This limit is rather narrow in width. It turns out that a shock wave will completely destroy a forest at a given distance, and slightly beyond this distance the effects will be small.

Observations indicate that when 5 calories/cm² of radiant energy, or more, are received from a nuclear weapon, we may expect incendiary action. As Table 2 shows, this will be received at a distance of 1.75 miles from a 20-KT bomb. The previously estimated energy of the Tunguska Meteorite explosion was

TABLE 2

RADIANT AND BLAST EFFECTS FROM NUCLEAR BOMBS

	LIMITING DISTANCES (MILES)	
	20 KT	1 MT
Fire effect............	1.75	10
Overturning of trees....	1.5	5.5

23.9 KT, and for this energy the limit of incendiary action would be about 1.9 miles. However, at the Tunguska fall the fire zone was 15 kilometers, or 9.3 miles, wide. To yield 5 calories/cm² at this distance on an average clear day, a much larger bomb, bursting in the air, would be required. This bomb is computed to be 781 KT, or a release of 3.3×10^{22} ergs.

For a nuclear explosion the effects of flash-flaming extend farther than the effects of the shock waves in breaking down trees. At the Siberian fall the reverse was true. It was observed that trees were felled for distances averaging 25 kilometers, or 15.5 miles.

It is not known just what type of scaling would be proper in intercomparing blast-wave effects of different-sized explosions, but an inverse-cube law will be used. Crater scaling tests indicate that in all cases the exponent is less than $\frac{1}{3}$. If that is so here, the energies to be derived will be too small.

Table 2 shows that a 20-KT bomb will break trees to an approximate distance of 1.5 miles. Scaling shows that a 23.9-KT explosion would do so to 1.6 miles. To overturn trees to the observed limit of the Tunguska Meteorite fall

Modern Terrestrial Meteoritic Craters

would require 22.2 MT, or about 9.3×10^{23} ergs. Several conclusions may be drawn:

1. The proportion of radiant energy, given off by a high-velocity meteorite on exploding, to the total energy is less than at a nuclear explosion.
2. Because of this, the energy derived from the limit of burning will be too low and the energy derived from the limit of tree damage will be too high unless the scaling exponent is too large. In this case the two effects work oppositely.
3. The energy released by the Tunguska Meteorite as calculated by Astapovich and Whipple is definitely too low.
4. The attentuation of shock and sound waves in air, which led to the calculated value of 10^{21} ergs, requires some modification.
5. The seismic records show that only a microscopic fraction of the energy went into ground waves.
6. The small seismic-wave energy and the large area of the fire zone are clear evidence that the meteorite burst in the air and did not strike the ground.
7. This conclusion is strengthened by the complete absence of a crater or craters.
8. The energy released by the explosion of the Tunguska Meteorite was probably in the range of 10^{23}–10^{24} ergs.
9. This is similar to the energy released below the ground at the 4,000-foot-wide Arizona Meteorite Crater.
10. Krinov's conclusion (72) that the object lost a million tons of dust in its passage through the atmosphere now appears reasonably consistent with the above determination of the energy released in the terminal explosion of the bolide.

It is entirely possible that the object was composed largely of ices, in conformity with Whipple's concept of comet heads. Even such bodies must contain a certain amount of rocky material.

THE 1947 FALL

The morning of February 12, 1947, was a brilliantly clear day in eastern Siberia. Suddenly, at about 10:38 A.M. local time, a scintillating ball of fire with a luminous tail and sparks sped across the sky. It was visible for several seconds and was so bright that it blinded the eyes of people who watched it, and it cast moving shadows. Its path was marked by a wide, gray band of dust, which remained visible for several hours. Some minutes after the fireball disappeared, powerful detonations were heard.

Pilots soon flew over the taiga and found clear evidence of the end points of the meteorite's path. Fresh craters among the trees stood out against the snow because of the yellow-brown color of the uncovered rock and clay.

There were more than 300 eyewitnesses to the event. From these observations, the path of the meteorite through the atmosphere and in its orbit in space were determined. It proved to be a tiny asteroid whose orbit crossed that of the earth. On February 12, 1947, the object and the earth came to the same point at the same time.

The total mass of the fall has been calculated to be about 70 tons (91). The velocity with which it encountered the atmosphere was approximately $14\frac{1}{2}$ km/sec. These values give an energy of 6.7×10^{20} ergs; but the losses in mass and energy due to air resistance were tremendous. From the polished and etched sections of the recovered meteoritic specimens, it was found to be a nickel-iron object, which appeared to be pressed together from separate pieces, ranging in size from less than a millimeter to several centimeters. Frequently, there were thin layers of schreibersite, a compound of iron and phosphorus, between the separate pieces. This structure and the existence of the layers weakened the meteorite so that it shattered in flight into at least 375 pieces. The bigger pieces split into thousands of fragments on striking the ground. About 23 tons of this material have been collected. Fine meteoritic dust, particles found when the big fragments broke up, was discovered with the help of a magnet in the ground layers in and around the craters.

The object shattered high in the sky—estimates say 5 miles high—but the fragments struck in an elliptical area of only 1.6 square kilometers. Consequently, the meteorite did not actually explode as did the larger 1908 object.

Expeditions have located 122 craters, ranging from $1\frac{1}{2}$ to 90 feet in diameter. The largest crater was 20 feet deep. There were also 78 holes less than $1\frac{1}{2}$ feet across, and 175 small specimens were collected on the surface.

The body came from the north, and the smaller fragments lagged behind the larger (92-95). The craters and larger fragments were found on the southern part of the impact area.

The actual impact velocity was extremely low, for no true explosion resulted, even at the largest crater. Air resistance had practically stopped the particles before they struck. Only the extreme weakness of the meteoritic structure permitted it to fragment. If the object had remained intact and struck the ground, it would have produced a crater smaller than Odessa No. 1.

THE MURGAB METEORITE CRATERS

Some 250 years ago, in the morning, Kirghiz nomads saw two large meteorites fall and blast craters in the ground. They were near the town and river called Murgab in the highlands of East Pamir, just north of India in Tadzhik S.S.R. (96). The region is almost uninhabited and is about 12,100 feet above sea level.

The larger crater, 260 feet in diameter and 50 feet deep, was first described by H. D. Klavins in 1926. In 1951 a special expedition led by S. H. Zaharov and

A. M. Bakharev (97–99) discovered the second crater 820 feet from the principal one. The lesser crater is 52 feet in diameter. The craters lie in horizontal river deposits interspersed with layers of diluvial material over limestone.

A meteoritic origin appears definite. The minerals of the inner walls of the main crater have been scorched by a heat of short duration. The limestone is still covered by a weathered scale, dark brown in color and with many surface blisters. This crust is about 2–5 millimeters thick and includes considerable amounts of iron oxide. Beyond the crater, there is little evidence of scorching. Zaharov and Bakharev concluded that an iron meteorite struck at an angle from the horizontal of about 60° and a velocity of 2–3 km/sec. The site of the craters is named "Chaglgan Toushtou," meaning "the place where lightning fell."

THE AFRICAN CRATERS

AOUELLOUL
(Longitude 12°14′ W., Latitude 20°13′ N.)

The airplane is the greatest instrument for discovering meteoritic craters. On North American Flight Chart FC-62 (5th ed., November, 1947), an irregular feature was indicated as a "crater" (100). This location is in the west-central Sahara about 250 miles from the Atlantic. Native Mauritanians knew the feature and called it the "hole" of Aouelloul (pronounced Ah-way-lewel). The area is an ancient sandstone plateau lightly covered with drifting sands. The rocks are of Ordovician age.

The crater bottom has been largely filled up with sand, but the crater is everywhere still more than 20 feet deep. The inner walls attain inclinations of 20°–30° and are more or less cluttered with rubble. The outer walls are much flatter, dropping off at angles of only 5°–15°. This is characteristic of meteoritic craters.

The sandstone layers in the wall show, in over-all plan, an annular anticline much like that at Odessa No. 1. These ramparts are made up entirely of sandstones from the plateau layers. Nowhere are found ejectamenta of volcanic nature or materials from deep underlying layers.

In 1951, T. Monod found numerous pieces of silica glass in and around the pit. Spencer, Hey, and Smith of the British Museum analyzed it and found that it contained traces of oxidized iron and nickel, as well as other oxides.

Monod concluded that "the hypothesis of meteoritic origin seems to find itself considerably strengthened," pointing out that "the absence of nickel in the sandstones of Wabar and Henbury, and its presence in the glasses arising from the fusion of those sandstones give rise to the thought that the nickel in the Aouelloul glass was not derived from the beds in place but constitutes a foreign, i.e., meteoritic element."

The Aouelloul crater, like the Arizona crater, is non-circular in outline (101). The northern half is roughly a semiellipse, but the southeastern, southern, and southwestern segments are approximately straight lines. The corners are very much rounded, and the effect is less pronounced than in the Arizona example. The crater is about 840 feet in diameter.

THE CRATER OF TALEMZANE

It is no longer safe to rate meteoritic craters on the basis of size. Ten years ago, the Arizona Crater was the world's largest known. Now several certain or probable meteoritic craters have been found to surpass it. One of the most beautiful examples is the very ancient crater of Talemzane, which is 5,600 feet wide (102).

It appears as a circular depression on the "Oued Attar" sheet of the 1/200,000 topographic map of Algeria. Like all such craters, it shows the raised rim and sunken basin, with steep inner and gradually sloping outer walls.

The Arab name is the Daïet el Maädna, but it is not a "daïa" or shallow, closed depression such as characterizes the area. It is a true meteoritic crater, and Karpoff (103) has given it its name because it is only 9 kilometers southeast of an old cistern and "bordj" called Talemzane. The pit is 400 kilometers south-southeast of Algiers and only 1 kilometer off the Laghouat to Guerrara desert road. Although it is in a dry area, the depressed basin often shows a beautifully green field of grain, supported by the occasional autumn rains.

Karpoff and Dubief found no general magnetic anomaly in the neighborhood of the crater and no meteoritic materials. This is not surprising from two points of view. The crater is almost exactly symmetrical. This suggests that the fall was nearly vertical or that the velocity of impact was rather high. In either case, we would not expect many spalled-off meteorites from the back side scattered over the landscape.

Even if there were originally such spalled-off meteorites, the great age of the crater could have permitted erosion, oxidation, and deposition to hide or eliminate them. Karpoff feels that the crater may have been formed by a stony meteorite, but he is worried about the ability of large stony meteorites to survive the passage through the air to the impact point.

The crater is so old that the huge ejected blocks of limestone lying on the rim have been "peneplaned" to the point where they can be mistaken for a natural, nearly level surface.

The highest limestone layers involved in the crater are from the base of the Eocene or the top of the Cretaceous, but lying uncomformably on the limestone on the outer slope to the north appears a reddish continental stratum covered by a calcareous crust, which, in the Sahara, is normally placed in the Continental

Pliocene. Though usually horizontal, it appears to be folded very locally to the north of the crater. The crater then would seem to date, at the earliest, from the latest Pliocene, or over a million years ago.

The extreme maturity of the relief is immediately apparent, as is the advanced state of the filling of the interior with alluvial deposits of clay and sand. These deposits are obviously of considerable thickness.

On the rim are found numerous Neolithic "workshops" and also circular pre-Islamic tombs.

References

1. Gilvarry, J. J., and Hill, J. E. "The Impact of Large Meteorites," *Ap. J.*, **124**, 610, 1956.
2. Watson, F. G. *Between the Planets*, p. 114. Philadelphia: Blakiston Co., 1941.
3. "Dust Cloud around the Earth," *Sky and Telescope*, **21**, 71, 1961.
4. Watson, F. G. *Between the Planets*, p. 151. Philadelphia: Blakiston Co., 1941.
5. Barringer, D. M., Jr. "The Most Fascinating Spot on Earth. I, II, III," *Sci. American*, **137**, 52–54, 144–46, 244–46, 1927.
6. ———. "A New Meteor Crater," *Proc. Acad. Nat. Sci., Philadelphia*, **80**, 307, 1928.
7. ———. "The Barringer Meteorite," *Science*, N.S., **73**, 66, 1931.
8. ———. "Meteor Craters," *Frontiers* (Philadelphia), **2**, 80, 1938.
9. Öpik, E. "Note on the Meteoric Theory of Lunar Craters" (in Russian with a French summary), *Bull. Soc. Russe des Amis de l'Étude de l'Univers* (*Mirové dénié*), **5**, 125, 1916.
10. Gifford, A. C. "The Mountains of the Moon," *New Zealand J. Sci. and Technol.*, **7**, 129, 1924.
11. ———. "The Origin of the Surface Features of the Moon," *Scientia*, **48**, 69, 1930. Also published in *Proceedings of the Royal Astronomical Society of Canada*, **25**, 70, 1931; *New Zealand J. Sci. and Technol.*, **11**, 319, 1930; French translation by H. de Varigny, *Scientia*, **48**, Suppl., 31, 1930.
12. Hager, D. "Crater Mound (Meteor Crater), Arizona: Is Its Origin Geologic or Meteoritic?" (abstr.), *Pop. Astr.*, **57**, 457, 1949.
13. ———. "Crater Mound (Meteor Crater), Arizona, a Geologic Feature," *Bull. Amer. Assoc. Petrol. Geologists*, **37**, 821, 1953.
14. ———. "Discussion. Notes on Crater Mound in Answer to Some Points Raised by H. H. Nininger," *Amer. J. Sci.*, November, 1954. [See H. H. Nininger, "Reply," same issue.]
15. Butcher, Devereux. *Exploring Our National Parks and Monuments*, p. 7. Boston: Houghton Mifflin Co., 1951.
16. Nininger, H. H. "Symmetries and Asymmetries in Barringer Crater," *Earth Sci. Digest*, **7**, 17, 1953.
17. Shoemaker, E. M. "Impact Mechanics at Meteor Crater, Arizona," prepared on behalf of the U.S. Atomic Energy Commission, Open File Rept., July, 1959.

18. MERRILL, G. P. "The Meteor Crater of Canyon Diablo, Arizona: Its History, Origin, and Associated Meteoric Irons," *Smithsonian Misc. Coll.* (quarterly issue), **50,** 461, 1908.
19. ———. *Ibid.,* p. 470.
20. ———. *Ibid.,* p. 464.
21. GOEL, P. S., and KOHMAN, T. P. "Cosmogenic Carbon-14 in Meteorites and Terrestrial Ages of 'Finds' and Craters," *Science,* **136,** 875, 1962.
22. FISHER, CLYDE. *Meteor Crater, Arizona* ("Sci. Guides," No. 92). 3d printing. New York: American Museum of Natural History, 1946.
23. NININGER, H. H. *Arizona's Meteorite Crater.* Denver, Colo.: World Press, Inc., 1956.
24. RINEHART, J. S. "A Soil Survey around the Barringer Crater," *Sky and Telescope,* **16,** 8, 1957.
25. NEVILL, EDMUND NEVILLE (EDMUND NEISON, pseud.). *The Moon, and the Condition and Configuration of Its Surface,* p. 576. London: Longmans, Green & Co., 1876.
26. NININGER, H. H. "Double Blast in Arizona," *Sky and Telescope,* **9,** 114, 1950.
27. JAKOSKY, J. J., WILSON, C. H., and DALY, J. W. "Geophysical Examination of Meteor Crater, Arizona," *Trans. Amer. Inst. Mining, Met. Engrs.,* **97,** 63, 1932.
28. MOULTON, F. R. Report (mimeograph copy) dated August 24, 1929, on file in Lowell Observatory Library, Flagstaff, Arizona.
29. FOOTE, A. E. "A New Locality for Meteoritic Iron with a Preliminary Notice of the Discovery of Diamonds in the Iron," *Proc. Amer. Assoc., Advance. Sci.,* **40,** 279, 1891.
30. NININGER, H. H. *Arizona's Meteorite Crater,* p. 139. Denver, Colo.: World Press, Inc., 1956.
31. ANDERS, E. "Diamonds May Be Formed as Meteorites Hit Earth," *Sci. News Letter,* October 29, 1960.
32. UREY, H. C. "Diamonds, Meteorites, and the Origin of the Solar System," *Ap. J.,* **124,** 623, 1956.
33. DECARLI, P. S., and JAMIESON, J. C. "Formation of Diamond by Explosive Shock," *Science,* **133,** 1821, 1961.
34. COES, L., JR. *Science,* **118,** 131, 1953.
35. NININGER, H. H., *Arizona's Meteorite Crater,* p. 50. Denver, Colo.: World Press, Inc., 1956.
36. *Science News Letter,* July 9, 1960.
37. CHAO, E. C. T., SHOEMAKER, E. M., and MADSEN, B. M. "First Natural Occurrence of Coesite," *Science,* **132,** 220, 1960.
38. MERRILL, G. P., *Proc. U.S. Nat. Mus.,* **32,** 547, 1907.
39. ROGERS, A. F. *Amer. J. Sci.,* **19,** 195, 1930.
40. DECARLI, P. S., and JAMIESON, J. C. *J. Chem. Phys.,* **31,** 1615, 1959.
41. BOYD, F. R., and ENGLAND, J. L. *J. Geophys. Res.,* **65,** 749, 1960.
42. DIETZ, R. S. "Astroblemes," *Sci. American,* **205,** 51, 1961.

43. Sellards, E. H., and Evans, G. "Statement of Progress of Investigation at Odessa Meteor Craters," University of Texas, Bureau of Economic Geology, September 1, 1941 (mimeographed).
44. ———. "Odessa Meteor Craters," Views in Texas Memorial Museum, *Mus. Notes*, **6**, 13, July, 1944.
45. Monnig, O. E., and Brown, R. "The Odessa, Texas, Meteorite Crater," *Pop. Astr.*, **43**, 34, 1935.
46. Leonard, F. C. "Magnetic Anomalies at the Ungava Crater," *Meteoritics*, **1**, 2, 229, 1954.
47. LaPaz, L. "Evidence on the Nature of the Ungava Crater Unobscured by Glaciation," *Meteoritics*, **1**, 2, 2, 1954.
48. Meen, V. B. "Chubb Crater, Ungava, Quebec," *J.R.A.S. Canada*, **44**, 169, 1950.
49. Millman, P. "The New Quebec Crater," *J. Brit. Astr. Assoc.*, **47**, 116, 1957.
50. Meen, V. B. "Chubb Crater, Ungava, Quebec," *J.R.A.S. Canada*, **44**, 7, 1950.
51. Harrison, J. M. "Ungava (Chubb) Crater and Glaciation," *J.R.A.S. Canada*, **47**, 16, 1954.
52. Meen, V. B. "Chubb Crater—a Meteor Crater," *J.R.A.S. Canada*, **51**, 137, 1957.
53. Nininger, H. H. *Out of the Sky*. Denver: University of Denver Press, 1952.
54. Nininger, H. H., and Figgins, J. D. "The Excavation of a Meteorite Crater near Haviland, Kiowa Co., Kansas," *Proc. Colorado Mus. Nat. Hist.*, **12**, 3, 1933.
55. Nininger, H. H. "Kansas Meteorites since 1925," *Trans. Kansas Acad. Sci.*, 1936.
56. Watson, F. G. *Between the Planets*, p. 161. Philadelphia: Blakiston Co., 1951.
57. Spencer, L. J. "Meteorite Craters as Topographical Features on the Earth's Surface," *Geog. J.*, **81**, 227, 1933.
58. Alderman, A. R. "Meteorite Craters at Henbury, Central Australia," *Mineralog. Mag.*, **23**, 19, 1932.
59. Spencer, L. J. "Meteorite Craters," *Nature*, p. 8, May 28, 1932.
60. Madigan, C. T. *Trans. R. Soc., South Australia*, **61**, 187, 1937; abstr., *J. Geomorph.*, **1**, 173, 1938.
61. Simpson, E. S. *Mineralog. Mag.*, **25**, 157, 1938.
62. Nininger, H. H. "Another Meteorite Crater Studied," *Science*, **130**, 1251, 1959.
63. Cassidy, W. A. "The Wolf Creek, Western Australia, Meteorite Crater," *Meteoritics*, **1**, 2, 197, 1954.
64. Guppy, D. J., and Matheson, R. S. "Wolf Creek Meteorite Crater, Western Australia," *J. Geol.*, **58**, 30, 1950.
65. Fisher, Clyde. *Nat. Hist.*, **38**, 292, 1936.
66. Reinvaldt, L., with Luba, A. "Bericht über geologische Untersuchungen am Kaalijärv (Krater von Sall) auf Ösel," *Tartu Ülikoolijuures oleva Loodusuurijate Seltsi Aruanded (Sitzsb. Naturforsch. Gesellsch., U. Tartu)*, **35**, 30–70, 1928.
67. ———. Published separately as *Pub. Geol. Inst. U. Tartu*, No. 11, pp. 1–42, 1928.
68. Kraus, E., Meyer, R., and Wegener, A. "Untersuchungen über den Krater von Sall auf Ösel," *Gerlands Beitr. z. Geophys.*, **20**, 312–78, 1928. Prepared according to information and drawings furnished by Reinvaldt.

69. Philby, H. St. J. "Rub' al Khali: An Account of Exploration in the Great South Desert of Arabia," *Geog. J.*, **81**, 1, 1933.
70. Chao, E. C. T., Fahey, J. J., and Littler, J. "Coesite from Wabar Crater near Al Hadida, Arabia," *Science*, **133**, 882, 1961.
71. Krinov, E. L. *The Tunguska Meteorite*, p. 196. Moscow: U.S.S.R. Academy of Sciences, 1949.
72. ———. "The Tunguska Meteorite," *Internat. Geol. Rev.*, **2**, 8, 1960.
73. ———. "Der Tungusker Meteorit," *Chemie der Erde*, **19**, 207, 1958.
74. Kulik, L. A. "On the Question of the Locality of the Fall of the 1908 Meteorite," *D.A.N. Proc., U.S.S.R. Acad. Sci.*, p. 393, 1927.
75. Vosnossenski, A. V. "The Fall of a Meteorite on June 30, 1908, in the Upper Reaches of the Chatanga River," *Mirowedenje*, **14**, 25, 1925.
76. Merrill, G. P. *Science*, May 11, 1928.
77. Astapovich, I. S. "New Material on the Flight of a Great Meteorite in Central Siberia on June 30, 1908," *A.J.*, **10**, 465, 1939.
78. LaPaz, L. "Contraterrene Meteorites," *Pop. Astr.*, **49**, 265, 1941.
79. Whipple, F. G. "The Great Siberian Meteor and the Waves, Seismic and Aerial, Which It Produced," *Quart. J. R. Meteorol. Soc.*, **56**, 236, 287, 1930.
80. Schönrock, A. M. "The Red Mornings of 17–30 June 1908," *Monthly Bull. Nikolajewsker Phys. Chief Obs.*, No. 6, 1908.
81. Rudnjer, D. D. "Light Night Clouds," work of the Scientific Students Union of Physics-Mathematics, St. Petersburg, **1**, 69, 1909.
82. Fesenkov, V. G. "The Dimness of the Atmosphere Caused by the Tunguska Meteorite Fall Which Occurred June 30th, 1908," *Meteoritika Issue*, **4**, 8, 1949.
83. Abbot, C. G. *Smithsonian Ann.*, **2**, 105, 1908.
84. Stanyukovich, K. P., and Fedynski, V. V. "On the Destructive Action of Meteorite Impacts," *D.A.N. Proc., U.S.S.R. Acad. Sci.*, **70**, 129, 1947.
85. Levin, B. Yu. "On the Question of the Speed and Path of the Tunguska Meteorite," *Meteoritika Issue*, **11**, 132, 1953.
86. Glasstone, S. *The Effects of Nuclear Weapons*, p. 19. Prepared by the U.S. Dept. of Defense, published by the U.S. Atomic Energy Commission, June, 1957.
87. ———. *Ibid.*, p. 20.
88. ———. *Ibid.*, p. 331.
89. Gilvarry, J. J., and Hill, J. E. "The Impact of Large Meteorites," *Ap. J*, **124**, 610, 1956.
90. Glasstone, S. *The Effects of Nuclear Weapons*, p. 85. Prepared by the U.S. Dept. of Defense, published by the U.S. Atomic Energy Commission, June, 1957.
91. Krinov, E. L. "The Siberian Fall of February, 1947," *Sky and Telescope*, **15**, 300, 1956.
92. Astapovich, I. S. "The Great Meteorite Fall on 1947, Feb. 12, Ussouri District, Far East, USSR," *Astr. Circ.* (U.S.S.R.), No. 62, May 25, 1947.
93. Fesenkov, V. G. "The Sikhotay-Alinsky Meteorite," *Russ. A. J.*, **24**, 318, 1947.
94. ———. *Astr. News Letter*, No. 37, p. 6, 1948.
95. ———. *Ibid.*, p. 8.

96. HOFFLEIT, D. "Murgab Meteorite Craters," *Sky and Telescope,* **12,** 1, 1952.
97. ZAHAROV, S. A., and BAKHAREV, A. M. *Astr. Circ. Russian Acad. Sci.,* January 7, 1952.
98. ———. "Meteor Crater in Eastern Pamir," *Rept. Tadzhik Acad.,* **6,** 3, 1953.
99. ———. *Astr. News Letter* (U.S.S.R.), **76,** 40, 1955.
100. LaPAZ, L., and LaPAZ, J. "The Adrar (= Chinguetti), Mauretania, French West Africa, Meteorite (CN = 0127,202)," *Meteoritics,* **1,** 2, 2, 187, 1954.
101. LEONARD, F. C. "Mineral Formulas, the Classification Sequence, and the Aouelloul Crater," *Meteoritics,* **1,** 3, 3, 1955.
102. BRADY, L. F. "The Crater of Talemzane in Algeria," *Sky and Telescope,* **13,** 9, 1954.
103. KARPOFF, R. "The Meteorite Crater of Talemzane in Southern Algeria," *Meteoritics,* **1,** 1, 1, 1953 (translated from the French and communicated by L. F. BRADY).
104. KENNEDY, G. C. Public lecture, reported by R. S. DIETZ, 1962.
105. BUNCH, T. E., and COHEN, A. J. "Precambrian Coesite," paper presented at first Western Meeting, American Geophysical Union, Los Angeles, Calif., December, 1961.
106. DIETZ, R. S. Personal communication, January 23, 1962.
107. BOON, J. D., and ALBRITTON, C. C., JR. "Deformation of Rock Strata by Explosions," *Science,* **96,** 402, 1942.
108. ———. *Field and Lab.,* **5,** 1, 1938.
109. ———. *Ibid.,* p. 53.
110. ———. *Ibid.,* **6,** 44, 1938.
111. DIETZ, R. S. "Geological Structures Possibly Related to Lunar Craters," *Pop. Astr.,* **54,** 1, 1946.
112. ———. "Meteorite Impact Suggested by the Orientation of Shatter-Cones at the Kentland, Indiana, Disturbance," *Science,* **105,** 2715, 1947.
113. "Hard Rock," *Sci. American,* **206,** 78, 1962.

3

PROBABLE AND POSSIBLE TERRESTRIAL METEORITIC CRATERS

Of the craters discussed in the preceding chapter, only the Arizona Crater has been so thoroughly studied that it passes every one of the tests for a meteoritic crater. Shatter cones have not been found at the Wolf Creek Crater. Meteoritic debris has not been found at the New Quebec Crater. Coesite has not been found at the Crater of Talemzane.

Nevertheless, these craters are of meteoritic origin. The youngest of the authenticated meteoritic craters is only 15 years old as of the date of writing. The next impact is 54 years old, then about 250 years. The others stretch back in time until the oldest, which is probably the Crater of Talemzane, goes back about a million years or to the end of the Pliocene.

There are many other structures known which are probably of similar nature or which may turn out to be meteoritic when enough facts have been determined. Craters in this category are described below.

The French Craters

Two probable meteoritic craters have been reported by Gèze and Cailleux (1) from the Department of Hérault in southeastern France. The larger is the depression of Cabrerolles, named "Le Clot." It was first discovered by stereoscopic observations on aerial photographs. It lies about 1 kilometer south-southeast of the village of Cabrerolles. The ground layers in this region are of insoluble schists and similar rocks below a Pliocene surface. The crater is tolerably circular, with a diameter of about 722 feet and an apparent depth of about 164 feet. It has an elevated rim.

The second crater is 5 kilometers to the east-northeast, or about 300 meters from the station of Faugères. It is smaller, about 197 feet in diameter and 75 feet deep. The excessive depth of the Faugères crater perhaps is to be explained by the fact that it was formed in a rather deep Quaternary ravine. The meteoritic nature of these two craters is deemed probable because of their shape and the type of rock from which they were blasted.

In the neighborhood of Faugères are five other similar, but smaller, craters. "Celles-ci se groupent grossièrement, avec celle de la gare de Faugères, sur un même alignement SSW-NNE comparable à celui d'un chapelet de trous de bombes, long de 2500^m, ce qui plaide en faveur d'une origine commune, par chute d'un bolide fragmenté" (1). Janssen (2) does not hesitate to call them meteoritic craters and says that they are about 10,000 years old.

The Pretoria Salt Pan

Obviously, not every hole in the ground was punched there by a meteorite. On the other hand, certain craters which have been called volcanic in origin sometimes have been proved to be impact structures. Others may well be of meteoritic nature.

One such feature is the Pretoria Salt Pan (Pl. V). This strange depression lies in the wooded, gently rolling Bushveld of the Transvaal, about 25 miles north-northwest of Pretoria. The surrounding country rock is the red Bushveld granite, nearly level, with isolated outliers of grit from the Coal Measure series of the Karroo system (3).

The great hollow is almost perfectly circular, with diameters north to south of 3,460 and 3,330 feet measured at right angles. The circumscribed ridge rises an average of 100 feet above the level of the neighboring country, with a peak of 200 feet on the northwest side. It is composed of coarse, uncemented breccia made up almost exclusively of large and small angular granitic blocks resting on Karroo grits. The inner slope is steep, the outer gradual. It is clearly in exactly the same position as that in which it fell after being blasted from the pit. Among the fragmental masses there is not to be found a single specimen of young volcanic rock. Many of the granite blocks are traversed by narrow fissures filled with a breccia composed of angular fragments of quartz and feldspar.

The floor of the crater lies 300 feet below the rim and presents a strange appearance. The outer portion is generally covered by a dazzling white saline incrustation, while a dark pool of soda-brine fills the center. Bore holes have been forced down through 230 feet of bedded layers. In the central portion, the 18 feet of the Trona-Mud zone are above the Gaylussite layer of permeable beds made up largely of Gaylussite crystals interbedded with and underlain by saline clays and marls. These layers are impregnated with clear, saturated soda-salt brine. It is upon this brine that the soda industry established at the Pan is based. There is evidence to show that the liquor has been and is being generated by underground water that has forced its way into the caldera deposits from above.

With an average apparent diameter of 3,400 feet and equations (7–2) and (7–6) of Chapter 7, we may solve for the expected apparent crater depth and the rim height on the assumption that this is a meteoritic crater. We find the following measurements: apparent depth, 516 feet; rim height, 134 feet; true depth, 382 feet.

The rim now averages about 100 feet high, and the present crater floor is 300 feet below the rim. This places the present floor about 182 feet above the predicted crater bottom. However, bore holes have been forced down through 230 feet of bedded layers. This leaves three possibilities: the structure may not be meteoritic; it may be a meteoritic crater of average form and the bore hole penetrated 48 feet through the top layers of pulverized and brecciated, brine-soaked materials below the original floor of the pit; or the crater was at least 48 feet deeper than an average meteoritic crater of this diameter. Such an extra depth would still be well within the normal scatter of depths of such craters relative to their diameters. The observed rim height is about right for a meteoritic crater 3,400 feet wide after a moderate amount of erosion had occurred.

An additional test is the relationship between the crater diameter and the width of the rim from the crest to where the change of curvature occurs when the rim proper reaches the normal ground level. This point is usually well marked at meteoritic craters, even though much scattered material is broadcast farther out. The average relationship is that the crater is about five times as wide as the rim. At the Pretoria Salt Pan, the ratio is 5.6, as nearly as can be told from aerial views.

About 1920, three nearly radial cuts were made through the rim into the crater to expedite the movement of materials into and out of the pit. The Mauss' Cutting breaches the wall on the south; the Wagon Road and Trona Haulage Cuttings are on the southeast. The sections afforded by the cuttings are complementary. The exposed sides of the Mauss' Cutting form a section through a ring-shaped cone composed of coarse, uncemented breccia made up almost exclusively of large and small angular blocks of red granite resting on a not inconsiderable thickness of grits, the boulders of granite projecting from the grass and scrub at the top of the ridge being merely portions of large blocks ejected by the explosion, which, by virtue of their size, have resisted denudation.

Wagner, in his difficult attempt to explain the crater by volcanic processes (1922), says: "The nature of the breccia makes it clear, moreover, that we have to do with a volcano of the explosive or phreatic type, or, in other words, with a volcano formed by one or more tremendous explosions apparently unaccompanied by the extrusion of lava or other magmatic material" (4).

The Mauss' Cutting was intended originally to serve as a direct haulage way between the caldera floor and the treatment plant. It is around 450 feet long and has a maximum depth of 45 feet. The section shows, in descending order, a maximum of 26 feet of breccia, grits, and native red granite *in situ*. The blocks are irregularly piled on one another, the interstices between the larger ones being filled with finer material. Some of them are grooved and striated, but they show no signs of chemical alteration. There is no break in the deposition of the ejected material, so that only one explosion is indicated. Rohleder found several shatter cones in the ejected breccia.

The grits, which have a thickness of 32 feet, are rudely stratified and are composed of subangular and rounded grains of quartz averaging about 3 millimeters in diameter, set loosely in a scanty matrix of white clayey matter. The layers range up to 2 feet in thickness.

The contact between the breccia and the underlying grits clearly marks the position of the surface at the time when the explosion occurred. This surface now dips radially away from the crater in all directions at an angle of about 7°. This is a measure of the updoming which occurred in the violent action. Wagner recorded the presence in the basal portion of the breccia of small blocks of grits of the same type as that by which it is underlain. This is exactly what would be expected at such an explosion site where some of the ejected fragments of the surface material would come down first as the overturning of the upper layers progressed. They would be found at the bottom of the rim. Those ejected from greater depths would be found in the upper portions of the rim. Shoemaker (5) showed this type of sequence beautifully at the Arizona Meteorite Crater.

The granite underlying the grits is, in contrast to that of the ejected blocks, much decomposed, being of a gray or grayish-pink color, owing to the kaolinization of its feldspar. This proved very useful in distinguishing the granite *in situ* from the breccia while the caldera was being mapped.

The surface layers of granite under the rim also show a radial dip of 7° away from the pit. Proceeding down the Mauss' Cutting, the dip of the divisional planes becomes steeper and steeper, eventually reaching 40° at the bottom of the cutting. At the same time, there is increasing evidence of shattering and compressional shear. The granite, indeed, presents all the phenomena characteristic of a massive, homogeneous rock that has been subjected to a strong compressive stress in the upper part of the zone of fracture. A similar, and in places even more pronounced, uptilting accompanied by the same shear phenomena is observable in the granite all around the rim.

Below the inner end of the cutting it is difficult to see just what has happened to the divisional planes of the rock. Where they can be seen clearly, these planes are found to dip inward toward the depression, evidencing a reversal in the direction of dip. The same phenomena are found in the Wagon Road Cutting. We then find a typical ring anticline surrounding the pit, with its crest slightly inside the crest of the rim. This description could have been written almost equally well about the Odessa No. 1 meteoritic crater, as shown in Figure 2 (p. 20).

On the inner slope of the caldera is a ring fault, sometimes double, which goes all the way around the pit. There is definite evidence of upthrusting of the region inside the ring fault, followed by a certain amount of subsidence. The close similarity of all features of this structure to those of a meteoritic crater is extremely strong but, of course, not conclusive evidence of its nature.

There is no obvious reason for the crater being situated where it is, for it is not visibly connected with any line of fissure or faulting. Apparently it was formed independently of any plane of structural weakness in the earth's crust. Rohleder concludes: "Nothing whatever suggests volcanic origin, but the parallelism with a meteor crater is obvious" (6, 7). Spencer (8) has also made the same suggestion independently. It seems highly probable that this crater, only slightly inferior to that in Arizona, was produced by meteoritic impact, possibly by a vast stony mass. No nickel-iron has yet been reported from this location. Whether or not it has been searched for is not recorded.

The photograph shown in Plate V was sent to me by H. B. S. Cooke, of the University of the Witwatersrand. In addition, he arranged for a series of photographs to be made of the Pretoria Salt Pan as a plane flew over it. The resulting pictures, when viewed in a stereopticon, show a beautiful three-dimensional view of a structure that has a striking resemblance to a meteoritic crater.

The Ashanti Crater

In the Gold Coast of Africa, now the new nation of Ghana, lies a peculiar structure which has caused many a scientific argument to rage. A vast crater, variously reported as $6\frac{1}{2}$ miles in diameter or 7 by 9 miles across, is partially filled by Lake Bosumtwi (Pl. VI). The lake measures 5 miles in diameter and has a maximum depth of 238 feet. The waters are self-contained, and there is no other lake within 500 miles.

The inner rim of the caldera is precipitous, rising from 1,000 to 1,500 feet from the lake. The walls are jungle-clad, and the pit is surrounded by a tropical forest which makes all study of the region difficult. The crater is bounded by a peripheral depression which separates the actual rim from a concentric ring of elevations. The width of the depression is $\frac{1}{2}$–2 miles.

Maclaren (9) originally suggested that the crater had a meteoritic origin, but this has been disputed by Rohleder (10) and Junner (11). No nickel-iron has even been found, with the possible exception of a specimen reputedly picked up over half a century ago on the shore of the lake.

Rohleder reported that the planes of schistosity of the metamorphic (pre-Cambrian) rocks on the inner walls of the caldera dip radially inward and that the dip gradually reverses itself until, beyond the concentric depression, the layers show a diminishing outward dip. He believes that the region was broadly uparched before the center collapsed to produce the caldera. Junner has cast grave doubts on these structural studies.

Both Junner and Rohleder base much of the assumption that the crater is of igneous origin on the finding of a few relic patches of pyroclastic debris on the inner walls. This debris, according to Junner, consists in large part of comminuted granite and phyllite, opaline quartz, and dacite pitchstone in a matrix of glassy pumice. Accompanying it are scattered masses of breccia supposedly produced by the shattering of the local rocks along faults. These extend to a distance of 7 miles from the center of the disturbance.

Junner concludes that the succession of events was as follows: First a doming of the crystalline crust by a laccolithic injection; then an upward and outward thrusting of the crushed rocks by gas pressure, culminating in several explosions, which blasted a funnel-shaped crater and ejected a little gas-rich magma, together with great quantities of lithic debris; and, finally, an engulfment caused by the loss of gas and consolidation of the magma body. That subsidence has continued since the catastrophe is probable from the fact that the lake sediments on the lower walls of the caldera show an inward dip.

Hence we are left on the horns of a dilemma. It is apparent that an explosion did occur, but the origin of the explosive force is under debate. It is also true that a subsidence followed the blast. Junner has outlined one reasonable hypothesis, but may it not be equally reasonable to expect a certain amount of settling within the ring faults produced by the impact of a meteorite and the subsequent rebound of the rock layers? This has been found at the Pretoria Salt Pan (3), the Steinheim Basin (12, 13), the Rieskessel (12), and others.

Shatter cones have been reported from the inner wall of the structure (14, 15). The lake hides the center in this case, and the crater is far larger than the others, where shatter cones have been limited to the central region.

Coesite has now been identified in suevite-like rock at the Ashanti Crater

(16). It now appears to be an authentic meteoritic crater of large size. It is to be hoped that the new nation of Ghana will encourage study of this natural wonder.

The Köfels "Crater"

In 1936 Suess (17) proposed a meteoritic origin for the Köfels "Crater," a curious widening in the Oetz Valley of the Tyrolian Alps. He found no meteoritic materials, but he did identify a peculiar pumice which apparently was derived by fusion of the gneissic country rock. On the suspicion that this pumice was actually meteoritic silica glass, Hammer (18, 19) analyzed samples of it but reported that there was no evidence of nickel-iron. The gneiss is locally fractured and brecciated, and the depression shows signs of having been formed by violent explosive forces. The form of the "crater" is irregular, which might have been expected from the normal rugged nature of the region. There are no volcanic rocks at Köfels and no recent volcanism. Schmidt (20) says, contrary to Suess, that there is evidence of tectonic movement in the region and that therefore the lack of such movement cannot be used as a criterion against the volcanic origin of the "crater."

Again there are not enough facts available to allow a decision to be made. It would be extremely interesting if the Köfels were to prove to be of meteoritic nature, for it would thus become the first recognized case in mountainous land.

The Franktown Crater, Ontario
(Longitude 76°3′.5 W., Latitude 45°03′ N.)

This is a feature about three-quarters of a mile in diameter, slightly depressed in the center. It occurs in Ordovician limestone, and its buried rim may still influence the attitude of the sediments. Beals, Innes, and Rottenberg (24) feel that it warrants a diamond-drilling program.

Labrador Crater
(Longitude 64°03′ W., Latitude 58°03′ N.)

Col. Arthur Merewether of the U.S. Air Force photographed the medium-sized Labrador Crater during World War II (21, 22). Later, V. B. Meen visited the spot in 1953 and 1954 to study the crater. It appears to be 525 feet in diameter and contains a small green lake. The rim is typical of a meteoritic crater. Meen is quoted as saying, in connection with his 1953 expedition: "There was no time for extensive exploration because the weather was closing in, but from all appearances this crater must be listed as of suspected meteoritic origin" (23).

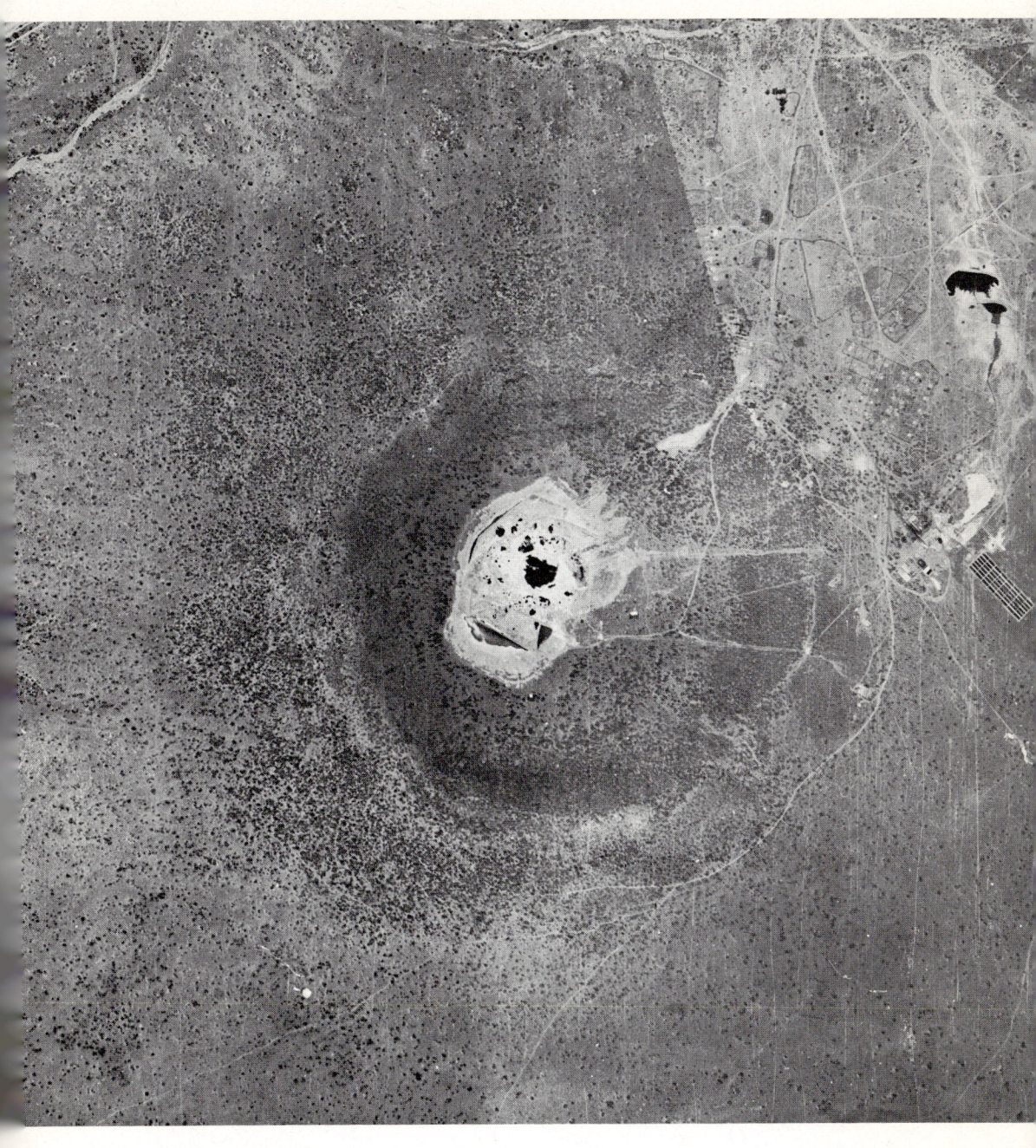

PLATE V.—Pretoria Salt Pan. (Photo of Pretoria Salt Pan reproduced under Government Printer's of the Union of South Africa Copyright Authority No. 2793 of 24/6/60.)

Plate VI.—Lake Bosumtwi in Ashanti Crater. (Courtesy Ghana Information Services Ref. No. R 15751/1.)

PLATE VII. Cratère de Tennoumer. (Courtesy M. Baussart, Secrétaire Général Adjoint de l'U.G.G.I.)

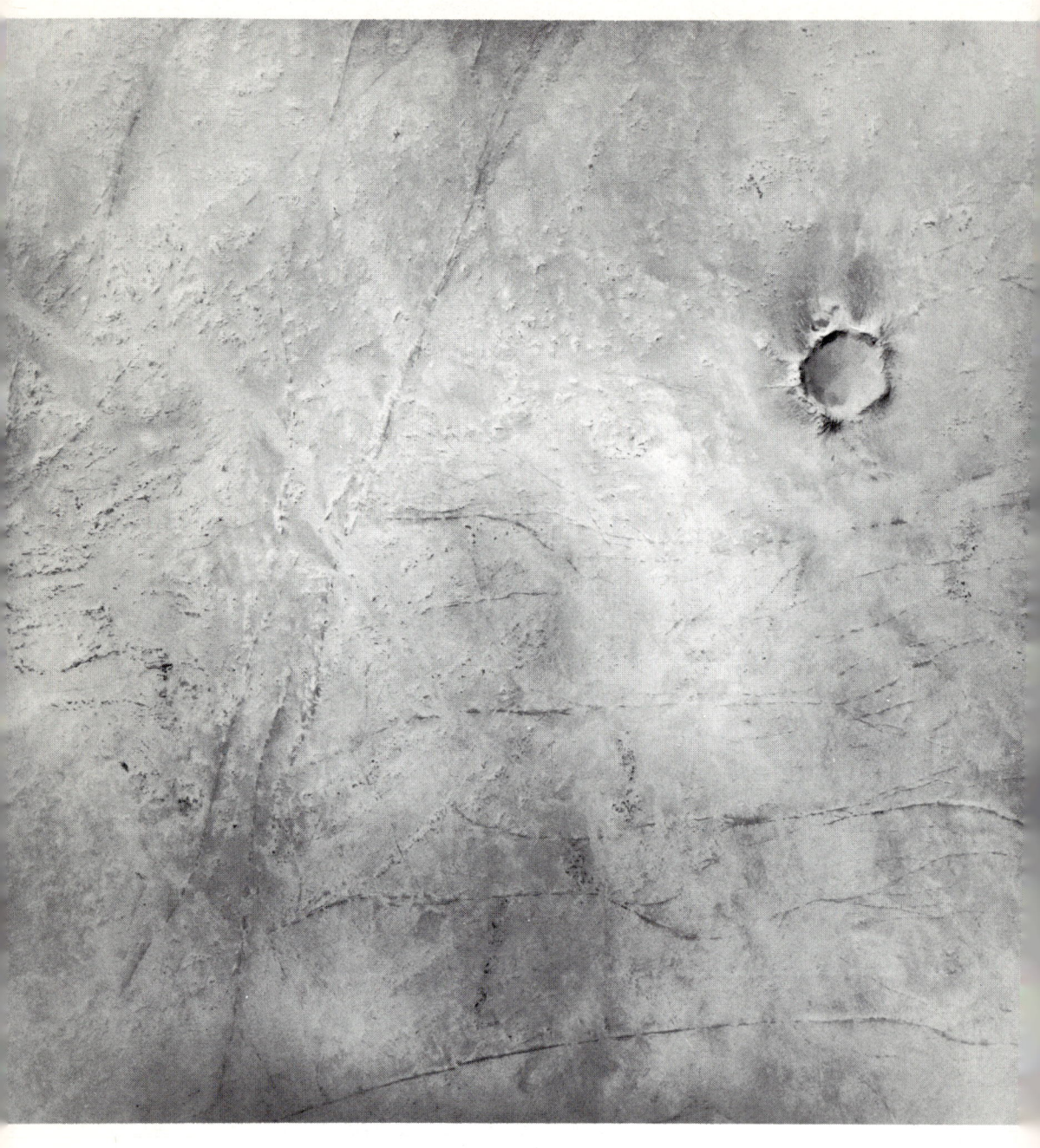

Plate VIII.—Cratère de Temimichat Gallaman. (Courtesy M. Baussart, Secrétaire Général Adjoint de l'U.G.G.I.)

The Clearwater Lakes, Quebec
(Longitude 74°20′ W., Latitude 56°10′ N.)

The Clearwater Lakes are two roughly circular lakes separated only by a screen of islands. They are 20 and 16 miles in diameter. All other lakes in the region are elongated, presumably because of glacial erosion.

In the larger lake is an approximately circular ring of islands concentric with the lake itself. Some of these islands are of considerable height. Geologic studies indicate that they are composed of lava.

Beals, Innes, and Rottenberg (24) suggest that this is a case of twin impact structures and that the islands were produced by the upward extrusion of lava in a ring dike.[1] Similar ring faults are known in several of the "cryptovolcanic" structures. Beals and his associates plan to drill and core the smaller crater lake during the winter of 1962–63.

The Manicouagan Lake Feature, Quebec
(Longitude 68°37′ W., Latitude 51°28′ N.)

An approximately circular area, inclosed by Lakes Manicouagan and Mushalagan, is a conspicuous feature of the map of Quebec (24). It is about 40 miles in diameter, and a mountain rises 3,000 feet high in the center. Geologic studies in the area (25) indicate that the central mountain is an igneous intrusion and that otherwise a large part of the area is covered by flat-lying lavas of somewhat different character. No definite conclusions have been reached, but it has been suggested that this is a very ancient, large crater whose rim has been removed by erosion, leaving the central mountain plus a lava floor.

Circular Structure, Carswell Lake Area, Saskatchewan
(Longitude 109°30′ W., Latitude 58°27′ N.)

In 1957 W. F. Fahrig (26), of the Geological Survey of Canada, discovered a feature approximately 18 miles in diameter, bounded on its circumference by concentric circles of rock outcrops consisting of sandstones and dolomitic sediments. These sediments are probably pre-Cambrian and were tilted and deformed in a manner reminiscent of those on the rim of the Arizona Crater. According to a sectional diagram provided by Fahrig, the strata give the impression of having been compressed along a radius and tilted more than 90° away from the center of the feature.

Beals, Innes, and Rottenberg (24) consider that this structure possibly has a meteoritic origin. Fahrig regards other explanations as more probable at this stage of the investigation.

---

[1] On October 24, 1962, C. S. Beals wrote that Shoemaker and Dence have identified the island ring as composed of breccia with columnar jointing, not lava.

The Nastapoka Islands Arc of Hudson Bay
(Longitude 80°02′ W., Latitude 57°40′ N.)

Beals and his associates (24) described a gigantic crater-like structure as follows:

These coordinates represent the center of curvature of an almost perfectly circular arc on the east coast of Hudson Bay, approximately 275 miles in diameter. This is a conspicuous feature even on a world map and many scientists [24, 27] and others have made the suggestion that it might have been due to the impact of a giant meteorite.

On a moderately large-scale map it is seen that over most of its length the arc is characterized by a screen of offshore islands of which the most important are the Nastapoka Islands, a chain over one hundred miles long of average latitude 57°. Geological studies of the Islands [28–30] have indicated that they are composed of Precambrian sediments, which sometimes extend to the mainland and throughout the length of the arc the sediments dip radially inward toward the centre at angles of a few degrees. Studies of aerial photographs have confirmed the radial direction of dip over the entire length of the arc and they have also confirmed that in many places the sediments extend to the mainland where it is often possible to see the contact between the sediments and the granitic rock of which the mainland is largely composed. When observed from a low flying aircraft, the seaward dip of the sediments is a very striking phenomenon and, considered in connection with the above geological and photographic evidence, suggests the existence of a deep circular basin in which great depths of sediments may well have been deposited. In addition to the offshore islands already mentioned, there are numerous other islands nearer the centre of the circle of which the most important are the Belcher Islands south and east of the centre. Where geological information is available the islands are composed of Precambrian sediments often capped or interbedded with lava flows. It appears that in contrast to observations on the border of the arc, the sediments on the more central islands are in general either flat lying or folded and do not correspond in dip to those on the arc [31]. In addition to the evidence for volcanism on the islands, lava flows are also a feature of certain areas of the mainland near the coastal arc.

On the landward side of the arc, hills normally rise to a height of several hundred feet; in places near Richmond Gulf the elevation is 1,500 ft. above sea-level and this is suggestive of an ancient and eroded crater rim. The incompleteness of the circle on the west is of course a handicap to interpretation and at present there is no evidence of a continuation, under water, of the visible features of the arc. It may be remarked however, having regard to the very great age of the feature (600,000,000 to 1,000,000,000 years) that it would indeed be surprising if it had remained completely intact over such an immense period of time. If this is truly a fossil meteorite crater we are fortunate in having such a substantial proportion of it remaining for study.

... Unfortunately the size of the Hudson Bay feature and its great age are formidable obstacles to investigation. It would appear logical to look for a lava floor under the sediments but their assumed great depth (3,600 ft. near the coast and presumably much greater farther out) would make drilling very expensive. It is also quite probable that consolidation and alteration of the sediments would make it difficult by geophysical methods to establish the existence of a boundary with the basement. In spite of these difficulties it is hoped to undertake gravity, magnetic and seismic work in the area as soon as facilities are available for making measurements of this kind at sea.

The offshore area of this possible crater will be drilled and cored by Beals and his associates during the summer of 1962.

THE GULF OF ST. LAWRENCE ARC
(Longitude 63°03′ W., Latitude 47°06′ N.)

The Gulf of St. Lawrence Arc is an incomplete circular arc about 180 miles in diameter and bounded by parts of Nova Scotia, New Brunswick, the Gaspé Peninsula, and perhaps Anticosti Island. Prince Edward Island and the Magdalen Islands lie within the circle. They seem to be counterparts, at least from their positions, of the ring of islands in the large Clearwater Lake and the Nastapoka Islands.

Seismic observations within the circle have indicated the presence of sediments at depths of approximately 6 kilometers. Much more extended observations will have to be made before conclusions can be drawn, but at least the configuration suggests a great crater (24, 27).

Beals, Innes, and Rottenberg (24) also list other circular features from their studies of Canadian aerial photographs, which, because of the stratification of the rock layers, somewhat similar to that at the Holleford Crater, or for other reasons—e.g., excessive depth, evidence of shattering around the shore line, or simply unexplained incongruity with their surroundings—are considered as possible meteoritic-impact structures. These features are listed in the accompanying table.

Name	Longitude	Latitude	Diameter (Miles)
Mecatina Crater	59°22′ W.	50°50′ N.	2
Lake Michikamau	64°27′ W.	54°34′ N.	3½
Menihek Lake	66°40′ W.	53°42′ N.	3
Menihek Lake	67°10′ W.	54°19′ N.	2½
Sault au Cochons	70°05′ W.	49°17′ N.	7
Lac Couture	75°20′ W.	60°08′ N.	10
West Hawk Lake	95°12′ W.	49°46′ N.	3
Keeley Lake	108°08′ W.	54°54′ N.	8
Ungava Bay	67°20′ W.	60°00′ N.	150

Beals, Innes, and Rottenberg point out that all the features listed in this chapter from the Franktown Crater to Ungava Bay are represented as interesting possibilities worthy of further investigation but cannot yet be presented as probable fossil craters to be included in the statistics of earthly, as compared with lunar, features.

The rest of the world has not been subjected to as thorough a search for features of this type as has Canada, but still there has been an impressive number of craters found which may be impact structures.

The French, through the Union Geodésique et Géophysique Internationale (32), have made an aerial survey of portions of the Sahara and have discovered several additional craters which may, on closer inspection, prove to be meteoritic.

Cratère de Tennoumer
(Longitude 22°55′ W., Latitude 10°25′ N.)

The beautiful object, the Cratère de Tennoumer, is well shown on a photograph taken in 1957 (Pl. VII), although A. Allix (33) suggested in 1951 that it was of meteoritic origin. In a letter from M. Baussart, Secrétaire Général Adjoint de l'U.G.G.I., the scale of the picture is given as 1 to 5,000. If correct, this means that the pit is 625 feet wide.

In spite of the dryness of the area, there have been considerable erosion and infilling, but an upraised rim is well shown. The center is still depressed. Radial markings beyond the rim are clearly seen.

A ring appears on the photograph concentric with the crater and of a diameter slightly more than three times the crater diameter. Its nature cannot be told from the picture. The appearance of the crater of Tennoumer is exactly that of a meteoritic crater of considerable age.

Cratère de Temimichat Gallaman
(Longitude 24°15′ W., Latitude 9°39′ N.)

The Cratère de Temimichat Gallaman is an exceptionally beautiful crater, much resembling Aouelloul. It has the same general shape as Aouelloul, including the two nearly straight sides. If the scale of the photograph (Pl. VIII) is 1 to 5,000, the crater is about 235 feet in diameter, but the picture detail suggests that it was taken from a higher elevation. Like Tennoumer, there is a definite ring surrounding and concentric with the crater. This ring is roughly four times the crater diameter. The impression is given that it is a sunken region. This is a probable meteoritic crater.

Other peculiar structures found on the French photographs are listed in the table on page 63.

Each of the foregoing five formations is of the same order of size as the Tennoumer, Temimichat Gallaman, Aouelloul group. The three named craters all appear to be about the same age and lie on a straight line, with Aouelloul considerably to the northeast of the others. Four of the other five features lie on, or close to, an extension of the same line to the northeast. They are bunched slightly beyond Aouelloul.

Position		Comments
Longitude	Latitude	
11°40′ W.	19° N.	A nearly round structure with raised rim
9°50′ W.	25°40′ N.	A nearly round feature with a suggestion of a raised rim and with a central area much lighter in color than the surroundings
12° W.	22°45′ N.	Two peculiar round features
1°30′ W.	18°15′ N.	Round feature, poorly defined

Close to Aouelloul is a most peculiar structure, considerably larger and probably considerably older than the three nearby craters. It is called the Cratère de Semsiyât. It is usually described as a highly eroded laccolithic dome. The rock layers are exposed in circular fashion, with all participating strata dipping radially away from the center. It is sufficiently unusual to warrant more study than has yet been accorded it.

Lonar Lake, India
(Longitude 76°51′ E., Latitude 19°59′ N.)

Lonar Lake is a circular feature just over a mile in diameter and 400 feet deep (24). Geologists (34–36) who have examined it have attributed it to a volcanic explosion, but specific evidence for volcanism appears to be lacking. It may well prove to be meteoritic, for it is circular and has a raised rim.

In addition to the above large craters, several others have been reported. In *Sternenwelt* (37, 38) such a crater was described in 1952. Nothing further has been published on it.

LaPaz (39), in 1947, described a possible meteoritic crater on Amak in the Aleutian Islands. It had first been found by Keenan in 1945. The dimensions are given as 350 feet wide by 50 feet deep, and it lies in soft dirt and gravel. It possesses a low rim.

In 1951 Rinehart and Elvey (40) described a possible crater of this type near Duckwater, Nye County, Nevada. The pit is 225 feet in diameter, round, and

10–15 feet deep relative to a low rim. They found that there were at least 25 feet of sediments in the crater. There may be an Indian legend concerning it. No meteoritic material was discovered.

In 1949 LaPaz (41) described a very small crater of unknown origin in northeastern New Mexico. It was only 30 by 3 feet, and the photograph showed that it resembled a surface-burst crater rather than a meteoritic crater.

In 1954 Knetsch (42, 43) reported the discovery of two new craters near Baghdad which possibly are meteoritic.

In 1957 Bonfanti (27) described the Hudson Bay, Gulf of St. Lawrence, and New Quebec structures.

Also in 1957 Sangster (44) wrote that P. S. Hossfield, senior lecturer in geology at the University of Adelaide, Australia, had discovered a peculiar crater about 2 miles across, in the Northern Territory. It may have several associated smaller craters. From the photograph it does not appear very similar to known meteoritic craters, but Hossfield believes that there is no geologic evidence for a terrestrial origin.

References

1. Gèze, Mrs. B., and Cailleux, A. "Existence probable de cratères météoritique à Cabrerolles et à Faugères (Hérault)," *C.R.*, April 24, 1950.
2. Janssen, C. Luplau. "The Meteor Craters in Hérault, France," *J.R.A.S. Canada*, **45**, 190, 1951.
3. Wagner, P. A. "The Pretoria Salt-Pan, a Soda Caldera," *Mem. Geol. Surv. South Africa*, Vol. **20**, 1922.
4. ———. *Ibid.*, p. 21.
5. Shoemaker, E. M. "Impact Mechanics at Meteor Crater, Arizona." Prepared on behalf of the U.S. Atomic Energy Commission, Open File Rept., July, 1959.
6. Rohleder, H. P. T. *Zs. f. Deutsch. Geol. Gesellsch.*, **85**, 463, 1933.
7. ———. *Geol. Mag.*, **70**, 489, 1933.
8. Spencer, L. J. "Meteorite Craters as Topographical Features on the Earth's Surface," *Geog. J.*, **81**, 227, 1933.
9. Maclaren, M. "Lake Bosumtwi, Ashanti," *Geog. J.*, **78**, 270, 1931.
10. Rohleder, H. P. T. "Lake Bosumtwi, Ashanti," *Geog. J.*, **87**, 51, 1936.
11. Junner, N. R. *Gold Coast Geol. Surv. Bull.*, No. 8, 1937.
12. Bucher, W. H. "Cryptovolcanic Structures in the United States," *Rept. 16th Internat. Geol. Cong.*, **2**, 1055, 1933.
13. Branca, W., and Fraas, E. "Das Kryptovulcanische Becken von Steinheim," *Akad. Wiss., Berlin*, Abh. I, p. 1, 1905.
14. Rohleder, H. P. T. *Centr. Mineral. Geol.*, **1934A**, 316, 1934.
15. Dietz, R. S. "Meteorite Impact Suggested by Shatter Cones in Rock," *Science*, **131**, 1781, 1960.
16. ———. "Astroblemes," *Sci. American*, **205**, 51, 1961.

17. Suess, F. E. *Neues Jahrb. Beil., A,* **72,** 98, 1936.
18. Hammer, W. *Verh. Austria. Geol. Bund.,* Nos. 9–10, p. 195, 1937.
19. ———. *Ibid.,* No. 12, p. 268.
20. Schmidt, W. *Zentralbl. Mineralog.,* B. 2, No. 5, p. 222, 1937.
21. Beals, C. S., Ferguson, G. M., and Landau, A. "A Search for Analogies between Lunar and Terrestrial Topography on Photographs of the Canadian Shield," *Contr. Dom. Obs., Ottawa,* **2,** 3, 1956.
22. ———. "A Search for Analogies between Lunar and Terrestrial Topography on Photographs of the Canadian Shield," *J.R.A.S. Canada,* **50,** 203–11, 250–61, 1956.
23. *Toronto, Globe and Mail Limited,* May 24, 1954.
24. Beals, C. S., Innes, M. J. S., and Rottenberg, J. A. "The Search for Fossil Meteorite Craters," *Contr. Dom. Obs., Ottawa,* **4,** 3, 1960.
25. Rose, E. R. "Manicouagan Lake, Mushalagan Lake Area," *Quebec Paper,* Geol. Surv. of Canada, No. 55-2, 1955.
26. Fahrig, W. F. Reported in Beals *et al.* (24), p. 26, 1960.
27. Bonfanti, N. "Una nuova ipotesi nella storia della terra," *Universo: Riv. bimestrale Ist. Geog. Militare,* Anno XXXVII-4, Luglio–Agosto, 1957.
28. Bell, R. "Report on the East Coast of Hudson Bay," *Geol. Surv. Canada, Rept. of Progress,* No. 128, 1877–78.
29. Low, A. P. "Report on East Coast of Hudson Bay," *Geol. Surv. Canada, Rept. D,* 13, 1900.
30. Kranck, E. H. "Geology of the East Coast of Hudson Bay," *Acta Geog.,* **11,** 1950.
31. Jackson, G. D. Private communication to C. S. Beals, reported in Beals *et al.* (24).
32. Communication from M. Baussart, April 29, 1959.
33. Communication from M. Baussart, June 16, 1959.
34. Medlicott, H. B., and Blanford, W. T. *A Manual of the Geology of India,* Part 1. 1879.
35. Blanford, W. T. *Records of the Geol. Soc. India,* **1,** 63, 1870.
36. Newbold. *J. R. Asiatic Soc.,* Vol. **9,** 1846–48.
37. "Meteor-Krater in Sudfrankreich," *Sternenwelt,* **4,** 174, 1952.
38. ———. Abstr. *Astr. Jahresb.,* 1952.
39. LaPaz, L. "A Possible Meteorite Crater in the Aleutians," *Pop. Astr.,* **55,** 156, 1947.
40. Rinehart, J. S., and Elvey, C. T. "A Possible Meteorite Crater near Duckwater, Nye County, Nevada," *Pop. Astr.,* **59,** 209, 1951.
41. LaPaz, L. "A Possible Meteorite Crater in Northeastern New Mexico," *Pop. Astr.,* **57,** 136, 1949.
42. Knetsch, G. "Zwei neue Meteoritenkrater?" *Sterne,* **30,** 66, 1954.
43. ———. Abstr., *Astr. Jahresb.,* 1954.
44. Sangster, R. L. "Another Meteorite Crater in Australia?" *Sky and Telescope,* **16,** 429, 1957.

4

ANCIENT METEORITIC CRATERS AND CRYPTOVOLCANIC STRUCTURES

In a few more million years, the great Crater of Talemzane will be difficult to recognize from the surface. Ancient as it is, the great structure is, geologically speaking, extremely youthful, for the latest figures indicate that the earth was formed 4,500,000,000 years ago. In this tremendous span, there must have been myriads of such craters formed, larger and smaller.

The problem is: When the craters are older than the Crater of Talemzane, how shall we be able to recognize them? Most of the remaining structures will have been lost to view, as thick layers of rock have been laid down over them. The oldest of the great shields on the surface of the earth (1) is perhaps 3,000,000,000 years old, and even it does not represent the primal crust; hence much of the record is permanently lost.

Often thousands of feet or even tens of thousands of feet of sediments have been eroded away from a given site. The earth's geologic history is a record of rising and falling lands, advances and retreats of vast epeiric seas, mountain building, faulting, folding, erosion, and deposition. All these must affect the record of ancient conditions as transmitted to us. Rare indeed is the surface of land which has descended without major change since even as recent a geologic time as the Cretaceous. What chance, then, is there for geologists to identify a meteoritic crater of Archeozoic or even of Cambrian age?

The answer to the foregoing is that it would have required an almost impossible series of events to have occurred for a complete fossil crater to have been preserved. All the normal features usually associated with a meteoritic crater would probably have vanished—the pit, rim, nickel-iron fragments, ejectamenta, the tilted and brecciated rocks. All that could reasonably be expected to remain would be the basement structures, the modifications produced deep beneath the earth's surface by the unimaginably great pressures and shock waves which developed as the intruding mass came to a halt and exploded, or a filled-in crater later exposed by erosion. These are the formations which must be present on and in the earth if the earth's craters are meteoritic in origin.

It is amply clear that the earth is being bombarded at the present time by masses large enough to produce small craters. It is highly illogical to assume that these bodies started falling only recently. It is highly probable that, throughout geologic history, impact craters were formed and that, mixed with the small meteorites, there were larger ones similar to, or part of, the lesser asteroids.

Recent astronomical discoveries have somewhat expanded our usual mundane viewpoint. Several small asteroids have been observed to approach perilously close to the earth. In 1898 Witt of Berlin found Eros 433, which at times comes within 14,000,000 miles. No more of this class were detected until 1911, when Albert 719 was discovered. It, too, comes well inside the orbit of Mars but was followed for so short a time that it has never been recovered on subsequent passages. Some day it will probably be found again. In 1932 the Belgian astronomer Delporte photographed Amor 1221, which comes nearer the earth than does either Eros or Albert. Apollo was found by Reinmuth at Heidelberg on April 27, 1932. It was shown to have been, on that passage, as close as 1,800,000 miles, to have a period of 1.8 years, and actually to go closer to the sun than does Venus. Apollo came within 84,000 miles of Venus on this passage. Hence it may often be in the earth's neighborhood.

On February 12, 1936, and October 28, 1937, Adonis and Hermes were discovered, the former by Delporte and the latter by Reinmuth, just to keep the contest even. Adonis passed the earth by about 900,000 miles, while Hermes was even closer. It missed by the astronomically small space of 600,000 miles, less than three times the distance of the moon—a definitely narrow escape. Unfortunately, Adonis, Apollo, and Hermes came so close to the earth that they flashed into view and quickly faded. Their orbits are very poorly known but, like Albert, may be picked up again on some future visit. Several other small neighboring asteroids have been detected.

Except for Eros, which is several times larger, these little bodies are perhaps 1 mile in diameter and hence can be seen or photographed only when they are very near, but, in order to pass close to the earth, their orbits must be so oriented that these asteroids cross the plane of the ecliptic when they are at the approximate distance of the earth from the sun. These cosmic cannon balls, each loaded, in effect, with high explosive, must be representative of thousands of such bodies. Moderately close approaches are rather frequent, as we are just coming to realize. Actual direct hits must be rather rare. Watson believes that "the earth probably goes at least a hundred thousand years between collisions with them" (2).

Each tiny asteroid would be capable of blasting a hole in the earth's crust entirely comparable to all save the very largest of lunar craters. A mass not much greater than that of Eros would be adequate to produce the largest crater.

The impact of a large meteorite against a rocky surface is an occurrence of such magnitude that it far transcends any and all the familiar terrestrial phenomena. Nevertheless, we may apply known physical laws and properties of matter to the problem and arrive rather closely at a broad picture of the major happenings consequent to such a landing.

Somehow or other, the tremendous momentum and kinetic energy of the rapidly moving meteorite must be accounted for after the explosion. There are only four possibilities: First, the impact will shift the earth, or moon, slightly in its orbit. Second, if the explosion is violent enough, matter will be "kicked back" into interplanetary space and thus depart with some of the original energy. Third, strains may be set up in the crustal layers, some of which will be displaced, others crushed and powdered. Fourth, heat will be generated.

Loss due to the first point will always occur. In the second case, if the impact is on the earth, the blanket of air and high gravitational pull will prevent matter from again reaching outer space, and the net result of the ejection of matter from the explosion pit will be the production of heat, often distant from the point of original impact. Matter from a meteorite crater could easily be hurled completely away from the airless moon. The semipermanent strains set up in the crust will account for a small part of the energy, as will the changes in potential energy due to displacement of matter, but, in the long run, most of the meteorite's energy will be turned into heat; the crux of the problem is to identify where and how this heat is distributed.

It is a well-known fact that, in all but three of the meteoritic craters recognized on earth, there is little evidence of any great amount of heat being released upon impact. The two exceptions are the Wabar craters in the sandy wastes of the Arabian Desert, where large amounts of fused silica coat the walls and bottoms of the craters, and the larger of the Murgab craters. In other meteoritic craters, fused silica glass is usually found, and occasionally certain rocks are metamorphosed, but evidences of great amounts of heat are lacking. From these facts, it seems apparent that much of the energy is transmitted in shock waves through the crust and air and thence gradually converted into heat.

Large meteorites do, undoubtedly, reach the surface of the earth or moon with essentially undiminished speed. Direct measurements of meteoric velocities give top figures of the order of 50 miles per second. Others move more slowly but, even so, must strike with stunning force if they are of a size capable of producing a crater. It can easily be calculated that at a speed of only 4 miles per second each gram of matter in a meteorite possesses energy equal to 5 grams of TNT, or about 5,000 calories per gram, while at 50 miles per second the energy available is nearly 800,000 calories per gram. On impact, this energy must be quickly dissipated.

Except for the very smallest of the relatively few modern meteorite craters known on earth, evidences of tremendous explosive activity are obvious. There is always a radial distribution of both meteoritic matter and crustal rock scattered over perhaps ten times the radius of the crater proper. The local brecciation of the nearby rock layers and their pulverization into rock flour, the occasional manifestations of intense, but localized, thermal metamorphism, and the radially outward dip of rim rocks all lead to the conclusion that tremendous explosions have occurred at these localities and that these explosions have been caused by the impacts of meteorites.

These data being granted, it remains to see how the energy is dissipated. For large bodies, only a very small amount of the available energy of motion will be lost during the flight through the air. The amount lost is measured by the difference in striking velocity and that at the instant the air resistance began, with proper corrections for the acceleration of gravity. This loss must be a small percentage.

When a meteorite strikes the earth, it encounters tremendous resistance. This is the resultant of four factors. The first is the rigidity of the surface layers, particularly against shearing forces; the second and third are resistance to compression and the amount of heat produced; and the last is the inertia of the atoms and molecules in the earth as they fight to keep from being displaced. For a collision with a low-velocity body moving perhaps 5 miles per second or slower, the resistance is largely controlled by rigidity; but, once the velocity becomes greater than that of the waves created in the earth's crust, the resistance due to inertia becomes far and away the most powerful factor, and all other resistances fade into insignificance. It is easy to visualize this condition. For an ordinary blow, such as a hammer striking an anvil or even a bullet striking a piece of steel, the elastic waves travel in advance of the impinging body. Hence the molecules are set in motion; and they have time to get out of the way. On the other hand, a high-velocity meteorite strikes with a higher velocity than the shock waves it produces, the molecules are given no warning of its coming, and consequently they are trapped in front of the meteorite. The accelerations of the particles and the forces necessary to produce them become enormous.

A calculation can be made which applies this principle to a hypothetical collision between the earth and a large meteorite. It is assumed that the body is a spherical mass of nickel-iron, 250 feet in diameter, moving with a velocity of 100,000 feet per second (19 miles per second), and that the earth encountered has a density of 170 pounds per cubic foot. The peak pressure created by inertia is calculated to be about 25,000,000 atmospheres, or roughly ten times the calculated pressure at the center of the earth. The kinetic energy at the time of impact is 3.2×10^{24} ergs.

At first, the meteorite will plunge into the earth, moving faster than the shock waves and pushing ahead of it an ever increasing plug of compressed rock and probably a similar plug of compressed air. When the speed of the meteorite becomes less than that of the elastic waves, the vast amount of compression produced finds a shoulder against which to push, and the mass is soon stopped.

As long as the velocity is greater than that of the shock waves, very little heat is generated, as heat is a measure of the random motions of molecules and atoms, and these random motions are temporarily stopped during the initial phases of the impact when the tremendous meteoritic velocity is imparted to the plug. It is only after the velocity drops below that of the shock waves that the phenomenon of heat enters the picture.

The explosion and the production of shock waves, of course, start the instant the meteorite touches the ground. At high velocities, the meteorite has effectively no strength and will deform almost instantly; but this does not change the over-all general picture. With the stoppage of motion, the highly compressed meteorite is sitting on top of a tremendously compressed, tremendously hot, plug of matter. Naturally, an explosion of the utmost violence follows. This explosion will be directed upward at first in the direction of least resistance, but, as it develops and the explosive focus moves upward, the blast will tend more and more horizontally.

By contrast, if any large mass of meteoritic material is found to remain in a crater, it implies that the body struck with a low velocity, so that the resultant explosion was not violent enough to eject the meteorite. The size of the meteorite necessary to account for any given crater increases as the striking velocity decreases. It may be noted that this inertial resistance is not a function of the state of matter; hence water will stop and eject high-velocity meteorites as well as will rock.

When a large meteorite strikes the earth, it deals a terrific blow to a medium which has a limited degree of freedom and a high degree of elasticity of volume. While some materials, such as clay, have little or no elasticity of shape, they all have great elasticity of volume. Brittle substances are not shattered by pressure, if the pressure is applied to all sides, but by tension. Hence, after compression, they all rebound. Therefore, as a result of impact and explosion, a series of concentric waves would go out in all directions, forming ring anticlines and synclines. These waves would be strongly damped by the overburden and by friction along joint, bedding, and fault planes. The central zone, completely damped by tension fractures produced by rebound, would become fixed as a structural dome.

It must be admitted, however, that we do not know too many of the details of the basement structures. Nevertheless, the general and simplest type of structure to be expected beneath large meteoritic craters would be a central dome surrounded by a ring syncline and possibly other ring folds, the whole resembling a group of damped waves.

These structures should not be radially symmetrical unless the falling meteorites struck the surface of the earth at right angles. Rims of known craters commonly show opposed points of maximum and minimum uplift, suggesting that the impacts were oblique rather than vertical. An oblique blow would be expected to impart bilateral, rather than radial, symmetry to resulting *structures*, although the *craters*, which result from up and outwardly directed explosions, should approach more nearly radial symmetry. Long after these craters had been destroyed by erosion, the underlying formations might be preserved.

In 1938, two men, Boon and Albritton (3–6), recognizing the difficulties inherent in finding and identifying ancient meteoritic-impact structures, decided to see whether they could deduce just what could happen at various depths when a great meteorite struck. The characteristics of the upper parts—the crater, the rim, and the upturned rocks—were familiar. Little was known about the lower parts. They recognized that, under the influence of the shock, the rock layers would behave as though they were fluid and would react against the impact thrust. They also saw that the instant the pressure was released, the rocks would freeze in whatever contorted position they might find themselves, except in the immediate neighborhood of the meteorite, where the explosion would occur.

With these basic premises, Boon and Albritton drew the amazingly modern Figure 5. The section from A to A would represent a new-born crater. BB illustrates the same structure after erosion had destroyed the crater, but before the rock layers below the pit were exposed. CC and DD are progressively deeper sections. These two men found that there was only one type of structure

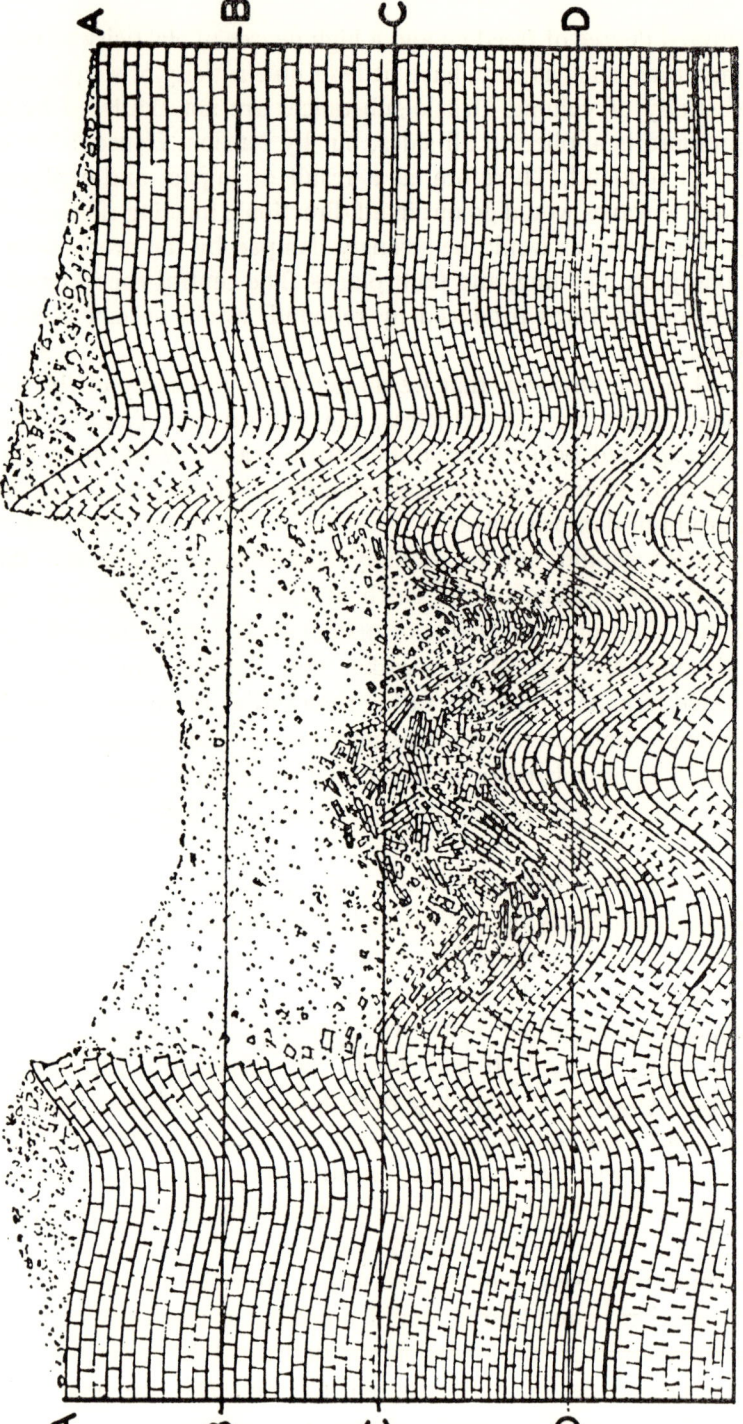

FIG. 5.—Attempt by Boon and Albritton to diagram the probable structure beneath a typical meteorite crater. The aspect of the scar at any time after origin depends on the level to which it has been denuded. The horizontal lines on the illustration indicate five possible dates: *AA*: modern, all features observed; *BB*: inconspicuous; *CC*: underlying strata begin to appear; *DD*: central uplift and ring folds become apparent; *Far below*: gone.

known on earth that fitted the predicted characteristics of their theoretical model. These were the so-called "cryptovolcanic" structures.

Distributed widely over the world, but not yet adequately studied except in scattered instances, are large disturbed regions of the earth's crust known as "cryptovolcanic structures." They are subcircular, complex domical structures characterized by intense deformation and brecciation within an area of a few square miles. Bucher (7) has cited the following characteristics common to six American objects that he believes to be cryptovolcanic. The same list applies equally well to similar objects anywhere on earth:

1. They show a tendency toward a circular outline.
2. A central uplift is surrounded by a ring-shaped depression, with or without well-developed marginal folds beyond it.
3. In the larger disturbances the area of the uplifted central part is small compared with the areas that sank.
4. Where the nature of the rock materials permits any judgment, evidence is found of violent action, such as seems explicable only as the result of sudden release of pressure—that is, of an explosive force.
5. Except in the Decaturville structure, no volcanic materials or signs of thermal action have been observed.

At the present time these cryptovolcanic structures are generally believed to have been formed by "disturbances produced by the explosive release of gases under high tension, without the extrusion of any original magmatic materials, at points where there had previously been no volcanic activity" (7).

The known cryptovolcanic structures did not all occur simultaneously or even close together in time. They range from early Paleozoic to late Tertiary, a ratio of perhaps 50:1 in age. It has been pointed out that the cryptovolcanic structures are all found in areas where bedrock layers are well exposed, unaffected by intense folding, which have been subjected to close geologic scrutiny. How many more lie under the immediate surface, hidden by layers of rock and soil deposited subsequently? What numbers of cryptovolcanic structures have been forever destroyed by the implacable forces of erosion or have been rendered unrecognizable by the processes of folding and mountain building? How many lie waiting on the surface for future geologists to discover?

It is not strange that geologists sought the origin of cryptovolcanic structures in known terrestrial processes. It is strange that they went so far astray. The cryptovolcanic structures are now recognized by many, though not all, authorities as being the results neither of hidden explosions nor of volcanic nature. They are almost certainly the marks showing where great meteorites came to their spectacular ends.

Dietz has coined the word "astroblemes" (8), meaning "star wounds," for such objects. This term is descriptive and may well be accepted, but I prefer to call these formations "impact structures." By this term is included all features by which the postimpact landscape and subsurface layers can be distinguished from the preimpact location.

The similarity of these formations to a theoretical impact structure is not enough to establish the identification. In the last few years, four additional tests have been discovered. The presence of coesite (9) or stishovite (76) in a crater proper now seems to be evidence that an explosion or shock of the necessary violence took place. They have never been found associated with volcanic explosions.

R. S. Dietz did not make the original discovery of shatter cones (Pl. IX) in these structures (10), but he alone is responsible for establishing the reason for their presence at formations which presumably are meteoritic in origin. He has long been an advocate (11, 12) of the theory that the astroblemes are old meteoritic landing places. He states (8):

> Shatter cones are striated cup-and-cone structures found usually in carbonate rocks, but they also have been noted in shale and chert. Presumably, a fine-grained homogeneous rock like dolomite favors their development, but it is not an absolute requirement. The striated surfaces radiate from small parasitic half-cones on the face of a master cone—a pattern which serves to differentiate these striations from the parallel grooving of slickensided fault planes. The apical angles of the cones are from 75 to nearly 90 degrees. The size of the cone apparently depends upon the thickness of the bed which yields as a unit. Some cones as small as 1 centimeter in height have been collected at the Crooked Creek (Missouri) deformation, while at Kentland (Indiana) cones as long as 2 meters have been seen in limestone and cones longer than 12 meters in shale. This coning is apparently a type of mechanical failure under percussion which causes a stratum to become thinner and slightly more elongate by normal faulting—that is, by downward slipping of the cup relative to the cone.

A pertinent question is: Do shatter cones unquestionably require a shock wave for their formation? This cannot be categorically answered, but it is clear that they are quite distinct and different from two geological structures which they vaguely resemble and with which they might be confused—namely, cone-in-cone structures and slickensides. In the United States, shatter cones have been found only very near the center of some of the structures identified in the 1940 edition of the *Structural Map of the United States* as cryptovolcanic structures—that is, deformations formed by a hidden explosion somehow considered to be related to volcanism although no direct evidence of this volcanism, such as volcanic rocks or hydrothermal alteration, is found. Thus, they are uniquely present in structures considered to have been caused by a natural explosion.

Shatter cones have never been reported from normal geological situations, so it would seem that they are not formed by tectonic stresses or by simple static loading. Nor have they been reported from rocks known to have been engulfed by volcanic explosions. So far as artificial detonations are concerned, low-velocity heaving explosives such as commercial dynamite (detonation wave velocity, about 5000 m/sec), which

PLATE IX.—Shatter cones collected near "ground zero" at Wells Creek Basin, Tennessee. Scale in centimeters. (Courtesy Robert S. Dietz.)

PLATE X.—Aerial mosaic photograph of the Steinheim Basin

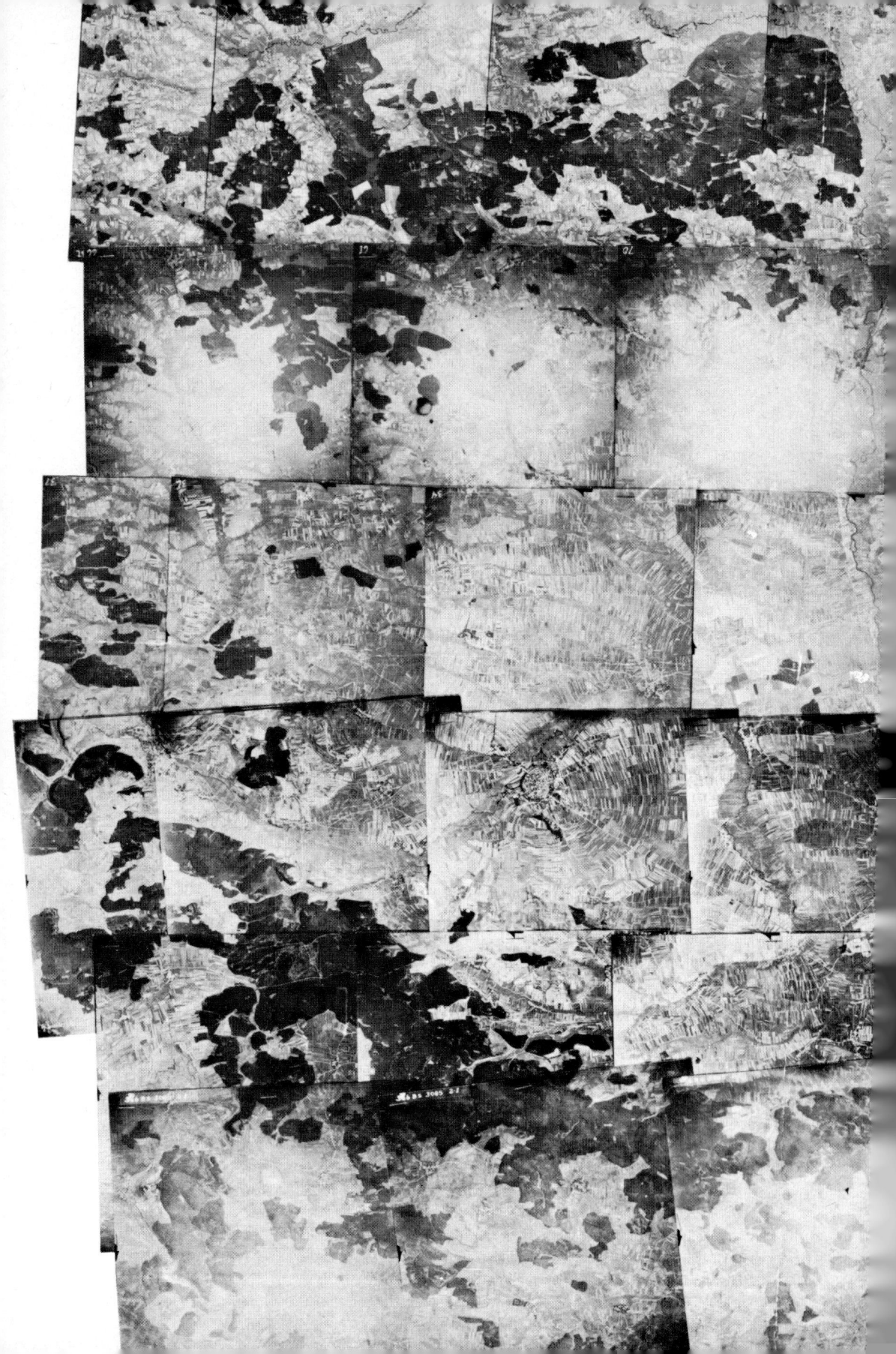

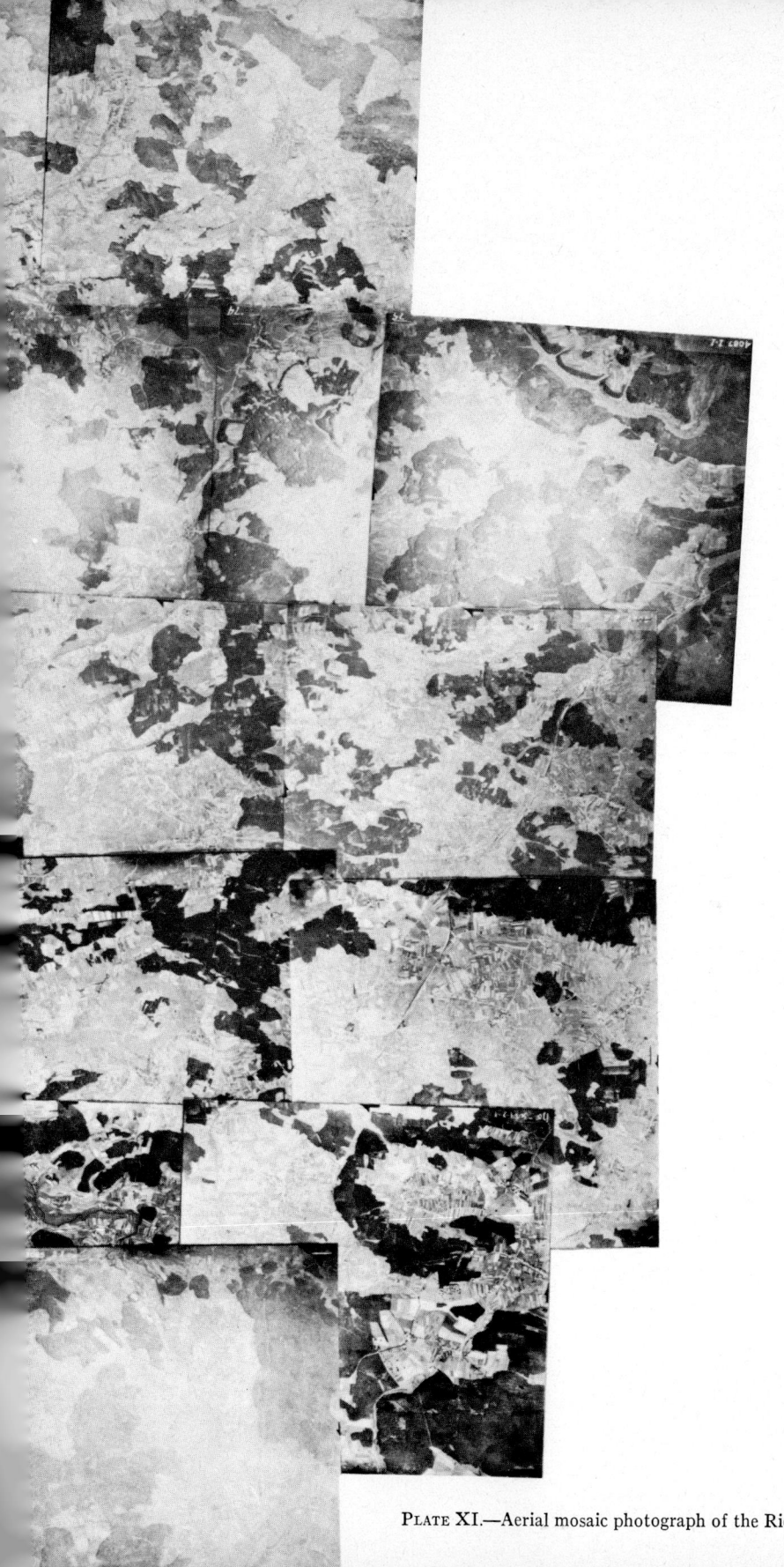

PLATE XI.—Aerial mosaic photograph of the Rieskessel

Plate XIII.—Aerial photograph of the Brent Crater. (Courtesy C. S. Beals, Dominion Observatory, Ottawa, Canada.)

Plate XII.—Aerial photograph of the Holleford Crater. (Courtesy of C. S. Beals, Dominion Observatory, Ottawa, Canada.)

PLATE XIV.—Aerial mosaic photograph of the Deep Bay Crater, Saskatchewan. (Courtesy C. S. Beals, Dominion Observatory, Ottawa, Canada.)

PLATE XV.—Aerial mosaic photograph of the Wells Creek Basin, Tennessee. (U.S. Department of Agriculture, Production and Marketing Administration.)

PLATE XVI.—Aerial mosaic photograph of the Sierra Madera Dome, Texas. (U.S. Department of Agriculture, Production and Marketing Administration.)

PLATE XVII.—Aerial mosaic photograph of the Decaturville Structure, Missouri. (U.S. Department of Agriculture, Production and Marketing Administration.)

Plate XVIII.—Oblique aerial photograph of the Crooked Creek Structure, Missouri

are almost exclusively used for quarrying and similar operations, commonly produce rude cones, but these lack the surface markings of shatter cones. On the other hand, military explosives of high detonation velocity and high *brisance* or shattering effect, like RDX (detonation wave velocity, 8000 m/sec), form cones with surface markings closely resembling those of shatter cones but not so perfectly formed as shatter cones. By extrapolation, it would seem that even greater *brisance* than that of RDX is needed to produce good shatter cones. Since the mean geocentric velocity of meteorites is about 15,000 meters per second, it is to be expected that extremely intense shattering will result for large bolides not appreciably cushioned by the atmosphere. One may conclude that rare natural conditions are required to produce shatter cones but that such conditions could be provided by a large meteorite impact. A visit to the site of an underground nuclear bomb test revealed portions of large shatter cones in volcanic tuff like those in Richmond shale at Kentland Quarry. Nearby, indurated clay beds beneath the sites of large test explosions of TNT displayed tent-shaped features very similar to shatter cones. The tent shape presumably resulted from a cylindrically spreading shock wave produced by the cylindrical high-explosive charge.

The orientation of shatter cones is useful for establishing the impact direction. In most cases the cones point opposite to the direction of shock-wave propagation—in other words, toward the locus of pulse source. It is to be expected that the simple spherical spread of the shock wave from its source will be complicated by reflections from interfaces and by other factors, so that shatter-cone orientation will be somewhat complex.

Shatter cones probably are a specific criterion for identifying the root structures of large fossil meteorite impacts, so geologists have a useful "index fossil" for astroblemes. These provide a site for the study of effects of hyper-velocity impacts which have released energies equivalent to several H-bombs.

Shoemaker, Gault, and Lugn, in an important paper (75), have demonstrated some of the mechanisms and conditions under which shatter cones may be formed by shock waves from hypervelocity impacts.

The fourth test for an impact structure is to find the eroded remnants of ring anticlines and synclines surrounding an upraised central dome. Some geologists are not yet willing to accept the tests as conclusive. One prominent geologist (76) has written me:

... as a geologist I find it impossible to accept the interpretation of origin by meteor impact for the three most emphasized examples of "astroblemes," viz., the Ries and Steinheim Basins, and the Vredefort Dome. The reasons lie in the regional relations and structural detail. As far as I know, the geologists familiar with these regions share my view. Please understand that we would be as quick as anyone to accept the meteor impact hypothesis since it would solve some of the worst difficulties we face in trying to explain the structures.

... wherever cryptovolcanic structures are sufficiently eroded the depression surrounding the central uplift is invariably shown to be due to downbending and downfaulting, not to explosion cratering. This is true of the larger features of the Ries Basin, where the depression caused by explosion is only a relative detail.

Finally, I believe it would be desirable to point out that not one of the terrestrial meteor craters shows any trace of central uplift, while all cryptovolcanic structures of similar sizes invariably show one.

In rebuttal it should be pointed out that the mechanisms of impact crater formation by shock and explosion as now understood require downbending, downfaulting, and a central rebound structure in the basement rocks. The absence of a central peak in the terrestrial meteoritic craters is not evidence of its absence in the rock layers below the pit proper. These are the layers exposed in cryptovolcanic structures. The terrestrial meteoritic craters that have been adequately studied are usually considerably smaller than the cryptovolcanic structures. Beals is studying the cores at the Brent Crater to see whether such an effect can be found. The Kaali Järv does show a small central peak.

With these four tests in mind, let us examine certain of the terrestrial features which do or may belong in the class of meteoritic-impact structures.

The Steinheim Basin, Germany

In 1905, Branca and Fraas (10, 13–16) coined the term "kryptovulcanic"—since Anglicized to "cryptovolcanic"—for the Steinheim Basin in southern Germany (Pl. X). This depression commences abruptly in a plain of flat Jurassic limestone. At present it measures about $1\frac{1}{2}$ miles in diameter[1] and is roughly 260 feet deep. Whatever rim it once had has been removed by erosion. In the center is a gently rounded hill, the Klosterberg, 130 feet high, composed of Jurassic rocks older than the other parts of the basin. The Klosterberg is the remnant of a domical structure from which the younger rocks have been stripped.

The original pit, formed in Miocene times, was filled with a Miocene lake without a drainage exit. Hence, even early in its history, the Steinheim Basin was a surface formation. Today the rim has twice been breached, so that the lake has been converted into placid farm land. It seems probable that the net effects of erosion have been simply to reduce the initial uplift of the Klosterberg and to remove the raised rim, meanwhile depositing sediments in the crater.

Several other factors lead to the same conclusions. No volcanic materials are present at Steinheim. The only extraneous objects are occasional large blocks of the usual Jurassic limestone which had been hurled from the pit at its forma-

---

[1] The generalized map of the Steinheim Basin shown by Bucher (7) scales almost exactly 2 miles in diameter measured from the outer edge of the syncline.

tion. The region of activity is sharply bounded by a rough circle. Inside this limit, the explosive forces operated with tremendous effects. Outside, apparently nothing was changed.

Rohleder says: "There is nothing whatever to prove the volcanic origin of the Steinheim Basin and the fact that the immediate neighborhood was in no way involved seems to point out that whatever caused the basin was an absolutely local event" (17), i.e., a meteorite.

The Klosterberg is the counterpart of the central peak so often found at the larger lunar craters. Its rocks have been uplifted 500 feet from their normal level, but, even originally, they did not extend above ground level. The limestone in the dome and other parts of the basin has been thoroughly shattered. They form a "confused structure that locally suggests a kneading together of heterogeneous units" (7).

Kranz and Gottschick (14) have found masses of breccia which make it probable that the basin was formed first as a very shallow, saucer-shaped explosion funnel. The best evidence of an explosion action is seen in the intense brecciation of the Upper Jurassic limestones along the outer edge of the disturbance. In some cases the beds are so shattered that every trace of the original bedding is lost. There are no traces of volcanic materials. Hot springs issued from the central hill, which was then an island in the lake, and from the eastern margin. Ring faults are well developed at the periphery of the ring syncline. A family of radial faults is also well developed, particularly in the northern and eastern sections.

R. S. Dietz, in 1956 and 1957, examined the shatter cones from this location in detail (11). They were originally discovered much earlier (10). The rock layers are poorly developed and are found mainly in excavations for houses. It was impossible to determine whether or not the cones exhibited a preferential orientation, as the limestone blocks were invariably not *in situ*. There were clear evidences of reflections of the shock waves from bedding planes in the rock.

The Steinheim Basin almost certainly is a meteoritic crater 15–20 million years old.

The Rieskessel, Germany

Not far from the Steinheim Basin is another great depression, the Rieskessel (7, 18, 19) (Pl. XI), whose floor is about 15 miles across and consists of a chaotic jumble of broken granite and Triassic and Jurassic sediments beneath the cover of Upper Miocene, fresh-water deposits. Werner (20), in 1904, suggested that it might be a meteoritic crater.

Williams (21), after rejecting the meteoritic theory of its origin without stating his reasons, described the Rieskessel as follows, implying a volcanic origin but continually emphasizing the complete inadequacies of the known terrestrial processes:

> Immediately surrounding the basin is a zone, several miles wide, made up of similar fragmented rocks in the form of large blocks and huge, imbricated thrust slices. This is the so-called *Schollen- und Schuppen-Zone*. Beyond this, and resting on the undisturbed limestones of the Alb Plateau, are rootless slices and smaller blocks of older rocks, thrust outward from the central area (*Zone der wurzellosen Schollen*). Still farther from the basin are scattered pieces of limestone, some of which, even 44 miles from the center of the Ries, measure 40 cm. across.

How are these peripheral zones of ejected blocks and thrust masses to be explained? Most authorities agree that during Miocene times the crystalline crust and its Mesozoic cover were strongly uparched, perhaps more than 400 meters, by the intrusion of a broad laccolith. Geophysical studies suggest that the laccolith is composed of basic rock and lies at a depth of only a few kilometers. Possibly, the central portion of the upraised area was punched up in bysmalithic fashion so that rocks formerly well below the surface were exposed. In any event, a steep doming of the laccolithic roof seems to have brought about widespread sliding on the surface, and this process was doubtless accentuated by repeated explosions. With good reason, Reck [22] doubts that the larger slices in the *Schollen- und Schuppen-Zone* could have been blown out. No masses even approaching them in size have ever been ejected in the most violent of historic eruptions. Had these huge slices of brittle limestone been hurled out by explosion, why were they not shattered as they fell, and why do they commonly show slickensides on their under surfaces? These features suggest that they were emplaced by sliding. It cannot be denied that much of the finer debris surrounding the Ries is truly pyroclastic, but Kranz's [23–25] view that the depression is a colossal explosion crater does not seem justified. True, torsion-balance measurements have indicated a funnel-shaped structure, approximately 12 km. in diameter, from 0.9 to 1.3 km. deep, and with sides sloping inward at 10° to 15°, beneath the center of the caldera, but that this is a gigantic explosion vent has yet to be proved. No known explosion craters exceed even a quarter of this in diameter. To produce a broadly flaring funnel of so great proportions, it is obvious that the explosion focus must be very shallow. But at shallow depths magma can hold little gas in solution and can only produce explosions of weak to moderate intensity such as would puncture the roof by a series of diatremes. No new magma was associated with the first explosions of the Ries; they seem to have been low-temperature, phreatic eruptions. All the more is there reason for doubting the theory of Kranz.

Bentz [18], Branca [10], Reck [22], and others agree that after the main explosive phase the summit of the dome collapsed along ring fractures, and that it was primarily this engulfment which formed the caldera. Rittmann [26] also seems to adopt this view when he refers to the Rieskessel as a "volcano-tectonic sink." After the collapse, dikes of suevite were injected close to the margins of the basin and mild explosions of suevite pumice ensued. Apparently the magma was produced by gas fluxing of the granitic basement.

Certain observations are pertinent. An uparching of 400 meters in a radius approximating 10 kilometers gives an average slope of 1 in 25, roughly 2°. Great masses certainly would not slide on such a slope; hence the overthrust masses must have been emplaced by an explosion or explosions acting nearly horizontally. This effect, in minor form, is well known at the small modern terrestrial meteoritic craters (see Fig. 2, p. 20). Williams' objection to the explosion hypothesis seems to be based entirely on the fact that no volcanic explosion of such a magnitude is known to have occurred anywhere on earth. The meteoritic-impact theory avoids these difficulties completely, for ample energy is available.

The incidence of a mild form of volcanism—the injection of suevite in ring dikes—is not surprising. An explosion of the obvious magnitude of this one might easily weaken the crustal layers, particularly in a volcanic area, whereas smaller explosions probably would not. Shoemaker (27) has recently suggested that the suevite is actually a glassy material which has been fused by shock. Ring fractures are typical at impact structures.

No central uplift is known, and no shatter cones have been reported, possibly because the central portion of the pit is not exposed. The deeper rocks of this region are a brecciated granite rather than a homogeneous layer, like limestone. Wilson (28) suggests that the close proximity of the Ries to the Steinheim Basin and probable similar age (Upper Miocene) suggest that the structures were formed simultaneously in a double holocaust.

When the debris is removed from the rim of the Ries, flat ground and striated surfaces are found on the beds, which are still more or less *in situ*. These structures point radially toward the center of the basin and thus indicate the locus of the explosion. Stutzer (29) and Werner (30) have studied this problem. Shoemaker and Chao (31) have announced the discovery of coesite in rock specimens collected near the rim of this great crater.

There seems to be very little question that the Rieskessel is a meteoritic crater 15–20 million years old.

The Holleford Crater, Ontario
(Longitude 76°30′ W., Latitude 44°47′ N.)

The Steinheim Basin and the Rieskessel still show significant surface craters. As they are only a few million years old, this is not overly surprising. The Holleford Crater is much older and has been preserved only because it was filled up with materials instead of being eroded.

C. S. Beals has led a systematic survey of aerial photographs to see how many Canadian meteoritic craters he can find. This ingenious and thorough study has been surprisingly successful,

The discovery of the Holleford, Ontario, Crater was announced in 1956 (32, 33). The aerial view (Pl. XII) shows a circular structure composed of a large number of concentric markings which can easily be seen, although most of the crater interior is farmland. The dark central area of the region is a wooded bog. There is still a surface suggestion of a crater, as the center is depressed about 100 feet and the rim is a moderately impressive cirque of nearly 180° when viewed from the center. Its shape appears to have been altered only slightly by erosion. It is of approximately the correct width for a meteoritic crater this size.

The area is covered with Paleozoic sediments, which are of the order of 50–100 feet thick. The thickness of the sediments increases markedly as the center of the circular marking is approached, indicating that the original crater still exists. Seismic soundings and measurements of gravity at once indicated the existence of a deep basin filled with lighter sediments. Without exception the infilling sediments dip radially toward the center.

The gravity contours (Fig. 6) are roughly circular and, in a general way, follow the outline of the depression. Correcting for regional effects, it was found that the crater produced a negative anomaly of about 2.2 milligals. It proved difficult to obtain an accurate measure of the depth of the low-density brecciated rock beneath the crater, but a minimum value of 1,000 feet would be consistent with the gravity measures. Beals and his associates (34) bored three holes at distances of 1,400, 2,500, and 3,750 feet from the center to depths of 1,128, 1,486, and 443 feet, respectively.

The diamond-drill cores obtained in hole No. 1 did not penetrate through the breccia, but the drill did go through 373 feet of it before it became stuck and had to be abandoned. Hole No. 2 penetrated the breccia, which was found to be 160 feet thick at that point. At hole No. 3 on the rim, the breccia was only 1 foot thick. A drawing of the cross-section suggests that the value of 1,000 feet of breccia at the center of the pit is reasonable. This leads to a value of 1,800 feet from the original ground level to the limit of major brecciation under the crater.

Known meteoritic craters of large size show a lens of breccia beneath the floor, as does the Holleford crater. The cores of the first hole showed the lens to be composed of broken, granulated, and powdered rock, weakly recemented. The powdered rock is the usual rock flour. The second hole went through fragmental material, which consisted of a mixture of various sizes of rock fragments, but below the brecciated layer the rock was uniform in character and appeared to be broken and fractured in place. Beals is continuing the study of these cores.

On the assumption that this was a meteoritic crater, Beals and his associates predicted, for the guidance of the drillers, that the original floor would be found at the first and second hole at 800 and 400 feet, respectively. Actually, at the first hole, the floor was encountered at 755 feet and in the second at 440 feet. The third hole was bored through sediments covering the vestiges of the original crater rim, which were encountered at a depth of 65 feet. Most of the original rim was eroded away before the deposition of sediments began (Fig. 7).

From the data of Chapter 7, we may make a check to see how closely the observed contours fit the theoretical average contour for a crater of this size. The diameter which fits best is 7,709 feet [Beals's value (34)]. The scaled depth of burst which is most probable for a crater of this size is $H/W^{1/3} = 0.091$. Interpolation between equations (7-1) and (7-2) in Chapter 7 gives an apparent depth of the original crater of 1,042 feet. Interpolation between equations (7-5) and (7-6) suggests a rim height of 276 feet, leading to a calculated

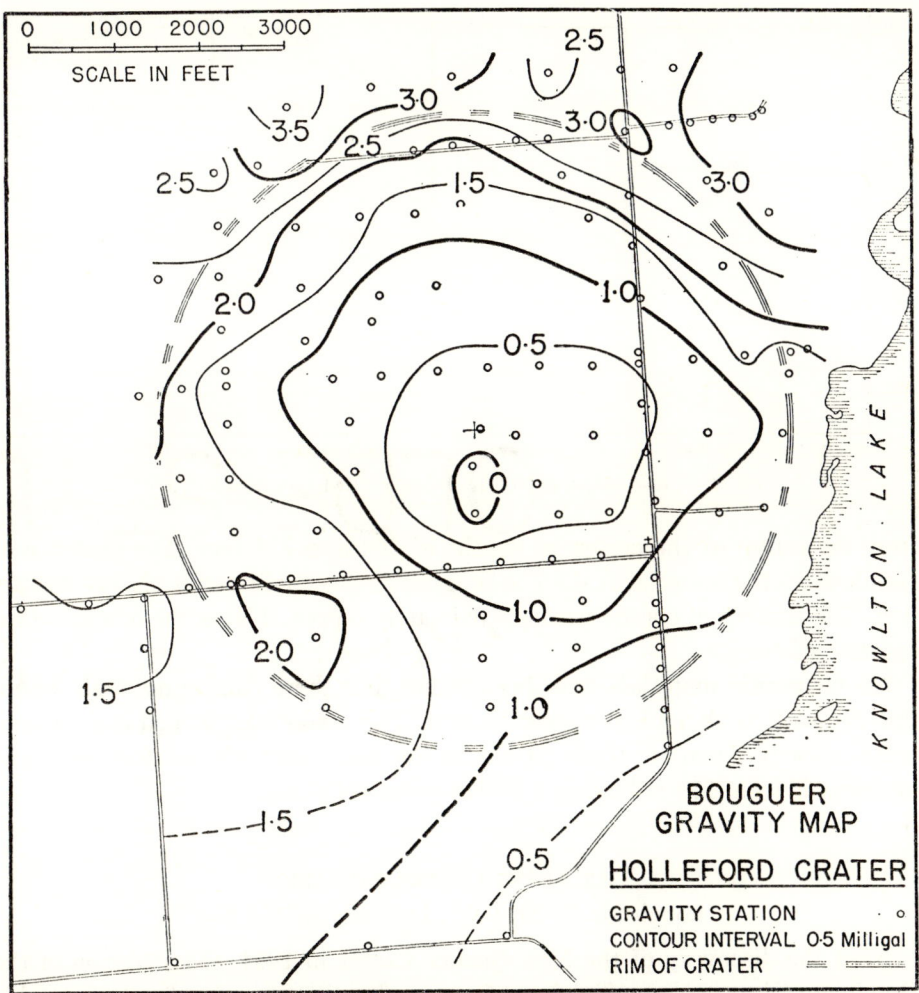

FIG. 6.—Bouguer gravity map, Holleford Crater. (Courtesy C. S. Beals, M. J. S. Innes, and J. A. Rottenberg, Dominion Observatory, Ottawa, Canada.)

true depth of 766 feet. The cross-section of the crater, defined by the three borings, gives a true depth of about 780 feet, on the assumption that the original ground level was at 480 feet. These data are plotted in Figures 19 and 20 (pp. 139, 141), and the Holleford point lies well within the lateral scatter.

Bunch and Cohen (70) have recently announced the discovery of coesite in materials from this site. This answers an important question as to the stability of this mineral. Clearly, it is stable and may be used with confidence to identify ancient impact structures. The coesite was discovered optically in siliceous fragments of brecciated rock in core sections from hole No. 2, drilled 2,500 feet

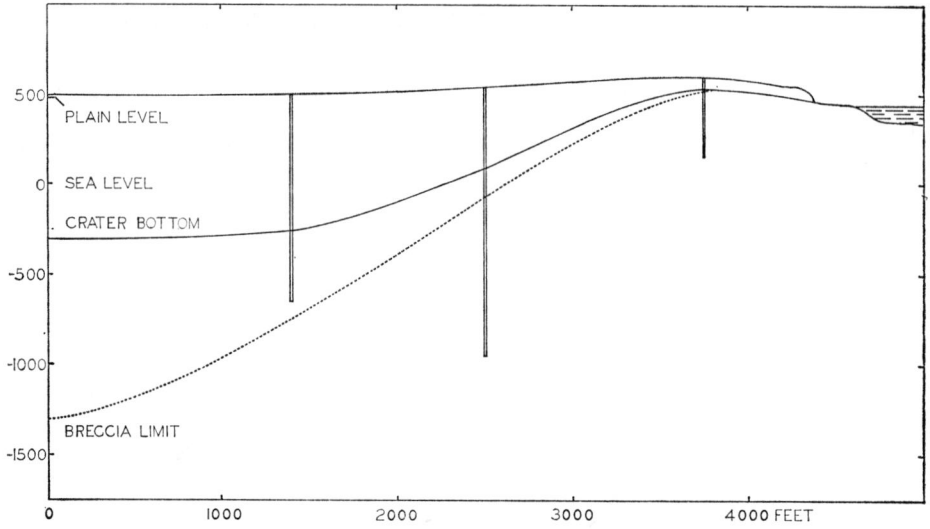

FIG. 7.—Cross-section of Holleford Crater. (Redrawn from Ref. 34.)

from the center of the crater, at depths of 704 and 737 feet. The presence of coesite from the material at the 704-foot depth was confirmed by X-ray diffraction. The content of coesite was approximately 6 ppm. This is the oldest known example of coesite.

No meteoritic materials have been found, but there can be no doubt about this crater. It is an ancient meteoritic scar. The crater was formed in pre-Cambrian or early Paleozoic times. It has an age of at least 450 million years and probably is from 500 million to 1 billion years old.

THE BRENT CRATER, ONTARIO
(Longitude 78°29′.5 W., Latitude 46°4′.5 N.)

After the discovery of the New Quebec Crater in 1951, the attention of the Dominion Observatory officials (35) was drawn to a circular topographic feature which appeared on aerial photographs of the northern part of Algonquin Park near Brent, Ontario (Pl. XIII). Stereoscopic studies showed that this feature was a relatively shallow depression, partly occupied by two lakes—Gilmour and Tecumseh—which, with their tributary streams, outlined a considerable part of the boundary. The central part of the feature lies about 300 feet below the hills that rise abruptly in places from the central floor.

The circle is nearly 2 miles in diameter. Seismic and magnetic observations by Innes, Willmore, and Clark (36) in 1954 and geologic studies by the Geological Survey of Canada indicate that this is a crater many hundreds of feet deep filled with Paleozoic sediments. Beyond the circle and for a distance of 2 or 3 miles, the drainage is, for the most part, radial and toward the depression.

Like the Holleford Crater, it is filled with Ordovician sediments, which thus define a minimum age of 400 million years. Most of the rocks in the surrounding area are pre-Cambrian. The trends of the pre-Cambrian rocks are generally westerly to northwesterly and appear to be terminated by the circle defining the present crater wall. In no way do these trends appear to conform to the circularity of the feature, which seems strong evidence that the folding and deformation of the gneissic rocks antedate the crater's formation.

Beals, Innes, and Rottenberg state: "It should also be remarked that any suggestion that the crater is the deeply eroded vent or caldera of an ancient volcano, finds no support from the surface geology" (34).

Several outcroppings of breccia were found along the circular drainage channel which separates the granitic rocks from the crater floor. Other blocks of breccia were found within the glacial drift.

Assuming an original diameter of 11,500 feet (34, 37)—equations (7-1), (7-2), (7-5), and (7-6)—and a scaled depth of burst of 0.076 lead to values of 1,438 and 381 feet for total depth and rim height, respectively and hence 1,057 feet for the depth of the crater floor beneath the original ground level (Fig. 8).

Millman, Liberty, Clark, Willmore, and Innes (34, 37) have made extensive geophysical investigations, employing gravity, seismic, and magnetic methods. The gravity anomaly map (34) shows contours following closely the circular outline of the structure (Fig. 9). There is a circular gravity minimum of 5–6 milligals concentric with the crater. These measures suggest that a lens of breccia extends 700–3,000 feet beneath the original crater floor.

The seismic tests indicate that the thickness of the sedimentary rock increases to about 1,000 feet near the center of the crater, confirming the predicted value for its depth. The estimated thickness of the brecciated zone is of the order of 4,000 feet as determined by this method.

In 1959 a hole was bored to a depth of 3,500 feet. Practically all the core was recovered. This hole was drilled as close to the center as possible, during the winter months. Paleozoic sediments were penetrated at a depth of 851 feet, in-

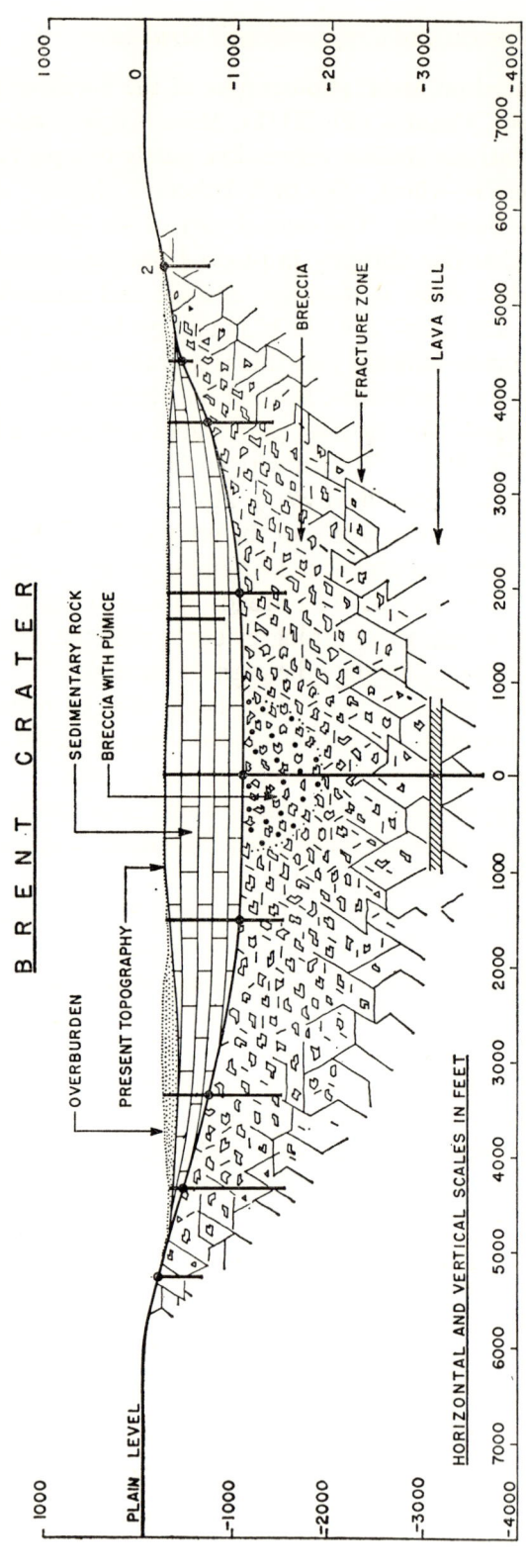

Fig. 8.—Cross-section of Brent Crater. (Redrawn from data furnished in letter of March 1, 1961, from C. S. Beals, Dominion Observatory, Ottawa, Canada.)

dicating that the floor of the crater lies 1,080 feet below the mean level of the surrounding country. This is reasonably close to the predicted true depth of 1,057 feet. These data are plotted in Figures 19 and 20 (pp. 139, 141), and the resulting point lies well within the lateral scatter.

Underlying the sedimentary column, fragmental material and gneiss breccia similar to that discovered near the rim were encountered and continued to a depth of more than 3,000 feet. The breccia consisted of fragments of all sizes, from minute particles to large blocks several feet in thickness. These blocks generally increased in size with increasing depth, so that no definite level could

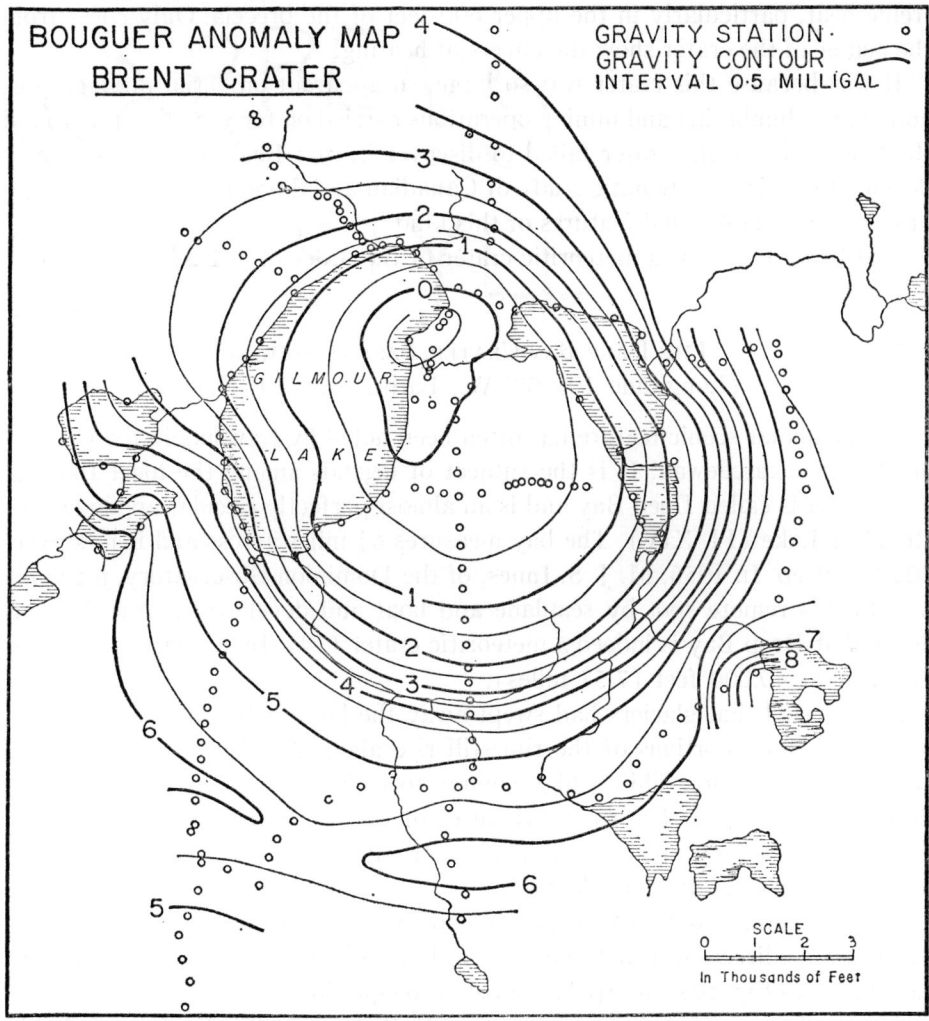

Fig. 9.—Bouguer gravity map, Brent Crater. (Courtesy C. S. Beals, M. J. S. Innes, and J. A. Rottenberg, Dominion Observatory, Ottawa.)

be established for the division separating the fractured zone from the undisturbed granitic rocks. Beals (38) writes:

> ... We believe that the fracture zone extended to a depth of about 3300 feet. It is possible, or even probable, that there was some cracking and fracturing of large blocks to a greater depth but we consider we were pretty well through at 3300 feet. The depth of the top of the hole was of the order of 250–300 feet below the surrounding ground level so this would bring the fracture zone of the order of 3600–4000 feet below the original ground level.

Emphasizing the meteoritic nature of the explosion are the evidences of extreme heat, particularly in the upper 600 feet of the breccia. Only cores from the center of the crater show the effects of heating.

It was because this crater was so large, so apparent on aerial photographs, and yet the lumbering and mining operations carried on for years in and around the boundaries of the crater failed to discover it, that Beals and his associates decided to begin a systematic study of Canadian aerial photographs in the hope of discovering additional features of this kind.

The Brent Crater is a meteoritic crater of the order of $\frac{1}{2}$–1 billion years old.

The Deep Bay Crater, Saskatchewan
(Longitude 103°00′ W., Latitude 56°24′ N.)

A spectacular scenic feature has often been noted by travelers in the wilds of northern Saskatchewan. It is the subject of legends among the local Indians. This crater is called Deep Bay and is an almost perfectly circular appendage of Reindeer Lake (Pl. XIV). The bay measures $6\frac{1}{4}$ miles across and is now 500–700 feet deep. In 1956, M. J. S. Innes, of the Dominion Observatory, made his way to this remote spot by seaplane and boat and developed impressive evidence that Deep Bay is indeed a meteoritic crater (39). Innes places the crater diameter at 40,000 feet (7.57 miles).

He reported that glaciers had swept away the loose debris of the explosion, but the bedrock portions of the rim still rise about 270 feet, on the average, above the lake shore. This is high enough to isolate the bay from the general drainage of the area. The rock structures of the region are not related to the crater. In fact, they are broken off abruptly at the rim, and the rim itself is fractured into huge angular blocks.

Measurements around the structure show gravity increasing away from the bay in every direction, which indicates a deep hole partially filled with loosely packed rock (Fig. 10). Interpolation between equations (7-1), (7-2) and (7-5), and (7-6) for a scaled depth of burst of $H/W^{1/3} = 0.046$, the most probable value for a $7\frac{1}{2}$-mile-wide meteoritic crater, yields an original depth of about 3,952 feet, with a rim height of 1,057 feet and a true depth of 2,863 feet. Innes suggests that glaciers have probably pushed in a great deal of material, as the present water depth averages 500 feet, with the maximum 720 feet (34). Beals (34) states:

That Deep Bay is the result of a tremendous explosion is clearly indicated by the intense fracturing and shattering of the granitic rocks which is most pronounced in the vicinity of the shore. Large-scale fracture and fault, zones of various widths, now partially obscured by glacial action and deposition, cut radially and obliquely across the rim and persist for several miles from the margin of the bay. A system of concentric fractures is also well developed particularly in that area less than 3 miles from the shore-line. Perhaps the most prominent feature, that may be the expression of such fracturing, is a narrow arcuate lake 3 miles in length located about 3 miles to the east of the crater. There is some evidence from the drainage pattern and dissected topography that this depressed zone is much longer and circumscribes the whole crater and has a diameter of about 12 miles. Within this area lie the rocks which form the now deeply eroded rim of the crater, with the general appearance of having been shattered into huge blocks by a process involving little or no horizontal movement.

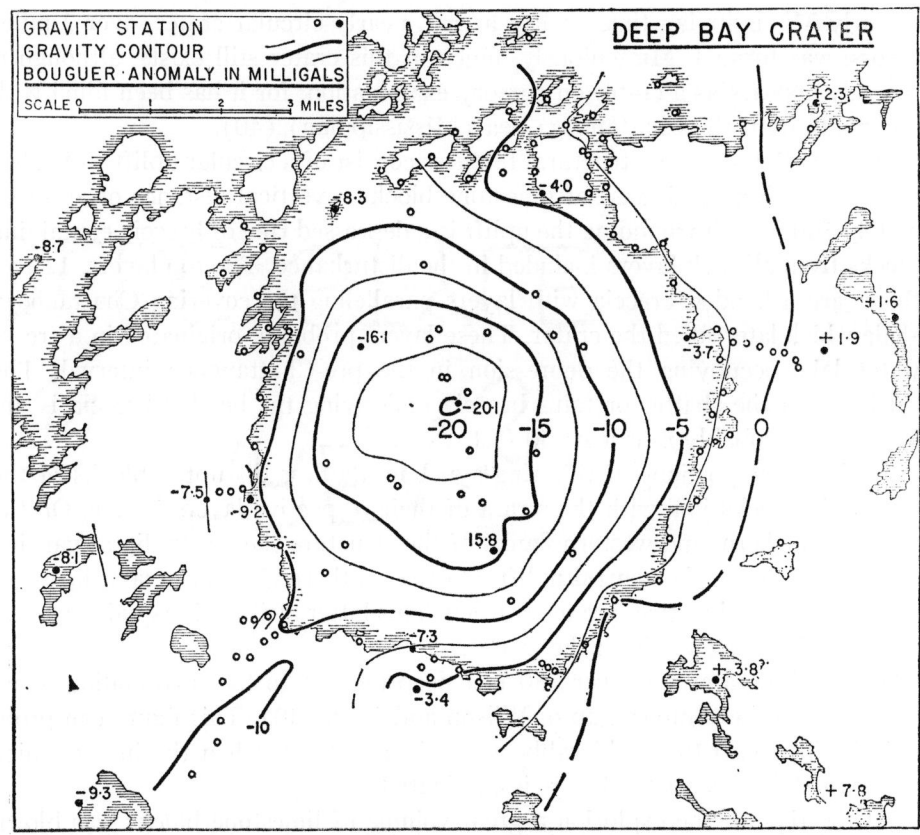

Fig. 10.—Bouguer gravity map, Deep Bay Crater. (Courtesy C. S. Beals, M. J. S. Innes, and J. A. Rottenberg, Dominion Observatory, Ottawa.)

The gravity anomaly reaches a maximum of 20 milligals near the center of the bay. The sedimentary material in the crater can account for about 3–6 milligals of the total anomaly, leaving the remainder to be explained by the underlying fragmentation from the explosion. The zone of deformation, under the original floor of the crater, extends to depths that may be as great as 10,000 feet below the original ground level. An aeromagnetic map of the area has been compiled by the Geological Survey of Canada, and the results are in qualitative agreement with the results obtained at other craters (34).

The extent of erosion indicates an age in millions of years, but probably it is not so old as Brent and Holleford, because it has not been so filled in. Fossils in the shales around the edge indicate a minimum age of 60,000,000 years.

This seems to be another definite identification of an ancient meteoritic crater.

The Flynn Creek Structure, Tennessee

A gigantic explosion occurred near Gainsboro, Tennessee, in late Devonian or early Mississippian times, when a vast, nearly circular crater, over 2 miles across, was formed. Miraculously, much of this crater still exists, although it scarcely appears as a crater on cursory examination, for it has been filled with sediments from the Chattanooga Sea (Mississippian) (40).

In general, the major structural features consist of a circular uplift which has raised a small central mass of limestone blocks a vertical distance of approximately 500 feet. Surrounding the uplift is a depressed ring of breccia containing blocks from all rock layers included in the disturbance. The top layers, 12 feet thick, are a bedded breccia with layers paralleling the covering Chattanooga shale which later filled the crater. These layers probably originated in a freshwater lake occupying the depression in the pre-Chattanooga interval. The thickness of the shatter or talus breccia underlying the bedded breccia is unknown but probably is at least 50 feet.

In the outer portions of this ring syncline, the less disrupted blocks of the Ordovician rocks join with the strata of their respective layers *in situ*. On the eastern, northern, and western flanks of the structure the strata dip at varying angles away from the central uplift. On the southern side the strata dip is toward the center, i.e., they are overturned. This latter attitude is explained as the result of thrusting out from the center.

The entire crater was over 350 feet deep relative to the surrounding plain, according to the contour map of Wilson and Born (40). This figure can probably be increased to considerably more than 400 feet when the inflow which forms the talus brecciated layers is eliminated.

At the time of the explosion a great volume of limestone blocks was blown out of the crater, some falling back into the pit and the majority being scattered around the margin of the crater within a radius of several miles. Since this rim was completely removed by erosion and yet the pit was not filled with air-borne sediments, the explosion is dated as shortly before the deposition of the Chattanooga shale, or in late Devonian time. Late Devonian time corresponds to an age of about 300,000,000 years.

Wilson and Born report the important discovery that there is a powder breccia, injected dikelike along fractures in the large blocks of limestone in the central part of the area. Stringers or veins of this breccia, which range in width from a feather edge to a foot, extend across the blocks. In some cases the injection of this rock flour into minor fissures took place on a microscopic scale.

In the original article by Wilson and Born (40), the emphasis was all upon a cryptovolcanic mode of origin. Wilson has now revised this opinion, attributing the origin of the structure to a meteoritic impact (28).

Wilson, Stearns, and Dietz (8) discovered shatter cones in 1959 in a thin bed along a new road cut not far from the structure's center.

The structure does not appear prominently on the aerial photographs of the United States Department of Agriculture, Production and Marketing Administration.

This is undoubtedly an ancient meteoritic scar.

The Wells Creek Basin Structure, Tennessee (7)

Aside from the normal faulting, the Wells Creek Basin structure reveals more clearly than most of the others the dominant pattern of an impact structure. It has the appearance of damped waves. A central uplift, $2\frac{1}{2}$ miles across, was pushed up at least 1,000 feet, although at present it rises but 75 feet from the surrounding landscape. Safford (41) considers the uplift to be not less than 2,500 feet. The whole central area is broken up into blocks, large and small. Surrounding it are two pairs of up-and-down folds with diminishing amplitude. The inner ring syncline, $2\frac{1}{4}$ miles from the center, was dropped at least 700 feet. A pattern of this type usually arises from a sudden impulse, such as that of an explosion. A similar structure may be seen on high-speed pictures of a drop of liquid falling into water. There are two main differences. The Wells Creek Basin structure became frozen in the disturbed position, and, instead of radial symmetry, it exhibits a distinct bilateral symmetry. The outer edge of the outer ring syncline is 8.5 miles in diameter (Fig. 11).

The Wells Creek Basin structure is not alone. During the post-Eutaw–pre-Wilcox (Cretaceous) interval, at least four basins were located in this region (28). The largest one was probably over 6 miles in diameter. This was the crater resulting from the explosion that formed the Wells Creek structure. The crater coincided with the inner depressed ring and contained a central hill coinciding with the central uplift. Five miles north-northeast of its northern rim was the much smaller Cave Spring Hollow Basin, the extent of which is unknown. Three miles farther north was the Indian Mound Basin, at least 2,000 feet in diameter and deeper than 263 feet, but with a central hill rising above the level of the floor of the basin. About 1,700 feet to the north was the very small Austin Basin, 375 feet in diameter and deeper than 40 feet.

It seems logical that the four basins or craters had a similar origin at the

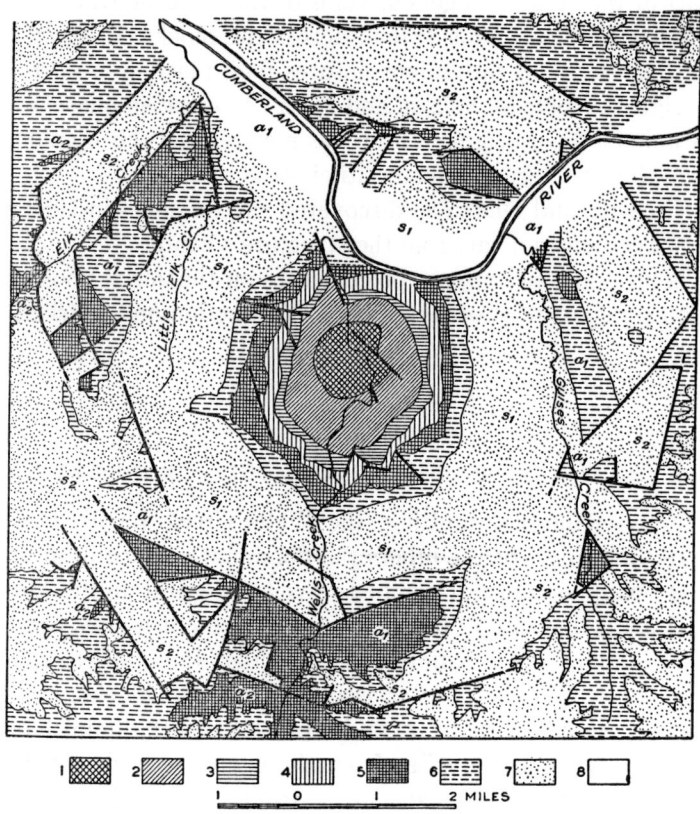

FIG. 11.—Geologic map of the Wells Creek Basin, Tennessee. (Courtesy Walter Bucher and Geological Society of America.)

		Feet	
8.	Alluvium.		
7.	St. Louis limestone (may include higher beds locally; thickness estimated, probably more)	200	Middle Mississippian
6.	Warsaw limestone	120	
5.	Fort Payne chert	110	
	Ridgetop shale (including Maury member)	35	Lower Mississippian
	Chattanooga shale	25	
4.	Harriman chert	20	
	Birdsong formation	20	Devonian
	Limestone series	110	
	Red shales with limestone layers	130	Silurian
3.	Hermitage formation	100	
2.	Post-Beekmantown, pre-Trenton limestones (estimated, probably more)	500	Middle Ordovician
1.	Wells limestone (estimated, probably more)	200	Lower Ordovician

s_1, Inner ring depression; a_1, inner ring in the anticlinal bulges; s_2, second depressed zone, a_2, incomplete second marginal zone of anticlinal bulges.

same time. That origin would have been related to the phenomenon that formed the Wells Creek Basin structure. Wilson (28) states:

> The writer believes only two known forces could account for the origin of Indian Mound crater: (1) a local, abnormally deep sink hole; (2) the depression ring of an explosion crater. It seems to the writer that the sink hole can be eliminated when ... it must have been cut: (1) 130 feet below the present level of bedrock in Cumberland River valley, and (2) through at least 200 feet of Fort Payne and Ridgetop beds. These relatively insoluble beds are underlain by the Chattanooga shale and about 50 feet of Devonian Harriman chert, a sequence that would have prohibited, or made improbable, the cutting of such a deep sink hole. The writer believes that this basin represents a doughnut-shaped explosion crater with a central hill; it is very similar in cross section to the Steinheim Basin and the Flynn Creek crater. Austin and Cave Spring Hollow craters represent small meteoritic pits, or craters, without known central hills.
>
> It is concluded that a swarm of meteors approached the earth's surface from the south, or a single meteor fragmented into at least four pieces before striking the surface. The largest fragment struck at the present position of Wells Creek Basin, and the second in size struck at the Indian Mound locality. Smaller fragments ploughed into the earth to form the Austin and Cave Spring Hollow craters. The Ries Basin and the near-by and much smaller Steinheim Basin are both post-Jurassic, pre-Miocene in age. If these were formed at the same time, which seems logical, they would represent twin craters similar to the Wells Creek Basin and Indian Mound craters.

Around the periphery of the central uplift of the main structure the beds dip away from the center, as would be expected, but along the northern periphery the rock layers dip steeply to the south or toward the center. This is consistent with the idea that the meteorites approached from the south.

The axis of asymmetry of the Wells Creek Basin structure points unerringly toward the Indian Mound Crater.

In 1947 the Ordman Company cored Wells Creek Basin on the interpretation that it was a salt dome. The core is essentially complete from 23 to 2,000 feet. It started and bottomed in Knox Dolomite, which has a great, but unknown, thickness in this area. Wilson (28) states:

> The examination of this core was an unusual privilege and in a way an eerie experience. The deep fingers of grotesque injection dikes and the intense, bizarre, ever-changing patterns of brecciation and deformation are awe-inspiring. Each new box of cores revealed new, strange, and different intricacies.
>
> This core presents three significant features. (1) Deformation was instantaneous, and did not result from normal tectonic stresses; (2) progressive downward dying out of deformation may be traced, in spite of the brecciation between 1743 and 1930 feet; (3) in the top 200 feet of the core, the shatter cones are all horizontal, except for some that point obliquely upward. The greatest concentration of horizontal cones is at a depth of 100 feet. Below 200 feet, shatter cones are rare, incomplete, and poorly defined with a single exception at 1237 feet. As the core was not oriented, it is impossible

to state in which direction these cones pointed. In an exposure about 650 feet south of the well, shatter cones are horizontal, and their axes strike N. 7° W. They occur in a vertical block of dolomite with strike of N. 20° E. These cones were not formed by the impact of the meteorite, as such should be normal to the bedding and oriented stratigraphically up, but rather by the explosion of the rocks compressed beneath the penetrating meteorite. This block must have been knocked out of position by impact and penetration a few moments before the explosion. It is believed that the cones in the core pointed northward as do those in the near-by exposure. At a depth of 1237 feet, an oblique shatter cone points upward at a 45 degree dip. Its orientation is likewise unknown, but, from the evidence furnished by the exposure, it is believed to have been the same. The orientation of the shatter cones in the exposure and the postulated similar orientation of those in the core suggest that the center of explosion was north of the well.

These features are believed to present definite evidence that the deformative force came from above and not from below. This evidence combined with the occurrence of four aligned craters, of which the Indian Mound crater has critical depth and cross section, and the southward dip of the Ross and Decatur limestones on the north periphery of the uplift of Wells Creek Basin all harmonize to tell the same story of meteoritic origin.

The main impact structure is clearly evident on the United States Department of Agriculture, Production and Marketing Administration, photographs (Pl. XV).

This is a spectacular group of four associated meteoritic-impact structures about 100,000,000 years old.

The Sierra Madera Dome, Texas (6)

Twenty-five miles south of Fort Stockton, Texas, is a large dome which forcibly suggests an impact scar from which the crater and other top layers have been so completely removed by erosion that the underlying structures are revealed. According to King (42), this dome is roughly circular and perhaps 3 miles in diameter. It is composed of abruptly updomed Permian strata which stand nearly vertically or dip radially outward at high angles. On the south these beds are overturned and incline toward the center of the dome at angles of 60°–70°. The central portions are composed of an unmetamorphosed dolomitic limestone, but these regions are highly fractured, jointed, and apparently "jumbled and twisted in hopeless disorder" (42). The dome has been drilled (43, 44) to a depth of over 2 miles (12,096 feet) without revealing the presence of an igneous core. In fact, the structure appears to die out with increasing depth, which emphasizes its non-volcanic origin.

Numerous radial tear faults displace the rocks in the center, while small folds flank the uplift to the east and west. Apparently, these are portions of nearly circular damped waves similar to those which appeared so strongly at the Wells Creek Basin structure. On the eastern flank, layers of Cretaceous rock lie uncomfortably on the Permian strata of the structure, thereby dating the explosion as post-Permian and pre-Comanchean. This gives an age from perhaps 135,000,000 to 200,000,000 years.

It is well to realize that, while this is the only method capable of dating these cryptovolcanic structures, the great discontinuities in geologic history as shown by the rock layers at any particular point leave tremendous spans of time unaccounted for. Hence the dates of formation of these objects are uncertain, usually by tens of millions of years and often by hundreds of millions. The usual shatter cones of such objects are plentiful, according to Dietz (8), who searched for them in October, 1959.

The structure shows up very clearly on the photographs of the United States Department of Agriculture, Production and Marketing Administration (Pl. XVI).

Shoemaker (45) has made an interesting tentative reconstruction of the original crater. He feels that the jumbled mass at the top of the dome is part of the lens of breccia normally found beneath the central sections of meteoritic craters. His suggestion has great merit, although it may prove that the erosion in the area was greater than he shows and the crater may have been at a considerably higher elevation.

The Vredefort Dome, Orange Free State

On a far larger scale, the Sierra Madera dome is reproduced in the northern part of the Orange Free State in Africa. There is found an almost circular uplift, approximately 75 miles across. It has a core one-third as wide, composed of non-intrusive granite. The dome is thus strictly comparable in size with many of the larger lunar craters; indeed, it is wider than Copernicus and approaches Ptolemaeus in diameter. Younger rocks which encircle the core are usually overturned so as to dip toward the center of the dome. Elsewhere they dip radially outward at high angles (46).

In the Vredefort memoir by Hall and Molengraaff, it was suggested that the updoming (not a rebound) was "initiated by centripetal pressure. The relief of load resulting from this movement caused a younger magma below the granite to become active and to rise and thus assist the updoming and the upward movement of the much heated, but passive, granite" (47). They proposed no motivating cause for the development of the centripetal pressure and emphasized the inadequacies of all existing theories other than the meteoritic by mentioning the extreme difficulty of accounting for an almost circular dome by appealing to tangential stresses.

The similarity between this and other cryptovolcanic structures is still further brought out by the presence of transverse and oblique faults around the margins of the core, much like those of the Sierra Madera dome. In addition, no volcanic materials directly associated with the dome have been identified.

Throughout the entire mass of the dome are evidences of unprecedented pressures. All the rocks, including the Vredefort granite, reveal under the microscope the effects of these powerful forces. They have been crushed and pulverized in many places. Evidence for the operation of these pressures appears principally in veins of flinty crush-rock which literally riddle the granite and adjoining rocks. The volume of the crushed and shattered materials has been estimated to be as high as 800,000,000 cubic meters, but even this must be too low.

Maree (48) has made gravimetric measures which outline the structure but show a possible extension to the southeast, where the dome is still covered by later sediments. Daly (49), Baldwin (50), and Dietz (11) have given the meteoritic origin of this peculiar feature some backing. Recently Dietz asked Robert Hargraves, of the University of the Witwatersrand, to search for shatter cones (51, 73). He found them in abundance and showed also that if the rocks were returned to their original positions, the cones would all point upward toward the center of the ring. Dietz (51) summarizes the event as follows:

Upon reconstruction, the event that produced this structure emerges beyond doubt as the greatest terrestrial explosion of which there is any clear geological record. Apparently an asteroid a mile or so in diameter plunged into the earth from the southwest, for the structure is overturned somewhat to the northeast. The huge object drilled into the earth and released enormous shock forces, causing a gigantic upheaval. Strata nine miles thick peeled back like a flower spreading its petals to the sun, opening a crater 30 miles in diameter and 10 miles deep. The shock must have reached with shattering force down through the entire 30-mile thickness of the earth's crust. Shock pressures of many millions of atmospheres spread through the collar, forming scattered pockets of pseudo-tachylite (fused rock) like raisins in raisin bread. Rock that had lined the cavity was melted and injected into the rock walls as great dikes of fused rock (of a type called enstatic granophyre) 100 feet across and several miles long. Except for these rocks, which remained molten until the shock had passed, the collar rocks are intensely and wonderfully shattered, and it is in these that the shatter cones abound.

Dietz (73) has recently described this structure and its origin in greater detail. This is an important paper.

My interpretation of the Vredefort event differs only in minor details from that of Dietz. The crater was at least 40 miles wide and had a true depth of not more than 3 miles.[2] Isostatic adjustments and erosion account for part of

---

[2] At similar, but smaller, structures, the diameter of the central rebound dome is substantially less than the apparent crater diameter.

Ancient Meteoritic Craters and Cryptovolcanic Structures

the apparent updoming. At a reasonable velocity of impact, the asteroid was perhaps 3 miles in diameter, between Hermes and Eros in size.

The structure is dated as post-Transvaal (late pre-Cambrian) and pre-Carboniferous.

THE DECATURVILLE STRUCTURE, MISSOURI

Near Decaturville in Camden and Laclede counties is a large impact structure (7). In general, it has been listed as being concealed beneath later layers of rock, but the aerial photographs of the United States Department of Agriculture, Production and Marketing Administration (Pl. XVII), show it to be beautifully developed. Actually, it has only a soil covering. It has a central uplift, over a mile across, surrounded by a ring syncline and a prominent ring anticline whose outer edge has a diameter of 3.7 miles.

The structure is dated as late Cambrian or early Ordovician. It is around 400 million years old and undoubtedly is very much eroded. The central uplift shows tremendous internal brecciation. John L. Rich has likened this internal brecciation to the effect of the sudden release of highly compressed steam on rice in the manufacture of "puffed rice" (52).

In October, 1961, I visited this site and, under the guidance of Mr. H. B. Hart, spent considerable time studying the structure. Mr. Hart owns or has under lease essentially the entire region affected. Because of his mining operations, he has drilled and cored numerous parts of the dome. These cores are remarkably similar to those from the Brent Crater. They show tremendous evidence of brecciation and jumbling of the strata. In several cores, rocks from the same level were found at two or three widely different depths. Often rocks would be found above others which normally are about 500 feet higher. Mr. Hart felt that an average of 530 feet would be reasonably close to the uplift of the dome. No estimates of the downthrow of the ring syncline were available, but, from the few cores, it appears to be about 200 feet. The surface expression of the rocks is limited to two exposures. A mining cut about 400 feet southwest of the center shows a fantastic jumbling of dissimilar rocks and breccia. There is an almost complete lack of order here.

A unique feature of this structure is the existence of a small outcrop of highly quartzose pegmatite (53, 54) a few hundred feet west of the center of the formation. In part this rock has the texture of graphic granite. It has an exposure of only a few square yards and is rapidly disintegrating. The importance of this observation is still unknown. Tarr (55) has maintained that the igneous rocks in the Decaturville structure are much older than the explosion which was the cause of the structure. Drill holes tend to bear out Tarr's hypothesis, for, in places a few hundred feet down, the pegmatite dike disappears where it would normally be expected to be found.

Probably nothing of the original crater remains, but the area still gives the impression of a crater. The dome is gently rounded and rises perhaps 40 feet above the ring syncline. The outer ring anticline is the highest part and in places rises substantially above the center.

At the outer edge of the dome is an incomplete ring composed of a rather poorly marked set of raised regions. The nearest parallel is the Crooked Creek structure, and perhaps the same phenomenon occurred here. When more data are collected, it may prove that the center of the dome collapsed and left a small apparent ring anticline toward the edge of the dome.

Partway down the dome toward the syncline may be found a series of at least six tiny sets of anticlines and synclines about 1 foot high and about 20 feet crest to crest. Mr. Hart reported that they go completely around the dome.

In the area of rock exposure where the great jumbling is noticed and in certain cores from the dome, there are evidences of heat. This is strongly reminiscent of the scorched breccia beneath the Brent Crater (38).

Similar letters of November 14, 1961, from E. C. T. Chao, of the U.S. Geological Survey, and H. B. Hart included the information that shatter cones had been found on November 12 by Heyl and Brock, of the same organization. This probably confirms the Decaturville structure as an ancient meteoritic scar. The shatter cones were found in cores. A very beautiful example came from hole No. 2 at 288 feet, while Hart found one at 700 feet in a different core.

THE KENTLAND DISTURBANCE, INDIANA

The Kentland disturbance is a well-known, localized area of intensely deranged Paleozoic beds in a region of flat-lying strata. Because of the overlying strata, the extent of the disturbance is not known from geologic evidence. It is exposed at its center in a large quarry near Kentland, Indiana, but does not appear on aerial photographs. An uplift of 1,500 feet is indicated (56).

Most writers agree—e.g., Bucher (7), Shrock (56), Dietz (11)—that the structure was formed by a violent natural explosion. The earlier writers called it a "cryptovolcanic structure." The violence of the shock is evidenced by the jumbling of the strata, the shattering of the limestones, and the pulverization of some of the sandstones into a rock flour resembling the comminuted Coconino sandstone at the Arizona Meteorite Crater.

In 1953, Pullen (57) investigated the transmission characteristics of the ground layers of the Kentland area for the radio waves of the 5,000-watt station WIND in Gary and found that the structure had a roughly circular shape at least 4 miles across, with the McCray Quarry in the middle (Fig. 12). The field-intensity contour map of the area determined from the radio station WAAF in Chicago suggests that the structure is somewhat larger, or about 8 miles in diameter. The intensity anomaly noted here may be explained by the reflection and excitation and/or signal reinforcement through Silurian rocks connecting Kentland and Chicago. Other similar radio stations underlain by Pennsylvanian or Devonian rocks show little signal reinforcement at the site of the impact structure.

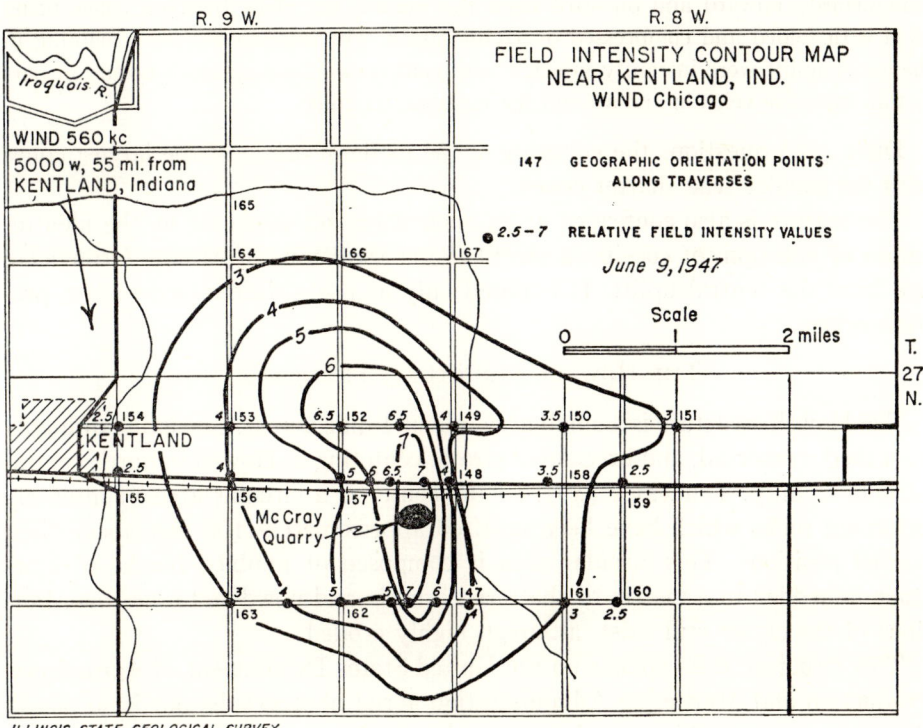

FIG. 12.—Radio field-density contour map near Kentland, Indiana (Station WIND). This map outlines the central portions of the Kentland impact structure. (From "Geologic Aspects of Radio Wave Transmission," by M. William Pullen in *Report of Investigations No. 162,* Illinois State Geological Survey, 1953.)

Many shatter cones, including some very large ones, were found by Dietz in 1945 (12). He reported a predominantly upward orientation for the vertices of the cones and concluded that the deforming stress must have come from above, as in a meteoritic impact. This preferred orientation of shatter cones was questioned by Gutschick (58), but Dietz (11) has reinvestigated the problem and finds that roughly 95 per cent are oriented normal to the strata and point upward. The rest of the cones are inverted as though by a directly reflected shock wave.

Cohen, Reid, and Bunch (71, 72) have identified "megashatter cones" at this impact structure, including one which is 160 feet high by 250 feet wide at the quarry floor. These gigantic cones are themselves made up of smaller parasitic cones. They suggest that the entire uplift may be one composite megamegacone, possibly containing a small central crater produced by the explosive jet of the meteorite when it reached its maximum penetration. They conclude:

> This uplift in the center of a large crater would have many megacones thrust simultaneously upward and outward from this center, the voids between cones being filled with folded and compressed overlying strata. The interstices then remaining in the entire uplift were instantly injected with compressed breccia which occurs in part as thin intrusive veins in and around the megashatter cones.

Dietz (74) questions the equating of the mode of origin of the largest cones with the smaller true shatter cones.

The feature is also confirmed as an ancient impact structure by the identification of 100 ppm of coesite in the St. Peter sandstone of Middle Ordovician age from the central uplift. It is dated only as post–Middle Ordovician, pre-Pleistocene.

The Howell Structure, Tennessee

The Howell structure is a small area, about 1 mile across, consisting of highly disturbed, contorted, and brecciated strata, centering at Howell in north-central Lincoln County, Tennessee (59). In the center is a circular area of intensely deformed rocks which have been uplifted approximately 100 feet above their normal positions. This circular area is composed of jumbled blocks of limestone imbedded in a matrix of shattered breccia and powdered limestone. Portions of the crater still exist, although highly eroded.

This structure does appear on the United States Department of Agriculture, Production and Marketing Administration aerial photographs but is so poorly shown that one would not pick it out unless he knew just where to look.

Dietz (11) reports that the rock outcrops at the Howell structure are too poorly developed to permit any intensive search there for shatter cones. Even so, this appears to be a small but ancient meteoritic crater.

Jeptha Knob, Kentucky

Jeptha Knob is a small cryptovolcanic structure with a central uplift about $1\frac{1}{3}$ miles across. As at other such structures, this uplift is completely bounded by normal faults. There are the usual irregularities in the direction and amount of dip of the rocks in this knob (7). Surrounding the central region, which has been raised by from 200 to 400 feet, there is a ring syncline which has been dropped 100–200 feet. As at the Wells Creek Basin structure and others, the

Jeptha Knob syncline shows two oppositely placed points of greatest depression. The entire structure is about $2\frac{1}{3}$ miles in diameter.

Immediately outside the ring depression are a few relatively small and local marginal synclines and anticlines. These are best developed in the north and south.

A cursory search by Dietz failed to locate any shatter cones (8).

The structure is of mid-Silurian age, or about 350,000,000 years old.

The Serpent Mound Structure, Ohio

The Serpent Mound structure is somewhat different from most of the others, in that the central portion consists of irregular, angular fault blocks which give that portion of the uplift a very angular outline (7). The rock layers in the central dome have been raised about 350–950 feet from their normal levels. This area is perhaps $1\frac{1}{2}$–2 miles wide as measured on aerial photographs. The 950-foot value is derived from a small area very near the center of the structure. Bucher has noted irregular and erratic dips and rapidly changing strikes in the Serpent Mound central area.

This structure shows a ring syncline surrounding the dome. It is about 4 miles in over-all width, and the downthrow is over 500 feet. The major fault system is one surrounding the outside of the depression. This syncline is bilaterally, not radially, symmetrical and shows two maximum and two minimum depths. Here, as at the Jeptha Knob, marginal anticlines surround the ring depression, with the largest marginal anticline adjoining the deepest synclinal depression on the northeast side. It was first mapped by Bucher in 1920 (60), who called it a cryptovolcanic structure similar to the Steinheim Basin.

Bucher makes the interesting point that ". . . over 60 per cent of its surface is occupied by strata that lie more than 100 feet (30 meters) below their original altitude; less than 30 per cent lie within 100 feet of their original altitude; and less than 10 per cent lie more than 100 feet higher than they did originally" (61).

Dietz (8) has found shatter cones here.

The Serpent Mound Structure is younger than 270,000,000 years, but much of the later geologic record is missing from this area. It is post–Lower Mississippian and pre-Pleistocene.

A concentration of 10 ppm of coesite has been identified in shatter cones of Lilley dolomite of Middle Silurian age from the central uplift (71, 72). This identification is still somewhat tenuous.

The formation appears moderately well developed on aerial photographs of the United States Department of Agriculture, Production and Marketing Administration.

Upheaval Dome, Utah

Upheaval Dome is a highly dissected cryptovolcanic structure with a central dome about $1\frac{1}{2}$ miles across. Its rock layers are intensely and minutely deformed. They have been uplifted about 1,200 feet. The adjacent ring syncline has an outer diameter of perhaps 3.4 miles, and the downthrow is roughly 500 feet. The formation terminates abruptly in all directions, perhaps because it was formed in massive sandstones (7). McKnight (62) suggests that it is not a cryptovolcanic structure, but rather is a salt dome.

It appears prominently on aerial photographs of Aero Service Corp., U.S. Department of Interior, Geological Survey.

Its age is post-Navajo (Jurassic).

The Kilmichael Structure, Mississippi

An area complicated by intense faulting and reverse dips has been identified as a cryptovolcanic structure by R. R. Priddy (63). Its center lies about 4 miles north of Kilmichael, and the entire formation is about 8 miles in diameter.

This diameter is not strictly comparable with certain of the other similar objects, for its outer structure is clearly shown. There is a central uplift about $3\frac{1}{2}$ miles wide, and the uplift is 800–1,200 feet. Outside the uplift is a surrounding graben averaging 6 miles in diameter; then there is a ring anticline and poorly defined series of graben and horsts involving successively younger beds.

At most cryptovolcanic structures the last clearly marked feature is the first graben and the inside edge of the surrounding ring anticline. As often appears, there is a noticeable bilateral symmetry.

The Des Plaines Structure, Illinois

An intensely deformed structure centered on the city of Des Plaines is considered by Shoemaker (45) as possibly an impact structure. Emrich and Bergstrom (64) question the capacity of the usual theories of the origin of cryptovolcanic structures to account for its formation.

The Crooked Creek Structure, Missouri

One of the best-known of the highly eroded cryptovolcanic structures is that in the Steelville Quadrangle in Missouri. The Crooked Creek structure was discovered in 1910 by a reconnaissance field party of the Missouri Geological Survey. Hughes (65) published a brief description of the formations involved and also a generalized geologic map.

Hinds mentioned the occurrence of Pennsylvanian coal deposits in sinks in southern Crawford County and observed that ". . . their slight extent and their crushed and disturbed condition make most of them of doubtful value" (66, 67).

The Crooked Creek structure is the dominant geologic feature of the Steelville Quadrangle. It consists of an essentially circular, highly deformed, and uplifted central area inside a peripheral series of high-angle normal faults. These faults form the inner margin of an encircling ring syncline. The whole structure is 3–4 miles in diameter. The oldest beds exposed are about 1,000 feet above their normal positions (68).

The uplifted central area is roughly $1\frac{1}{2}$ miles in diameter. It is a definite dome, but it is not a simple dome as it is in other such structures now known. It is a collapsed dome. The center of the uplift is slightly depressed. The diameter of the central peak crater is close to a mile, and the crater was about 500 feet deep. Of course, all surface exposures of the crater have been eroded away, so that this conclusion is based on the observed trends of the remaining rock layers. Hendriks' work and charts show this collapsed dome beautifully.

The beds on the outer flank of the ring anticline portion of the dome dip slightly outward and end against the upthrown side of the peripheral faults. The beds in the central section dip gently inward into the central basin. Faults, drag faults, and small irregular undulations cause many local variations in the direction and amount of dip.

Hendriks has found small but numerous shatter cones in the central region, particularly in the edge of the basin. These shatter cones are oriented upward in intensely fractured beds of the Cambrian period. They generally plunge in a south to southeasterly direction and form angles of approximately 70° with the recognizable bedding.

The uplifted central area is surrounded by a ring syncline from 1 to 2 miles wide. It lies between the peripheral series of high-angle normal faults that bound the central dome and the high-angle normal faults that separate the structure from the essentially horizontal bedrock of the surrounding area. The Roubidoux rocks of Ordovician time dip gently inward toward the center line of the ring syncline. However, along the margins of the ring the Roubidoux has been intensely fractured and dragged up. In places it dips as much as 70°–80° away from the faults.

The aerial photograph of the United States Department of Agriculture, Production and Marketing Administration shows a slight evidence of the structure. It certainly is not prominent. However, a later photograph (Pl. XVIII) taken in the autumn shows the formation beautifully developed. The vegetation in the area must be strongly affected by the nature and orientation of the rock layers. Hendriks (69) states:

It is believed that the structural pattern of the Crooked Creek area and other cryptovolcanic structures is similar to that which may be expected beneath a large meteorite crater. Features of the central uplift which are in accord with a meteoritic origin are: (1) the circular shape; (2) the size comparable to the dome under Meteor Crater; (3) the jumbled, brecciated, and shattered character of the rock; and (4) the upward orientation of the shatter cones. The character of the encircling synclinal graben, with its two areas of maximum depression, could be the result of the meteorite striking the earth at an angle of less than 90°, as was probably the case at Meteor Crater. A meteoritic impact origin does not have to explain the lack of magmatic materials or the lack of evidence for hydrothermal action. The absence of meteoritic material is to be expected, as the explosion of the meteorite would scatter it over the adjacent area, where it would be removed by erosion.

Certainly, this object came into existence after the deposition of the Jefferson City beds of Lower Ordovician age, for these beds are involved in the disturbance. Unfortunately, the sedimentary record is missing between the Jefferson City and the Pennsylvanian in this area. Certain Pennsylvanian age residual boulders of sandstone lie with equal distribution over the entire structure. Closer dating is not possible at present. The crater thus was formed between 400,000,000 and 240,000,000 years ago.

References

1. SCHUCHERT, C., and DUNBAR, C. O. *Outlines of Historical Geology*, p. 69. 4th ed. New York: John Wiley & Sons, Inc., 1945.
2. WATSON, F. *Between the Planets*, p. 29. Philadelphia: Blakiston Co., 1941.
3. BOON, J. D., and ALBRITTON, C. C., JR. "Deformation of Rock Strata by Explosions," *Science*, **96,** 402, 1942.
4. ———. *Field and Lab.*, **5,** 1, 1938.
5. ———. "Meteoritic Scars in Ancient Rocks," *Field and Lab.*, **5,** 53, 1938.
6. ———. *Ibid.*, **6,** 44, 1938.
7. BUCHER, W. H. "Cryptovolcanic Structures in the United States," *Report of the Sixteenth International Geological Congress*, **2,** 1055, 1933.
8. DIETZ, R. S. "Meteorite Impact Suggested by Shatter Cones in Rock," *Science*, **131,** 1781, 1960.
9. CHAO, E. C. T., SHOEMAKER, E. M., and MADSEN, B. M. "First Natural Occurrence of Coesite," *Science*, **132,** 220, 1960.
10. BRANCA, W., and FRAAS, E. "Das kryptovulcanische Becken von Steinheim," *K. Preuss. Akad. Wiss. Abh.*, 1. Berlin, 1905.
11. DIETZ, R. S. "Shatter Cones in Cryptoexplosion Structures (Meteorite Impact?)," *J. Geol.*, **67,** 496, 1959.
12. ———. "Meteorite Impact Suggested by the Orientation of Shatter-Cones at the Kentland, Indiana, Disturbance," *Science*, **105,** 76, 1947.

13. KRANZ, W. "Wanderungen, Probleme und weitere Forschungen im Becken von Steinheim am Albuch," *Schwäbischer Albver. Blätter, Jahrg.*, **38,** 68, 1926.
14. KRANZ, W., and GOTTSCHICK, F. "Zur Tektonik des Steinheimer Beckens," *Zs. f. deutsch. geol. Gesellsch.*, Vol. **77,** Monatsber., p. 37, 1925.
15. KRANZ, W. "Begleitworte zur geognostischen Spezialkarte von Württemberg," *Atlasblatt Heidenheim*, Vol. **77,** Monatsber., p. 37, 1925.
16. KRANZ, W. "Ein neuer Aufschluss des württembergischen statistischen Landesamts im Steinheimer Becken," *Zs. f. deutsch. geol. Gesellsch.*, Vol. **78,** Monatsber., p. 215, 1926.
17. ROHLEDER, H. P. T. "Steinheim Basin and the Pretoria Salt Pan; Volcanic or Meteoric Origin," *Geol. Mag.*, **70,** 489, 1933.
18. BENTZ, A. "Das nördlinger Ries-Problem und seine Deutungen," *Preuss. geol. Landesanstalt Sitzb.*, Heft 3, p. 85, 1928.
19. LÖFFLER, R. "Beiträge zur Ries Entstehungshypothese," *Oberrhein. geol. Ver. Jahresber.*, N.F., **14,** 73, 1926. Contains exhaustive bibliography.
20. WERNER, E. "Das Ries in der schwäb.-fränk. Alb," *Blätter d. Schwäb. Albver.*, 1904.
21. WILLIAMS, H. "Calderas and Their Origin," *U. California Pub., Bull. Dept. Geol. Sci.*, **25,** 239, 1941.
22. RECK, H. "Zur Deutung der vulkanischen Geschichte und der Calderabildung auf der Insel La Palma," *Zs. f. Vulk.*, **11,** 217, 1928.
23. KRANZ, W. "Aufpressung und Explosion oder nur Explosion in vulkanischen Ries bei Nordlingen und im Steinheimer Becken," *Zs. f. deutsch. geol. Gesellsch.*, **66,** Monatsber., p. 9, 1914.
24. ———. "Vulkanexplosionen, Sprengtechnik, praktische Geologie und Ballistik," *ibid.*, **80,** 257, 1928.
25. ———. "Fünfte Fortsetzung der Beiträge zum nördlingen Ries-Problem," *Centralbl. f. Min. u. Päl.*, **13,** 262, 1934.
26. RITTMANN, A. *Vulkane und ihr Tätigkeit*. Stuttgart, 1936.
27. SHOEMAKER, E. "Geological Interpretation of Lunar Craters." Administrative report prepared for the U.S. Geological Survey, May, 1961.
28. WILSON, C. W., JR. "Wilcox Deposits in Explosion Craters, Stewart County, Tennessee, and Their Relations to Origin and Age of Wells Creek Basin Structure," *Bull. Geol. Soc. America*, **64,** 753, 1953.
29. STUTZER, H. " 'Meteor Krater' Arizona und nördlinger Ries," *Zs. f. deutsch. geol. Gesellsch.*, Vol. **88,** 1936.
30. WERNER, E. "Das Ries in der schwäbisch-fränkischen Alb," *Zs. f. deutsch. geol. Gesellsch.*, Vol. **88,** 1936.
31. SHOEMAKER, E. M., and CHAO, E. C. T. "New Evidence for the Impact Origin of the Ries Basin, Bavaria, Germany" (abstr.), presented at A.A.S. Meeting, December, 1960.
32. BEALS, C. S., FERGUSON, G. M., and LANDAU, A. "A Search for Analogies between Lunar and Terrestrial Topography on Photographs of the Canadian Shield,"

Contr. Dom. Obs. Ottawa, **2,** 3, 1956; also *J.R.A.S. Canada,* **50,** 203–11, 250–61, 1956.

33. BEALS, C. S. "A Probable Meteorite Crater of Great Age," *Sky and Telescope,* **16,** 11, 1957.
34. BEALS, C. S., INNES, M. J. S., and ROTTENBERG, J. A. "The Search for Fossil Meteoritic Craters," *Contr. Dom. Obs. Ottawa,* **4,** 3, 1960; also *Current Sci.,* **29,** 205–18, 249–62 (Bangalore, India, 1960).
35. MILLMAN, P. M. "Editorial Note," *Sky and Telescope,* **11,** 9, 1951.
36. INNES, M. J. S., WILLMORE, P. L., and CLARK, J. F. *R. Soc. Canada, Minutes and Proc.,* June, 1954.
37. MILLMAN, P. M., LIBERTY, B. A., CLARK, J. F., WILLMORE, P. L., and INNES, M. J. S., "The Brent Crater," *Pub. Dom. Obs.,* **24,** 1, 1960.
38. BEALS, C. S. Personal letter to author, January, 1961.
39. INNES, M. J. S. "A Possible Meteorite Crater at Deep Bay, Saskatchewan," *J.R.A.S. Canada,* **51,** 235, 1957.
40. WILSON, C. W., JR., and BORN, K. E. "The Flynn Creek Disturbance, Jackson County, Tennessee," *J. Geol.,* **44,** 815, 1936.
41. SAFFORD, J. M. *Geology of Tennessee.* Nashville: S. C. Mercer, Printer to the State (accompanied by *Geological Map of Tennessee*).
42. KING, P. B. "The Geology of the Glass Mountains, Texas," *Descriptive Geol.,* Part 1, No. 3038, p. 123, 1930.
43. PHILLIPS PETROLEUM COMPANY. Letter to author. Two wells drilled in Sierra Madera dome.
44. Ref. 28, p. 767. Similar letter to C. W. Wilson, Jr.
45. SHOEMAKER, E. M., and HACKMAN, R. J. "Interplanetary Correlation of Geologic Time" (preprint copy), Seventh Annual Meeting, American Astronautical Society, Dallas, Texas, January 16–18, 1961.
46. NEL, L. T. "The Geology of the Country around Vredefort: An Explanation of the Geological Map," *Pretoria, Geol. Survey South Africa,* 1927.
47. HALL, A. L., and MOLENGRAAFF, G. A. F., "The Vredefort Mountain Land in the Southern Transvaal and the Northern Orange Free State," *K. Akad. Wetensch., Amsterdam, Abdeeling Natuur. Verh., 2d sec.,* **24,** 1, 1925.
48. MAREE, B. D. "The Vredefort Structure as Revealed by a Gravimetric Survey," *Trans. Geol. Soc. South Africa,* **47,** 183, 1944.
49. DALY, R. A. "The Vredefort Ring Structure of South Africa," *J. Geol.,* **55,** 125, 1947.
50. BALDWIN, R. B. *The Face of the Moon.* Chicago: University of Chicago Press, 1949.
51. DIETZ, R. S. "Astroblemes," *Sci. American,* **205,** 51, 1961.
52. Ref. 7, p. 1071.
53. WINSLOW, A. "Lead and Zinc Deposits," *Missouri Geol. Survey,* Vol. **7,** 1894.
54. SHEPARD, E. M. "Spring System of the Decaturville Dome, Camden County, Missouri," *U.S. Geol. Survey, Water Supply Paper 110,* p. 113, 1905.
55. TARR, W. A. Discussion in Ref. 7, p. 1084.

56. Shrock, R. R., and Malott, C. A., "The Kentland Area of Disturbed Ordovician Rocks in Northwestern Indiana," *J. Geol.*, **41**, 337, 1933.
57. Pullen, M. W. *Geologic Aspects of Radio Wave Transmission.* Report of Investigations, No. 162, State of Illinois, Dept. of Registration and Education, Div. of State Geol. Survey, 1953.
58. Gutschick, R. Personal communication to R. S. Dietz, Ref. 11, p. 503.
59. Born, K. E., and Wilson, C. W., Jr. "The Howell Structure, Lincoln County, Tennessee," *J. Geol.*, **47**, 371, 1939.
60. Ref. 7, p. 1062.
61. Ref. 7, p. 1063.
62. McKnight, E. T. "Geology of Area between Green and Colorado Rivers, Grand and San Juan Counties, Utah," *U.S. Geol. Survey Bull.*, No. 908, 1940.
63. Priddy, R. R., and McCutcheon, T. E. "Montgomery County Mineral Resources."
64. Emrich, G. H., and Bergstrom, R. E. "Intense Faulting at Des Plaines, Northeastern Illinois," *Bull. Geol. Soc. America*, **70**, 2, 1596, 1959 (abstr.).
65. Hughes, V. H. *Reconnaissance Work: Missouri Bur. Geol. and Mines Bienn. Rpt., 1909–10.* 1911.
66. Hinds, H. *The Coal Deposits of Missouri.* Missouri Bur. Geol. and Mines, 2d Ser., Vol. **11**, 1912.
67. Ref. 68, p. 9.
68. Hendriks, H. E. *The Geology of the Steelville Quadrangle, Missouri.* State of Missouri, Dept. of Bus. and Admin., Div. of Geol. Surv. and Water Resources, 1954.
69. Ref. 68, p. 69.
70. Bunch, T. E., and Cohen, A. J. "Precambrian Coesite." Paper presented at the first western meeting, American Geophysical Union, Los Angeles, Calif., December, 1961.
71. Cohen, A. J., Reid, A. M., and Bunch, T. E. "Central Uplifts of Terrestrial and Lunar Craters, I. Kentland and Serpent Mound Structures." Paper presented at the first western meeting, American Geophysical Union, Los Angeles, Calif., December, 1961.
72. Cohen, A. J., Bunch, T. E., and Reid, A. M. "Coesite Discoveries Establish Cryptovolcanics as Fossil Meteorite Craters," *Science*, **134**, 1624, 1961.
73. Dietz, R. S. "Vredefort Ring Structure: Meteoritic Impact Scar?" *J. Geol.*, **69**, 499, 1961.
74. ———. Personal communication, January 23, 1962.
75. Shoemaker, E. M., Gault, D. E., and Lugn, R. V. "417. Shatter Cones Formed by High Speed Impact in Dolomite," *Short Papers in the Geologic and Hydrologic Sciences,* Articles 293–435, D-365, Geology and Hydrology Applied to Engineering and Public Health. 1961.
76. Bucher, W. H. Personal communication, February 2, 1962.

5

CRATER FREQUENCY ON EARTH AND MOON

The Central Plains area of the United States has been rather thoroughly explored geologically. In this area there are known to be at least eleven cryptovolcanic structures. In the same general area is the Odessa group of terrestrial meteoritic craters and the Haviland pit. In Canada, particularly on the Great Shield, we recognize three very old craters—Brent, Holleford, and Deep Bay. In addition, there are the modern New Quebec Crater and the Labrador Crater. Other areas of the United States and the world are less well known, or the nature of the surface is such that the records of these structures are not recognized.

Several other Canadian structures show some good, but not definite, evidence of having had a meteoritic origin. Among these we include the Franktown Crater, the two Clearwater Lakes, the Mecatina Crater, and the structure near Carswell Lake. Certain others must be considered as possible craters of this type.

It may be considered presumptuous by some to identify the cryptovolcanic structures as old meteoritic craters, but at the present writing the evidence is so strong that no other conclusion seems tenable. The complete solution of the problem of the cryptovolcanic structures and their correlation with ancient meteoritic craters should be a prime objective. There are now sufficient data on cryptovolcanic dimensions that we may intercompare them and show that these formations all belong to one family. The known facts are listed in Table 3.

Basically, the amount of erosion which has occurred at these impact structures is widely different. At the Flynn Creek and Howell disturbances, as well as at the German Steinheim Basin and the Rieskessel, there are remnants of the original surface crater. Most of the structures exhibit bilateral symmetry. The main pattern is the typical shock-wave rebound pattern. All show a central dome when they are sufficiently eroded to expose the basement structures. In most cases this dome possesses a central region in which the rocks are hopelessly jumbled. On our model, the dome is a rebound, and the central jumble

TABLE 3

IMPACT STRUCTURES OF THE CRYPTOVOLCANIC TYPE

	Dome	Diameter in Miles of						Dome Uplift (Feet)	First Syncline Downthrow (Feet)
		Syncline		Anticline		Syncline			
		Middle	Outer Edge	Middle	Outer Edge	Middle	Outer Edge		
Howell............	0.4	(0.7)	1	....	....	....	....	100	Small
Flynn Creek.......	0.6	(1.4)	2.2	....	....	....	....	500	350
Steinheim.........	0.7	(1.35)	2	....	....	....	....	500	260
Sierra Madera.....	0.9	(1.5)	2.1	(2.55)	3.0	....	....	....	....
Decaturville.......	1.3	(1.75)	2.2	(2.95)	3.7	....	....	530	(200)
Jeptha Knob.......	1.4	(1.85)	2.3	....	....	....	....	200–400	100–200
Crooked Creek.....	1.5	(2.5)	3.5	....	....	....	....	1,000	....
Upheaval Dome...	1.6	2.7	(3.05)	3.4	(3.8)	....	....	1,200	....
Serpent Mound....	2.2	(3.1)	4	4.5	(5.0)	....	....	950	500
Wells Creek.......	3.2	4.5	6	6.3	7.2	8.1	9	2,500	700
Kilmichael........	3.5	(4.75)	6	7	8	....	....	800–1,200	....

corresponds to the shattered and brecciated layers which normally lie beneath the terrestrial meteoritic crater.

Beals is studying the cores from the Brent Crater to see whether there is any evidence of the rebound dome under this crater. The determination will be made on the levels at which brecciated rocks from different layers are actually found. The Brent crater does not show a central peak in the original apparent crater.

The portion immediately surrounding the dome of a cryptovolcanic structure is a ring syncline. Its area is greater, but the downthrow is less than the upthrow of the dome. The syncline is considered to correspond in position with the main crater but, of course, is not usually the crater depression itself but is a basement structure brought into view by extensive erosion. At the smaller impact structures, this is all that appears. The outer edge of the syncline joins with the normal area rocks *in situ*. The smashing of the rock layers decreases with increasing distance from center. The shatter cones are normally found in the central areas only.

As the objects become larger, a ring anticline appears surrounding the syncline. At still larger impact structures, the anticline is itself bordered by a second ring syncline. There are suggestions of this at Kilmichael, and it is well developed at the Wells Creek Basin. Table 3 shows this expanding trend of complexity versus size, and, in addition, we note that there is a decided tendency for the uplift at the dome to become greater at the larger formations and a similar increase in the magnitude of the downthrow of the first syncline.

The fact that all the highly eroded impact structures show a central rebound dome does not imply that all the original craters exhibited central peaks. The peaks may have been visible, but in many cases they may not have extended through the breccia at the crater bottom. There are no visible peaks in the Arizona and New Quebec craters. Only the Kaali Järv is found to have such a central peak. In the lunar craters the frequency of central peaks is found to increase to a maximum as the crater diameter becomes larger, while there seem to be no central peaks in the small pits.

These peculiar formations are thus one family, and all evidence implies that we may consider them with the Canadian meteoritic craters in any statistical study. By comparison with known terrestrial meteoritic craters, we find that the position of the first ring anticline is approximately under the rim of the original crater. Odessa No. 1 suggests that the rim is actually slightly beyond the middle of the first anticline. At the Arizona meteoritic crater, the surface layers at or near the anticline have hinged and overturned. In such a case the anticline center is perhaps just inside the crater rim. Probably a coincidence of rim and anticline is a good first approximation. On this basis, these ancient objects once sported craters ranging from 1 to perhaps 7 miles in diameter.

At the New Quebec Crater there are portions of two ring anticlines outside of and parallel to the one imbedded in the rim. The crater is 2.14 miles in average diameter, or about the same size as the Flynn Creek structure. The incomplete ring anticlines are thus about 1.6 and 2.0 times larger than the crater.

At Deep Bay Crater the surrounding terrain is deeply eroded, but there is a definite rim with a diameter of about $7\frac{1}{2}$ miles. At a diameter of 11 miles, there is what almost certainly is a reasonably well-developed ring anticline. It appears most clearly on the east through the south, but suggestions of it can be found in other sections. Surrounding the anticline is a lower zone marked by an arcuate lake in the east and other lakes throughout more than 180°. The northern section merges with Reindeer Lake, and so the structure there is unclear. Inasmuch as these symmetrical lakes are all bounded within about $1\frac{1}{2}$ miles, there is indirect evidence of a third ring anticline with a diameter of perhaps 14 miles. The relative dimensions here are slightly less than at the New Quebec Crater, namely, 1.5 and 1.9 times the crater diameter, respectively. The smaller Holleford, Brent, and Arizona craters do not seem to have produced these surface ripples in the rock beyond the rim.

The Wells Creek Basin structure probably once had a crater about 6.3 miles in diameter. It shows a definite ring syncline around it, and fragmentary indications of a ring anticline about 10 miles in diameter, or 1.6 times the crater diameter.

It would appear that it takes a large impact to produce the second and third ring anticlines, with accompanying synclines. It also seems that these outer structures are more closely confined to the surface layers than are the first syncline and anticline. This is consistent with the observed fact that the crater dimensions vary as functions of applied energy to an exponent higher than an inverse cube of the distance from the explosion.

The second and third ring anticlines do not seem to be present at Theophilus or most lunar craters. When Aristillus, Bullialdus, and Bailly are seen just at sunrise or sunset, a faint raised rim may be seen around each at approximately twice the crater radius from the center. At least one such anticline has appeared prominently at each of the larger impact maria.

Shoemaker and Hackman (2) have made an interesting attempt to see whether the frequency of the cryptovolcanic structures is consistent with the frequency with which the small craters appear on the lunar maria and also whether the frequency of either is consistent with the present rate of infall of meteoritic bodies to the earth. They used several assumptions and limits, but only one typical set will be used here.

Harrison Brown (3) has derived an expression for the frequency with which such bodies strike the earth, based on the maximum estimates for the frequency of falls of stone and iron meteorites. There is absolutely no information on how well this formula represents the infalls of meteorites throughout geologic history or how well it represents the very large masses which occasionally strike the earth, but, even so, it is interesting to apply the formula to see what results it predicts:

$$f = \left[\frac{6.9 \times 10^{11}}{(10^6 \text{ km}^2)(10^9 \text{ years})(\text{gm}^{-0.80})} \right] M^{-0.80}, \qquad (5\text{-}1)$$

where f = the total impact frequency of bodies of mass $> M$ and M is the mass of the impacting body in grams.

Shoemaker used the scaling relationship between crater diameters and expended energy as derived from nuclear explosion craters. He finds the scaling exponent to be $1/3.4$, a constant.

From the more extensive work in this volume, the exponent is found to be variable, and the average size of a chemical explosion pit is very slightly larger than one produced by a nuclear explosion of the same yield. Shoemaker's procedure, slightly modified, is as follows:

The energy released by the impact of a meteorite is simply the geocentric kinetic energy:

$$\left(4.185 \times 10^{19} \frac{\text{ergs}}{\text{KT}} \text{ TNT equivalent}\right) W = \tfrac{1}{2} M V_r^2, \qquad (5\text{-}2)$$

where W is the number of kilotons of TNT equivalent released, M is the meteoritic mass in grams, and V_r is the impact velocity in cm/sec.

The new scaling equation—from equation (8-10) (p. 164)—is

$$\text{Diameter} = \frac{88 W^{1/y}}{(\text{KT})^{1/y}}, \qquad (5\text{-}3)$$

where Diameter is the apparent crater diameter in meters, and 88 is the apparent crater diameter in meters for a 1 metric kiloton burst at a scaled depth of burst of $H/W^{1/3} = 0.10$; y is the variable exponent. It ranges from 3.2 to 3.6 as the applied logarithmic energy in calories varies from 2 to 20. One metric kiloton has $E = 12$. This equation is reasonably accurate for craters up to 10 or 20 miles in diameter.

The exponent y varies according to

$$y = \frac{E - 12}{D - 2.4584}, \qquad (5\text{-}4)$$

where D is the logarithm of the apparent crater diameter in feet. With equations (5-1), (5-2), (5-3), and (5-4), we find

$$f = \left[\frac{7.96 \times 10^3}{(10^6 \text{ km}^2)(10^9 \text{ years})}\right] 88^{0.8y} V_r^{1.6} D_m^{-0.8y}, \qquad (5\text{-}5)$$

where f is the cumulative frequency of impacts per 10^6 km^2 per 10^9 years which have produced craters whose logarithmic apparent diameters in meters are greater than D_m. This equation is very similar to Shoemaker's equation (15). It predicts slightly more small craters than does Shoemaker's curve and slightly fewer large craters.

The predicted rate of production of meteoritic craters larger than 1 kilometer in diameter is 1 in 60,000 years according to equation (5-5). Shoemaker's equation (15) would yield one such crater in 90,000 years. These figures are for the earth. If we are dealing with the moon, the frequency would be slightly less but would still be of the same order of magnitude. If the size of such a crater is 10 kilometers or larger, the rates become 1 in 70,000,000 years and 1 in 50,000,000 years, respectively. For larger craters, both equations predict impossibly low frequencies.

For craters of the order of about 1–3 kilometers in diameter, both equations yield results about equal to the observed frequency of cryptovolcanic structures in the central United States. There are ten such structures known in 770,000 square kilometers. They have an average age of 235,000,000 years.

Shoemaker and Hackman (2) have made a re-evaluation of the frequency of craters in the various lunar maria. Their table is reproduced in Table 27 (p. 161). They have made a strong attempt to eliminate any craters which might have been caused by ejecta from nearby large impacts, such as Copernicus. On the assumption that the maria are 4.5×10^9 years old, they find that the observed frequencies of the lunar postmare craters are in agreement with the extrapolated rate of equation (5-5) (or Shoemaker's eq. [15]) only for the 1–3-kilometer-wide craters and that there is an excess of one or two orders of magnitude in the abundance of the larger craters on the lunar maria. They conclude that the observed slope of the cumulative frequency distribution curve of the craters on the maria is less than the slope predicted from the mass distribution curves for meteorites observed to fall on the earth.

Inasmuch as the extrapolation is long here, it is not too surprising that the discrepancy is substantial. It is probable that the mass distribution of large objects in the neighborhood of the earth and moon over the last several billion years has been significantly different from the calculated distribution for recent meteorites. It is also possible that they were unable to eliminate many of the secondary craters caused by ejected fragments from the impacts which produced large postmare lunar craters. If the angles of fall were steep, the resulting craters would be circular and might not be closely distributed near to the parent crater. The moon's gravity is so low that widespread scattering of such fragments might be possible. Shoemaker and Hackman (2) conclude:

> If we reverse our viewpoint by adopting the age of the maria as 4.5 billion years and take the observed frequency distribution of the craters as the best data on the mass distribution of objects in the Earth's neighborhood, then the predicted frequency of impact of the larger objects on the Earth may be increased. An increase of an order of magnitude in the probabilities of finding an impact structure the size of the Vredefort Dome would relieve the necessity of attributing the exposure and recognition of this structure to good luck. Such an increase probably would also be in better accord with the fact that 13 asteroids have now been discovered in the neighborhood of the Earth, considering that their discovery has largely been accidental and many more are likely to be found. If the probable incompleteness of our information on terrestrial impact structures is taken fully into account, it would appear that the geologic record of impact is consistent with the hypothesis that the lunar maria are about 4.5 billion years old, that the craters assumed to be of primary impact origin are indeed formed by impact, and that the space density and mass distribution of large solid objects in the neighborhood of the Earth has remained nearly constant over most of geologic time.

A corollary of this hypothesis is that the rate of formation of craters on the Moon in the period prior to the development of the maria was much greater than during subsequent geologic history.

It is also very probable, as will be shown later, that the maria are younger than 4,500,000,000 years and that the observed excess of such craters on the maria is greater than has been calculated. A very extended period of time elapsed between the formation of the lunar surface and the genesis of the maria.

A slight correction factor must be applied to lunar impact rates as calculated from terrestrial impact rates, because the vacuum-sweeper effect of the planet is larger than that of the satellite.

The correlation of the numbers of craters on the lunar maria with the numbers of cryptovolcanic structures in the central United States and the estimates of the numbers of falls of meteorites tell us that the lunar maria are old. They cannot tell us how old they are, but an age of less than 4.5×10^9 years gives a closer agreement than does the full accepted value for the age of the moon.

References

1. BALDWIN, R. B. *The Face of the Moon*. Chicago: University of Chicago Press, 1949.
2. SHOEMAKER, E. M., and HACKMAN, R. J. "Interplanetary Correlation of Geologic Time," Seventh annual meeting of the American Astronautical Society, Dallas, Texas, January 16–18, 1961.
3. BROWN, H. "The Density and Mass Distribution of Meteoritic Bodies in the Neighborhood of the Earth's Orbit," *J. Geophys. Res.*, **65**, 1679, 1960.

6

CHARACTERISTICS OF EXPLOSION CRATERS, TERRESTRIAL METEORITIC CRATERS, AND LUNAR CRATERS

Many books and articles have been written on the general subject of the moon. Some of these papers are purely factual and descriptive. They form an invaluable resource library. Others are frankly speculative. In them the authors have attempted to explain various features of the moon and even the origin of the moon itself. Some of these speculative hypotheses have had little to recommend them, save only their novelty.

Out of this welter of conflicting ideas, one theory of the origin of the lunar-surface features now has apparently gained the ascendancy. This is the idea that, very early in the moon's history and to a lesser extent throughout the moon's life, great meteorites[1] crashed into the lunar surface and generated the various markings we now know.

Although it is clearly evident from the recent published and unpublished comments of a large number of astronomers and others interested in the moon that the meteoritic theory of the origin of lunar structures is greatly favored, it should not be supposed that this attitude is universal. Far from it, for re-

---

[1] Purists may question the usage of the word "meteorite" in this connection. If they would prefer "planetesimal," "planetoid," or other verbiage, I could not stand in opposition. The word "meteorite" has often been used in the past, and the connotation is well understood. It will generally be used here.

nowned men such as Wilkins (1), Moore (2), Marshall (3), Escher (4, 5), Spurr (6–10), Alter (11–16), Miyamoto (17, 18), Green (19, 20), and Firsoff (21), to name but a few, have argued for their own versions of the various volcanic theories or combination volcanic-meteoritic theories. This is a healthy situation, as supine acceptance of a theory which can be settled absolutely only by going to the moon or by seeing a great crater formed there can quickly lead to sterility. Considered opposition will often bring out ideas which will lead to further advances. No student worthy of the name would argue that the evidence so far gained from studies of the lunar surface shows that only one process was operative.

In *The Face of the Moon,* published in 1949 (22), was presented a summary of the various theories and hypotheses which had been advanced up to that time. Since then, Urey (23–29) and Kuiper (30–32) have offered modifications of the sequence of happenings and slight differences in the interpretation of evidence. These valuable works will be discussed later. Except for these two modifications, no new evidence has been found or interpretations made which would invalidate the conclusions reached in *The Face of the Moon,* namely, that the great majority of lunar-surface features were formed very early in the moon's history by the impacts and subsequent explosions of myriads of meteorites, great and small. There is clear evidence of some sort of igneous action occurring as a secondary phenomenon in the chain craters and possibly in lava-filled Wargentin, in the great domes on the lava flows, the lava flows themselves, and possibly in the occasional central peak craters. Yet, even if these are all granted to be igneous structures, the dominant process appears to be meteoritic impact.

In *The Face of the Moon* the arguments were primarily qualitative. Some attempts to inject quantitative measures and interpretations were made, but at that stage these were necessarily crude. At present, it seems possible to refine and extend some of these earlier efforts.

Inasmuch as a meteoritic impact results in an explosion, much energy will be expended here in establishing the characteristics of crater formation as brought about by explosions. Unfortunately, we have no ready-made source of meteorites which could be used to produce nice, new terrestrial meteorite craters. Many terrestrial meteorite craters are known, and more are being discovered, but usually erosion and infilling have somewhat modified their original contours. Still, many facts have been discovered about them, and certain generalities may be accepted.

A larger source of data on craters comes from man-made explosions. Descriptions of many hundreds of large and small explosions and craters are available. Most of these man-made explosions were from TNT and similar chemical explosives.

The energy densities in these detonations, while high, were somewhat less than in most meteoritic-impact explosions. Nevertheless, there is absolutely nothing in the now existing data which indicates that this possible difference in energy density has caused significant differences in results. It follows that, to a first approximation, meteoritic craters and chemical explosion pits are essentially identical for equal expenditures of energy. This may probably be attributed to the fact that practically all the energy in each case is released as mechanical energy and very little as radiant energy. Little has been published on similar applications of atomic energy, but, judging from the accounts of the tremendous floods of radiant energy from fusion and fission bursts and the penetrating power of atomic fragments, it may confidently be expected that nuclear explosions would be considerably less efficient in forming craters than chemical or meteoritic explosions. Both the nuclear test shots, Teapot Ess and Jangle U, reported by Shoemaker (33, 34) are slightly too small for the expended energy, although the crater shapes are correct.

Table 4 (Appendix 2) lists the raw data on man-made explosion pits from chemical and nuclear explosives. Table 6 (Appendix 2) lists similar data from terrestrial meteoritic craters, but only in those cases where reasonable approximations are known to the original crater dimensions, either because the crater is not much eroded or from excavation and drilling. Table 7 (Appendix 2) summarizes the known facts concerning dimensions of the lunar craters. The lunar craters are divided arbitrarily into five classes. Class 1 is composed of those craters which appear to be most nearly in their original condition. They are newer-appearing than most and less damaged by impositions of subsequent craters. They often show ray systems. The Class 2, Class 3, and Class 4 craters are progressively older in appearance. The Class 5 craters have been filled or partially filled with some material. All craters produced by bursts of any kind of explosive are similar. Variations are due to several factors, such as the position of the explosive center relative to the ground level and the absolute magnitude of the energy release and the soil or rock characteristics, including density.

A crater may or may not have a central peak or peaks. The larger lunar craters generally do have peaks. Terrestrial explosion pits may have them, but usually do not. The size, complexity, and presence or absence of a central peak do not distinguish two different classes of craters but merely offer an additional variable.

When one examines the charts from which are derived the curves relating the various crater parameters, he cannot fail to be impressed by the scatter of the points. This scatter is real. It is observed at all controlled tests of explosive craters.

If one takes several identical charges of a given explosive and runs tests under supposedly identical conditions, there often will be variations in the crater dimensions of the order of ±25 per cent. The different crater dimensions seem to vary almost independently.

On occasion, the explosions seem to be directed much more in one direction than in another. I well remember a test conducted on the sandy shores of Lake Michigan with quarter-pound blasting-powder charges. The charges were spherical. The procedure was to bury the charge to a predetermined depth, fill in the narrow hole above the charge, light a long, slow fuze, and get out of the way. Several tests were moderately concordant, and the last charge was planted just like the others and lighted by Mr. A. W. Hewitt, the dignified president of the local school board. He lit the fuze and ran to a distance of about 50 feet. When the explosion came, it was not symmetrical but was directed almost entirely at Mr. Hewitt, who was drenched with sand. The point is that, in dealing with explosive craters, we must be careful to deal in averages only, as the individual crater may deviate widely from the norm.

Experimental work by Lampson (35) and others has shown that a surface burst yields a broad, shallow crater, with a relatively low rim. When the burst occurs immediately below the surface, the picture changes. The crater becomes slightly wider, the depth increases markedly, and the rim height increases. The ratio of diameter to depth for the apparent crater drops quickly to a value of 3–4 and remains very approximately at that value for a wide spread in depths of burst. This is because both diameter and depth of the resulting crater increase at about the same rate as the depth of burst increases. Finally, a point is reached where burying the explosive charge more deeply results in wider, but shallower craters. At a still greater depth, a point of maximum apparent crater diameter is reached, and thereafter the surface crater becomes rapidly smaller as the depth of burst is further increased, until a camouflet is reached.

Only in the very shallowest explosions is the crater cleaned out of loose detritus. The volume of loose material remaining in the pit increases steadily as the explosive focus is moved deeper.

This sequence is very informative. It tells us that, for those depths of penetration of meteorites into the earth or moon that we may logically expect, the form of the visible crater is almost completely independent of the depth of explosion. This accords with observation, for the form of terrestrial explosion pits, uneroded terrestrial meteoritic craters, and the undistorted lunar craters is purely a function of diameter. A second important conclusion is that, for bursts centered at the surface or very slightly below, the crater produced is relatively shallow.

Recognition of this effect allows a minor correction to be made in the basic relationship between diameters and depths of explosive craters. The equation derived in *The Face of the Moon*, Figure 12, was

$$D = 0.1083 d^2 + 0.6917 d + 0.75 , \qquad (6\text{-}1)$$

where

$D =$ the logarithm of the diameter (feet),

and

$d =$ the logarithm of the depth (feet),

both for the apparent crater. This equation represents the observed points beautifully, yet it contains a disturbing element. For the very smallest craters, a few feet in diameter, the depth was observed to increase more rapidly than the diameter. This did not seem to be logical, and so an attempt was made to find out why the observed points defined a curve of this type in this region. Through the co-operation of Mr. Earl Picard, a local fireworks display expert, a series of test explosions was set off under carefully controlled conditions. These subsurface bursts yielded craters from about 2 inches in diameter to pits several feet across. Data were obtained roughly corroborating Lampson's work. When the diameters and depths were plotted on a log-log scale against the curve defined by equation (6-1), they were found to give points lying below the curve and gradually approaching it with increasing diameter until they merged at a diameter of about 35 feet. With this information, the parts of the puzzle fell into place. The lower portion of the original curve was defined by craters formed from shell bursts. We were dealing only with subsurface bursts, and data were gathered for this purpose from military tests of delay-action fuzes in shells. Nevertheless, the center of a shell explosion is always several inches behind the nose. When such an explosion is initiated by impact with the ground, the explosive focus either is very much closer to the ground level than the nose of the shell or may even be at ground level, even though the fuze were a delay-action type. Thus shell explosion pits should be shallower than those from equivalent energies buried and detonated more deeply below the surface.

The shell craters reported in *The Face of the Moon* were essentially all from low-angle firings. Therefore, the explosion centers were always closer to ground level than if the shells had fallen at high angles. This aggravates the above-mentioned distortion of the D versus d curve and clearly indicates that the shell craters must be rejected in the derivation of a definitive relationship between diameters and depths of craters from subsurface bursts.

The same arguments should hold for bomb craters, but, because the bomb mass and the angle of fall are each much greater than for the shells studied, the bombs with delay-action fuzes are able to penetrate sufficiently below ground to yield results consistent with bursts from buried explosives. The bomb data are retained.

The portions of the original curve describing craters between about 35 feet in diameter and about 100 miles in diameter are not significantly affected by this correction.

This change in the small-diameter end of the D versus d curve does not affect in the slightest degree the conclusions reached in *The Face of the Moon*, but the present improvement defined in equations (7-1) through (7-4) does permit various basic relationships to be found between crater forms and sizes and energies needed to produce the wide range of observed terrestrial and lunar craters.

Man has set off millions of explosions. Modest numbers of these have been assigned to beneficial uses. A tiny few of these explosions and their effects have been reported in more or less detail. An infinitesimal number of controlled tests have been run, mostly by the military. An even smaller set of data has been released in non-classified form; and most of the data processed by the military, both classified and unclassified, are nearly valueless as far as shedding any light on the mechanics of crater formation and establishing valid relationships covering wide ranges of applied energies and different locations of the bursts above and below ground level.

An explosion acts so nearly instantaneously that it is difficult to determine the sequence and nature of the events which result in the familiar crater pit. Even so, certain simple tests may be designed to yield information on the mechanics of crater formation. One such test was conducted by Lieutenant Fred Olsen, of the United States Army Service Forces, at Arco, Idaho, in 1944. His unclassified report, a typewritten document, came to light as a by-product of a discussion on a different subject held with Dr. Ralph Ilsley in Washington, D.C. In this test, a 100-pound block of TNT was placed on the ground and detonated. It may be pointed out in passing that this, technically speaking, means that the explosion was an air burst and not a surface burst, inasmuch as the burst center was above ground. The following observations were made by Lieutenant Olsen (36):

> The visible subsurface crater was 7′ in diameter, about 10″ deep. The ground around the crater was covered by fine silt as well as a layer of 5″ of silt in the crater itself. There were no lumps or larger pieces scattered about. There was an even cone of hard rocklike material (5′ × 5′ × 10″), precisely in the center of the crater which had worked its way through the layers of silt. The caked material extended to a depth of 30″ and was permeated with fine root systems. The root system in normal soil extended to about 20″.
>
> Because the normal soil is loosely packed loam with a noticeable moisture content, it was possible to define the limit of the compressed soil. Primarily it can be stated that the crater is purely the result of compression. The shape of the crater is roughly parabolic as opposed to conical. There was apparently no material thrown out of the crater which may be deduced from the lack of small pieces of hard compressed ground in the area surrounding the crater.
>
> The cone in the crater is not made by pieces of earth falling back into the crater. When the shock wave first compresses the ground under the crater, there is a displacement of ground roughly equivalent to the final apparent crater dimensions. However,

the dry nature of the soil, as well as the packing of the loam, coupled with the intermittent layers of lava, makes the ground structure comparatively resilient. On restoration after the compression the dense compressed material from the bottom of the crater is broken up and piled in a cone. This would account for the precise location of the cone in the center of the crater. The finding of root structures in the cone material confirmed this.

The silt found in the crater was drawn into the crater from the surrounding ground by the thermal up-draft which occurred over the center of the crater. This silt did not derive from the soil in the crater volume prior to the detonation.

The lip of the crater evidently is not composed of materials scoured out of the crater, but rather is displaced and distorted by the compressive forces which find relatively less resistance on the surface of the ground. In a sense, the surface immediately surrounding the crater "buckles" from a combination of vertical and horizontal forces.

The lumps of compressed ground were hot and showed evidence of baking, although the imbedded roots were not charred. There was evidently no heat transfer from the hot explosion products to the compressed earth. The baking was rather the heat effect of mechanical non-elastic compression to form the crater. This "potential energy of compression" is released in situ as heat energy, which is responsible for the baking or sintering. It is interesting to note that the roots which did not offer the same resistance to compression (such that only a small energy is expended in compression) were not charred, even though sufficient oxygen was available.

Thus it seems that the crater resulting from a high-order surface or near-surface explosion is a product of simple compression by the shock or blast wave. There are no digging or scouring effects. The form of the crater is approximately that of a paraboloid of revolution, except for the rebound cone in the center. Lieutenant Olsen also makes the point in his article that "craters normally have a lip (rim) and frequently have well developed cones in the bottom." It must be granted that the conditions of this test were particularly suited to the formation of a central cone, because the ground layers were "comparatively resilient." Under similar circumstances, any terrestrial or lunar material would behave in a more or less comparable fashion, and thus we have a clue which may indicate how the great, low, central peaks were formed in the craters of the moon.

Shoemaker (39) has made an assumption which can be correct only in part. He says:

The final phase of displacement of material in fairly large impact craters is the slumping back of the breccia and part of the crater walls toward the center of the crater. At Meteor Crater, Arizona, this displacement appears to have preceded the showering down of ejecta thrown to great height, but at the Ries basin it may have occurred later. The centripetal movement of material from the crater walls produces a rise in the level of the central part of the crater floor. No sharply defined central hump is present on the floor of Meteor Crater, though a subdued off-center ridge may be buried under the Pleistocene lake beds below what Barringer [37, 38] referred to as Silica Hill. The convergent movement resulting from slumping, however, would appear to be entirely adequate to explain the formation of a central hill such as observed in the Steinheim basin.

This mechanism is quite distinct from that of a rebound after compression. It would appear that a rebound is necessary opposite to the direction of approach of the meteorite after the shock wave has left the impact point. Slumping may well operate to aid in the production of a central peak, but a rebound would operate far more rapidly than the slumping and appears to be the dominant mechanism.

After the passage of a shock wave, the rock layers revert to their original strength, or approximately so, except in the immediate subcrater breccia layer.

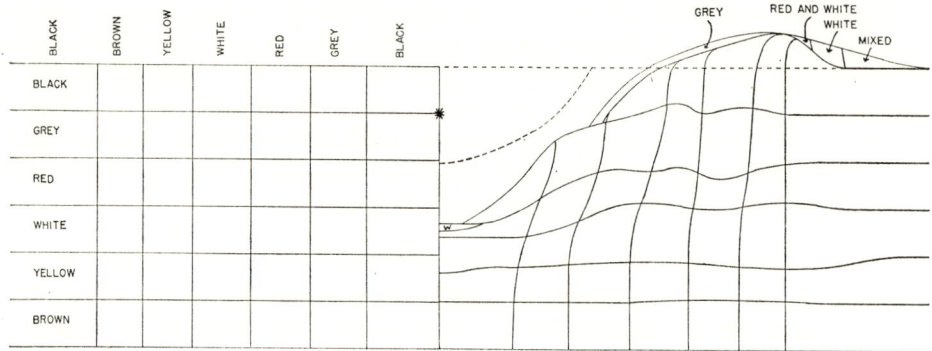

Fig. 13.—Section across crater formed by 40-grain dynamite charge illustrating mechanics of crater formation.

It is difficult to picture a massive slumping mechanism as aiding the production of a central peak. There may indeed be a certain amount of slumping such as might produce the terraced walls of Copernicus, but here, too, as Shoemaker has shown, there is a radial motion of the rock layers away from the explosive center, and we should expect a certain amount of rebound centripetally from the walls. This rebound, coupled with some slumping, should produce terraced walls.

Because the results reported by Lieutenant Olsen referred only to a burst just above the surface, a similar test was devised on a smaller scale. A 40-grain dynamite charge was detonated 1 inch below the surface of a specially built-up volume of soil. A large box 3 feet square and 1 foot deep was filled with soil. The bottom 6 inches consisted of ordinary garden dirt. Above this were six horizontal layers of colored soil, each 1 inch thick. They were, in ascending order, brown, yellow, white, red, gray, and black. Each layer was lightly tamped into place. The horizontal lines on the left side of Figure 13 show this arrangement.

The star on the center line gives the burst center. The right-hand portion of Figure 13 shows the sectioned crater produced by the explosion. The roughly horizontal lines are the boundaries between the different-colored materials.

A second test designed to complete the picture was then run and the results incorporated in Figure 13. A series of vertical layers of colored soil was placed in the box. The center black section was 3.2 inches wide, and the pattern indicated by the vertical lines in the left half of the Figure 13 shows the before-explosion arrangement. The distorted up-and-down lines in the right-hand half give the results found by a sectioning of the crater on a vertical plane through the burst center and perpendicular to the vertical soil layers.

The data from two explosions are thus incorporated in one crater drawing. With very minor modifications, the two sets of data are mutually consistent. Several interesting conclusions may be drawn.

A. Most of the materials above the burst center, within the crater limits, were ejected from the pit. The curved dotted line gives the limit of the volume from which the soil was actually blasted and scoured out of the crater.

B. Some materials were scoured out of the crater by the blast, even below the burst.

C. Most of the volume of the crater was produced by permanent compression and displacement of the soil, not by ejection of materials. The zone between the dotted line and the solid line corresponding to the true crater represents the amount of compression and permanent distortion of the soil.

D. The amount of compression of the soil was greatest close to the burst and was directed radially away from the burst.

E. A marked crater rim was formed. From above, it looked as though it were produced by scoured-out materials. Sectioning showed that it was produced primarily by the compressional effects which squeezed the materials up above the original surface.

F. The upper third of the crater—the inside of the rim up to and a little beyond the rim crest—was covered by a thin wash of the material from the layer immediately below the burst.

G. A portion of the rim beyond the crest was composed of red and white materials which did not come from below the burst but from the portions of the rim which were being squeezed upward at the same time as the black and gray materials were being ejected.

H. The ejected materials departing close to the surface interacted with the rising rim materials, and the top of the rising layers were stripped off and deposited in the lee of the rim.

I. The chief evidence for the interpretation of H is the fact that the top of the squeezed-up rim layers have been pushed over sharply away from the burst center.

J. Complicated shock waves were sent pounding through the soil, and several of them were damped so rapidly that the distorted waves froze in position. This is most clearly shown by the distortion of the horizontal gray, red, and white layers.

K. The first ring anticline appeared inside the crest of the rim, not under it.

L. The density of some of the materials below the burst was markedly increased.

M. An incipient central cone was formed, as is shown by the flat area in the center of the crater.

N. This flat area was a rebound phenomenon. Some of the horizontal white layer was found above the red layer and partially under the gray layer. It was of lower than normal density as contrasted with the increased density of the yellow and white layers below it. Apparently, the rebound had either flipped a portion of the red and white layers completely over or had driven the white material back up through the red so that it lay clearly exposed in the center of the crater.

Bearing in mind the difference in scale between the two tests and the larger test at Arco, the results are logically consistent. A crater is largely a compressional structure rather than one which is formed by ejection of material, although some ejection usually occurs. The raised rim is primarily produced by upward and outward squeezing of the ground layers rather than the piling-up of ejected crater materials, although the top layers are composed of broadcast debris.

The frozen shock waves in the ground layers of this small-scale test brought to mind forcibly a test made at the Aberdeen Proving Ground during World War II. A group of us was in a flimsy wooden lookout, high above the east shore of Chesapeake Bay, conducting a test of our own. Some distance away, on Abbey Point, the Army scientists detonated a large bomb. It might have been as large as 12,000 pounds. It made a tremendous flash, and then, seconds later, our lookout was damaged and nearly torn apart by the blast. Some weeks later I had the opportunity of visiting Abbey Point and made an effort to find where the bomb was set off. It obviously had been an air-burst, probably with the bomb suspended from a post. The crater was a simple depression where the ground had been pressed downward, and the depression was permanent. There was no central peak. The most interesting features, however, were two ring synclines about 6 inches and 1 foot deep, respectively, completely circling the crater. They were only about 3–4 feet wide and were many feet from the center. The corresponding ring anticlines were not very high—perhaps an inch or two—but were much wider. These structures were formed by shock waves, which forced the ground to roll and writhe almost as though it were liquid.

The above conclusions concerning explosive craters, while not previously

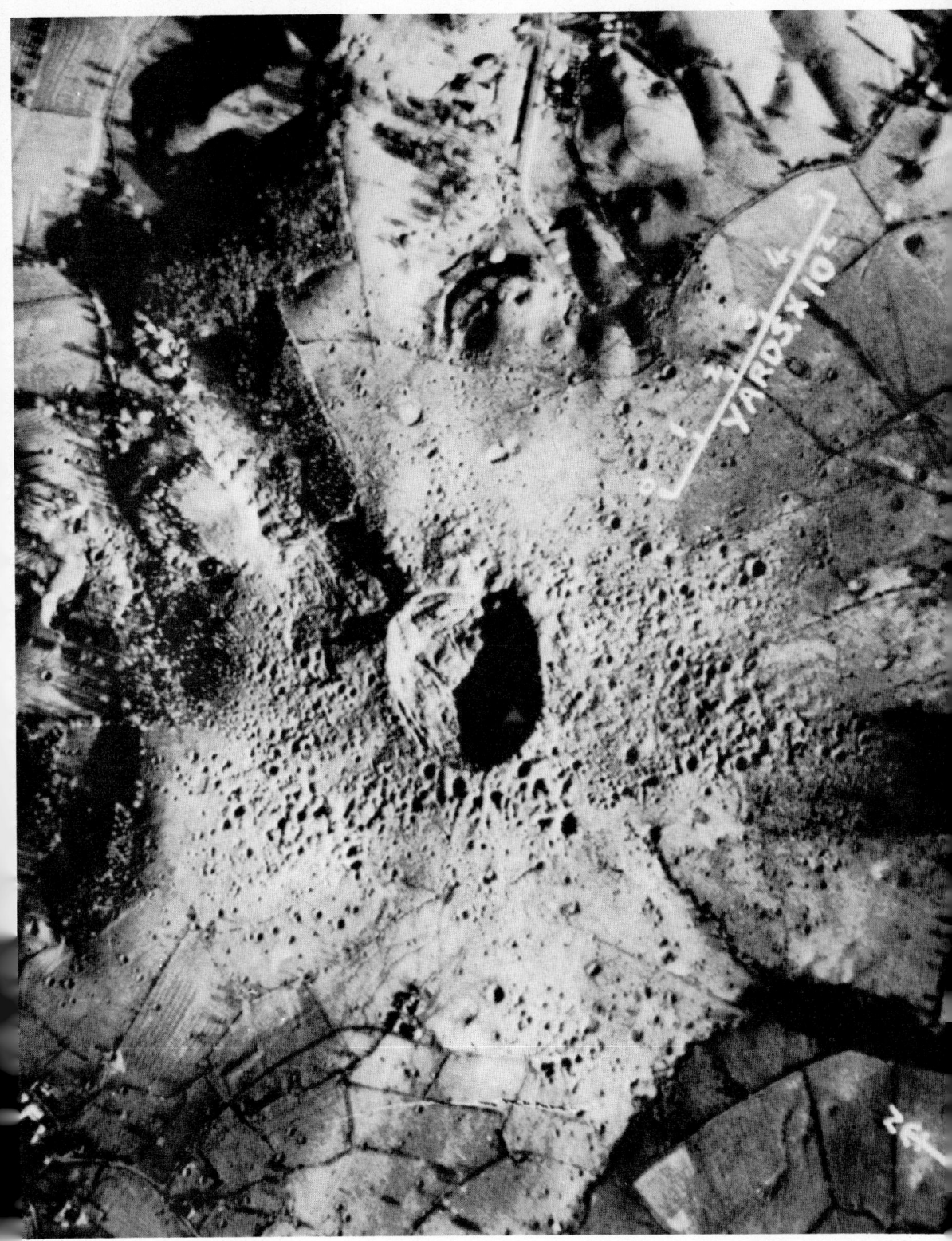

PLATE XIX.—Burton-on-Trent Crater. This crater (No. 362), produced by the explosion of stored ammunition, is typical of a deep burst. Much material was ejected at high angles and fell near the main crater. Some of the smaller craters produced in this fashion are over 100 feet across. (Courtesy D. E. Jarrett, of Ministry of Supply, Armament Research and Development Establishment, Woolwich, S.E. 18, England.)

published, have been available, had we only read the information correctly. At all well-observed terrestrial meteoritic craters, the rim strata are highly distorted and usually are pushed up so that they slope away from the crater. This upthrusting is observed primarily under the rim. Hence the largest part of the rim structure is composed of ground layers pushed out and up, and the rough and jagged surface layers on the rim are only that—surface layers. These layers do, indeed, come from the material ejected from the pit, but the rims are mainly compression phenomena.

The cross-sections of the Arizona Meteorite Crater are not complete enough to give all desired information, but they, too, indicate that much of the rim was pushed into place. This is particularly evident in the southern wall; a wedging action has raised many layers about 100 feet.

The key bit of information is the cross-section of the Odessa No. 1 Crater by Sellards and Evans (40) (Fig. 2, p. 20). This crater is only 550 feet across, and the energy expended was only slightly greater than that of the Burton-on-Trent explosion (Pl. XIX). Even so, the energy density was so high that the rock layers, normally rigid, acted as though they were extremely plastic for an instant and were then folded, wrinkled, displaced, and faulted. The layers on the left side of the section were crowded upward and outward. The layers at the top and bottom of the right side were also shifted upward and outward, but the layers in the middle depths were bodily compressed about 50 feet; this compression led to a tremendous buckling of the limestone strata so that they changed from horizontal to effectively vertical at one point. While other sections would be needed to give a fuller picture, it seems probable that the meteorite came from the left at a considerable angle to form the Odessa No. 1 Crater.

It is clear that the usual ideas concerning crater formation are seriously in error. Such mechanisms as those advocated by Gold (41) and Öpik (42) are only partially correct. Conclusions based on such incomplete descriptions must be revised. In Gold's hypothesis, the crater is essentially a product of the scouring action of the gases generated by the impact, a process which "sandpapers" materials out of the crater and into the rim. These gases cannot push straight down successfully; hence they eject the material above the explosion center and scour an increasingly deep trench surrounding a central cone. The formation of the central cone is aided to some extent by rebound phenomena.

Certainly some scouring must occur at subsurface bursts, but the evidence at man-made explosion pits and terrestrial meteoritic craters is conclusive that it is not the major mechanism. Öpik (43) postulates:

From the centre of impact the shock is propagating laterally; the radial momentum of the shock is at first transmitted inelastically, vaporizing concentric layers of the surface material that are nearest to the centre, and crushing the more distant layers....

At a certain critical distance from the centre, which may be called the radius of destruction, the elastic forces of the material will be able to absorb the energy of the shock and carry it further as a regular vibration propagating with the velocity of sound. The surface where destruction ends corresponds to the crater bowl.

Öpik (44) also states:

Near incompressibility actually holds in the outer portions of the bowl. Compressibility would alter somewhat the course of calculations for the central region of the collision, affecting flattening and penetration [of the meteorite].

Öpik's calculations undoubtedly are reasonably accurate for the central region of the crater. This is the section bounded by the curved, dotted line in Figure 13 and marks the volume in which the material has been shattered by the explosion and then ejected. But between this dotted line and the crater inner walls is a major zone which is clearly a compression effect. Öpik's assumptions of near-incompressibility of the rock strata is observed to be incorrect. The slight compressibility of such materials is magnified by the ability of the layers to extrude upward into the above-ground-level regions in the instant when the explosive forces have overridden the bonds normally holding the rocks together. Öpik correctly points out that a relatively small percentage of the energy available is expended in these directions and that most of it is used for shattering and pulverizing the central region or is wasted harmlessly into the void above ground. Nevertheless, the assumption that the "radius of destruction" and the crater bowl are equivalent is not borne out by the observations. Judging by the smaller craters in soil and the Odessa No. 1 and larger craters in rock, the "radius of destruction" is significantly smaller than the final crater radius. Consequently, Öpik's calculations of the energy needed to produce meteoritic craters should be substantially too high, and this is indeed the case (see Chap. 8, p. 169, where Öpik gives the highest value of the energy needed to form the Arizona Meteoritic Crater; his value of the energy is nearly an order of magnitude too high).

At any crater-forming explosion, the resistance to scouring, ejection, or compression is less in a horizontal direction than in the vertically downward direction. This effect will increase with increasing angular depth measured from the burst center. The ground immediately beneath the burst offers the greatest resistance to scouring and compression, and thus the rebound will be a maximum from this direction. Any rebound will tend to make the crater more shallow.

Shoemaker (33) has given detailed cross-sections of craters from two nuclear explosions set off below ground at the Nevada Testing Site of the Atomic Energy Commission. In each case, the released energy was 1.2 ± 0.05 kilotons

Explosion, Terrestrial Meteoritic, and Lunar Craters

—i.e., energy equal to that which would be released by 1,200 ± 5 metric tons of TNT. The data on these tests are given in Table 4 (Appendix 2).

In the first test, named Teapot Ess, the charge was detonated 67 feet below ground level. The resulting crater was about 339 feet wide.

In the second test, named Jangle U, the charge was detonated 18 feet below ground level. The resulting crater was about 275 feet wide.

The Teapot Ess charge was relatively deeper than the center of the explosion at the Arizona Crater or the Odessa No. 1 Crater. The Jangle U charge was roughly at the same relative depth as the explosion center at the two meteoritic craters.

The two nuclear craters were slightly smaller than the Odessa No. 1 Crater and less than one-tenth as wide as the Arizona Crater.

The ground layers at both Jangle U and Odessa No. 1 were folded into a ring anticline beneath the rim. At the Arizona and Teapot Ess Craters, the anticline is missing, although the layers at the upper edges of the craters are raised and, in some instances, overturned.

Shoemaker points out that the deeper explosions will raise the rock cover and, at a specific distance from the center, a hinge will develop, and the layers toward the crater will be overturned and scattered onto the rim. If the explosion is shallow, the nearly horizontal thrust of the explosion will compress the layers into a ring anticline.

Odessa No. 1 is clearly of the shallow-burst type. The ring anticline is well developed. There are indications that the anticline on one side was overturned, and the top of the anticline on the opposite side is completely missing. The cross-section shows that the thrust forced the rock layers, even below the burst center, to buckle into an anticline.

We really do not know much about the contortions of the rock under the rim of the Arizona Crater. Where the rocks are exposed, they commonly dip outward, and there are some positive evidences of the hinge mechanism operating, as some layers were completely overturned. Shoemaker suggests that the evidence calls for a depth of burst center of about 400 feet. Later in this volume, a depth of burst of 261 feet will be found for this crater. It does not appear that there is a need to postulate as deep a penetration by the Arizona Meteorite as 400 feet. The Odessa No. 1 meteorite probably burst about 30 feet below ground.

References

1. WILKINS, H. P., and MOORE, P. *The Moon*. A complete description of the surface of the moon, containing the 300-inch Wilkins lunar map. New York: Macmillan Co., 1955.
2. MOORE, P. *Guide to the Moon*. London: Collins Paper Books, Ltd., 1953.

3. MARSHALL, R. K. "The Origin of the Lunar Craters (a Summary)," *Pop. Astr.*, **51,** 415, 1943.
4. ESCHER, B. G. "Moon and Earth," *Koninklijke Nederlandsche Akademie van Wetenschappen, Proceedings of the Section of Sciences*, **42,** 127, 1939.
5. ———. "Three Caldera-shaped Accidents; Volcanic Calderas, Meteoric Scars, and Lunar Cirques," *Bull. Volcanol.*, Union Géodesique et Géophysique International, Ser. 2, **16,** 55, 1955.
6. SPURR, J. E. *Geology Applied to Selenology.* Lancaster, Pa: Science Press, 1944–49.
7. ———. *The Imbrian Plain Region of the Moon*, Vol. **1.** Lancaster, Pa.: Science Press, 1944.
8. ———. *Features of the Moon*, Vol. **2.** Lancaster, Pa.: Science Press, 1944.
9. ———. *Lunar Catastrophic History*, Vol. **3.** Concord, N.H.: Rumford Press, 1949.
10. ———. *The Shrunken Moon*, Vol. **4.** Concord, N.H.: Rumford Press, 1949.
11. ALTER, D. "Evolution of the Moon," *Griffith Observer*, **12,** 114, 1948.
12. ———. "The Nature of the Typical Lunar Mountain-walled Plains," *Pub. A.S.P.*, **67,** 404, 437, 1956.
13. ———. "Explosion Craters of the Moon," *ibid.*, **69,** 411, 533, 1957.
14. ———. "The Nature of the Domes and Small Craters of the Moon," *ibid.*, pp. 408, 245.
15. ———. "Peculiar Features of the Lunar Surface," *ibid.*, **70,** 416, 489, 1958.
16. ———. "Scientific Aspects of the Lunar Surface," *Proceedings of the Lunar and Planetary Exploration Colloquium*, **1,** 3, 1958.
17. MIYAMOTO, S. "A Geological Interpretation of the Lunar Surface," *Contr. Inst. Ap. and Kwasan Obs.*, No. 90, 1960.
18. ———. "Magmatic Boiling and Underground Structure of the Moon," *ibid.*, No. 96, 1960.
19. GREEN, J. "Geochemical Implications of Lunar Degassing," *Proceedings of the Lunar and Planetary Exploration Colloquium*, **1,** 1, 1959.
20. ———. "Lunar Physics and Topography," *ibid.*, **2,** 7, 1961.
21. FIRSOFF, V. A. *Surface of the Moon.* London: Hutchinson & Co., Ltd., 1961.
22. BALDWIN, R. B. *The Face of the Moon.* Chicago: University of Chicago Press, 1949.
23. UREY, H. C. *The Planets: Their Origin and Development.* New Haven: Yale University Press, 1952.
24. ———. "The Surface of the Moon," *A.J.*, **57,** 27, 1952.
25. ———. "Some Criticisms of 'On the Origin of the Lunar Surface Features' by G. P. Kuiper," *Proc. Nat. Acad. Sci.*, **41,** 423, 1955.
26. ———. "The Moon's Surface Features," *Observatory* (London), **76,** 232, 1956.
27. ———. "The Origin and Significance of the Moon's Surface," *Vistas in Astronomy*, ed. ARTHUR BEER, **2,** 1667–80. London: Pergamon Press, 1956.
28. ———. "The Origin of the Moon's Surface Features," *Sky and Telescope*, **15,** 108, 160, 1956.

29. ———. "Meteorites and Origin of the Solar System," *Physical Society, Yearbook*, Vol. **14,** 1957.
30. KUIPER, G. P. "On the Origin of the Lunar Surface Features," *Proc. Nat. Acad. Sci.,* **40,** 1096, 1954.
31. ———. "The Lunar Surface, Further Comments," *ibid.,* **41,** 820, 1955.
32. ———. "Exploration of the Moon," *Vistas in Astronautics Symposium,* Vol. **2.** New York: Pergamon Press, 1959.
33. SHOEMAKER, E. M. "Impact Mechanics at Meteor Crater, Arizona." Open file report, prepared on behalf of the U.S. Atomic Energy Commission, July, 1959.
34. ———. "Penetration Mechanics of High Velocity Meteorites, Illustrated by Meteor Crater, Arizona," *Twenty-first Internat. Geol. Cong., Copenhagen,* Rept. 18, p. 418, 1960.
35. LAMPSON, C. W. "Effects of Atomic Weapons," ed. S. GLASSTONE, p. 410. Washington, D.C.: Combat Forces Press, 1950.
36. Unclassified report from files of Dr. Ralph Ilsley, Washington, D.C.
37. BARRINGER, D. M. *Meteor Crater (Formerly Called Coon Mountain or Coon Butte) in Northern Central Arizona.* Philadelphia: The Author, 1910.
38. ———. "Volcanoes—or Cosmic Shell Holes: A Discussion of the Origin of the Craters on the Moon and of Other Features of Her Surface," *Sci. American,* **131,** 10, (specifically p. 11), 62, 102, 142, 1924.
39. SHOEMAKER, E. M. "Geological Interpretation of Lunar Craters," *Adm. Rept., U.S. Dept. Int., Geol. Survey,* May, 1961.
40. SELLARDS, E. H., and EVANS, G. *Statement of Progress of Investigation of Odessa Meteor Craters.* Austin: University of Texas, Bureau of Economic Geology, Sept. 1, 1941.
41. GOLD, T. "The Lunar Surface," *M.N.,* **115,** 6, 1955.
42. ÖPIK, E. J. "Meteor Impact on Solid Surface," *Irish A.J.,* **5,** 14, 1958.
43. ———. *Ibid.,* p. 20.
44. ———. *Ibid.,* p. 22.
45. *Work of the Royal Engineers in the European War, 1914–1919, Military Mining.* Chatham: Published by the Secretary, Institute of Royal Engineers. W. & J. McKay & Co., Ltd., 1922.
46. SHELTON, A. V., NORDYKE, M. D., and GOECKERMAN, R. H. "The Neptune Event: A Nuclear Explosive Cratering Experiment," *UCRL-5766, Nuclear Explosions, Peaceful Applications, TID-4500, UC-35.* 15th ed., 1960.
47. ROBINSON, C. S. *Explosions, Their Anatomy and Destructiveness.* New York: McGraw-Hill Book Co., 1944.

7

RELATIONSHIPS BETWEEN CRATER PARAMETERS

To a first approximation, the dimensions of craters vary in proportion to the cube root of the expended energy. This is only an approximation but is reasonably valid over a range of a hundred or so times in energy; this permits general relationships to be set up between craters of different sizes.

The terms "apparent crater" and "true crater" are not the same as used by the military.[1] They are best shown by reference to Figure 14. Essentially, the

---

[1] For comparison purposes, this footnote is included. An Army report (1) gives the following data:
"The cratering effect which would be produced by certain explosives can be estimated from the following formulae:

$$V = 0.4Q^{8/7},$$

where V is the crater volume in cubic feet, the soil is average, the explosive is commercial dynamite, and Q is the weight of dynamite in pounds.

"Since military explosives such as TNT are more effective cratering agents than commercial blasting explosives, the above formula does not apply to them. If a proper charge of TNT is buried so that a common crater is formed, the ratio of the true diameter of the crater in feet, D, to the depth of burial in feet, H, will be two. Crater formation can then be estimated from

$$H^3 = \frac{2W}{Z} \quad \text{and} \quad D^3 = \frac{16W}{Z},$$

where W is the weight of TNT in pounds and Z is a constant having values of 0.054, 0.066, 0.084, and 0.10 for light earth, common earth, hard sand, and hardpan or heavy clay, respectively.

"The cratering effect from 500-pound and 1,000-pound general-purpose bombs loaded with TNT can be estimated from the formula

$$V = 4.13W,$$

where V is the volume of the crater in cubic feet and W is the weight of the explosive component of the bombs in pounds."

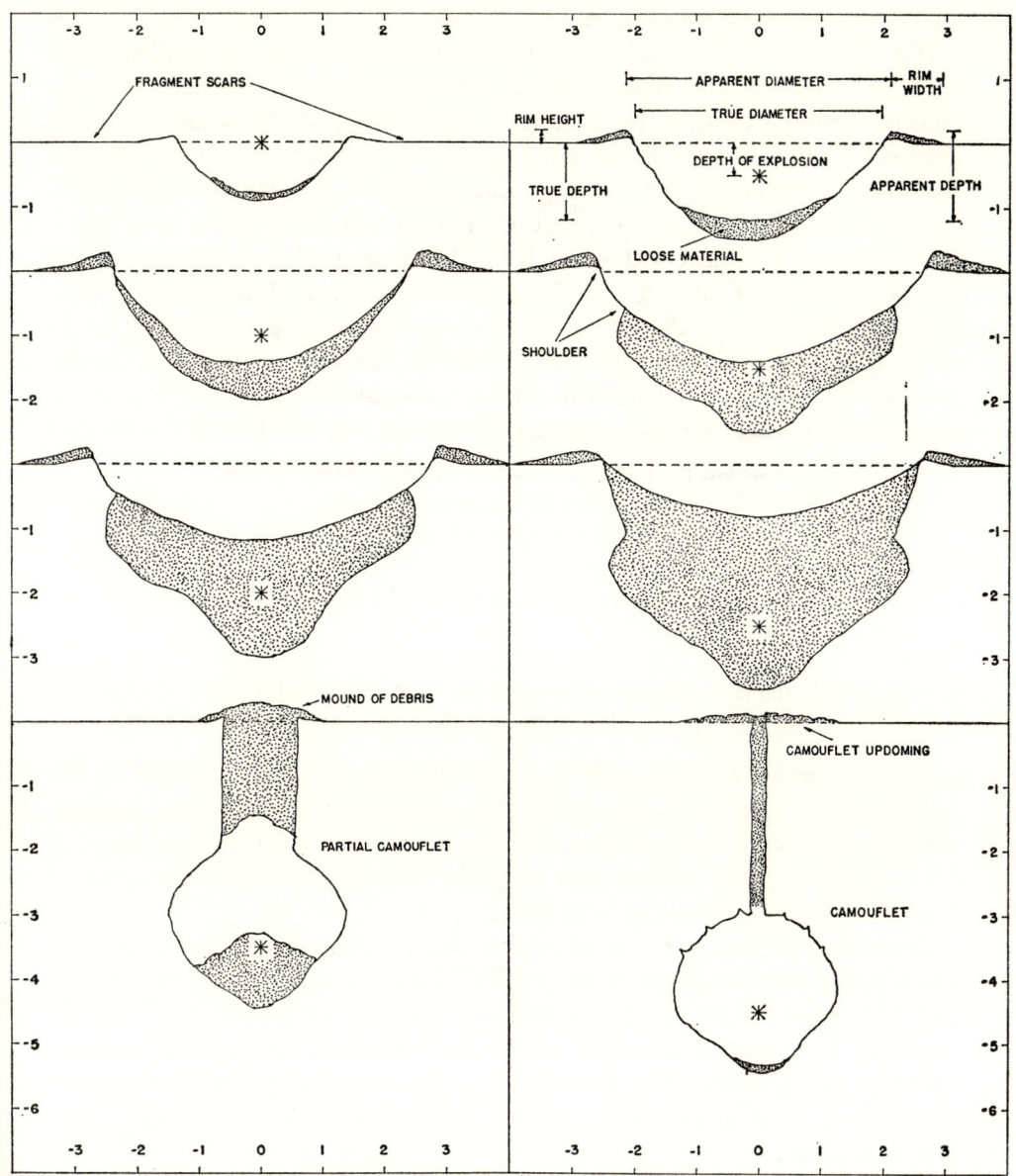

Fig. 14.—Schematic relationships between dimensions and depths of explosions for terrestrial craters

apparent crater and the true crater, as there defined, are the only ones we can examine in two of the types of objects we wish to study—the terrestrial meteoritic craters and the lunar craters.

If a given size explosive charge is detonated at various distances above and below ground level, the dimensions of the resulting craters are definite functions of the burst location, the nature of the ground, and the nature of the explosive. The heat of explosion is not a completely reliable basis from which to calculate the cratering effect from various explosives. An explosive with a low rate of detonation, one which produces a large volume of gases, may dig a larger crater than a faster explosive, if the charge is well below the surface and the ground is not too hard. If the charge is on the surface of the ground, a slow explosive produces a smaller crater than a faster one because of the lesser action of the gases resulting from their lower velocity. We may assume that, on the average, the explosives involved in the many man-made craters will have the same cratering effect as TNT and that an average soil was involved. If we call

$H =$ height $(-)$ or depth $(+)$ of the burst center in feet,
$R =$ crater radius in feet for the apparent crater,
$S =$ crater depth in feet for the apparent crater,
$T =$ rim height in feet,
$W =$ explosive energy in terms of pounds of TNT equivalent,

then

$H/W^{1/3}$ is the scaled depth of burst, $R/W^{1/3}$, $S/W^{1/3}$, and $T/W^{1/3}$ are the scaled radius, depth, and rim height, respectively, of the apparent crater.

We may derive $R/W^{1/3}$, $S/W^{1/3}$, and $T/W^{1/3}$ as functions of $H/W^{1/3}$. We find that $R/W^{1/3}$, $S/W^{1/3}$, and $T/W^{1/3}$ are smallest for air-bursts and increase through the ground-burst position to a maximum at a scaled depth of burst of 1.5 to 2. This type of scaling may be called "Lampson scaling," inasmuch as Lampson (2) established the basic laws.

However, true Lampson scaling implies that the exponent is one-third for all energies and that the shape of the craters for a given scaled depth of burst remains the same for all energies. As will be shown, neither implication is strictly true even for a small difference in energy, and the effects become quite marked for large energy differences. Nevertheless, Lampson scaling is a good approximation over small energy ranges and will be used as a starting point in the following analysis.

The exponent for the diameter (or radius) equations is slightly smaller than one-third and is itself variable. This will be discussed in Chapter 8. This means that when the exponent is used as one-third, we cannot strictly compare craters of widely differing energies. For smaller ranges of energies, the scaling differences are not great. The exponent is still smaller for the depth equations. It does not appear to be variable for the rim-height equations.

Lampson further discussed the effects of varying types of soil on the size and form of explosion craters. This effect is very real but is very small compared with the range of crater sizes that we shall be dealing with; and, in fact, the effect of soil type on craters is not much, if any, greater than the scatter in crater dimensions produced by a series of a given size of explosive charges in a single type of soil. The effect of soil constants systematically shifts the observed relationships between crater dimensions and crater scaling. In the following discussion, we shall consider typical average soils and later discuss deviations from these conditions.

The data on explosive craters are usually very carefully guarded by the military. It is apparently a great secret that a certain-sized hunk of TNT will produce a certain size or shape of crater. One is tempted to ask how long it will be before potential enemies or even friends of the United States will discover TNT for themselves. The net result is that the vast majority of craters discussed in non-classified sources come from accidental explosions and were not tested under controlled conditions. This introduces unavoidable errors into the data; but, inasmuch as we shall work with averages, the data are sufficient.

In the laboratory, we cannot yet impel projectiles with velocities above 20,000–25,000 feet per second. Theory suggests that it is only when a projectile strikes with a velocity above about 6 miles per second that a true, violent explosion occurs. In the lower-velocity regime, the explosion is only partial; or in the very low-velocity range, there may not be an explosion. Even in these cases, a projectile has a definite amount of kinetic energy and momentum, and, when stopped suddenly on impact, the effect is similar to an explosion, and craters may be formed.

The nature of the resulting crater is very sensitive to the material of the projectile and particularly to the material of the target. Obviously, an ounce of feathers striking at 5,000 feet per second would give a different result than an ounce of tungsten carbide. Equally, steel balls striking a malleable lead target at low velocity will produce craters which differ from those that would be found in a rock target. As the velocity of impact increases, the natures of the missile and target have less and less effect on the result.

In an extremely interesting article, Charters (3) described the sequence of events involved in low-velocity impact. His results illustrate three of the five possible modes of impact.

In the first case, not discussed by Charters, we have the simple elastic collision where target and projectile are not permanently deformed.

When steel spheres are shot into lead at velocities below about 1,500 feet per second, the projectile is undeformed and simply punches a narrow, deep cavity in the target. The penetration increases linearly with the velocity.

Above this limit and up to 8,000 feet per second, the projectile is deformed or shattered. The penetration is high at the low-velocity end of this regime, then falls sharply and later rises slowly until, at 8,000 feet per second, it is about the same as at 1,600 feet per second.

At higher velocities, both the projectile and the neighboring target flow like liquids. Up to at least 20,000 feet per second, the penetration slowly increases.

A different choice of target and projectile material would have changed the amount of penetration and the limiting velocities for each type of impact, but the general principles would be the same. It is in this fluid-flow range that most of the small terrestrial meteoritic impact craters are formed.

Unfortunately, it is not yet possible to produce in the laboratory velocities sufficiently high to yield the fifth type of impact, in which the projectile and part of the target are vaporized. The larger meteoritic-impact craters are formed by explosions of this type.

In the range from 8,000 to at least 20,000 feet per second, a projectile strikes, producing a cavity. The projectile effectively turns itself inside out, and the crater is lined with the remnants of the missile. It is easy to understand this result when we reflect that, in this velocity range, the forces set up are very much greater than the mechanical strengths of either projectile or target. For example, when a steel ball hits a lead target at 20,000 feet per second, the pressure in the ball early in the collision is greater than 30,000,000 pounds per square inch, more than one hundred times the ultimate strength of the steel. Under such conditions, the materials flow as if they had essentially no strength.

It is recognized that these low-energy experiments cannot be considered as duplicating the conditions obtaining in a meteoritic impact, even though the low end of the meteoritic-impact velocity range lies within the scope of the laboratory experiments. A meteoritic crater is formed by very considerably larger bodies. The action takes place farther down, in an absolute sense, within a rocky layer. The explosion takes longer, and the energies are greater. The net result is that the larger explosions are somewhat muffled by the overburden, and the rock layers can flow and distort. A raised rim is produced in rock, as well as in the more plastic metals. It is encouraging to note that a tiny lens of breccia is found in the small laboratory impact craters in rock, as well as in nuclear craters and in terrestrial meteoritic craters.

In the fifth regime—that where the velocity of impact is so high that the projectile is largely vaporized—we would expect much the same sort of phenomena to occur, although on a more violent scale. Except in the very slowest of collisions, the projectile is backfired out of the pit. This agrees with the observations regarding meteoritic craters.

If a prediction may be made at this point, it is that, for a given size and composition of a projectile and a given type of target and for impact velocities of above 6 miles per second, the amount of penetration will not increase as a linear function of the velocity, as is suggested by the lower-velocity tests. It

Relationships between Crater Parameters

will change very little with velocity and may even decrease as the impact velocity is increased. Calculations based on cratering by jets from shaped charges will usually give penetrations into the ground that are too great.

In this chapter we wish to derive relationships which will define the size and shape of craters produced by given applications of energy at different heights or depths of bursts in soil of average composition. It will develop that families of curves are necessary to relate each set of crater parameters. The derivation of these final families of curves into charts and equations, which can be used easily, required the processing of a considerable amount of data and the use of a series of intermediate families and curves. These will be developed in detail in the remainder of this chapter. Suffice it to say that each step proved to be necessary and resulted in equations which could be applied successfully, with minor interpretations, to the terrestrial man-made explosion and meteoritic craters and the lunar craters.

Table 4 (Appendix 2) summarizes the known data on craters formed by chemical and nuclear explosions. The first 128 tests were run by the author. The remaining craters were found in the literature or were from the few non-classified tests described by the Armed Forces.

The logarithms used in each case are to the base 10. For logarithms of quantities less than 1, the minus 10 has been omitted. The energy unit is the calorie. Logarithms are generally used because our derived relationships must cover, equally well, craters less than a foot across and those over 100 miles across.

For the scientist it would perhaps have been more convenient to use the c.g.s. system with the erg as the energy unit; but, because the present system was used in *The Face of the Moon* and because the terrestrial meteoritic craters and the lunar craters are customarily described in feet and miles, it was felt wise not to change. The conversions are simple. From the values of Table 4, the quantities

$H/W^{1/3}$ = scaled depth of burst,
$R/W^{1/3}$ = scaled radius of apparent crater,
$S/W^{1/3}$ = scaled depth of apparent crater,
$T/W^{1/3}$ = scaled rim height,

were computed.

These values are listed in Table 8 (Appendix 2) in columns 2, 3, 4, and 6, respectively. Column 1 lists the crater identification number from Table 4.

A first attempt was made to plot the scaled crater radii and depths against the scaled depths of bursts for all craters. This did not prove possible. There are sufficient departures from Lampson's scaling to force us to divide the craters into four groups according to the energy of the crater-forming blasts. The groups selected were those craters for which the logarithm of the explosive energy $= E = 2.88$ and 3.31; those craters with E ranging from 3.78 to 4.91, centering about $E = 4.5$; and the remaining craters with E ranging from 7.58 to 12.39, centering about $E = 10$.

Within each of the four energy ranges, the data are sufficiently homogeneous that we may plot $H/W^{1/3}$ versus, respectively, $R/W^{1/3}$ and $S/W^{1/3}$ for the apparent crater. The results are shown in Figure 15, and $H/W^{1/3}$ versus $T/W^{1/3}$ is shown in Figure 16.

In Figure 15, in each case, the points indicate the same trend, yet are so few in number and the intrinsic scatter is so great that it would be possible to draw inconsistent pairs of curves. In other words, when dealing with the data which show as large real deviations from the average as do the values for individual craters, it is necessary to draw curves which not only represent the plotted data, but must correctly show the relationships between all crater parameters.

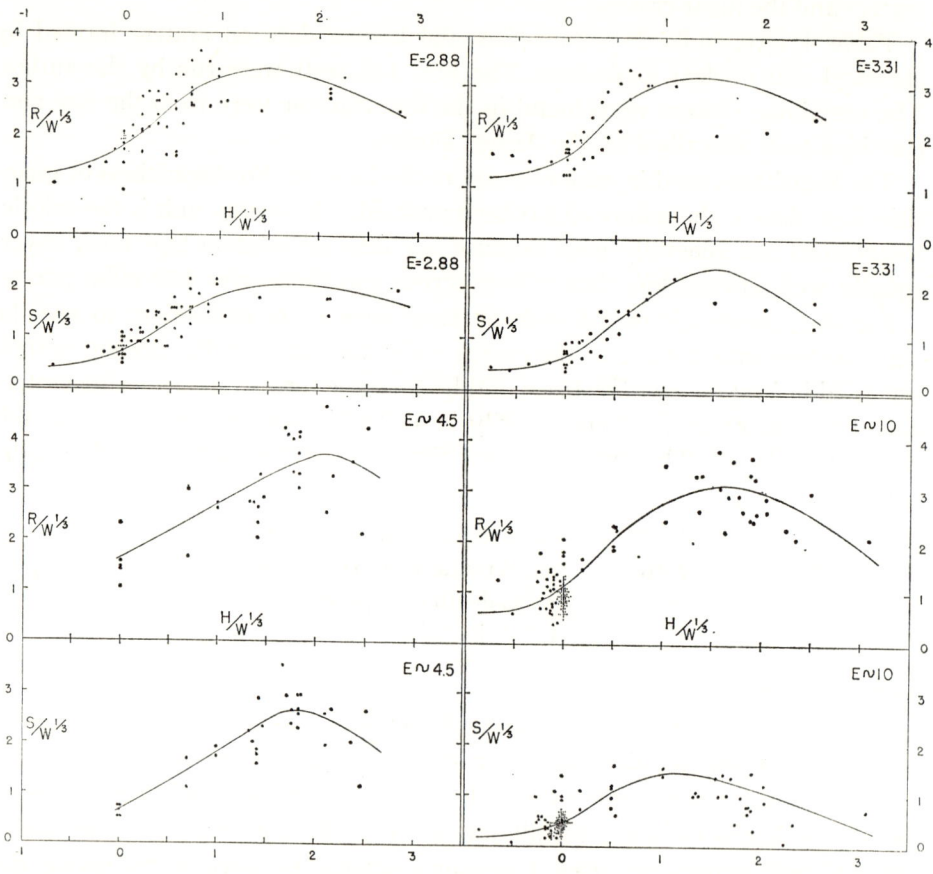

Fig. 15.—Scaled radius and scaled depth for apparent crater versus scaled depth of burst of explosive charge for four energy ranges. $E =$ logarithm of explosive energy (calories); $R =$ apparent crater radius (feet); $S =$ apparent crater depth (feet); $H =$ distance of explosion center from ground level (feet); $W =$ pounds equivalent of TNT.

To aid in plotting reasonably consistent curves in Figure 15, another observed relationship was used. The ratio R/S was computed for each crater and plotted against the scaled depth of burst separately for each energy range. A typical curve is shown in Figure 17. In this curve, for the apparent crater, the ratio of crater radius to crater depth grows less with increasing depth of burst to a minimum at about a scaled depth of burst of 1.5–2.0 and thereafter in-

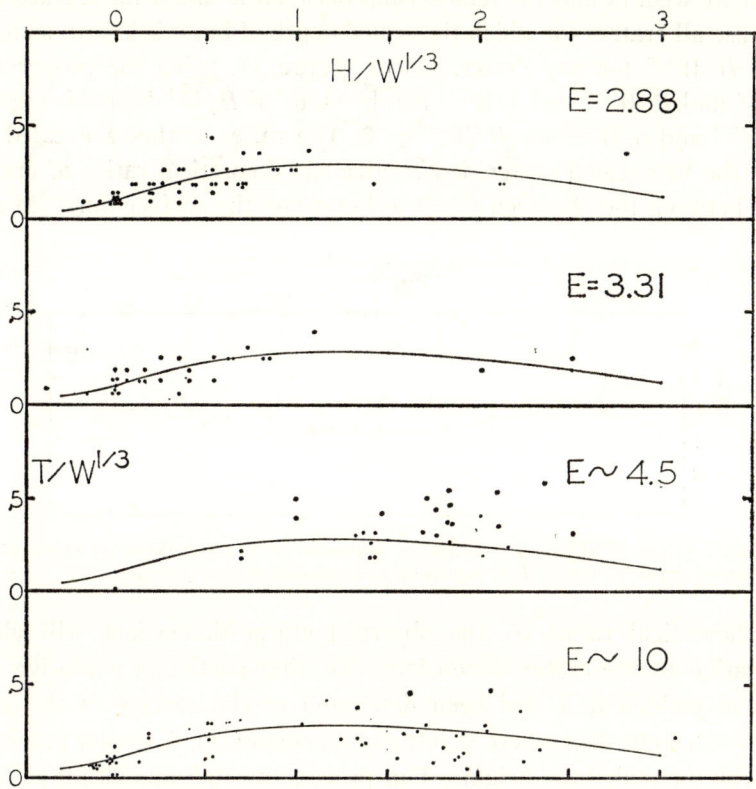

FIG. 16.—Scaled rim height for apparent crater vs. scaled depth of burst. T = rim height (feet); H = distance of explosion center from ground level (feet); W = pounds equivalent of TNT; E = logarithmic energy (calories).

creases. In each case, this type of curve is very well defined and aided greatly in determining the precise position of the curves of Figure 15.

The 365 craters described in Table 4 come from explosions whose scaled heights of burst above ground are as high as -0.76 and whose scaled depths of burst below ground are as low as $+3.08$—a truly tremendous range. Consequently, a curve such as log apparent diameter ($=D$) versus log apparent depth ($=d$) derived from the unsegregated data would be seriously in error. For example, on the average, the larger explosions occurred at systematically smaller average scaled depths of bursts. Because of this, the upper section of the curve would lie higher, because D would become systematically larger relative to d. Extrapolation into the terrestrial meteoritic-crater range and particularly to the lunar-crater range would become progressively more in error. Similar errors would be encountered in all the relationships between crater parameters.

It is possible, however, to derive consistent relationships for these diverse craters. If we wish to find the relationship between D and d for surface bursts, we could use all craters for which the scaled depth of burst is known as follows:

Derive $H/W^{1/3}$ for any crater. Go to Figure 15, using the proper energy range, and find $R/W^{1/3}$ and $S/W^{1/3}$ for the value of $H/W^{1/3}$ for this crater and also $R/W^{1/3}$ and $S/W^{1/3}$ for $H/W^{1/3} = 0$. The ratios of the two scaled crater radii and the two scaled crater depths give the theoretical ratios of crater dimensions between the observed depth of burst and the surface burst. Multiply-

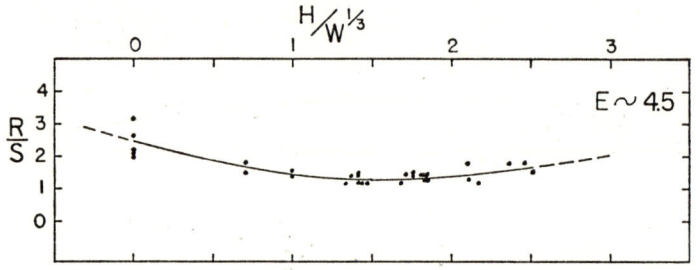

Fig. 17.—R/S versus $H/W^{1/3}$. R = radius of apparent crater; S = depth of apparent crater; $H/W^{1/3}$ = scaled depth of burst; E = logarithm of explosive energy (calories).

ing the theoretical ratios by the observed crater dimensions will yield an approximation to the crater dimensions that that particular explosive charge would have yielded if it had been detonated at the surface of the ground, $H/W^{1/3} = 0$, rather than where it actually was exploded. A similar process may be used with the scaled rim heights. This process thus permits all fully observed crater data to be used as corrected to any scaled depth of burst.

It is not known just how deeply meteorites penetrate into the ground. It is known that the penetration is slight and the released energies enormous. This means that the scaled depths of burst of the meteorites capable of causing explosive craters must be small. For this reason, four scaled depths of burst, $H/W^{1/3} = 0.00, 0.10, 0.25$, and 0.50 were selected. By the above process, the data from each crater were successively converted into equivalent values of D and d for each of the four scaled depths of burst and reported in Table 8 (Appendix 2) and the results separately plotted.

We have now arrived at the point where consistent equations may be derived relating the logarithms of the diameter and depth of the apparent crater, D and

d, for specific values of scaled depth of bursts. Equations (7-1) through (7-4) are for the English system of units:

For $H/W^{1/3} = 0.00$ (surface bursts),

$$D = 0.0315 d^2 + 1.0038 d + 0.6386 . \tag{7-1}$$

For $H/W^{1/3} = 0.10$,

$$D = 0.0256 d^2 + d + 0.6300 . \tag{7-2}$$

For $H/W^{1/3} = 0.25$,

$$D = 0.0234 d^2 + 0.9970 d + 0.5876 \tag{7-3}$$

For $H/W^{1/3} = 0.50$,

$$D = 0.0225 d^2 + 0.9907 d + 0.5367 . \tag{7-4}$$

Equations (7-1A) through (7-4A) are for the metric system of units:

For $H/W^{1/3} = 0.00$ (surface bursts),

$$D_{km} = 0.0315 d_m^2 + 1.0363 d_m - 2.3520 . \tag{7-1A}$$

For $H/W^{1/3} = 0.10$,

$$D_{km} = 0.0256 d_m^2 + 1.0264 d_m - 2.3461 . \tag{7-2A}$$

For $H/W^{1/3} = 0.25$,

$$D_{km} = 0.0234 d_m^2 + 1.0211 d_m - 2.4087 . \tag{7-3A}$$

For $H/W^{1/3} = 0.50$,

$$D_{km} = 0.0225 d_m^2 + 1.0139 d_m - 2.4630 . \tag{7-4A}$$

The data and corresponding curves for equations (7-1) through (7-4) are shown in Figure 18. These charts show that equations have been derived which are consistent with the crater dimensions found for terrestrial explosion craters.

It now remains to see whether they are consistent with the dimensions of terrestrial meteoritic craters. Figure 19 shows the extended curves from equations (7-1) through (7-4) corresponding to surface bursts and scaled depth of bursts of 0.10, 0.25, and 0.50, respectively. The data for the ten terrestrial meteoritic craters are from Table 6 (Appendix 2) and are plotted as dots. They represent only those pits which have been excavated or closely measured to give original dimensions. The Holleford and Brent points are also shown.

It is clearly seen that the terrestrial meteoritic craters fall systematically below the surface-burst curve. This may be interpreted to mean that the meteorites penetrated an appreciable but small distance into the ground before exploding. At first glance, the plotted curve for $H/W^{1/3} = 0.50$ seems to represent the ter-

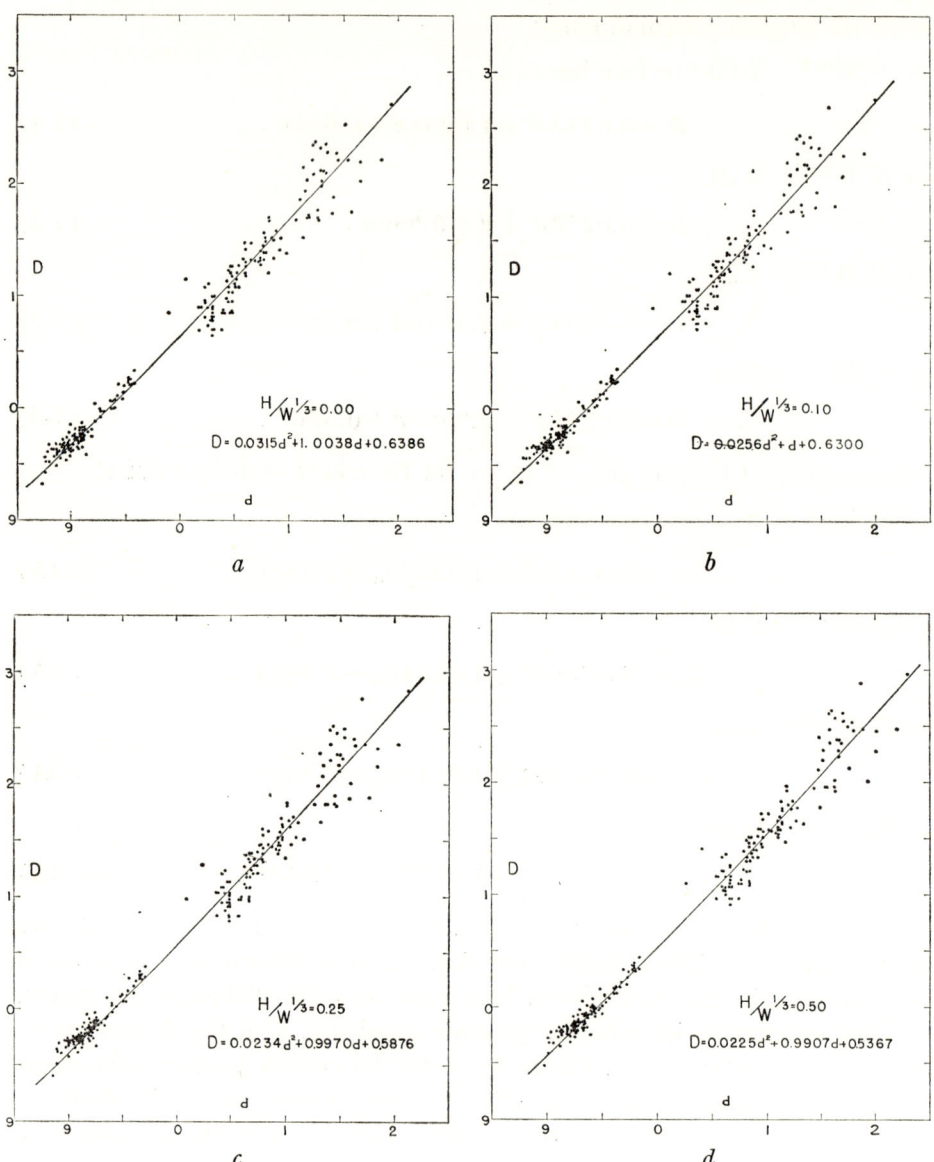

Fig. 18.—Logarithmic diameter (D) versus logarithmic depth (d) for apparent crater for four scaled depths of burst.

restrial meteoritic craters well, but a closer examination reveals that three of the four points below this curve are from the least well-established crater data, i.e., Henbury No. 13, Le Clot de Cabrerolles, and Faugères. Henbury No. 13 represents a special case, in that the impinging body did not explode. The crater was dug simply by the kinetic energy of the meteorite applied mechanically without vaporizing the body. The striking velocity must have been very low.

The two French craters have been tentatively classified as terrestrial meteoritic craters, but this is not certain. Faugères, particularly, is rather deep for its

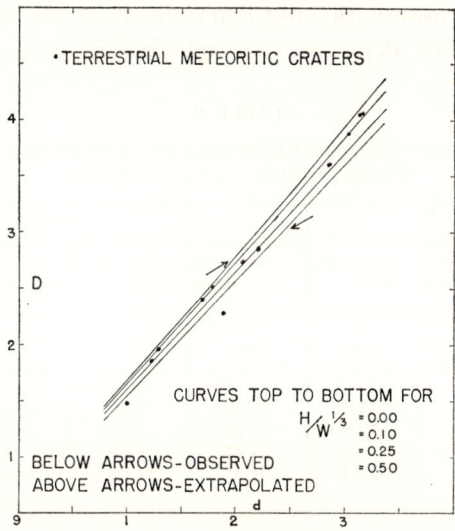

Fig. 19.—Logarithmic diameter (D) versus logarithmic depth (d) in terrestrial meteoritic crater range. Curves from Fig. 18.

diameter if it is a meteoritic crater, but this discrepancy may be due to its location in a rather deep Quaternary ravine.

All the remaining terrestrial meteoritic craters on this list have been thoroughly studied, and the dimensions are closely defined. They yield points which lie very close to the scaled depth-of-burst curve for $H/W^{1/3} = 0.10$, with the possibility that a slightly larger value may be appropriate for the smaller craters.

If the scaled depth of penetration is given by $H/W^{1/3} = 0.25$ or 0.10 over the entire range of explosive crater pits and the impinging meteorite is composed of stone of density 2.5 or nickel-iron of density 7.9, then Table 9 shows the corresponding depth of penetration in terms of the meteoritic diameter. The stony meteorite would be greater in diameter than the metallic body by a factor of about 1.38. The penetration is defined as that of the center of the meteorite and hence the center of explosion. For a given scaled depth of burst, the metallic meteorite would penetrate farther in terms of its diameter than the less dense, larger, stony object. It is not certain that this effect would be a simple proportion.

It is quickly seen that the penetration increases rapidly with increasing impact velocity for a given scaled depth of burst. Conversely, if the meteoritic penetration remains low for any impact velocity, the scaled depth of burst must decrease and approach zero as the impact velocity increases.

The establishment of the exact scaled depth of burst for meteorites is not possible by this method, but it is highly probable that the correct order of penetration has been observationally determined for the terrestrial meteoritic craters. The best scaled depth-of-burst relationship to use for general cases

TABLE 9

PENETRATION OF CENTERS OF METEORITES FOR SPECIFIC SCALED DEPTHS OF BURST

IMPACT VELOCITY (MILES/SEC)	PENETRATION IN DIAMETERS FOR SCALED DEPTH OF BURST = 0.10		PENETRATION IN DIAMETERS FOR SCALED DEPTH OF BURST = 0.25		SCALED DEPTH OF BURST FOR PENETRATIONS OF ONE DIAMETER	
	Stone Meteorite	Iron Meteorite	Stone Meteorite	Iron Meteorite	Stone Meteorite	Iron Meteorite
2........	0.50	0.68	1.24	1.71	0.202	0.146
10........	1.45	2.00	3.62	5.00	.069	.050
25........	2.67	3.68	6.67	9.21	.037	.027
50........	4.23	5.84	10.58	14.61	0.023	0.017

would seem to be that for $H/W^{1/3} = 0.10$. The amount of penetration definitely seems to be somewhat greater than a radius, so the explosion focus is below ground.

The observations of the terrestrial meteoritic craters and explosion pits lead to the following conclusions:

1. The terrestrial meteoritic craters fall among the extensions of the D versus d curves for subsurface bursts and lie below similar curves for surface bursts.

2. The terrestrial meteoritic craters and subsurface-burst pits show prominent rims; the surface-burst craters do not.

3. The terrestrial meteoritic craters and the subsurface-burst craters show evidences of upthrusting and overturning in the rim layers. The surface-burst craters do so only to a minor extent.

4. The form of the terrestrial meteoritic craters and the subsurface-burst craters are similar. Except for the rebound-induced central peak, these pits are always concave. The inner walls of craters from surface bursts are often slight-

Relationships between Crater Parameters

ly convex, except very close to the center, where a funnel form is sometimes found.

5. The energy needed to form surface-burst craters is more than an order of magnitude greater than that needed to produce most subsurface-explosion pits.

6. Meteoritic material has been found deep below the Arizona meteoritic crater.

7. Some of the Öesel meteoritic craters, upon excavation, showed definite subsurface-explosion foci.

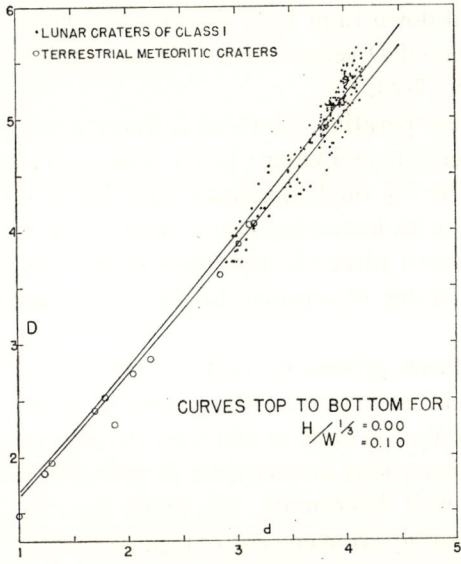

FIG. 20.—Logarithmic diameter (D) versus logarithmic depth (d) in lunar crater range. Curves from Fig. 18.

The remaining step is to extrapolate still further the curves above the terrestrial meteoritic craters through the region defined by the lunar craters. This is done in Figure 20. The data are from Table 7 (Appendix 2) and differ substantially from data represented in *The Face of the Moon*. This is primarily due to new measures of crater diameters made by J. Young and published by D. W. G. Arthur (4). These improved dimensions lead to a reduced scatter of the D versus d points and also help verify and define the curvature of the D versus d relationship in the larger-diameter region.

Throughout most of the range, the curve for surface bursts, $H/W^{1/3} = 0.00$, is definitely too high, and the curves for $H/W^{1/3} = 0.25$ and 0.50 are too low to fit the points for the lunar craters of Class 1. The curve which best fits the terrestrial meteoritic craters, that for $H/W^{1/3} = 0.10$, also seems a good fit to the smaller lunar craters with diameters of 1 mile to possibly 20 miles. On the logarithmic scale used, this means those craters where $D = 3.723$ to nearly 5. The larger craters, where D is between 5 and 5.65, tend to become progressively shallower. This trend may be interpreted in either or both of two ways. Probably the scaled depth of burst for the larger lunar craters, following the trend suggested by the terrestrial meteoritic craters, continues to approach and perhaps even to reach zero. This means that the explosion center for the large lunar craters was approximately at ground level. Conversely, some effect, greatest for the largest craters, could have operated to reduce the apparent depth of the craters since they were formed. From Figure 29 (p. 166) we would expect this effect to be found down to at least the $D = 4$ crater, but between $D = 4$ and $D = 4.9$ the observed D versus d curve is not significantly displaced from the curve for equation (7-2).

Whichever effect was operative, the result is that the terrestrial explosion and terrestrial meteoritic craters and lunar Class 1 craters up to about $D = 5$ can be uniquely represented by the single quadratic equation (7-2), but a discrepancy appears and increases with increasing crater sizes above that point. This problem can best be discussed after the equations of the next chapter have been derived. They relate crater dimensions to the energy needed to produce the craters.

For each of the three groups of craters—the explosion pits, terrestrial meteoritic craters, and lunar craters—the scatter of points around its mean is large; but it is essentially the same in all three. In spite of the inherent scatter, the range of crater dimensions in each case is sufficiently large to permit the average trends to be well determined. The trend lines are all mutually compatible and extend smoothly from one crater type to another.

Equations (7-1) through (7-4) for the apparent craters each have the quadratic form with the coefficient of the linear term equal to unity or nearly so. This indicates that, to a first order, D varies in proportion to d; and, therefore, scaling of the type proposed by Lampson is a good first approximation.

With the data from Tables 4–8, we may construct several important curves in addition to those relating D and d. The first step was to ascertain the scaling laws for the rim heights of explosive craters as functions of the scaled depth of burst of the explosion. As expected, the shallow craters formed by surface bursts have low rims, and the deeper bursts yielded higher crater rims. It was entirely unexpected, however, to find that the relationships $H/W^{1/3}$ versus $T/W^{1/3}$ were apparently the same for all energy ranges of man-made pits. Figure 16 (p. 135) shows this composite chart. Without proof, it has been assumed that the same law holds in the terrestrial meteoritic-crater range and that for the lunar craters of Class 1. The consistency of the results of this assumption in this and the next chapter lends confidence in its validity.

Over the range of scaled depths of burst from $H/W^{1/3} = 0.00$–0.50, the scaled crater rims increase in maximum height from $T/W^{1/3} = 0.100$ to 0.226.

Relationships between Crater Parameters

The finally determined equations for the English system of units relating the logarithmic apparent crater diameters and logarithmic rim heights, D and R_H, are as follows:

For $H/W^{1/3} = 0.00$,

$$R_H = 0.004366\,D^3 - 0.054571\,D^2 + 1.1316\,D - 1.4940 . \quad (7\text{-}5)$$

For $H/W^{1/3} = 0.10$,

$$R_H = 0.004366\,D^3 - 0.054571\,D^2 + 1.1316\,D - 1.3800 . \quad (7\text{-}6)$$

For $H/W^{1/3} = 0.25$,

$$R_H = 0.004366\,D^3 - 0.054571\,D^2 + 1.1316\,D - 1.2640 . \quad (7\text{-}7)$$

For $H/W^{1/3} = 0.50$,

$$R_H = 0.004366\,D^3 - 0.054571\,D^2 + 1.1316\,D - 1.1400 . \quad (7\text{-}8)$$

In the metric system of units, the above four equations become:

For $H/W^{1/3} = 0.00$,

$$R_{H_m} = 0.004366\,D_{km}^3 - 0.008506\,D_{km}^2 + 0.9098\,D_{km} + 1.4847 . \quad (7\text{-}5A)$$

For $H/W^{1/3} = 0.10$,

$$R_{H_m} = 0.004366\,D_{km}^3 - 0.008506\,D_{km}^2 + 0.9098\,D_{km} + 1.5987 . \quad (7\text{-}6A)$$

For $H/W^{1/3} = 0.25$,

$$R_{H_m} = 0.004366\,D_{km}^3 - 0.008506\,D_{km}^2 + 0.9098\,D_{km} + 1.7147 . \quad (7\text{-}7A)$$

For $H/W^{1/3} = 0.50$,

$$R_{H_m} = 0.004366\,D_{km}^3 - 0.008506\,D_{km}^2 + 0.9098\,D_{km} + 1.8387 . \quad (7\text{-}8A)$$

These curves have to be represented by cubic equations. They are complicated by the fact that at the low end the tiny man-made pits show incomplete rims because the materials are scattered broadcast rather than being concentrated in the rim.

In establishing these relationships, the points for the lunar craters represent only the Class 1 craters. The craters of Classes 2, 3, 4, and 5 give progressively lower average rim heights for a given diameter. The rims of the Class 4 craters average about 24 per cent lower than those of Class 1.

Equation (7-6) will uniquely represent the terrestrial meteoritic craters and smaller lunar craters; the larger lunar craters can better be represented by equation (7-5). The transition zone is the same as for the D versus d curves (Fig. 21).

In all craters, the measured width of the rim is an apparent width, which is less than the real rim width. It is also true that the crater rims have no definite outer boundaries. Figure 22 shows the observed relationship between D and the logarithm of the width (feet) of the apparent rim, R_W. The latter was measured from the well-defined crest to the point where the outer rim reached approximately the level of the undisturbed surface. In most craters this point is easy to determine, as a rather abrupt change of slope occurs. At the Arizona Meteorite Crater, meteoritic particles have been hurled at least 6 miles from the pit (23), yet the apparent rim proper is only about 800 feet wide, on the average. This is much less than the quarter-mile usually quoted but may easily be veri-

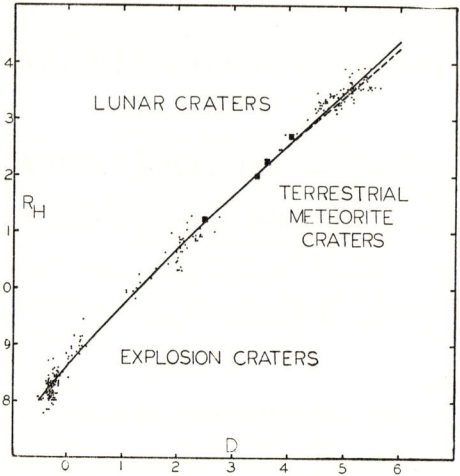

FIG. 21.—Logarithmic rim height (R_H) versus logarithmic apparent crater diameter (D). Solid line from eq. (7-6) for scaled depth of burst $H/W^{1/3} = 0.10$. Dashed line interpolated between eqs. (7-5) and (7-6).

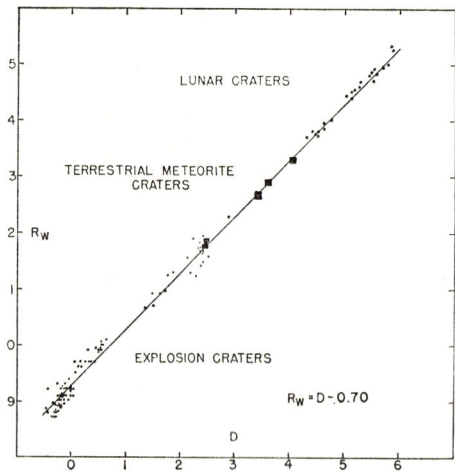

FIG. 22.—Logarithmic apparent rim width (R_W) versus logarithmic apparent crater diameter (D)

Relationships between Crater Parameters

fied by scale measurements on a photograph of the rim of the crater taken from a distance. A thin, continuous layer of rim materials often extends hundreds of feet farther from the crater.

In Figure 22 a straight line,

$$R_W = D - 0.70, \qquad (7\text{-}9)$$

adequately represents the terrestrial explosion and meteoritic pits and the lunar craters in either the English or the metric system of units, provided that R_W and D are in the same unit.

A key point here, which has previously been overlooked, is that all the curves relating crater dimensions may be extended without discontinuities from the terrestrial explosive craters into the lunar craters of Class 1. This presents a difficulty, inasmuch as the surface gravity on the earth is six times that on the moon. This can only mean that the crater-forming energy must be expended in some manner where the mass of the displaced ground materials is involved and not weight. The rims of the craters are not formed primarily by flying debris. Such objects, once started, would fly much farther on the moon than on the earth. The rims of explosive craters on the moon would thus be much lower and much wider than the terrestrial crater rims. As they are not lower and wider in proportion than terrestrial explosive crater rims, the crater-forming energies must have pushed much of the rims into place instead of throwing them into place.

With the aid of two simple assumptions, we may now solve for the dimensions of the true crater. By definition, the true crater is the subsurface portion of the apparent crater. It is not concerned with the depth of loose materials still remaining in the pit.

The first assumption is that the central peak may be neglected in these computations. The form of the apparent crater is nearly that of a parabolic surface of revolution. Therefore, the second assumption is that the true crater is in the form of a smaller parabolic surface of revolution.

The true crater depth is easily found. It is simply the apparent depth minus the rim height, both of which are known. The true crater diameter is less easily found, since few data are available. The observed contour of the lunar crater Theophilus by McMath, Petrie, and Sawyer (5) yields one approximate point. The apparent crater diameter is 65 miles, and the true crater diameter is measured from their cross-section to be 57.5 miles. The true crater diameter is thus about 0.885 times that of the apparent crater.

Table 10 summarizes all known data on the relationship between true and apparent crater diameters. The small explosion craters were tested by the author; the large man-made pits either were from World War I military mines

TABLE 10
Relationship between True and Apparent Crater Diameters for Explosion Pits

Crater No.	Apparent Diameter	True Diameter	True Diameter ÷ Apparent Diameter	Notes*
8	5 inches	4 inches	0.800	1
9	5.5	4.5	.818	
10	5.75	4.75	.826	
24	6.5	6	.923	
25	6.75	5.75	.852	
26	7.75	7.12	.919	
30	6.25	5.5	.880	
32	7.5	6	.800	
33	7.5	5.75	.767	
39	9	7	.778	
40	9.25	7.5	.750	
41	9.75	7	.718	
46	10	8	.800	
47	10.25	9	.878	
48	11	9.75	.886	
53	10	8	.800	
56	10.5	9	0.857	
Average	....	....	0.827	
310	139 feet	105 feet	0.755	
311	159	141	.887	
316	235	195	.830	
323	242	202	.835	
324	263	217	.825	
325	237	175	.738	
326	279	217	.778	
327	233	183	.785	
328	198	182	.919	
330	270	220	.815	In chalk
331	268	228	.851	
332	177	130	.734	
333	285	235	.824	
334	252	210	.833	
335	253	204	.806	
336	206	200	.971	Omit, partial camouflet
337	340	273	.803	
338	102	80	.784	
339	115	88	.765	
340	268	240	.896	
341	306	250	.817	
342	261	205	.785	
343	224	176	.786	
346	127	112	.882	
347	159	122	.767	
348	180	137	.761	
349	257	240	.934	In clay
350	207	194	.937	In sandstone
352	165	142	.861	
353	183	146	.798	
354	184	138	.750	
355	201	165	.821	
356	210	177	.843	
359	338	315	.932	
360	428	398	.930	
362	800	741	.926	
363	339	289	.852	
364	275	248	0.902	
Average	....	....	0.829	
Kaali Järv	319 feet	263 feet	0.826	Refs. 9–11
Arizona	3,850	3,380	.878	2; Ref. 12
New Quebec	11,290	9,115	.810	Ref. 13
Theophilus	65 miles	57.5 miles	.885	Ref. 5
Average	....	....	0.850	

Grand average for all 58 craters = 0.830

*Notes to Table 10:
1. Crater numbers are from Table 4 or Table 8. 2. From Shoemaker's drawing of original crater.

Relationships between Crater Parameters

set off by the British (6) or were from later tests run at Arco, Idaho (7, 8), by the United States in disposing of surplus World War II explosives.

The straight line,

$$\frac{\text{True diameter}}{\text{Apparent diameter}} = 0.830, \quad (7\text{-}10)$$

seems a satisfactory relationship, as Figure 23 shows. There is no significant variance from this relationship as a function of depth of burst for any value of energy of man-made explosion pits.

Many years ago Schröter (14) announced a rule which stated that the volume of the rim of a crater on the moon was equal to the volume of material dug out of the ground to produce the crater. This must be true for all explosive

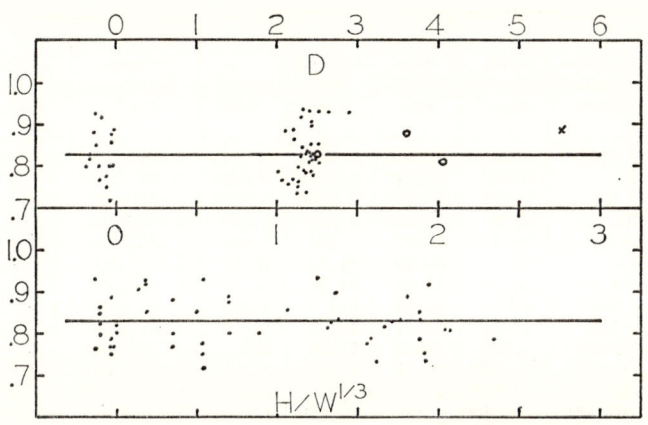

Fig. 23.—Ratio of true crater diameter to apparent crater diameter as function of: (*upper*) logarithmic apparent crater diameter (D); (*lower*) scaled depth of burst ($H/W^{1/3}$).

craters if we include the debris scattered widely by the explosion. Because of this material ejected far beyond the rim, the subsurface-crater volume should always be more than the volume of the rim. This discrepancy should be most pronounced for small craters and should decrease as the crater size increases and it becomes harder to eject material from the pit.

This relationship may be tested by using equations (7-1), (7-2), (7-5), (7-6), (7-9), and (7-10) and the simplifying assumptions that the volume of the true crater may be represented as the volume of a paraboloid of revolution and the volume of the rim may be represented as the difference of two frustums of cones. Table 11 summarizes the results. It may be seen that the expected relationship is confirmed. There is less volume in the rim than in the true crater, but the discrepancy becomes less with increasing diameter. A modifying influence is that material shattered and hurled from the pit will be of lower density than the original material. To a first approximation, Schröter's rule is confirmed, for he was studying the larger lunar craters. The variations in the scaled depths of bursts listed in Table 11 came from the analysis in the next chapter.

One of the interesting results of the work by Beals (15) and his associates and Innes (16, 17) is the determination of the nature of the rocks in the immediate subcrater region. They are highly brecciated. The central portion of the shattered lens is thickest and is also most highly fragmented and pulverized. Sometimes it is even burned. The lens seems to fade out in all directions as the fractures become farther apart. It is effectively gone horizontally at the distance of the rim from the center; and below the crater the jumbling of the rocks

TABLE 11*

INVESTIGATION OF SCHRÖTER'S RULE: CLASS 1 CRATERS

D	$H/W^{1/3}$	Apparent Diameter (Feet)	True Diameter (Feet)	Rim Height (Feet)	Apparent Depth (Feet)	True Depth (Feet)	Volume True Crater (Feet)^3	Volume Rim (Feet)^3	Volume Rim/Volume Crater
9.5...	0.10	0.3162	0.2624	0.0110	0.0684	0.0574	1.549×10^{-3}	4.090×10^{-4}	0.264
0.0...	.10	1.0	0.830	0.0417	0.2236	0.1819	4.911×10^{-2}	1.576×10^{-2}	0.321
0.5...	.10	3.162	2.624	0.1488	0.7406	0.5918	1.598×10^{0}	5.710×10^{-1}	0.357
1.0...	.10	10	8.30	0.5028	2.326	1.823	4.925×10^{1}	1.960×10^{1}	0.398
1.5...	.10	31.62	26.24	1.619	7.103	5.484	1.481×10^{3}	6.408×10^{2}	0.433
2.0...	.10	100	83	5.010	21.14	16.13	4.357×10^{4}	2.014×10^{4}	0.462
2.5...	.10	316.2	262.4	15.00	61.39	46.39	1.253×10^{6}	6.118×10^{5}	0.488
3.0...	.10	1,000	830	43.81	174.3	130.5	3.524×10^{7}	1.812×10^{7}	0.514
3.5...	.10	3,162	2,624	125.7	484.6	358.9	9.687×10^{8}	5.275×10^{8}	0.544
4.0...	.083	10,000	8,300	341.4	1,289	947.6	2.557×10^{10}	1.453×10^{10}	0.568
4.5...	.053	31,620	26,240	894.1	3,255	2,361	6.375×10^{11}	3.855×10^{11}	0.605
5.0...	.020	100,000	83,000	2,335	7,838	5,503	1.486×10^{13}	1.020×10^{13}	0.686
5.5...	.00	316,200	262,400	6,389	18,650	12,261	3.310×10^{14}	2.829×10^{14}	0.855
6.0...	0.00	1,000,000	830,000	18,800	45,680	26,880	7.263×10^{15}	8.430×10^{15}	1.161

* Data for this table were derived from eqs. (7-1), (7-2), (7-5), (7-6), (7-9), and (7-10). These equations are valid only to $D = 5$ or a little beyond. Hence the data for $D = 5.5$ and 6.0 are only approximate. There are no Class 1 craters larger than $D = 5.65$.

ceases at a reasonably sharp surface, even though the fracturing of the rock in position seems to continue downward for an unknown distance.

There are fairly consistent data for the three Canadian craters, all of which are very old. Shoemaker (12) has summarized earlier data on the Arizona Meteorite Crater and on the Jangle U nuclear explosion; Sellards and Evans (18, 19) have sectioned the Odessa No. 1 Crater; and Reinvaldt (9–11) has given a cross-section of the Kaali Järv. Shelton, Nordyke, and Goeckermann (20) have described the Neptune atomic burst. The pertinent facts on these eight craters are given in Table 12.

When B is plotted against d_1 (Fig. 24), it is evident that the depth to the limit of major brecciation below the original ground level increases more rapidly than the true depth of the meteoritic crater. A straight line,

$$B = 1.2658 d_1 - 0.3417 .\qquad(7\text{-}11)$$

Relationships between Crater Parameters

gives an approximation to this relationship. B and d_1 are logarithms of the dimensions in feet. In the metric system, where the quantities are logarithms of the dimensions in meters, we find

$$B_m = 1.2658 d_{1_m} - 0.2045 \; . \qquad (7\text{-}11\text{A})$$

The right-hand panel of Figure 24 shows the equivalent relationship between D, the logarithmic apparent crater diameter, and B, as a dotted line. The solid line is D versus d_1, the true depth of the crater. The D versus B curve is below the D versus d_1 curve for all crater diameters greater than $D = 2.06$, or 115 feet. Consequently, we should not expect to find a significant lens of breccia beneath the smaller terrestrial meteoritic craters, nor is one found.

In extrapolation, the lens of breccia under a 10-mile-wide crater should

TABLE 12
LIMIT OF BRECCIATION OF ROCK BENEATH EXPLOSION CRATERS

Crater	Diameter (Feet)	D^*	True Depth (Feet)	d_1^*	Depth of Breccia† (Feet)	B^*
Neptune..........	255	2.407	35	1.544	80	1.90
Jangle U..........	275	2.439	44	1.643	78	1.89
Kaali Järv........	319	2.504	36.4	1.561	57	1.76
Odessa No. 1......	550	2.740	81.5	1.911	106	2.02
Arizona...........	4,000	3.602	500	2.699	1,100	3.04
Holleford.........	7,700	3.887	780	2.892	1,800	3.26
Brent............	11,500	4.061	1,080	3.033	3,800	3.58
Deep Bay.........	40,000	4.602	2,863	3.457	10,000	4.00

* D, d_1, and B are logarithms of the preceding respective column figures.
† From original surface.

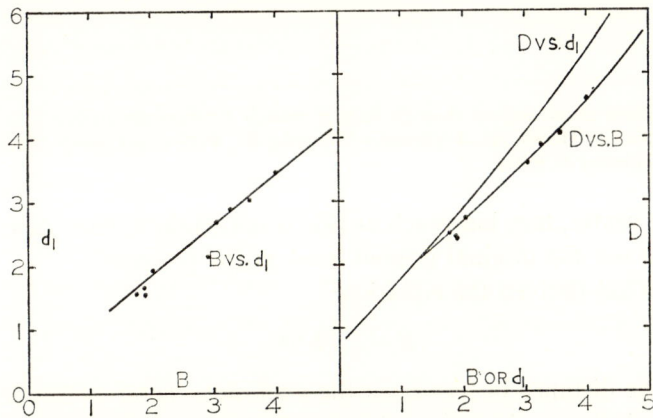

FIG. 24.—Relationships between depth of brecciation from ground level and true crater depth (*left*) and apparent crater diameter (*right*), logarithmic scales.

stretch to 3 miles below ground level; at 50-, 100-, and 150-mile craters the rock layers should be smashed and jumbled to vertical distances of the order of 11, 21, and 30 miles. Of course, the data are few, and the extrapolation is large, but the general order of this effect would seem to be established.

Presumably, the depth to the limit of brecciation should be measured from

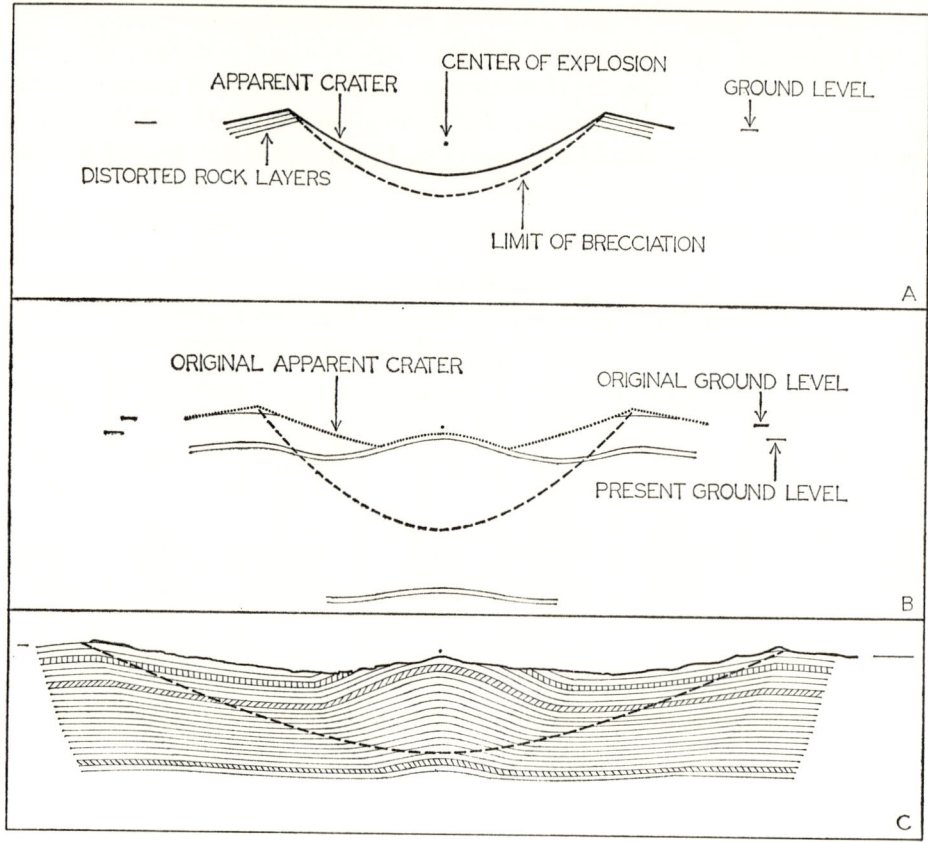

Fig. 25.—Idealized cross-sections through impact craters showing distortions of rock layers and zone of brecciation. *A:* Odessa No. 1, diameter 550 feet; *B:* Wells Creek Basin, diameter 6.5 miles; *C:* Theophilus, diameter 65 miles.

the explosive center, but, inasmuch as this is uncertain in most cases, it is more convenient to use the original ground level as the referent.

Shoemaker has derived the equation

$$R = 5.7W^{1/3}, \qquad (7\text{-}12)$$

where R is the distance in feet from the explosion center to the limit of brecciation and W is the number of metric tons of TNT equivalent. This equation fits the smaller craters tolerably well but becomes increasingly in error at the larger pits. As in the crater scaling equations, the exponent appears to be less than one-third.

With the data of the preceding chapters, we may reconstruct average cross-sections in depth of a small meteoritic crater about the size of Odessa No. 1, a large crater such as Wells Creek Basin as it was originally, and a lunar crater, Theophilus. The surface contour of Theophilus is that derived by McMath, Petrie, and Sawyer (5) from their analysis of motion pictures taken at the McMath-Hulbert Observatory.

The basis for showing the rebound dome as a diminished, but still real, structure, even below the limit of brecciation, is the test well sunk to 12,096 feet at the Sierra Madera Dome (21, 22). The dome there had not yet disappeared at this depth, but in all probability, the zone of major brecciation had been penetrated. In this brecciated zone, the general trend of the rock layers is indicated, but a tremendous jumbling of the materials in this lens is normally observed (Fig. 25).

References

1. Private communication from Mr. ROBERT FRYE, assistant, Samuel Feltman Laboratories, Picatinny Arsenal, Dover, N.J.
2. LAMPSON, C. W. *Effects of Atomic Weapons*, ed. S. GLASSTONE, p. 410. Washington, D.C.: Combat Forces Press, 1950.
3. CHARTERS, A. C. "High Speed Impact," *Sci. American*, **203**, 128, 1960.
4. ARTHUR, D. W. G. *The Diameters of Lunar Craters*. Private publication.
5. MCMATH, R. R., PETRIE, R. M., and SAWYER, H. E. "Relative Lunar Heights and Topography by Means of the Motion Picture Negative," *Pub. Obs. U. Michigan*, **6**, 67, 1937.
6. *Work of the Royal Engineers in the European War, 1914–19, Military Mining*. Chatham: Published by the Secretary, Institute of Royal Engineers, W. & J. Mackay & Co., Ltd., 1922.
7. "Igloo Tests," *Technical Paper No. 3*. Washington, D.C.: Army-Navy Explosives Safety Board, 1945; revised November 6, 1947.
8. "Igloo and Revetment Tests," *Technical Paper No. 5*. 1946.
9. REINVALDT, L., with LUBA, A. "Bericht über geologische Untersuchungen am Kaalijärv (Krater von Sall) auf Ösel," *Tartu Ulikooli juuresoleva Loodusourijate Seltsi Aruanded (Sitzb. Naturforsch. Gesellsch. U. Tartu)*, **35**, 30–70, 1928.
10. ———. Separate as *Pub. Geol. Inst. U. Tartu.*, No. 11, pp. 1–42, 1928.
11. KRAUS, E., MEYER, R., and WEGENER, A. "Untersuchungen über Krater von Sall auf Ösel," *Gerlands Beitr. z. Geophys.*, **20**, 312–78, 1928. Prepared according to information and drawings furnished by REINVALDT.
12. SHOEMAKER, E. M. "Impact Mechanics at Meteor Crater, Arizona." Open file report, prepared on behalf of the U.S. Atomic Energy Commission, July, 1959.
13. MILLMAN, P. "The New Quebec Crater," *J. Brit. Astr. Assoc.*, **67**, 116, 1957.

14. SCHRÖTER, J. H. *Selenotopographische Fragmente.* 2 vols. Göttingen: Lilienthal & Helmst, 1791–1802.
15. BEALS, C. S., INNES, M. J. S., and ROTTENBERG, J. A. "The Search for Fossil Meteoritic Craters," *Contr. Dom. Obs., Ottawa,* **4,** 3, 1960.
16. INNES, M. J. S. "The Use of Gravity Methods To Study the Underground Structure and Impact Energy of Meteorite Craters," *J. Geophys. Res.,* **66,** 3, 1961.
17. ———. "The Use of Gravity Methods To Study the Underground Structure and Impact Energy of Meteorite Craters," *Contr. Dom. Obs., Ottawa,* **5,** 3, 1961.
18. SELLARDS, E. H., and EVANS, G. *Statement of Progress of Investigation at Odessa Meteor Craters.* University of Texas, Bureau of Economic Geology, September 1, 1941.
19. ———. "Odessa Meteor Craters," *Views in Texas Memorial Museum, Museum Notes,* **6,** 13 (July, 1944).
20. SHELTON, A. V., NORDYKE, M. D., and GOECKERMANN, R. H. "The Neptune Event: A Nuclear Explosive Cratering Experiment," *UCRL-5766, Nuclear Explosions, Peaceful Applications, TID-4500, UC-35.* 15th ed. 1960.
21. PHILLIPS PETROLEUM COMPANY. Letter to author. Two wells drilled in Sierra Madera Dome.
22. WILSON, C. W., JR. "Wilcox Deposits in Explosion Craters, Stewart County, Tennessee, and Their Relations to Origin and Age of Wells Creek Basin Structure," *Bull. Geol. Soc. America,* **64,** 767, 1953. Letter from Phillips Petroleum Company to C. W. WILSON, JR.
23. NININGER, H. H. *Arizona's Meteorite Crater.* Denver: World Press, Inc., 1956.

8

DETERMINATION OF ENERGIES NEEDED TO PRODUCE METEORITIC CRATERS

The equations and charts of the preceding chapter are interpreted to mean that all three categories of craters—explosion pits, terrestrial meteoritic craters, and lunar craters of Class 1—are of the same basic type of origin. The scaling of the terrestrial meteorite craters is fairly closely established, but the scaling for the lunar craters is still to be pinpointed. Inasmuch as the first two groups are known to be produced by explosions, it follows that the third type—the lunar craters—are also of explosive origin. If this is granted, then it follows either that the lunar craters were produced by meteoritic impact and subsequent explosions or that they were formed by explosions of unknown origin. The former interpretation is usually accepted because only vast meteorites are known to possess sufficient energy to produce the lunar craters. No other known source of energy has been shown to be capable of accounting for the structures of the craters of the moon. The meteoritic-impact theory has been generally, but not universally, accepted. It is here assumed to be correct, as it most satisfactorily accounts for the observed lunar features.

Various terrestrial craters have been proved to be of meteoritic origin. They are also known to be of explosive origin. Therefore, it follows that meteorites can and do explode on impact. To substantiate this, various authors have made theoretical calculations to verify the conditions under which a meteorite could explode on impact.

Gold, in a public lecture (1), tried to solve this problem in shock hydrodynamics by setting up an equation of state corresponding to an ideal gas of dissociated nuclei and electrons. For a large and fast-moving body, he found a temperature of about 1,000,000° C to be produced by sudden stopping.

Gilvarry and Hill (2, 3) have extensively studied this problem. They have published several papers presenting the results of computations on the pressures and temperatures attained in the impact of large meteorites on the lunar surface based on the equations of state as obtained from the Thomas-Fermi statistical model of the atom.

As a gross simplification,[1] they reduced the problem to one-dimensional form. The meteorite is considered as a right cylinder whose base impinges on the lunar surface. A plane shock wave enters the cylinder along its axis, and a plane shock wave enters the lunar surface. Behind each shock the flow patterns and the distribution of pressure and temperature are uniform up to the interface between meteoritic and lunar material. For simplicity, the meteorite and the lunar surface are assumed to be composed of the same pure element. This assumption yields the conclusion that the pressure, temperature, and density are the same behind both shocks; the velocity of the material between the two shocks is uniform and is equal to $v/2$ directed into the moon, if v is the velocity of the impinging meteorite. This idealized model will apply to the impact on the lunar surface of a large meteorite of roughly spherical shape for a time sufficiently short that the distances traversed by the shocks in the meteorite and the lunar crust are small compared with a linear dimension of the impinging mass. For times sufficiently long that the shock in the meteorite reaches the rear surface (from which a rarefaction wave is reflected in this case), the idealization loses physical validity, since appreciable lateral motion of the meteoritic material can exist and the shock wave can no longer be regarded as plane.

Gilvarry and Hill find that the highest pressures possible are over 100 megabars; in general, the pressures attained are nearly of the order of those reached in the explosion of an atomic bomb. Similarly, the temperatures may reach values larger than 100,000° C and thus are comparable with those attained in an atomic explosion.[2] In general, the temperatures are far above the range (of the order of 5,000° C) corresponding to a chemical explosion but are below the value (1,000,000° C) estimated by Gold. The possibility of such

---

[1] Shoemaker (4) later amplified the work of Hill and Gilvarry and improved it substantially. He reached essentially the same conclusions.

[2] Brickwedde (5) estimates that temperatures of the order of 10,000,000° C are attained in an atomic explosion.

Energies Needed To Produce Meteoritic Craters

high pressures and temperatures stems from the fact, of course, that the kinetic energy per unit mass of a body moving at the higher speeds considered is far in excess of the explosive energy (about 2×10^{13} ergs) per pound of TNT.

Initially, these pressures and temperatures exist in a volume of the order of the initial volume of the striking meteorite, if the volume swept out by the shock wave in the moon's crust is included. The explosive release of the energy, initially confined in this volume, blasts out the lunar crater. Since the velocity of the shock wave in the meteorite relative to the meteorite itself is of the order of the impact velocity, it follows that the explosive pressures and temperatures are created in a time of the order of that required for the colliding mass to traverse a distance equal to its diameter. The velocity of a shock wave is a function of the pressure, which is proportional to the velocity of impact. Therefore, the effective center of the explosion must be within a depth below the surface of the moon of the order of a linear dimension of the striking mass, unless the explosive velocity is considerably smaller than the meteoritic striking velocity. In this case, the momentum of the meteorite will carry the explosive focus down into the ground farther than Gilvarry and Hill estimated. They (6) state:

> It should be emphasized that the results presented are valid only when the shock formation is the dominant mechanism of energy transfer, that is, for impact velocities greater than the order of, say, six miles per second for large bodies. For the lower impact velocities possible on the earth, because of atmospheric deceleration, a continuous gradation exists between this regime and that corresponding to plastic yield or elastic strain. In spite of their qualitative nature, the results tend to bear out the fundamental correctness of Baldwin's picture of the origin of lunar craters.

If Gilvarry and Hill are correct—and it appears that they must be reasonably close to a true picture—then the scaled depth of burst from a meteoritic impact must be rather small. Let H equal 1 diameter of the meteorite, and let the striking velocity vary. The results are shown in Table 9 (p. 140).

If, on the contrary, the penetration of a meteorite is somewhat greater than 1 diameter, the scaled depth of burst will be proportionately greater. Because we do not yet know the exact degree of penetration, we shall concern ourselves in this chapter with relationships valid for all scaled depths of burst 0.0–0.50, recognizing that the greater scaled depths of burst cannot be possible for the higher-velocity meteorites.

In setting up the values of the logarithms of the diameters and depths of true and apparent craters in Table 4 (Appendix 2), we have also made possible the derivation of explicit correlations between the energies needed to blast the craters and the sizes of the craters. This can be done for any scaled depth of burst, because the diameter and depth of each crater have been converted into the corresponding dimensions for four different scaled depths of burst. At the same time, we have the amount of energy that was actually applied to the formation of the original craters covering approximately 50 per cent of the logarithmic energy range.

In the calculations of this chapter, the attempt will be made to determine the relationships between the energy released and the crater dimensions as functions of the depth of burst. Inasmuch as theory is not yet able to tell us what these relationships should be over wide ranges of crater sizes, the approach here is purely empirical. The only tests of man-made explosions are those from charges which, when compared with the sizes of the resulting craters, are essentially point energy sources.

It is readily admitted that meteoritic-impact explosions are not identical with point-source bursts. A meteoritic explosion starts at the instant of impact, and during the early stages of penetration it is roughly a line or cylinder explosion. Very quickly the meteorite flattens out and stops. In this, the major part of the burst, the effect is more like an explosion from a surface. Shockwave phenomena become very important here.

However, the meteorites capable of producing the terrestrial pits and smaller lunar craters are quite tiny relative to the crater size, and so the difference between a point-source model for a chemical or nuclear explosion and the complicated burst of an impacting meteorite is very small. The observations indicate that a point-source model will represent rather well all impact bursts up to those which will produce a $D = 4$ crater, i.e., a crater 10,000 feet wide. For another factor of 10 in diameter, the approximation is rather good.

In order to reach certain necessary conclusions, extrapolations will first be made to determine the relationships between dimensions and energies for point-source models of craters up to 189 miles in diameter or $D = 6$. No Class 1 crater is this large. These extrapolations will be shown to be rather closely limited. Then the resulting equations will be modified to account for the deviations from the point-source model which are found at the larger impact craters on the moon. This may be called the "surface-source explosion regime."

Let us arbitrarily assume a scaled depth of burst of $H/W^{1/3} = 0.10$ to be true of all sizes of craters. Plots of D versus E and d versus E in the energy range of $E = 2.88$–12.39 show rather well-defined relationships, but still the scatter of points will permit serious errors if the curves are extrapolated through the lunar crater region. If this were all the information we had available, we could not establish accurately the energies needed to produce the lunar craters. Fortunately, the problem may be limited by a completely different approach.

Whatever the mechanism by which the rim for a given lunar crater was produced, the energy must have been greater than a calculable minimum value.

Energies Needed To Produce Meteoritic Craters

It has been shown that the volume of the true crater is somewhat larger than the volume of the rim for all craters of this type. Therefore, the starting point of this calculation is the mass of material displaced to produce a true crater.

This minimum rim-forming energy may be found by taking a cross-section of the crater and finding the center of gravity of the rim and the center of gravity of the near half of the true crater and the distance between these two points. For an assumed density of surface materials of 2.0 grams per cubic centimeter and the known crater dimensions, we may solve for the total mass moved out of the true crater and hence the energy necessary to deposit it in

TABLE 13*

Crater-Forming Energies (Terrestrial Craters)

(Scaled Depth of Burst $H/W^{1/3} = 0.10$)

D	Log Volume	Log Mass	Average Distance	Log Velocity^2	Log Minimum Energy	Log Observed Energy	$E_{obs} - E_{min}$
9.5	7.198	1.951	0.134	3.603	7.933	2.50	4.567
0.0	8.692	3.445	0.411	4.089	9.913	4.08	4.167
0.5	0.198	4.951	1.33	4.599	1.929	5.66	3.731
1.0	1.680	6.433	4.22	5.101	3.913	7.28	3.367
1.5	3.155	7.908	12.2	5.562	5.849	8.88	3.031
2.0	4.622	9.375	37.5	6.050	7.804	10.48	2.676
2.5	6.083	10.836	117	6.544	9.759	12.13	2.371
3.0	7.537	12.290	373	7.047	11.716	13.83	2.114
3.5	8.982	13.735	1,210	7.558	13.672	15.55	1.878
4.0	10.418	15.171	3,700	8.044	15.594	17.35	1.756
4.5	11.845	16.598	12,300	8.565	17.542	19.18	1.638
5.0	13.262	18.015	37,300	9.047	19.441	21.09	1.649
5.5	14.665	19.418	117,000	9.544	21.341	23.06	1.719
6.0	16.055	20.808	379,000	10.054	23.241	25.147	1.906

* The data are for point-source bursts, and D = logarithm of diameter (feet) of apparent crater. The volume of true crater is expressed in cubic feet. The mass displaced in producing the true crater is expressed in grams. The distance between the center of gravity of a section of the crater rim and the center of gravity of the near half of the true crater is expressed in feet. The velocity is expressed in centimeters per second. E represents the logarithm of the energy in calories. Values above the heavy line are observed; values below are extrapolated.

the rim along a minimum-energy path. These calculations are summarized in Table 13.

No matter what rim-forming process is assumed, it will require more energy than the above calculation yields. Let us assume a newly formed crater with $D = 6$. This is larger than any Class 1 lunar crater and about equal in diameter to ruined Bailly. The logarithm of the mass in grams of the material displaced to form the true crater is 20.808. The distance moved is 379,000 feet, and the minimum needed energy is found to be $E = 22.940$ (calories). However, this must be increased by 0.301, inasmuch as only half the explosive energy at a maximum can be utilized by such a surface explosion. At least half the energy will be wasted into air or space. The minimum is thus raised to $E = 23.241$.

Now, in the smaller-energy ranges, there is a series of different-sized craters which can be produced by explosions of a given energy, depending on how deeply the explosive charge was buried. The largest craters at all observable energy ranges are found at a scaled depth of burst of $H/W^{1/3} = 1.50$–2.00. Consequently, we assume a similar relationship to hold for the postulated very large craters. Therefore, if $E = 23.241$ is the minimum logarithmic energy needed to produce a $D = 6$ crater at $H/W^{1/3} = 0.10$, we might also assume that this same amount of energy would produce a $D = 6.477$ crater at $H/W^{1/3} = 1.50$. However, this larger crater would be formed by displacing more material and moving it a greater distance. Therefore, the above-mentioned minimum energy for the $D = 6$ crater could not produce a $D = 6.477$ crater. It would take at least $E = 25.06$ to produce the larger crater. Therefore, the minimum logarithmic energy needed to produce the $D = 6$ crater would also be $E = 25.06$, because the two craters would have to be formed by the same amount of energy released at two different depths of burst.

In Table 13, we have calculated minimum energies derived by moving the crater materials into the rims along minimum-energy paths and also the observed energy which actually produced the craters for logarithmic energies less than $E = 12.39$. The difference in the sense $E_{obs} - E_{min}$ grows steadily less as the craters grow larger. This means that the crater-forming process utilizes an increasing percentage of the available energy as the magnitude of the burst increases. At the tiny craters, less than a foot across, the available energy is several thousand times the theoretical minimum energy. By the time the largest chemical explosive pits are reached, the ratio is reduced to many hundreds to one. Extrapolation suggests that, for $D = 6$ and $H/W^{1/3} = 0.10$, the needed logarithmic energy can be estimated at about 25.147, or about 80 times the theoretical minimum energy; but if the explosive is set off deeper in the ground, the ratio comes down to about 1.2 to 1. It may be concluded that the relationship between diameter and energy is reasonably closely defined throughout the energy range of explosive crater sizes. For a lunar-type crater with $D = 6$ produced by a point burst at a scaled depth of burst of $H/W^{1/3} = 0.10$, the logarithmic energy cannot be less than 25.06 for physical reasons, and the observed trends in the lower-energy regions suggest that it cannot be significantly greater. Figure 26 shows this clearly. We shall adopt $E = 25.147$ for a $D = 6$ crater at a scaled depth of burst of $H/W^{1/3} = 0.10$ as reasonably close to the correct figure for a point-source burst.

Öpik (7) has correctly pointed out that there is a real difference between terrestrial explosions and explosions from meteoritic impact:

The terrestrial explosion is the result of expansion of a gas ball without initial translational motion; its efficiency depends on the arbitrarily chosen depth of a planted charge, and on the energy released in the explosion which in underground nuclear explosions determines the initial mass of the gas ball, without regard to the mass of the

Energies Needed To Produce Meteoritic Craters

charge. In meteorite impact, the primary agent is the translational motion of the meteoric body which itself, from mere inertia, is a powerful cratering factor and determines also the depth of penetration; at this predetermined depth, a gas ball develops as a secondary agent of cratering which may be missing at low velocities. The action of the gas ball is similar in both cases but its mass, in the case of impact, is not simply proportional to the kinetic energy of the projectile. There is no direct way of applying quantitative results of explosions to meteorite impact, although the geometry of the craters may be similar.

Öpik, in effect, says that a gas ball containing a certain number of ergs produced from a terrestrial explosion below ground level, which produces its

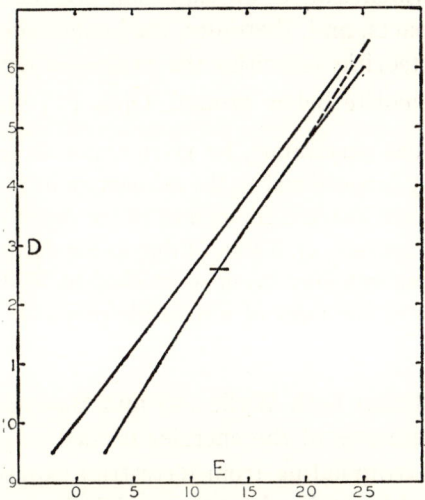

FIG. 26.—Crater-forming energies. *Left line:* minimum E needed to produce a crater of the indicated D. *Right line below bar:* observed E versus D. *Right line above bar:* solid line is extrapolated E versus D for point-source bursts. Dashed line is E needed to produce very large lunar craters and circular maria, surface-burst regime. $E =$ logarithmic energy (calories); $D =$ logarithmic apparent crater diameter (feet); $H/W^{1/3} =$ scaled depth of burst as indicated in Table 16.

own shock waves, will not yield the same effect as a meteoritic impact which produces its own shock waves and also a gas ball which then contributes to the shock waves. The effective center of this action is at the same distance below ground as the terrestrial explosion, and the total energies are identical in the two cases.

It is important to note that experiments show that when a missile strikes at a reasonably high velocity, the explosion starts at the instant of impact, but the body continues down into the target at a rapidly reducing velocity. All the time it is moving at velocities above the shock-wave velocity, the expanding shock front in the meteorite and in the ground is generating a very highly compressed gas ball, which, when the motion is essentially stopped, will result in a terrific explosion. While still moving at a high velocity, the meteorite acts as though it had no strength as it flattens and literally turns itself inside out (8). The flattening definitely limits the amount of penetration. In this connection it appears that, for velocities above perhaps 6 miles per second, essentially the entire meteorite is converted into gas. Undoubtedly part of the ground is also vaporized. Öpik also feels that the momentum of the body determines the amount of penetration.

Theoretical work by Gilvarry and Hill (2) and later by Shoemaker (4) clearly shows that the velocities of the shock waves generated in the meteorite and in the ground are roughly proportional to the velocity of the impact, and hence the conditions which lead to an explosion are reached more quickly at the higher-velocity impacts, and, therefore, the penetration is not a function of momentum, but at meteoritic velocities the explosion will occur within about 2 diameters of the meteorite below ground. Öpik (7) states:

> Because of predetermined penetration, for given materials the volume efficiency of impact cratering is mainly proportional to the momentum of the projectile, and not to its kinetic energy, with some increase (doubling) of the factor of proportionality with increasing velocity (5–30 km/sec, *cf*. Table 2) due to the development of the gas ball. This circumstance has been unknown to, or overlooked by Wylie, Baldwin and others concerned with estimates of the mass of a meteoric projectile required to produce a certain size of crater.

It is readily granted that both Wylie (9) and Baldwin (10) were in error in their preliminary estimates of the energies needed to produce the explosive meteoritic craters by extrapolating from terrestrial explosions. The very large effects of scaled depth of burst for these terrestrial pits were not properly taken into account, and also the scaled depths of burst for the meteoritic explosions were not known previously. These defects are remedied in this chapter.

The effect of this assumption of "predetermined penetration" by Öpik leads inevitably to a larger value for the energy of the meteorites needed to produce specific meteoritic craters than is consistent with the present work and that by Johnson (11), Shoemaker (4), and others, and the discrepancy increases as the size of the crater decreases.

Let us now leave Öpik's theory for a little while and continue the argument of this chapter. We shall return to it later. With the available information, we may now derive D and d as functions of $E = \log$ energy (calories) and develop a family of curves for each scaled depth of burst. We already have four equations, (7-1) through (7-4), relating D and d for four scaled depths of burst.

Now, for each scaled depth of burst, we plot D and d, respectively, versus E. Examples are shown in Figures 27 and 28. The necessary condition is that, for example, if we derive an equation, D versus E, for a scaled depth of burst of

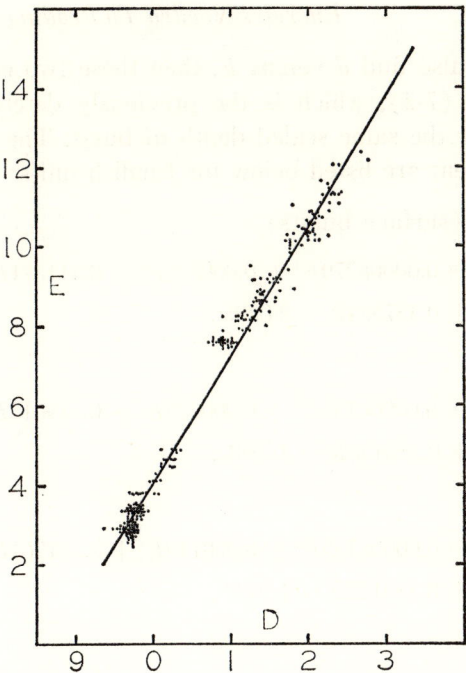

Fig. 27.—Observed logarithmic energy (*E*) versus logarithmic apparent crater diameter (*D*) for a scaled depth of burst, $H/W^{1/3} = 0.10$.

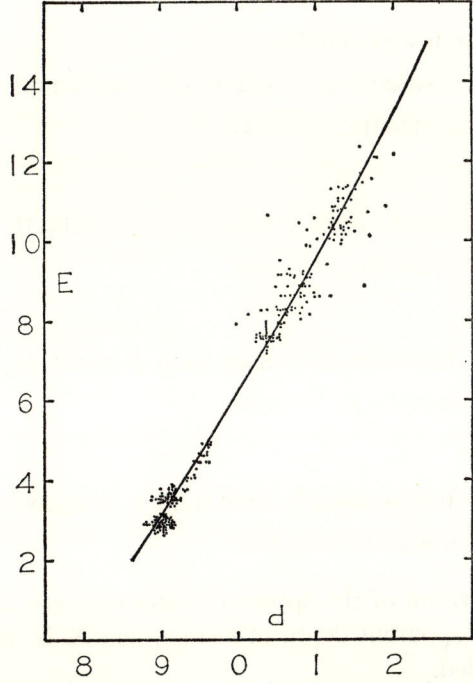

Fig. 28.—Observed logarithmic energy (*E*) versus logarithmic apparent crater depth (*d*) for a scaled depth of burst, $H/W^{1/3} = 0.10$.

$H/W^{1/3} = 0.10$ and also find d versus E, then these two equations combined must yield equation (7-2), which is the previously determined relationship between D and d for the same scaled depth of burst. The eight equations so found and so consistent are listed below for English units:

For $H/W^{1/3} = 0.00$ (surface bursts),

$$D = 0.00000043E^4 - 0.00008279E^3 + 0.000823E^2 + 0.311871E - 1.3213, \quad (8\text{-}1)$$
$$d = -0.003675E^2 + 0.357584E - 2.0895. \quad (8\text{-}2)$$

For $H/W^{1/3} = 0.10$,

$$D = 0.00000029E^4 - 0.00006155E^3 + 0.000225E^2 + 0.319015E - 1.3018, \quad (8\text{-}3)$$
$$d = -0.003375E^2 + 0.356183E - 2.0381. \quad (8\text{-}4)$$

For $H/W^{1/3} = 0.25$,

$$D = 0.00000022E^4 - 0.00005139E^3 + 0.000149E^2 + 0.321761E - 1.2614, \quad (8\text{-}5)$$
$$d = -0.003092E^2 + 0.355122E - 1.9432. \quad (8\text{-}6)$$

For $H/W^{1/3} = 0.50$,

$$D = 0.00000022E^4 - 0.00005009E^3 + 0.000090E^2 + 0.326579E - 1.1908, \quad (8\text{-}7)$$
$$d = -0.003098E^2 + 0.359330E - 1.8188. \quad (8\text{-}8)$$

The same eight equations converted into the metric system are as follows:

For $H/W^{1/3} = 0.00$ (surface bursts),

$$D_{km} = 0.00000043E_e^4 - 0.00009590E_e^3 + 0.002866E_e^2 + 0.284136E_e - 7.1293, \quad (8\text{-}1A)$$
$$d_m = -0.003675E_e^2 + 0.413604E_e - 5.5444. \quad (8\text{-}2A)$$

For $H/W^{1/3} = 0.10$,

$$D_{km} = 0.00000029E_e^4 - 0.00007039E_e^3 + 0.001733E_e^2 + 0.304345E_e - 7.2089, \quad (8\text{-}3A)$$
$$d_m = -0.003375E_e^2 + 0.407630E_e - 5.4649. \quad (8\text{-}4A)$$

For $H/W^{1/3} = 0.25$,

$$D_{km} = 0.00000022E_e^4 - 0.00005810E_e^3 + 0.001401E_e^2 + 0.310144E_e - 7.1986, \quad (8\text{-}5A)$$
$$d_m = -0.003092E_e^2 + 0.402255E_e - 5.3455. \quad (8\text{-}6A)$$

For $H/W^{1/3} = 0.50$,

$$D_{km} = 0.00000022E_e^4 - 0.00005680E_e^3 + 0.001312E_e^2 + 0.316088E_e - 7.1687, \quad (8\text{-}7A)$$
$$d_m = -0.003098E_e^2 + 0.406555E_e - 5.2535. \quad (8\text{-}8A)$$

Here D_{km} is the logarithm of the apparent crater diameter in kilometers, d_m is the logarithm of the apparent depth in meters, and E_e is the logarithm of the number of ergs applied.

Energies Needed To Produce Meteoritic Craters

Equations (7-1) through (7-4) are valid from the tiny man-made explosion pits beyond the largest Class 1 lunar crater, provided that they are produced by point-source bursts at the indicated scaled depths of burst. Equations (8-1) through (8-8) are consistent with equations (7-1) through (7-4) and were derived from observations covering 50 per cent of the logarithmic energy range. Therefore, we may, with reasonable certitude, use the above equations to obtain approximations to the energies which would be needed to produce explosion craters on earth and moon for bursts at specific scaled depths of burst. This does not mean that equations (8-1) through (8-8) can be applied directly to the terrestrial meteoritic craters and the lunar craters. It does mean that equations have been derived which give the crater dimensions that one could reasonably expect if certain energies were released explosively at certain distances below the surface of the earth and moon.

It must again be pointed out that these are extrapolations, but, in spite of these limitations, the conditions to be met may be approximated reasonably closely; and it is probable that the energies so derived are of the correct order of magnitude in the higher ranges and more accurate at the smaller craters.

It is usual to attempt to relate crater diameters (or depths) to the weight of TNT equivalent, or energy, by an equation of the form

$$\text{Diameter} \sim W^{1/y} . \tag{8-9}$$

In Lampson (12) scaling, y would be exactly 3. Johnson (13) finds $y = 3.4$ from large chemical and nuclear explosions. The results of the present investigation show that y cannot be a constant over the entire range of crater diameters studied. It varies from approximately 3.2 at the smallest craters studied to 3.4+ in the range of diameters from 3,000 to 10,000 feet and up to 3.6+ at a 10-mile crater, then very slowly drops as the point-source type of meteoritic explosion gradually shifts to a surface-source type of burst. It would go to more than 3.8 for very large point-source bursts, while at Mare Imbrium the denominator appears to be about 3.38 as y becomes smaller at the larger craters of the surface-burst regime.

The first three values were computed for a scaled depth of burst of $H/W^{1/3} = 0.10$ (eq. [8-3]). Values for larger craters were taken from Tables 18 (p. 177) and 30 (p. 315). The use of the exponent, 1/3, in deriving equations (8-1) through (8-8) was made necessary by the fact that, until equations accurately relating E and D were found for specific conditions, the exponent would not be known. Consequently, a scaled depth of burst of $H/W^{1/3} = 0.10$, for example, does not mean the same thing for large craters as for small, but, as defined, the equations are simple and usable.

If E is the logarithm of the number of calories released according to equation (8-3) and $E = 12$ corresponds to 1 metric kiloton of TNT equivalent, then the apparent crater which would be produced by a kiloton of TNT equivalent detonated at $H/W^{1/3} = 0.10$ would be 287 feet wide on the average. Thus

$$\text{Diameter (feet)} = 287 \text{ feet } \frac{W_{KT}^{1/y}}{(1\text{KT})^{1/y}} \qquad (8\text{-}10)$$

or

$$\text{Diameter (meters)} = 87.5 \text{ meters } \frac{W_{KT}^{1/y}}{(1\text{KT})^{1/y}}, \qquad (8\text{-}10\text{A})$$

where

$$y = \frac{E - 12}{D - 2.4584}. \qquad (8\text{-}11)$$

with E as logarithm of the number of calories and D as logarithm of apparent diameter in feet. At 1 KT, $y_0 = 3.336$.

In the metric system

$$y = \frac{E_e - 19.6218}{D_m - 1.9424}, \qquad (8\text{-}11\text{A})$$

where E_e is the logarithm of the number of ergs and D_m is the logarithm of the apparent diameter in meters.

It has already been shown that equations (7-2), (8-3), and (8-4) will represent the terrestrial meteoritic craters well, and they should be used in that range of energy. It has also been shown that the smaller lunar craters are rather closely represented by the same equations but that the larger lunar craters do not agree well with equation (7-2), and the trend is toward, or even beyond, equation (7-1) at the high end. Fortunately, it can be shown that there is a physical reason for this shift in the D versus d relationship among the larger lunar craters.

Let us first assume that equations (7-2), (8-3), and (8-4) are valid throughout their entire observable range. Let us further assume that the meteorite is composed of nickel-iron of density 7.9 grams per cubic centimeter. A scaled depth of burst of $H/W^{1/3} = 0.10$ is postulated.

Table 14 then gives the logarithmic energy and the logarithmic mass and the diameter for meteorites capable of producing certain craters for four different velocities of impact. The data of this table are derived from equation (8-3), which relates apparent crater diameter to explosive energy for a scaled depth of burst of $H/W^{1/3} = 0.10$.

Let us assume an impact velocity of 10 miles per second. At this scaled depth of burst, this implies that the center of the nickel-iron meteorite penetrated the ground to a depth of 2 of its own diameters and then exploded. This does not mean that the equations say that such meteorites will or can penetrate this far but simply implies that if they did, the results would be as indicated if the explosion were approximately from a point source.

Energies Needed To Produce Meteoritic Craters

The meteoritic diameters so determined range from 157,000 feet, or about 30 miles for one capable of producing a 189-mile crater, to one 7 inches in diameter capable of producing a 31.6-foot crater.

The left side of Figure 29 is a key figure in determining the actual scaled depth of burst of those explosions which actually produced the lunar and terrestrial craters. Seven craters are shown, all reduced to the same width in the drawing. Each crater is drawn to actual scale from the data from equation (7-2). The top drawing is a small crater 1,000 feet across, or $D = 3.0$. The others are successively larger to the last, which is 189 miles across, or $D = 6.0$. The decreasing relative depth with increasing size is shown, as is the decreasing

TABLE 14*

SIZES OF NICKEL-IRON METEORITES WHICH COULD PRODUCE CERTAIN CRATERS IF THEY EXPLODED AT A SCALED DEPTH OF BURST OF $H/W^{1/3} = 0.10$
(POINT-SOURCE MODEL)

D	E	LOGARITHMIC MASS (POUNDS) FOR VELOCITIES OF				METEORITIC DIAMETER IN FEET FOR VELOCITIES OF			
		2 m.p.s.	10 m.p.s.	25 m.p.s.	50 m.p.s.	2 m.p.s.	10 m.p.s.	25 m.p.s.	50 m.p.s.
1.5	8.88	3.13	1.73	0.94	0.33	1.7	0.6	0.3	0.2
2.0	10.48	4.73	3.33	2.54	1.93	5.9	2.0	1.1	0.7
2.5	12.13	6.38	4.98	4.19	3.58	21	7.2	3.9	2.5
3.0	13.83	8.08	6.68	5.89	5.28	78	26	14.4	9.1
3.5	15.55	9.80	8.40	7.61	7.00	290	99	54	34
4.0	17.35	11.60	10.20	9.41	8.80	1,160	395	215	135
4.5	19.18	13.43	12.03	11.24	10.63	4,700	1,610	875	550
5.0	21.09	15.34	13.94	13.15	12.54	20,400	7,000	3,800	2,400
5.5	23.06	17.31	15.91	15.12	14.51	92,500	31,600	17,200	10,800
6.0	25.147	19.397	17.999	17.203	16.60	459,000	157,000	85,200	53,700

* D = logarithm of apparent crater diameter; E = logarithm of needed energy (calories) derived from eq. (8-3). Mass is the mass in pounds of the nickel-iron meteorite having the kinetic energy indicated in the second column at selected velocities.

relative rim height. The rim width remains proportionately constant. For each crater, the size of the meteorite which could produce the crater is pictured at a distance below ground level for the meteoritic center of 2 diameters.

The relative size of the needed meteorite is shown, and it is clearly evident that the larger craters require proportionately larger meteorites for their genesis. This increasing size of the meteorites needed for large craters is a measure of the failure of the Lampson type of scaling (12). This trend is real.

It will be noted that, for the smaller craters, the meteorite which is 2 diameters down is still well above the crater bottom. This would be even more obvious at still smaller, not charted, terrestrial meteoritic craters. But, as the crater size increases and the meteorite size increases, a spherical meteorite 2 diameters down would first touch bottom and then for larger craters would be completely buried beneath the visible crater bottom.

Reference to Figure 14 (p. 129) will show that craters of the type where the explosive energy is released below the bottom of the resulting visible crater can occur; but, in every case, this requires a scaled depth of burst of considerably more than $H/W^{1/3} = 0.50$. A penetration of 2 diameters at 10 miles per second implies a scaled depth of burst of $H/W^{1/3} = 0.10$; and in all except the very largest craters, when the scaled depth of burst is less than $H/W^{1/3} = 0.50$, the explosive focus is above the crater bottom, and for $H/W^{1/3} = 0.10$ it is far

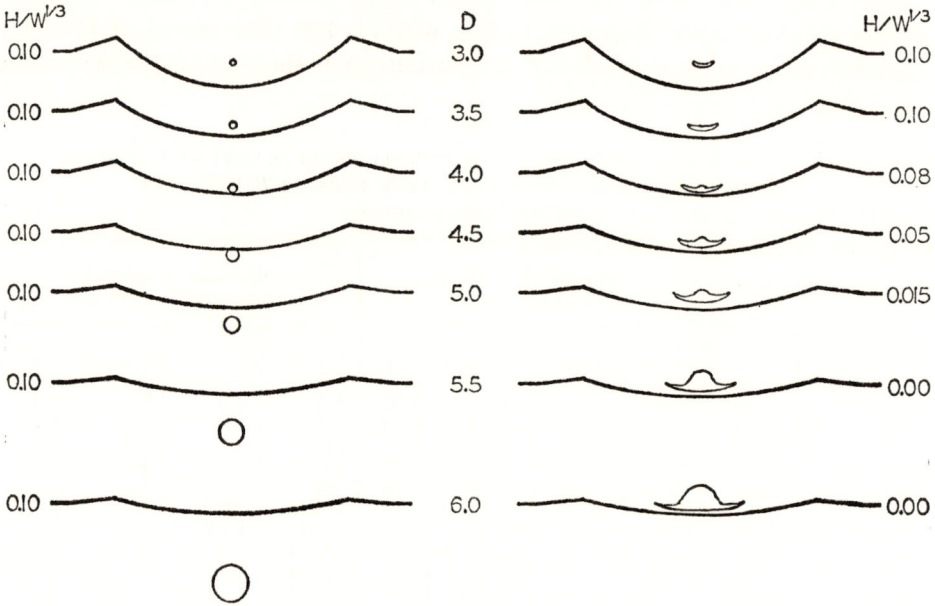

Fig. 29.—Illustration of relationship between meteoritic size, penetration into ground, and crater size (point-source model). *Each panel:* seven craters ranging from 1,000 feet to 189 miles in diameter. The meteorites are in correct proportion to craters. *Left panel:* meteoritic penetration of the 2 diameters shown for scaled depth of burst of $H/W^{1/3} = 0.10$ at 10 m.p.s. *Right panel:* realistic representation of meteorite at approximately maximum penetration, showing flattening.

above the crater bottom. It may confidently be stated that the scaled depth of burst for the top two craters of Figure 29 cannot be more than about $H/W^{1/3} = 0.10$. For the third crater, the scaled depth of burst cannot be greater than $H/W^{1/3} = 0.10$ and more probably is of the order of $H/W^{1/3} = 0.08$. For all larger craters, the minimum scaled depth of burst must approach closer to zero as the crater size increases. At the $D = 5.5$ or 60-mile crater, the scaled depth of burst must be essentially equal to zero and similarly for larger craters.

Here another phenomenon is made manifest. At the 60-mile crater, the meteoritic diameter is twice the true crater depth. Hence, if the burst were a surface burst, the meteoritic bottom (assuming no deformation) would be at the final crater bottom. This does not seem possible. At the 189-mile crater, the undeformed meteorite would extend well below the final crater bottom.

An explanation such as the following seems in order. The velocities of small meteorites are so high in terms of their diameters that they are stopped in the ground very quickly and almost as a unit. The velocity of a shock wave in the ground or meteorite is roughly the same as the impact velocity. If the impact velocity is 10 miles per second, the shock wave would go completely through a small body before it had penetrated far into the ground; but at 10 miles per second, the 189-mile crater requires a meteorite of the order of 30 miles in diameter. It would require at least three seconds for the shock wave to travel through this body. Hence the explosion would last for at least 3 seconds, and, for all of that time, the back of the meteorite would not know that the front end had struck and was being violently slowed. The necessary corollary to this is that the larger meteorites will flatten out on contact and will maintain the explosion center close to the ground level. Others, such as Urey (14) and Öpik (8), have mentioned this flattening-out before but have not connected it with a determination of the location of the explosive focus and the scaled depth of burst. Shoemaker (4) also discussed this phenomenon but limited his discussion to terrestrial meteoritic and man-made explosion pits.

Reference to Table 9 (p. 140) indicates that, for impacts of velocity higher than 10 miles per second, the scaled depth of burst for any reasonable penetration, such as zero to 2 diameters, also approaches zero, or the burst center comes closer and closer to the ground level. Hill and Gilvarry (15, 16) have theoretically demonstrated that penetration of the order of 2 meteoritic diameters or less is all that can be expected.

If the impact velocity is less than 10 miles per second, the required size of the object needed to produce a given crater becomes larger, and the above arguments are even more valid. The explosion will last longer, because the shock-wave velocity will be less and the diameter of the impinging body will be greater. Low-velocity impacts will produce explosions significantly different from point-source energy releases, but, in any event, for the larger objects, the center of the explosion will be close to, at, or possibly just above, the ground level. The equations for a zero scaled depth of burst should reasonably approximate the conditions of the large-scale, low-velocity impacts.

It is realistic to think that small meteorites penetrate a short distance into the ground. The Oesel craters show this for certain, but, as the meteoritic diameter increases, the penetration probably approaches some sort of maximum in terms of absolute penetration or in terms of meteoritic diameter. When we say that an explosion lasts for several seconds, it implies that the explosion begins with the instant of impact. Materials must be ejected horizontally from the impact region at the bottom of the meteorite. In addition, the back sections of the meteorite will slide off the rising shock wave, and this section will tend to radiate more or less toward the horizontal. When the shock wave reaches the back surfaces of the body, a rarefaction or tension wave will be reflected, and fragments will be spalled off, again more or less horizontally. The momentum of the meteorite will be transmitted to the moon largely in the direction of motion of the meteorite. The explosive forces will act effectively more and more toward a horizontal plane as the size of the projectile increases. Perhaps this is one reason why the percentage of available energy actually utilized in forming such craters increases with increasing crater size. The tiny meteorite is almost instantaneously stopped, it explodes as a unit, and blows a thin surface layer above the burst up into space or air and wastes a great part of its energy against the non-resisting open surface. The large impacts send a bigger proportion of the available energy directed downward and particularly sideward into the ground.

It may also be concluded that if, as suggested here, the central peaks of lunar craters are rebound phenomena, the deep explosions with a scaled depth of burst of more than $H/W^{1/3} = 0.50$ are excluded. In such cases the rebound would exist, but it would be buried under a tremendous amount of loose material left in the crater. In order for a rebound to be visible, it must come from a very shallow explosion from the impact of a high-momentum but low-penetration meteorite.

The moon is a relatively small body, and it is conceivable that the surface curvature in the region occupied by a large crater would cause a crater to be of a different depth from an otherwise identical crater formed on a plane. This effect can be checked on the assumption that the crater on the moon is reduced in depth by the same amount as the center of a chord lies below its arc when the length of the chord is equal to the crater diameter. Table 15 summarizes the results of this calculation where all crater dimensions have been derived from equation (7-2) for a scaled depth of burst of 0.10.

Inspection of Table 15 and comparison with Figure 20 (p. 141) show clearly that any effect of curvature of the moon's surface on crater depths is very small for craters less than 100 miles in diameter. At no point is this correction large enough to make it a principal factor in the departure of the upper end of the observed D versus d curve from the curve for $H/W^{1/3} = 0.10$. It is concluded that the effective burst center actually does approach ground level as the size of the crater increases. The curvature correction cannot be the cause of this shift.

With these points in mind, let us derive our relationships between diameter, depth, and necessary energies for terrestrial meteoritic craters and lunar craters on the point-source model. These cannot be expressed in simple quadratic form, inasmuch as the scaled depth of burst is a function of crater size. The results are summarized in Table 16. We set a scaled depth of burst of $H/W^{1/3} = 0.10$ as representing the terrestrial meteoritic craters and smallest lunar craters. For diameters above about 10,000 feet, the scaled depth of burst slowly decreases and reaches zero, or a ground-level burst at $D = 5.5$ or a 60-mile-wide crater. All data of Table 16 are derived for the lower region from equation (8-3). In the upper regions they are found by interpolation between equations (8-1) and (8-3).

This table contains the energies which would be required to produce each of a selection of terrestrial meteoritic craters and lunar craters if the meteoritic explosion is from a point source. From these energies, we may then calculate the masses of the assumed spherical bodies which might have produced the craters by exploding when they struck the earth or moon. To do this, it is neces-

TABLE 15

EFFECTS OF CURVATURE OF MOON'S SURFACE ON CRATER DEPTHS

Diameter (Miles)	D	d	Depth to Chord (Feet)	Corrected d
10	4.723	3.735	61	3.731
20	5.024	3.987	244	3.976
40	5.325	4.235	977	4.213
60	5.501	4.380	2197	4.338
100	5.723	4.560	6106	4.480
150	5.899	4.703	13747	4.564
200	6.024	4.803	24463	4.592

sary to assume a striking velocity in each case. Four such velocities were chosen arbitrarily, inasmuch as we do not know the actual impact velocities. Two miles per second is approximately the lowest velocity with which a body could strike the moon. Fifty miles per second is approximately the highest impact velocity. The objects were assumed to be nickel-iron with a density of 7.9 grams per cubic centimeter. If the meteorites were of the stony type, the diameters would have to be multiplied in each case by 1.38.

Let us now return to Öpik's hypothesis that the meteoritic crater would be formed in proportion to the momentum of the colliding body and not the kinetic energy. In two of his papers (7, 8) he derives the following figures for the Arizona Meteoritic Crater. With a lip-to-lip diameter of 1,200 meters (3,924 feet) the penetration of the object would be 320 meters (1,046 feet) at an impact velocity of 16 km/sec (range 11.5–21 km/sec). The size of the meteorite would be 81 ± 8 meters (265 ± 26 feet), and the mass would be 2.2×10^6 tons, with a range of 1.6–2.9 million tons. This value of the mass is essentially one order of magnitude higher than is derived in this book. It is the highest found by anyone who has studied this problem.

His value for the amount of penetration is such as to bring the body down almost to the bottom of the lens of breccia below the floor, if the impact were vertical, or about 740 feet below ground level if the collision path were tilted 45° from the vertical. In either case the explosive focus is assumed to be from 240 to 546 feet below the original apparent crater bottom, while the lens of breccia ends about 600 feet below the original floor.

My own studies made at the crater and buttressed by the preceding discussion of the depths at which meteorites explode indicate that the burst center was about 261 feet below ground level or 239 feet above the original apparent crater floor.

Shoemaker (4), from the most detailed work at the crater, suggests 400 feet as a maximum depth of explosion, while Johnson (11) finds 210 feet. Nininger (17) also claims that the explosive focus was shallow.

It is not possible that the explosion could have occurred as far below ground as Öpik derives. The configuration of the rock layers precludes this.

TABLE 16*

Relationships between Crater Diameters and Meteoritic Parameters for Certain Lunar and Terrestrial Meteoritic Craters on the Assumption of a Point-Source Explosion Model

Name	D	Apparent Crater Diameter	$H/W^{1/3}$	E	Logarithmic Mass (Lb.) for Velocities of				Meteoritic Diameter in Feet for Velocities of			
					2 m.p.s.	10 m.p.s.	25 m.p.s.	50 m.p.s.	2 m.p.s.	10 m.p.s.	25 m.p.s.	50 m.p.s.
Clavius†	5.881	144‡	0.000	25.07	19.32	17.92	17.13	16.52	433,000	148,000	80,300	50,600
Pythagoras†	5.626	80	.000	23.95	18.20	16.80	16.01	15.40	183,000	62,700	34,000	21,400
Tycho†	5.455	54.0	.000	23.21	17.46	16.06	15.27	14.66	104,000	35,500	19,300	12,100
Aristillus†	5.268	35.1	.005	22.44	16.69	15.29	14.50	13.89	57,300	19,700	10,600	6,710
Kepler†	5.028	20.2	.014	21.44	15.69	14.29	13.50	12.89	26,300	9,010	4,890	3,080
Bessel	4.723	10.0	.033	20.18	14.43	13.03	12.24	11.63	10,100	3,460	1,880	1,180
Bullialdus F	4.422	5	.056	18.98	13.23	11.83	11.04	10.43	4,010	1,380	744	469
Piton A	4.138	2.6	.073	17.88	12.13	10.73	9.94	9.33	1,810	620	337	212
In Purbach	3.723	1	0.097	16.35	10.60	9.20	8.41	7.80	536	184	99	63
New Quebec	4.06	11,290‡	0.076	17.58	11.83	10.43	9.64	9.03	1,375	470	255	162
Arizona	3.60	4,000	.100	15.91	10.16	8.76	7.97	7.36	381	130	71	45
Wolf Creek	3.43	2,700	.100	15.31	9.56	8.16	7.37	6.76	242	83	45	28
Odessa No. 1	2.74	550	.100	12.95	7.20	5.80	5.01	4.40	39.5	13.5	7.3	4.6
Kaali Järv	2.50	319	.100	12.18	6.43	5.03	4.24	3.63	22.0	7.5	4.1	2.6
Chaglgan Toushtou No. 1	2.41	260	.100	11.84	6.09	4.69	3.90	3.29	6.8	5.8	3.1	2.0
Sikhote Alin	1.95	90	.100	10.31	4.56	3.16	2.37	1.76	5.2	1.8	0.97	0.61
Odessa No. 2	1.85	70	.100	9.99	4.24	2.84	2.05	1.44	4.1	1.4	0.76	0.48
Henbury No. 13	1.48	30	0.100	8.79	3.04	1.64	0.85	0.24	1.6	0.55	0.30	0.19

* D = logarithm of diameter (feet) of the apparent crater; $H/W^{1/3}$ = scaled depth of burst accepted as best representing data for each crater; E = logarithm of energy (calories) released by meteoritic impact. Masses in pounds are those of meteorites required to produce observed results at specific impact velocities and specific scaled depths of burst.

† The data of this table are accepted as reasonable only for those craters 10 miles in diameter and below. At the larger craters the point-source model fails and must be replaced by a surface-source model. See Table 18 for accepted data for craters larger than 10 miles in diameter.

‡ In this column the upper nine values are miles; the lower nine values are feet.

Energies Needed To Produce Meteoritic Craters

We may apply Öpik's theory to the case of the small terrestrial meteoritic craters. It will be shown later in this chapter for Odessa No. 1 that a velocity of 10 miles per second, a projectile diameter of 13.5 feet, and a depth to burst center of 30 feet are consistent with the observations. Using Öpik's formulae at the same velocity of impact, the crater diameter should be 13.9 times greater than the projectile diameter. Inasmuch as the crater diameter is 550 feet, the nickel-iron meteorite should be nearly 40 feet in diameter, and the released energy should have been 25 times greater than the value accepted in this study. Similarly, Öpik's formulae predict that the center of the burst should be 155 feet below ground level. This is completely below the top of the undisturbed materials below the crater.

Other examples could be cited, but it is clear from the above that Öpik's mechanism is not valid at the terrestrial meteoritic craters up to at least 1 mile in diameter. At the larger craters the case is somewhat different. Öpik's values for the size of the objects which produced the larger terrestrial meteoritic craters and the lunar craters are less than here predicted for diameters of point-source craters above about 6 miles at an impact velocity of 10 miles per second. It is probable that above this size the crater-forming mechanism depends on a power of the velocity which is somewhat smaller than 2 but must still be considerably larger than 1.

Bjork (37) has done a major calculation of a complete two-dimensional hydrodynamic theory of crater formation. He gets a three-dimensional picture by reason of symmetry. His results are tied in to the Thomas-Fermi statistical model of the atom at the larger end and into experimental results at the small crater end. The results appear to be reasonably good, particularly at the low end of the scale.

Shoemaker (18) has recently revised earlier work concerning the mechanics of large meteoritic impacts in rock. He emphasized that shock waves will be produced in the ground and in the meteorite and that most of the ground engulfed by shock is accelerated downward and outward away from the oncoming body and that most of the meteorite engulfed by shock is decelerated and flows in directions paralleling the flow of the underlying shocked rock. Part of the kinetic energy of the meteorite is converted into internal energy of the meteorite, but most is transferred as kinetic and internal energy to the shocked rock ahead of the meteorite.

His analysis was simplified so that he considered the meteorite as an infinite plate of iron striking the ground. Such a plate differs in important fashions from a meteorite, but Shoemaker felt that his numerical results could establish several major qualitative facts about hypervelocity impact. First, they show that, by compression alone, an iron meteorite at typical geocentric velocities will penetrate in its entirety below the surface of a target composed of ordinary silicate rocks. This is undoubtedly so for a small meteorite, but it is not at all clear that a meteorite or planetoid of such size that the shock wave takes an appreciable time to go through the meteorite will remain sufficiently close to its original form that it will penetrate completely below the ground.

Second, Shoemaker's calculations suggest that, even after reflection of a rarefaction from the back side, the meteorite will not necessarily fly apart but that the whole meteorite may continue to move into the ground. The same question applies here as at his first point. One of the difficulties is that Shoemaker's calculations were made for an infinite plate and not for a roughly spherical object which necessarily will flatten more quickly and will penetrate a lesser distance into the ground and will not have smoothly symmetrical shock waves moving through it.

Third, Shoemaker points out that a major part of the original energy of the meteorite is transferred to the shocked rock ahead of the meteorite at a very early stage of penetration; and, finally, for the velocity (15 miles/sec) and conditions illustrated, the internal energy of the meteorite never exceeds one-seventh of the original kinetic energy, and only a fraction of this internal energy will be trapped thermally. A major part of this internal energy is released by the expansion of the meteorite behind the rarefaction and contributes to the further propagation of the shock into the rock. Shoemaker (19) says:

At a more advanced stage in the opening up of the crater the shock and pattern of flow produced by impact begin to resemble more and more closely the shock and flow produced by shallow nuclear explosions. The shock propagates away from the immediate vicinity of the cavity and is followed by the rarefaction reflected from the free surface of the ground. Material engulfed by the shock is accelerated in the directions of shock propagation, which at some distance from the cavity will be approximately along the radii of a sphere. Momentum is trapped in part of the material above the rarefaction wave and it will move upward and outward, individual fragments following ballistic trajectories. As the shock engulfs an ever-increasing volume of rock the shock strength will decrease until ultimately the shock decays to an ordinary elastic wave. The margin of the crater is determined primarily by the radial distance, at the surface, at which there is just sufficient kinetic energy in the rocks behind the reflected tensional wave for fragments to be torn loose and lofted over the rim.

In effect, Shoemaker is saying that the production of a hypervelocity-impact crater differs from an ordinary explosion crater in the sense that most of the velocity is imparted to the rocks in the former case by the shock mechanism rather than by the production of a great amount of vapor. This is very probably true. Experiments with nuclear charges seem to require such a mechanism. It is not the purpose of this book to establish the theoretical means of formation of terrestrial and lunar craters. It is one of the purposes to establish limitations and empirical relationships between energy and dimensions that will define how a crater may be formed without giving a mathematical formulation of the exact mechanism.

All authorities seem to agree that the object which produced the Arizona Meteorite Crater struck with a relatively low velocity. Shoemaker (20) states:

Meteoritic iron is dispersed in the breccia (beneath the crater) chiefly as microscopic spheres in drops of sintered dolomite, which appear to be most abundant near the base of the breccia. The sintered material constitutes not more than a few percent of the breccia.

And also (21):

From the fact that much of the meteorite now appears to be dispersed as minute spherical drops in fragments of sintered dolomite in the breccia under the crater, it was concluded that the impact velocity was greater than 9.4 km/sec. No evidence has been recognized, however, by which it could be shown that more than a small fraction of the meteorite or rocks at Meteor Crater ever behaved as vapor.

The samplings of Rinehart (22) and Nininger (23) indicate that at least 12,000–15,000 and probably considerably more tons of meteoritic materials were vaporized, then condensed, and *still exist* in the general neighborhood of the crater in the form of these tiny spherical drops. How much more was immediately oxidized or was condensed into droplets too small to fall near the crater or has been oxidized in the thousands of years which have elapsed is not known, but in all probability the amount was very large.

The percentage of the nickel-iron still remaining as droplets beneath the crater must be considered to be a relatively small part of the original body. Droplets of this type have been found at the Wabar Craters and at Odessa and others. It is extremely doubtful that all these objects struck within such a narrow velocity range that the impinging mass liquefied but did not vaporize. The nature of the droplets and their wide distribution clearly point to the fact that a moderately high-velocity meteorite will almost completely vaporize and that the craters produced by such impacts owe much of their size to a true explosive action supplemented by a shock mechanism of ejection.

Shoemaker's conclusion—"In the final analysis we should not expect an especially close relation between the depth-diameter ratios of craters produced by high explosives and impact craters. Because a large amount of gas is produced, the cratering mechanics of high-explosive detonation are substantially different from the cratering mechanics of high-velocity impact" (24)—does not appear to be correct. The relationships between diameter and depth of explosive and meteoritic and lunar craters do form consistent families. The two types of craters cannot be distinguished from their dimensions in the range of craters with which we are familiar on the earth.

Shoemaker's excellent study (25, 26) of the ejectamenta from Copernicus shows that the very great majority of masses expelled from the explosion site were moving at velocities of 0.4 km/sec or less. Such a velocity is many times less than the velocity of expansion of a gas ball at a high-explosive, nuclear, or meteoritic-impact explosion.

Regardless of the mechanism by which the kinetic energy of the meteorite is transformed into lateral motion of surface materials, the efficiency turns out to be essentially that of the pure gas ball, and equations derived from experiments with gas balls will yield results that can be applied with relatively little error to the formation of explosion pits, terrestrial meteoritic craters, and lunar craters at least up to the size of Copernicus.

Indeed, it will be shown in the chapter on rays that the explosive gas-ball mechanism will give a better approximation to the velocity distribution of materials ejected from small lunar craters than does the pure shock-wave mechanism.

Regardless of the nature of the mechanism which produced the larger craters, the energy limitation derived earlier in this chapter must be respected. The minimum logarithmic energy needed to produce a $D = 6$ crater at a scaled depth of burst of $H/W^{1/3} = 0.10$ at an efficiency of 50 per cent was found to be 23.241 (log calories). On the point-source model the most probable value for the logarithmic energy was found to be 25.147.

At 10-, 20-, 30-, and 50-km/sec striking velocity, Öpik (7) would find that nickel-iron objects 96,000, 57,500, 42,500, and 33,500 feet in diameter were needed. These objects would have logarithmic kinetic energies of 24.105, 24.027, 23.987, and 24.117, respectively. These values are all more than the theoretical minimum by less than one order of magnitude but are less than required by a point-source model by more than an order of magnitude.

Now the very large craters we have been discussing are, in the main, theoretical craters. They are probably quite close to the craters which would be produced by tremendous point-source explosions, but they differ in a definite fashion from meteoritic-impact bursts, and the difference increases as the size of the colliding body becomes larger.

Let us assume a striking velocity of 10 miles per second and that each nickel-iron object penetrated 2 of its own diameters into the ground and that its deceleration was constant in the ground. Under these assumptions, each body would be stopped according to Table 17. The meteoritic diameters are from Table 14.

Until the duration of the object's path in the ground becomes appreciable—let us say 0.01 second—the stoppage occurs so suddenly that we have essentially a point-source explosion. Actually, of course, the small missile will flatten and turn inside out and will explode from a surface and not a point, but the difference is small for the smaller objects, and hence at the smaller meteoritic craters, up to about $D=4$ or a 10,000-foot crater, the explosions may be considered to be effectively from point sources.

At impacts of the meteorites whose diameters are measured in hundreds or thousands of feet, the back side of the meteorite continues in motion for a definite length of time before it knows that the front has struck. The front in the meantime is being violently slowed, compressed, and flattened. The net effect is that, at the larger impacts, the explosion lasts for a considerable time and is confined between the resisting ground and the advancing meteoritic rear. The meteorite is thus flattened and extruded at high velocity parallel to the ground. The penetration is not large, and the effective burst center is near ground level. Much momentum is transmitted vertically into the ground, but the major explosive force operates more nearly horizontally in a plane. The craters so produced are relatively more shallow than those from smaller im-

TABLE 17

TIME NEEDED TO STOP 10-M.P.S. METEORITES WHICH PENETRATE 2 DIAMETERS INTO THE GROUND

D	Meteoritic Diameter (Feet)	Time to Stop (Seconds)	D	Meteoritic Diameter (Feet)	Time to Stop (Seconds)
1.5	0.6	0.000045	4.0	395	0.0299
2.0	2.0	.000150	4.5	1,610	0.122
2.5	7.2	.000545	5.0	7,000	0.530
3.0	26	.00197	5.5	31,600	2.39
3.5	99	0.00750	6.0	157,000	11.9

pacts, and, because the explosion acts more nearly horizontally, a given crater diameter will require less energy for its formation than is released by a point-source model which yields a deeper crater with the same diameter. At the very large lunar craters and circular maria the explosions must have lasted many seconds.

It is probable that a mechanism of this sort requiring smaller energies than point-source models in the range of possibly $D=4$ to $D=6$ or more, and where the burst center approaches ground level at the larger bursts, is the reason why the lunar craters of Class 1 are suddenly seen to become relatively more shallow above $D=4.75$ (see Fig. 20, p. 141). It is an observed fact that the larger lunar craters of Class 1 become systematically more shallow than a simple quadratic, equation (7-2), would indicate. As this is so, they may have required less energy for their formation than a point-source model, such as was given in Table 16. It is difficult to determine the magnitude of this effect, but the relationship can be graphically shown. On Figure 26 (p. 159) the left-hand line shows the minimum energy required to produce craters of the logarithmic diameters as read on the left-hand scale, on the assumption that the efficiency was 50 per cent. The right-hand line up to the horizontal bar shows the observed logarithmic energies needed to produce the corresponding craters at a scaled depth of burst of $H/W^{1/3} = 0.10$ and point-source explosions. Above the bar the right-hand curve is extrapolated, still for point-source models, and the top point of $E = 25.147$ at $D = 6$ is reasonably close to correct for such a crater. Above roughly $D = 4$, where the impact explosions begin to last a considerable time and the point-source model begins to change to a surface-source model, a dashed line gives an approximation to the logarithmic energies needed to produce the very large lunar craters. This line has been extended up into the region of the circular lunar maria.

While there is at present no known theoretical way of defining the curve in this range, it cannot be very much in error because of the nature of the limiting conditions. The scaled depth of burst must approach and probably reach zero in this range. From this cause the energy required must be higher than for the deeper bursts. The observed craters become relatively more shallow than those from deeper bursts. This leads to the conclusion that less material was displaced and hence less energy was required. The latter effect must be dominant.

The dashed line on Figure 26 permits us to complete the empirically determined relationship between meteoritic size, impact velocity, and crater diameter. All data in Table 16 from Bessel downward are considered to be representative and are accepted. The point-source model is reasonably accurate in this range.

For craters larger than $D = 4.75$, the straight line,

$$D = 0.3284E - 1.9041, \qquad (8\text{-}12)$$

is accepted as a good first approximation for the relationship between the logarithmic crater diameters (feet) and the logarithmic energies (calories) of the great meteorites which produced the larger lunar craters. This represents approximately the surface-source model which must obtain at these large impacts. In the metric system, we have,

$$D_{km} = 0.3284 E_e - 7.9240, \qquad (8\text{-}12A)$$

where D_{km} is the logarithm of the diameter of the apparent crater in kilometers and E_e is the logarithm of the number of ergs.

Using equation (8-12), the accepted data for the large lunar craters are listed in Table 18.

The data for Table 18 are approximately in agreement with Öpik's for the object which produced Clavius if the impact velocity were low. For all craters, Öpik's method yields larger meteorites than does the empirical method here described, and, at the sizes of the terrestrial meteoritic craters, it requires impossibly large meteorites and depths of penetration. It is apparent that a mechanism by which the crater volume is assumed to be proportional to the meteoritic momentum is not tenable, except possibly at the lunar craters of the size of the larger circular maria.

From the accepted lower part of Table 16 and Table 18, two conclusions immediately appear. First, the sizes of the meteorites necessary to produce the various craters were amazingly small, unless the impact velocity were extremely low. The great lunar crater Clavius was caused by a mass about 29 miles in diameter if it were produced by a 2-mile per second impact of a nickel-iron object and about 40 miles in diameter if it were of stone. Conversely, the necessary size of the impinging object decreases very rapidly with increasing impact velocity. If the meteorite which caused Clavius had struck at maximum veloc-

TABLE 18

RELATIONSHIPS BETWEEN CRATER DIAMETERS AND METEORITIC PARAMETERS FOR CERTAIN LUNAR CRATERS LARGER THAN 10 MILES IN DIAMETER ON THE ASSUMPTION OF A SURFACE-SOURCE EXPLOSION MODEL

NAME	APPARENT CRATER DIAMETER				LOGARITHMIC MASS (LB.) FOR VELOCITIES OF				METEORITIC DIAMETER IN FEET FOR VELOCITIES OF			
	D	Miles	$H/W^{1/3}$	E	2 m.p.s.	10 m.p.s.	25 m.p.s.	50 m.p.s.	2 m.p.s.	10 m.p.s.	25 m.p.s.	50 m.p.s.
Clavius	5.881	144	0.000	23.706	17.956	16.558	15.762	15.160	151,900	51,950	28,200	17,770
Pythagoras	5.626	80	.000	22.930	17.180	15.782	14.986	14.384	83,700	28,620	15,540	9,789
Tycho	5.455	54.0	.000	22.409	16.659	15.261	14.465	13.863	56,120	19,190	10,420	6,564
Aristillus	5.268	35.1	.005	21.840	16.090	14.692	13.896	13.294	36,250	12,400	6,730	4,240
Kepler	5.028	20.2	.014	21.109	15.359	13.961	13.165	12.563	20,690	7,076	3,841	2,420
Bessel	4.723	10.0	0.033	20.180	14.430	13.032	12.236	11.634	10,140	3,468	1,883	1,186

ity, 50 miles per second, it could have been nearly as small as 3 miles in diameter.

The energies required for the production of small craters are rather closely defined by experiment. In the middle range the scaling factors limit the energies within a small range. At the very large craters the minimum energies are known from the sizes, masses, and locations of the crater rims, while the maximum energies are established by the fact that the meteorite must be substantially smaller than the crater. For example, if the energy required to produce Clavius were higher than given in Table 18 by a factor of 10, a nickel-iron object 62 miles in diameter or a stony object 86 miles in diameter would be required for an impact velocity of 2 miles per second. The diameter of Clavius is 144 miles, and these values for the meteoritic diameter are far too large to be realistic. It is concluded that the relationships between crater dimensions and energy of impact are defined everywhere well within a factor of 10.

It is interesting to compare the energies found in this manner to be capable of producing the Arizona and New Quebec meteoritic craters with those found by different authors. In 1929, Moulton (27) derived 1.7×10^{24} ergs. He considered the energy necessary to shear the rocks at the edge of the crater, to crush and pulverize the rock within the limits of the crater, and the heating of the meteorites and adjacent masses of rock. He considered that the object was a densely packed swarm of meteorites rather than a single body. In 1943, Wylie (9), using data from chemical explosives, extrapolated a value of 9.4×10^{21} ergs as necessary to produce the Arizona Meteorite Crater. In 1949, in *The Face of the Moon* (10), I found 3.3×10^{21} ergs by a similar process. Neither Wylie nor I properly corrected for the effects of variations in depth of burst. This led to too small values of the energy.

In 1950, Rinehart (28) derived 1.4×10^{22} ergs from experiments at Aberdeen Proving Ground, which indicated that 2×10^{-2} cubic inches of rock could be displaced by the application of 1 foot-pound of energy. In 1956, Gilvarry and Hill (29), by a unique manipulation of the basic relationship between crater diameters and depths as given in *The Face of the Moon*, derived a value of 2×10^{23} ergs. In 1958, Öpik (8) found a value of 2.7×10^{24}, the highest yet estimated. Rostoker's determination (38) was nearly as high.

McPhail (30), basing his argument on the energy required for crushing and ejection of rocks, derived a value of 5×10^{23} ergs for the Arizona Crater. He used data from ore-dressing mills as to the amount of energy required to pulverize large quantities of rock. In 1959 Shoemaker (4) derived values ranging around 6×10^{22} ergs by extrapolating from nuclear explosion craters. In 1961 Innes (31, 32) found a value of 9.44×10^{22} ergs from the amount of work required to crush and fragment the observed amount of rock plus the assumption that the partitionings of the energy in impact explosions do not differ significantly from those in confined nuclear explosions. Also in 1961, Bjork (37) made his theoretical study of the cratering process on the assumption that the process is hydrodynamic in nature. His work as given in the preliminary report suggests that the meteorite had a mass between 30,000 and 194,000 tons, the range being due to the uncertainties in the impact velocity, 72–11 km/sec. The analysis in this form implies that the crater-forming ability of a meteorite is essentially proportional to its momentum and not to its kinetic energy. It is encouraging to note the close agreement with the values given below, which resulted from an entirely different approach to the problem. In the present book, the figure 3.4×10^{23} ergs is derived at a scaled depth of burst of $H/W^{1/3} = 0.10$.

For an impact velocity of 10 miles per second, which seems reasonable for the Arizona Meteorite Crater, the diameter of the nickel-iron meteorite is found to be 130 feet, and the mass 288,000 tons.

The corresponding energies necessary to produce the New Quebec Crater,

Energies Needed To Produce Meteoritic Craters

according to Gilvarry and Hill (2), are 3 or 4×10^{24} ergs, while Innes (31, 32) finds 1.95×10^{24} ergs. The present data suggest 1.7×10^{25}. The crater had not been discovered when several of the other estimates had been made concerning the Arizona Crater. The New Quebec Crater was not formed by a surface burst, but if it had been so formed, the energy here computed would have been 2.5×10^{25} ergs, which is thus an upper limit.

Innes has also derived the requisite energy for the Odessa No. 1, Holleford, Brent, and Deep Bay craters. They are, respectively, 2.19×10^{20}, 6.49×10^{23}, 2.06×10^{24}, and 8.69×10^{25} ergs. The corresponding results calculated from the equations of this chapter are 3.72×10^{20}, 3.67×10^{24}, 1.69×10^{25}, and 2.04×10^{27} ergs.

There is a progressive and systematic discrepancy between Innes' figures and mine. The ratios of energies are 1.70, 5.65, 8.20, and 23.5, with Innes' values being smaller. Such a progression, carried up into the range of the largest lunar craters, would lead to Innes' formulae, yielding energies so low that the crater materials could not be transported into the rim.

It must again be pointed out that the energy equations are extrapolations, that we do not know the nature of the lunar surface, and that we do not know the exact scaled depth of burst; but, in spite of these limitations, the conditions to be met may be approximated reasonably closely, and it is probable that the energies so derived are essentially correct.

The data of Table 16 suggest that the objects which produced the New Quebec and Arizona craters penetrated the ground so that the explosion centers were about 714 and 261 feet below the surface. Johnson (13) finds 540 and 210 feet, respectively, for the two craters. Shoemaker (4) calculated that the Arizona meteorite exploded 400 feet down.

Over the years the formation of meteoritic craters has been investigated by two different methods. Most studies have compared cratering by impact with cratering by chemical or nuclear explosion. Others have treated the problem mainly in terms of flow of incompressible fluids, a method that has been used successfully to analyze the penetration of metal targets by very high-velocity metallic jets.

The theory of the latter method as developed by Öpik (8) has proved to be rather insensitive to variations in striking velocity. He has found an energy for the Arizona meteorite considerably too high. As treated by Shoemaker (4), the success of the method depends on the exponents by which crater diameters are linked to energy expended. This exponent cannot be determined theoretically. It is a variable which can be determined only by experiment.

Any method which does not determine the exponent over a wide range of energies and does not take into account the changing relationship between diameter and depth of craters and the physically possible range of depths of burst for all sizes and velocities of the meteoritic bodies can yield results valid for only a narrow range of energies. Extrapolations in either direction will be increasingly in error.

At the present state of knowledge, the method used in this book is the only one that appears to be capable of defining the entire sequence of large and small craters and the energies needed to produce them.

The theoretical discussion of Gilvarry and Hill becomes indeterminate for impact velocities less than about 6 miles per second. For velocities above this approximate limit, masses certainly will explode on impact. Allen, Rinehart, and White (33) have accelerated steel pellets with masses of a few grams to velocities of 3.6 miles per second. They did not explode on impact but were heated and were seriously deformed. Evidently, the critical velocity is of the order of 4–5 miles per second.

Even though a mass strikes with a velocity somewhat below the critical velocity, it will produce a crater very similar to one formed by an application of equal explosive energy. Consequently, we may expect that the atmosphere of the earth will slow smaller masses below the critical speed; and yet these objects may still produce the usual crater forms. As an example, reference is made to the small Henbury No. 13 crater. This structure was 30 feet across and about 10 feet deep. In the crater, 10 feet down, were found four masses of nickel-iron totaling 441 pounds. A mass of 441 pounds moving at a certain speed possesses a definite amount of kinetic energy. This relationship is shown in the lower of the two ascending curves of Figure 30, where $E = \log$ kinetic energy in calories is the ordinate and velocity is the abscissa.

The lower descending curve, which intersects the E versus velocity curve for Henbury No. 13, gives the relationship between E and scaled depth of burst for a crater the size of Henbury No. 13. The abscissae of these two curves have no immediate obvious relationship; yet they tell us that the maximum possible scaled depth of burst for this crater is 0.13 and probably less, or around 0.10. The maximum possible impact velocity is about 5 miles per second and is probably around 3 miles per second. The mass did not explode; yet a crater was formed. In a slightly different approach, we may enter Table 16 with log mass $= 2.644$ for Henbury No. 13; and we can, by interpolation, conclude that this meteorite struck with a velocity of 3.2 miles per second.

This tiny projectile struck the earth and immediately began to slow down. It reached zero velocity in 10 feet. A rough approximation can be made to the interpretation of events, if we assume that the velocity reduction was linear, that is, that it lost one-tenth of its striking velocity for every foot it penetrated vertically. On this assumption, half its kinetic energy had been dissipated by the time the meteorite had penetrated 2.2 feet. This depth represents a rough approximation to the center of the energy release, or the center of the "explosion." The depth of burst corresponding to the most probable figure derived from Figure 30 is 1.3 feet, which checks reasonably closely.

The only other meteoritic crater to which we can apply this method is Odessa No. 2. This crater represents a transition type. The crater is larger than Henbury No. 13; and a form of explosion did take place, for the meteorite was highly fragmented. The component parts were found buried in the sides and bottom of the crater. An estimate of 6 tons of meteoritic material has been made, but this seems somewhat excessive. The upper two curves of Figure 30

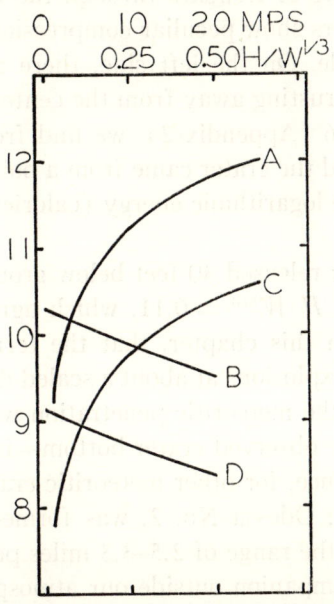

Fig. 30.—Determination of scaled depth of burst for two terrestrial meteoritic craters. Logarithmic kinetic energy (E) as a function of velocity for: A: 6-ton meteorite (Odessa No. 2); C: 441-pound meteorite (Henbury No. 13). Logarithmic kinetic energy (E) as a function of scaled depth of burst for: B: Odessa No. 2; D: Henbury No. 13.

are for Odessa No. 2. The ascending curve shows the increase in kinetic energy for increasing velocities of a 6-ton mass. The descending curve shows the relationship between log energy and scaled depth of burst for a crater of the size of Odessa No. 2. As at Henbury No. 13, the impact velocity must have been low, 5 miles per second or less. The effective scaled depth of burst was also low, of the order of 0.10 or less. Even if the mass at Odessa No. 2 were only 3 tons, the above conclusions are not changed. As at Henbury No. 13, we may enter Table 16 with log mass $= 4.078$, and, by interpolation, we find that the meteorite struck with a velocity of 2.5 miles per second. If the mass were as low as 3 tons, or $\log M = 3.778$, the striking velocity is found to be changed only to 3.3 miles per second.

These two examples show that modest masses, striking at 5 miles per second or less, will produce normal meteoritic craters, even though they do not vaporize or explode violently on impact. In no other terrestrial meteoritic craters do we know the mass of the meteorite. The Sikhote Alin craters may be exceptions to this statement, but the published data do not afford the information.

Of all the remaining terrestrial meteoritic craters, the published data permit the determination of the scaled depth of burst only for Odessa No. 1 and Oesel No. 4. Figure 2 (p. 20) shows a cross-section of the former crater as determined from actual digging of trenches through the rim and sinking of pits in the crater. The rock layers show peculiar compression effects on both sides but greatest on the right side. On the left side, there are two nearly horizontal faults which show overthrusting away from the center. Using the crater dimensions as given in Table 6 (Appendix 2), we find from the thrust planes that the forces which produced the crater came from a point about 30 feet below the normal ground level. The logarithmic energy (calories) derived for Odessa No. 1 is 12.95 (Table 16).

This amount of energy released 30 feet below ground level corresponds to a scaled depth of burst of $H/W^{1/3} = 0.11$, which agrees very closely with the observed fact, derived in this chapter, that the terrestrial meteoritic craters seem to be formed from explosions at about a scaled depth of burst of 0.10. The alternate solution—that the meteoritic penetration was so great that the burst center was well below the observed crater bottom—is completely ruled out for this crater and, by inference, for other meteoritic craters.

The companion crater, Odessa No. 2, was formed by a smaller meteorite traveling at a velocity in the range of 2.5–3.3 miles per second. The velocity of the main body and its companion outside our atmosphere must have been the same, but the smaller body would have been slowed down more by air resistance than would the larger object. Most of this slowing down occurred in the terminal parts of their paths; hence, even if the smaller object struck at a considerably lower velocity than the primary, the effect of gravity could not shift the impact points far apart. The time that either object was in the dense part of the atmosphere was relatively short, a time measured in seconds.

The major object thus struck at a higher velocity than the smaller body, and it might have been going as fast as 10 miles per second. Because it exploded violently, it certainly was going at least as fast as 6 miles per second. At 10 miles per second, the nickel-iron meteorite would have been 13.5 feet in diameter. It burst with its center 2.2 diameters below ground level, and, if the angle of fall were 45°, the actual penetration would have been 3 diameters. If the velocity had been less, the meteoritic diameter must have been greater and the absolute penetration the same as before, but the relative penetration would have been smaller.

A similar analysis may be made of the 65-foot crater No. 4, on Oesel, excavated by Reinvaldt (34–36). His cross-section, Figure 4 (p. 33), suggests that it was about 10 feet deep. Equation (8-3) indicates that the logarithmic energy was about $E = 9.87$. The crater was formed in soil above solid rock. The crater center immediately below the bottom shows a funnel-shaped hole. The explosion center must have been above this funnel. Consequently, the scaled depth of burst must have been less than 0.30. This is consistent with the conclusions reached in Chapter 7.

The data from these four small impacts (Henbury No. 13, Odessa Nos. 1 and 2, and Oesel No. 4) are remarkably consistent and rather closely define the conditions for formation of small meteoritic craters from low-velocity impacts. It would be very interesting if we had similar observations on a high-velocity impact.

A legitimate challenge may be raised against equations (8-1) through (8-8) on the grounds that they have been derived for "average soil conditions." This is true, and obviously some of the terrestrial meteoritic craters are formed from various rocks. Certainly, a small charge of explosive set off in a tiny shaft in a hard rock will not form a crater. Equally true, a charge detonated in very porous materials will yield somewhat different results than in soil.

Lampson (12) studied five different soil types and established constants for each, which could be substituted in his equations to yield crater dimensions for specific explosive charges at specific scaled depths of burst. He found, for a given charge and depth of burst, that the following soil types gave increasingly larger craters: loess, loam, silty clay, unsaturated clay, and saturated clay. The range of sizes was less than a factor of 1.5 over the entire range of soils. Army data indicate a maximum variation in crater diameter due to differences in soil characteristics of less than a factor of 1.25. This would introduce an error into the energy equations by a factor of, roughly, 3.

Lampson noted that clay soils yielded larger craters than did loam. Conversely, in the clays of Kent and Ottawa counties, in Michigan, my own experiments indicated that sandy loam gave larger craters than clay for equal expenditures of energy. My experiments with clay were usually made with smaller charges than Lampson used. Small to medium bombs exploded in chalk yield smaller craters than do similar bombs in soil; yet, by the time the great military mine crater, La Boisselle (No. 330) is reached, it falls nicely on the derived curves. Johnson (13) says that craters produced in basalt by 40,000-pound charges of TNT are about 20 per cent smaller than if they were produced in desert alluvium. Crater No. 350 formed in sandstone is slightly smaller in diameter than such an expenditure of energy would call for, but the depth is normal for a burst in soil.

From the foregoing, it seems probable that the effect of soil and rock nature on the dimensions of craters produced by specific charges of explosive is a variable which is greatest for low energies and decreases rather rapidly for increasing explosive violence. The effect of soil and rock nature on crater diameter may be somewhat slower in dying out than the effect on depth; but, in any case, by the time an energy of 10^{13} calories ($E = 13$) is reached, the effect appears to die out.

The relationships between log energy and crater dimensions are thus families of curves approaching coincidence at higher energies. The curves derived for "average soil conditions" are somewhere in the middle range and will thus be used as approximately typical.

This effect could have been predicted from the work of Gilvarry and Hill (2). As the violence of the explosion increases, the state of matter behind the shock front comes to resemble more and more a perfect fluid, and the initial characteristics of the material of the ground and meteorite have less and less bearing on the problem.

It is believed that the various equations and tables derived are reasonably valid and useful over a range of crater diameters from less than 3 inches across to well beyond 100 miles across.

Inasmuch as the energy equations were derived from log energies lying between 2.88 and 12.5 and were extrapolated to $E = 25.2$, the equations apply with decreasing rigor in the range of the larger lunar craters.

In 1936, the earth came perilously close to the tiny asteroid Hermes. If Hermes had struck the land area of the earth with a velocity of 25 miles per second, it would have duplicated the lunar crater, Kepler.

References

1. GOLD, T. Public lecture, University of California, Los Angeles, 1955.
2. GILVARRY, J. J., and HILL, J. E. "The Impact of Large Meteorites," *Ap. J.*, **124**, 610, 1956.
3. ———. "The Impact Theory of the Origin of Lunar Craters," *Pub. A.S.P.*, **68**, 402, 223, 1956.
4. SHOEMAKER, E. M. "Impact Mechanics at Meteor Crater, Arizona." Open file report, prepared on behalf of the U.S. Atomic Energy Commission, July, 1959.
5. BRICKWEDDE, F. G. *Temperature,* ed. H. C. WOLFE, Vol. **2.** New York: Reinhold Publishing Corp., 1955.
6. GILVARRY, J. J., and HILL, J. E. "The Impact of Large Meteorites," *Ap. J.*, **124**, 612, 1956.
7. ÖPIK, E. "Notes on the Theory of Impact Craters." Prepublication copy, p. 2, 1961. The substance of these notes was presented at the Cratering Symposium held at the Geophysical Laboratory, Washington, D.C., March 28–29, 1961.
8. ———. "Meteor Impact on a Solid Surface," *Irish Astr. J.*, **5,** 14, 1958.

9. WYLIE, C. C. "Calculations on the Probable Mass of the Object Which Formed Meteor Crater," *Pop. Astr.*, **51,** 97, 1943.
10. BALDWIN, R. B. *The Face of the Moon*. Chicago: University of Chicago Press, 1949.
11. JOHNSON, G. "Note on Estimating the Energies of the Arizona and Ungava Meteorite Craters," Prepublication copy, December 1, 1960.
12. LAMPSON, C. W. *Effects of Atomic Weapons*, ed. S. GLASSTONE, p. 410. Washington, D.C.: Combat Forces Press, 1950.
13. JOHNSON, G. W. Personal communication.
14. UREY, H. C. "The Origin of the Moon's Surface Features," *Sky and Telescope*, **15,** 108, 160, 1956.
15. HILL, J. E., and GILVARRY, J. J. "Application of the Baldwin Crater Relation to the Scaling of Explosion Craters," *J. Geophys. Res.*, **61,** 501, 1956.
16. ———. *Rand Rept.*, p. 801, January 27, 1956.
17. NININGER, H. H. Personal opinion expressed to me at the Arizona Crater, 1957.
18. SHOEMAKER, E. M. "Geological Interpretation of Lunar Craters," Administrative Report (for N.A.S.A.) of U.S. Dept. Interior, Geological Survey. Prepared for publication in *The Moon: Its Astronomy and Physics*. New York: Academic Press, 1961.
19. ———. *Ibid.*, p. 67.
20. ———. *Ibid.*, p. 46A.
21. ———. *Ibid.*, p. 65.
22. RINEHART, J. S. "A Soil Survey around the Barringer Crater," *Sky and Telescope*, **16,** 8, 1957.
23. NININGER, H. H. *Arizona's Meteorite Crater*. Denver, Colo.: World Press, Inc., 1956.
24. Ref. 18, p. 75.
25. ———. *Ibid.*, pp. 74 ff.
26. SHOEMAKER, E. M. "Ballistics of the Copernican Ray System," *Proceedings of the Lunar and Planetary Exploration Colloquium*, **2,** 7, 1960.
27. MOULTON, F. R. Report (mimeograph copy), dated August 24, 1929, on file in the Lowell Observatory Library, Flagstaff, Arizona.
28. RINEHART, J. S. "Some Observations on High-Speed Impact," *Pop. Astr.*, **58,** 458, 1950.
29. GILVARRY, J. J., and HILL, J. E. "The Impact of Large Meteorites," *Ap. J.*, **124,** 620, 1956.
30. Information in letter from C. S. BEALS, 1960.
31. INNES, M. J. S. "The Use of Gravity Methods To Study the Underground Structure and Impact Energy of Meteorite Craters," *J. Geophys. Res.*, **66,** 3, 1961.
32. ———. *Ibid., Contr. Dom. Obs.*, Ottawa, **5,** 3, 1961.
33. ALLEN, W. A., RINEHART, J. S., and WHITE, W. C. *J. Appl. Phys.*, **23,** 132, 1952.
34. FISHER, CLYDE. *Nat. Hist.*, **38,** 292, 1936.

35. REINVALDT, L., with LUHA, A. "Bericht über geologische Untersuchungen am Kaalijärv (Krater von Sall) auf Ösel," *Tartu Ülikooli Juures oleva Loodusuurijate Seltsi Aruanded (Sitzb. Naturforsch. Gesellsch. U. Tartu)*, **35,** 30–70, 1928.
36. ———. *Ibid.*, Separate as *Pub. Geol. Inst. U. Tartu*, No. 11, pp. 1–42, 1928.
37. BJORK, R. L. *J. Geophys. Res.*, p. 3379, October, 1961.
38. ROSTOKER, N. *Meteoritics,* **1,** 11, 1953.

SUMMARY INSTRUCTIONS FOR USE OF EQUATIONS WHICH REASONABLY REPRESENT TERRESTRIAL METEORITIC CRATERS AND LUNAR CRATERS OF CLASS 1

In each case the dimensions are statistically valid for the original craters. Where overlapping instructions are given the results will be similar for the two equations.

D = logarithmic apparent crater diameter (feet),
D_1 = logarithmic true crater diameter (feet),
d = logarithmic apparent crater depth (feet),
d_1 = logarithmic true crater depth (feet),
R_W = logarithmic apparent crater rim width (feet),
R_H = logarithmic crater rim height (feet),
E = logarithmic energy (calories),
B = logarithmic distance from original ground level to bottom of breccia lens below crater.

For craters up to $D = 4$, use

(7-2) for D versus d,
(7-6) for R_H versus D,
(8-3) for D versus E,
(8-4) for d versus E.

For craters between $D = 4$ and $D = 5.4$, interpolate between

(7-1) and (7-2) for D versus d,
(7-5) and (7-6) for R_H versus D,
(8-1) and (8-3) for D versus E,
(8-2) and (8-4) for d versus E.

APPROXIMATE RELATIONSHIP BETWEEN SCALED DEPTH OF BURST AND D

D	$H/W^{1/3}$	D	$H/W^{1/3}$
Below 3.60	0.10	4.66	0.04
3.89	.09	4.80	.03
4.02	.08	4.94	.02
4.20	.07	5.12	.01
4.36	.06	Above 5.40	0.00
4.51	0.05		

For craters larger than $D = 4.75$, use

$$(8\text{-}12) \text{ for } D \text{ versus } E,$$
$$D = 0.4831d^2 - 1.9941d + 5.4212 \text{ for } D \text{ versus } d.$$

For craters larger than $D = 5.4$, use

$$(7\text{-}5) \text{ for } R_H \text{ versus } D.$$

For all impact craters,

True diameter (at ground level) = 0.830 times apparent crater diameter, or $D_1 = D - 0.0809$.
True depth = apparent depth minus rim height.
Use (7-9) for R_W versus D,
    (7-11) for B versus d_1.

9

VARIATIONS IN LUNAR CRATERS AS FUNCTIONS OF THEIR AGES

The lunar craters of all sizes are of varying ages. Some craters are beautifully sharp and clear. They give the impression of having been formed yesterday. Excellent examples are Tycho and Aristillus. Craters of this type are called Class 1. All lunar craters tabulated have been arbitrarily divided into five classes on the basis of appearance.

Except as modified by the effects of the great dark areas, the impact craters seem to be distributed at random over the lunar disk. This distribution is random in space and in time of origin of individual craters. This does not imply that there have not been real variations in the rate at which craters were formed, but it does indicate that the craters were formed over a considerable period of time, a period which has not yet ended. All adequate studies so far made confirm the random characteristics of crater genesis.

Certain craters seem to be in alignment as though they were somehow associated. Often, but not always, in these cases the craters are found to be of different ages. Proponents of various volcanic hypotheses have avidly seized on these instances as proofs of volcanic modes of origin.

To check on the reality of these alignments and whether or not they added strength to the volcanic arguments, tests were made by firing shotguns of three bore sizes, each with four different size loads, against blank paper targets. The range to the target was such that the pellets had scattered to about a 2-foot circle. In every single case, there were numerous straight and curved lines of holes in the target, often from five or six, or even more, pellets. The similarity to the alignments of craters on the moon was striking. That the patterns of the shotgun pellets were random can scarcely be questioned.

In some cases, two or more craters on the moon probably came from the same fall. Several of the terrestrial meteoritic-crater groups are known to be from multiple impacts of swarms of meteoritic bodies traveling together through space and held close to each other by their own tiny gravitational forces. Presumably, some of the larger bodies may also have traveled in associations, and some of the lunar craters may thus be associated with their neighbors. This would be very hard to prove or disprove, but the possibility exists. Now, if craters were formed at random over the surface of the moon and were formed at different times, we would expect that some craters would be formed in the same areas as others.

If a large crater were formed last, it would obliterate all evidences of smaller craters formed in the same area, but earlier, unless the smaller crater projected out somewhat beyond the rim of the larger. A few cases of this are known.

If the large crater were formed first, then all subsequent smaller craters formed in this area would still be visible as markings which distort the larger object. Numerous examples of this sequence may be found. The relative frequency of the two types seems to be about as expected. Arthur (1) reports that "the statistical relation indicated is that the number of craters inside a larger crater is proportional to the square of its diameter, i.e., to its area. This result is precisely what we would expect of a random distribution."

The first four classes of craters as given in Table 7 (Appendix 2) are actually age classifications. The Class 1 craters are the newest-appearing craters. Classes 2, 3, and 4 are successively older in appearance, having been modified by increasing numbers of later craters. It seems to be correct that these four classes represent an age segregation, but the placing of a given crater in a given class has been done without absolute criteria. It can be only statistically correct. Class 5 craters are those which have been invaded by the dark matter similar to that of the maria and hence appear much shallower and often more deformed than craters which have not been so attacked.

Bearing these comments in mind, it is interesting to find that wide differences appear in the relationships between the crater parameters, $D = $ log diameter, $d = $ log depth, and $R_H = $ log rim height.

Figures 18, 19, and 20 show that terrestrial explosion pits, terrestrial meteoritic craters, and the smaller lunar craters of Class 1 can be perfectly represented by equation (7-2), which is for a scaled depth of burst of $H/W^{1/3} =$

0.10. The larger lunar craters depart systematically from this simple quadratic curve. Reasons for this departure have been discussed in Chapter 8.

The left panels of Figure 31 show the relationship between D and d for each of the succeeding four classes of lunar craters. In the upper panel, for Class 2 craters, almost all the points lie to the left of the curve for the Class 1 craters. The average difference in d between Class 1 and Class 2 craters is about 0.15. Similarly, the Class 3 craters are about 0.20 to the left of the Class 1 craters, and Class 4 perhaps 0.8 to the left. This means that the older classes of craters are, respectively, 21, 37, and 84 per cent shallower than the newer craters. The effect is obvious to the eye upon examination of photographs and is progressive, in that the older craters are considerably shallower than the younger craters, on the average. As might be expected, the Class 5 craters are also very much shallower than the Class 1 craters, but this is not necessarily an age effect. Some material has invaded these craters and reduced their apparent depths.

In the right-hand panels of Figure 31 are the similar charts relating the rim height as functions of apparent crater diameter. The solid line in each case is that found for the Class 1 craters. Statistically, the Class 2 craters seem to be about 0.05 to the left, at least at the larger or upper end of the curve. The Class 3 and 4 craters are about 0.10 and possibly 0.12 to the left. This is interpreted to mean that the three age groups show rims which are lower than the Class 1 craters by 11, 21, and 24 per cent, respectively. The Class 5 craters also seem to have lower rims than normal, indicating that they probably were older than Class 1 when invaded by the dark matter.

How can these data be interpreted? If we assume that all craters are formed as of Class 1—and this would be true instantaneously, even in water—then the older craters have been eroded and filled in, or they have been physically changed in shape after formation, or both effects have been operative.

Let us compare four great craters: Theophilus, Clavius, Maginus, and the great unnamed crater east of Walter, which have been assigned to Classes 1, 2, 3, and 4, respectively.

The contour chart of Theophilus, Figure 32, shows that the bottom is nowhere flat. The inner walls follow a parabolic curve down until suddenly they are interrupted by the upward sweep of the central peak. The peak in this case is tremendous. The outer walls are clear and sharp.

Clavius shows descending inner walls for roughly 15 miles, and then a rather abrupt change of slope occurs. The entire central area of Clavius is a plain whose contour approximates the curvature of the moon's surface. The form is not that of Theophilus but has been distinctly leveled out. The central peak is small, multiple, eccentric, and definite. The walls are massive but not as obvious as at Theophilus.

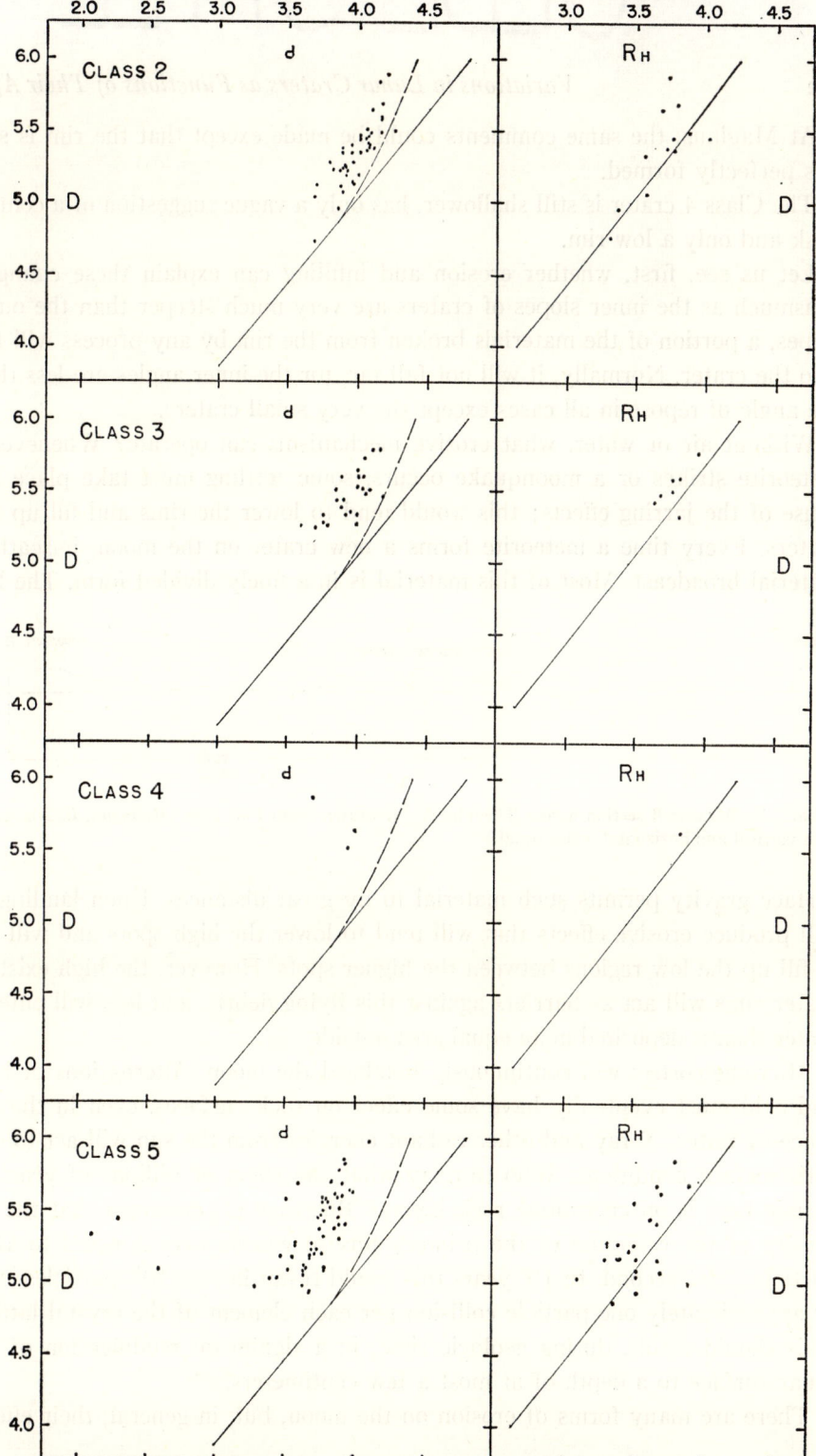

FIG. 31.—*Left panels:* relationships between logarithmic apparent crater diameter (D) and depth (d) for lunar craters of Classes 2, 3, 4, and 5. Classes 2, 3, and 4 are progressively older. Class 5 craters are lava-filled. Solid line from equation (7-2) for Class 1 craters for scaled depth of burst of $H/W^{1/3} = 0.10$. Dashed line is curve best representing larger lunar craters of Class 1 (see Fig. 20). *Right panels:* relationships between logarithmic apparent crater diameter (D) and rim height (R_H) for lunar craters of Class 2, 3, 4, and 5. Solid line from equations (7-5) and (7-6) for Class 1 craters.

At Maginus, the same comments could be made except that the rim is still less perfectly formed.

The Class 4 crater is still shallower, has only a vague suggestion of a central peak and only a low rim.

Let us see, first, whether erosion and infilling can explain these changes. Inasmuch as the inner slopes of craters are very much steeper than the outer slopes, a portion of the materials broken from the rim by any process will fall into the crater. Normally, it will not fall far, for the inner angles are less than the angle of repose in all cases except the very small craters.

Without air or water, what erosive mechanisms can operate? Whenever a meteorite strikes or a moonquake occurs, some settling must take place because of the jarring effects; this would tend to lower the rims and fill up the craters. Every time a meteorite forms a new crater on the moon, it scatters material broadcast. Most of this material is in a finely divided form. The low

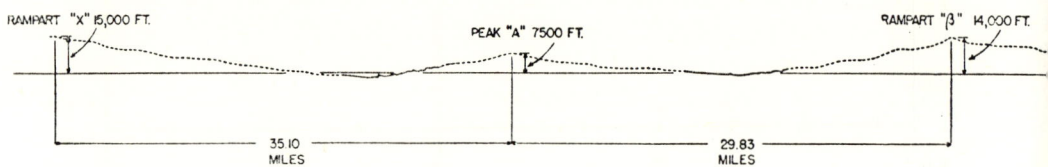

FIG. 32.—Measured section across Theophilus. (Redrawn from *Pub. Obs. Michigan*, **6**, No. 8, 67, with vertical and horizontal scales equal.)

surface gravity permits such material to fly great distances. Upon landing, it will produce erosive effects that will tend to lower the high spots and will act to fill up the low regions between the higher spots. However, the high existing crater rims will act as barriers against this flying debris, and less will enter a crater than is deposited in an equal area outside.

Micrometeorites will continuously bombard the moon. Alternations of heat and cold must eventually have some effect on rock surfaces, even in the absence of water. X-ray and other radiant energies from the sun will act to disturb surface conditions. Who can say what the effect of billions of years of cosmic rays is on crystalline rock layers? Buettner (2) assumed that the intensity of the radiation of the primary flux of cosmic rays is equal to 10^{-1} particles/cm^2/second. In 10^9 years this would result in 3×10^{15} particles/cm^2, or approximately one particle collision per each element of the crystal lattice. This should result, during geologic time, in a significant modification of the lunar surface to a depth of at most a few centimeters.

There are many forms of erosion on the moon, but, in general, their effects are small and slow. There does not seem to be any erosive process that would tend to fill up the crater faster than it reduced the height of the rim. The charts in Figure 31 show clearly that the crater depth is reduced with increasing crater age far more rapidly than the rim height is reduced.

The volume of a crater rim is proportional to the rim height. The volume of a crater is proportional to the crater depth. Let us take a $D = 5.5$ crater, which is a crater roughly 60 miles wide. The rim height of a theoretical Class 1 crater (eq. [7-5]) is 6,389 feet. The Class 2, 3, and 4 craters would show rims averaging 700, 1,300, and 1,500 feet lower than this. The apparent depth of this Class 1 crater is 18,650 feet. Statistically, the volume of the rim is observed to be 0.855 times as large as the volume of the true crater, or the volume of the apparent crater is 2.58 times the volume of the rim.

If all the materials lost from the rim were spread uniformly in a paraboloid over the low parts of the crater as an infill, the apparent crater depth would be reduced, to a first approximation, by the loss in rim height times 2.58, or 1,800, 3,400, and 4,000 feet, respectively. These values correspond to decreases in apparent depths of Class 2, 3, and 4 craters of 10, 18, and 21 per cent. As these values are very much lower than the observed reductions in apparent crater depths of 21, 37, and 84 per cent, respectively, it is concluded that erosive forces or infalling from above must be rejected as primary causes of the changes in shape of craters as functions of increasing age. Geometrically, it would seem impossible that all the lost rim material could find its way into the pit. Most of a crater's rim slopes away from the crater, not into it. This consideration increases the discrepancy.

The corollary is that most of the changes in crater shapes are real and are the results of distortions which occurred during the years after the craters were formed. The oldest, or Class 4, craters are the most distorted, and the youngest, or Class 1, craters are apparently not distorted at all except perhaps in the largest examples. It would seem logical that we are dealing here with a form of isostatic adjustment.

When a crater is formed, it suddenly puts drastic strains on a surface which formerly was in perfect balance. At large craters, there is a sudden relief of load as rock layers thousands of feet thick are removed from an area of hundreds of square miles. In the center of this area where the central peak lies, the relief is less than nearby. Surrounding the gaping hole is a rim, often thousands of feet high and miles wide, which represents a large increase in static load.

The mechanism is well known. The questions involved here are whether or not the moon's surface and subsurface layers were ever rigid enough to allow the craters to be formed and to remain and then plastic enough to permit slow adjustments to the over- and underloads on this crust. Apparently this is what happened. The rims were less high than the craters were deep. They were distributed in rings. They gradually sank somewhat over the years but never became markedly lower than originally.

The craters were round areas, and their centers were pushed up, with the central peaks becoming less prominent. Because of the nature of the forces involved, the crater bottoms rose more than the rims sank. Neither rim nor crater ever reached the original level. This is characteristic of an isostatic adjustment. When a load is superimposed or a relief of load occurs, the adjustments allow the surface to approach, but never to reach, its original condition.

Because erosive forces and infilling fail to explain the variations in crater shape with the age of the crater and isostatic adjustments appear to be consistent with these changes, the latter mechanism is provisionally accepted. If isostatic adjustments do account for these changes in crater shape, then there are three corollary conclusions which must be drawn.

The surface layers of the moon must have been hard and rigid so that the craters could be formed. The subsurface layers had little real strength. If they had been really strong, these adjustments could not have occurred. The strength of the subsurface layers increased with time. The oldest craters show the modifications of shape most clearly, and the youngest, Class 1, craters show it little or not at all.

Undoubtedly this process extended even farther back into the past than we have visible record remaining. At very early dates, the craters then formed were quickly almost erased, and the traces were often blotted out by subsequent crater formation.

It was reasoned that only a very large primitive crater could have been formed and still have remnants of its walls remain visible. A search was made for such objects, and several craters older than Class 4 were found.

Immediately west of Hipparchus is a still more archaic ring of equal size. It has been modified by the superposition of two smaller craters, each of which appears older than ancient Hipparchus. Somewhat farther west is another similar unrecognized crater. In this category are two tremendous objects extending into and under Mare Nubium. One includes the Straight Wall, and the other is east of Alpetragius. These last two have occasionally been noted before.

The prize discovery is an object which can only be considered in the category of a primitive mare, although it does not seem ever to have contained the dark material. It is about 350 miles in diameter and shows low walls which are almost complete. The central area is not appreciably depressed. It looks like the classic example of an isostatically adjusted crater which has reached its terminal condition.

The walls of this object pass just north of Almanon. Thence they go eastward to a point slightly south of Donati. Curving to the south and west, they pass northwest of Werner and Aliacensis, almost touching each. The line then goes to the north wall of Gemma Frisius and crosses the Altai Mountains northeast of Polybius. After a short arc inside the Altai ring, it curves out south of Tacitus and returns to the starting point. There is no question but that this is an ancient, gigantic crater ring. It is better seen in waning phases of the moon, and not all photographs show it clearly. It is close to the highest part of the moon's surface.

On earth it is recognized that large areas are isostatically compensated, even though extensive anomalies may exist for considerable periods of time, as witness the Himalaya Mountains. Given sufficient time, even small areas will become fully compensated. De Sitter (3) has expressed the opinion that full compensation certainly applies to areas as small as 100 × 100 km and probably to any area greater than 20 × 20 km. These dimensions are strictly comparable with the sizes of the older lunar craters, which have been observed to become more shallow and to possess lower rims with increasing age.

Haskell (4) has derived a convenient method of determining the viscosity of the asthenosphere of the earth from the measured rate of rise of the Fenno-Scandian area after the melting of the Pleistocene ice. From this he developed a formula relating the time needed for a full isostatic adjustment to the kinematic viscosity, the gravitational acceleration, and the linear dimensions of the area involved:

$$t_e \sim \frac{20\eta}{\rho g l},$$

where t_e = time in seconds, η/ρ = kinematic viscosity = 2.9×10^{21} cm^2 sec^{-1}; g = gravitational acceleration = 980 cm/sec^2; l = one-half of effective width in centimeters; η = viscosity; and ρ = density in c.g.s. units.

Use of this approximate formula applied to a crater 200 km in diameter on the moon suggests that it would become completely adjusted isostatically in about 10^6 years if the lunar viscosity were as low as that observed in the earth.

Inasmuch as the lunar craters are believed, with reason, to extend back in time to the terminal phase of the moon's period of accretion, this implies that the outer layers of the moon are now, and throughout most of its history have been, hard and rigid. Otherwise, most of the ancient lunar craters would have vanished long before now.

Conversely, there are positive evidences of modest isostatic adjustments which decrease with time, and hence the moon did become hot deep in its interior in the early stages.

References

1. Arthur, D. W. G. "The Distribution of Lunar Craters," *J. Brit. Astr. Assoc.*, **64**, 127, 1954.
2. Buettner, K. "Lunar Erosion," *Pub. A.S.P.*, **64**, 11, 1952.
3. De Sitter, L. U. *Structural Geology*, p. 333. New York: McGraw-Hill Book Co., Inc., 1956.
4. Haskell, N. A. "The Viscosity of the Asthenosphere," *Am. J. Sci.*, **33**, 22, 1937.

10

THE PROBLEM OF THE MOON'S MOTION AND SHAPE

One by one, the physical "constants," long believed to be invariant, are shown to be variable. Perhaps the most outstanding of these discoveries is that the length of the day—the period of time against which all clocks are regulated—is slowly increasing. To ascertain a change in the length of the day, observations must be made on celestial objects whose motions are independent of the rotation of the earth. The most convenient bodies are the sun and the moon, particularly the latter. A decrease in the earth's rate of rotation causes apparent secular accelerations of the sun and moon. Observations of the time of passage of the sun across the equator, when the precession of the equinoxes is known, yield directly its secular acceleration. Fotheringham (1, 2) has discussed the numerous observations of ancient Greek, Babylonian, Chinese, and Egyptian astronomers, in order to determine both secular accelerations. His decision was that, judging purely from observational evidence, the most probable values are $21''.6/(\text{Century})^2$ for the moon and $3''.0/(\text{Century})^2$ for the sun. Jeffreys (3) showed that the latter value might have to be somewhat reduced, but, even so, the values are closely limited. Schoch (4), De Sitter (5, 6), and H. Spencer Jones (7) have slightly improved the earlier results without changing the conclusions.

The lunar theory demands a secular acceleration of the moon of $12''.2/(\text{Century})^2$. The excess—about $9''/(\text{Century})^2$—is the observational fulcrum on which the lever of analysis in the brilliant hands of Darwin and Jeffreys operated to divulge so much information on the past history of the earth and moon. This excess secular acceleration, unaccounted for by gravitational theory, implies an increase in the length of the day by 1 second in the last 120,000 years.

It is usually stated that the source of this loss is not to be found in bodily tides, for the earth behaves like an elastic ball, or in tidal friction in the open sea. Taylor (8), Heiskanen (9), and Jeffreys (10) determined the rate of dissipation of energy in the tides in various shallow seas over the earth and found that the agreement between the loss observed from the motions of the sun and moon and that calculated from the water velocities was indeed excellent. The friction of water against sea bottom in the partially bounded shallow seas of the earth, where the currents are tidally induced, reacting against the ocean as a whole, is sufficient to account for the observed increase in the length of the day.

Munk and MacDonald (11) have recently reopened the question of the dynamic history of the earth-moon system. From the astronomical data they found that dissipative processes remove energy from the mechanical motion at a rate of 3.2×10^{19} ergs per second, a figure about triple that obtained by Jeffreys. The difference between the two values lies in the method of reduction, not in the data.

MacDonald (12) points out that this new value for the rate of dissipation of energy is somewhat greater than present estimates of the rate of release of seismic energy in the earth. Munk and MacDonald feel that the energy dissipated in the shallow seas is insufficient by at least a factor of 3 to account for the observed change in kinetic energy of the earth-moon system. Much of the energy must be dissipated within the body of the earth. They find that this suggestion is consistent with the observed damping of seismic waves and that the phase lags of these bodily tides, as observed at isolated continental stations, lead to the conclusion that energy is released within the body of the earth.

A loss in the angular rate of rotation of the earth is a loss in angular momentum; yet angular momentum cannot be destroyed, but only transferred. Since the moon is the most important body in producing tides, its action on the earth results in a reacting transfer of angular momentum of rotation into the moon's angular momentum of revolution. The moon is thus gradually receding from the earth, and the length of the month is slowly increasing.

This process may be reversed, mathematically, and the past history of the earth-moon system investigated. As the total amount of angular momentum in the system is constant, it can be shown that the month and the day may once have been equal at about 4.8 hours each. At that time, the moon's distance would have been about 9,000 miles, essentially at Roche's limit.

If the present angular momentum of the earth, P_E, is unity, its original angular momentum was 5,000. The present and original values for the moon, P_M, are 4.964 and 0.964. Tidal friction in the past has practically reversed the primal quantities. The value of P_M varies as the square root of the moon's distance from the earth, or

$$P_M = 4.964 a^{1/2}. \qquad (10\text{-}1)$$

These data give the relationships between the distance of the moon, the month, and the day but do not define the rapidity with which these changes occurred.

Jeffreys (13) calculated that it would have taken approximately 4 billion years for tidal friction to slow the earth down to the present observed value. This figure is in excellent agreement with the currently accepted age of the earth of 4.5 billion years.

The work of Munk and MacDonald implies that the time needed by the moon to recede from 10^5 kilometers to its present distance is 1.3×10^9 years if their estimate of the energy dissipation is correct and has been constant in the past. If non-linear terms are considered, the time of recession would be a few hundred million years. This period is so short that they suggest that the earth-moon system is very considerably younger than the earth.

The whole problem of the early history of the earth-moon system is in a very unsatisfactory state at present. If Munk and MacDonald are correct, their solution implies that the earth captured the moon relatively late in geologic history. Pending clarification of this problem, the following discussion is based on the assumption that the earth and moon are twin planets. The capture of a satellite late in the history of the earth would undoubtedly be a catastrophic event whose records would remain in the rocks of that era. No such record has been detected.

How far the moon was from the earth when it was first formed is not known. Darwin (14) carried the extrapolation backward to the point at which the month and the day were equal and then made the remarkable assumption that the moon once formed part of the earth. The slowest natural free vibration of a homogeneous earth is nearly half the original length of the day, or about 2 hours. Darwin's hypothesis was that the solar tides, coming in resonance with the free oscillations of the earth, resulted in monstrous tidal bulges, one of which ultimately broke off to form the moon. Jeffreys (3) found that closer agreement would be reached between the length of the day and the free vibration of the earth, if the earth, even then, had been non-homogeneous. Still later he had to abandon the theory (15) as contradicted by the facts, for he found that the friction engendered in the still liquid earth would have been too great to permit the resonance tidal bulges from ever reaching the necessary heights. Consequently, it is possible, but not certain, that the earth and moon are twin progeny, born nearly in contact and at the same time, but always having been separate.

In the future the moon will recede until the day and the month are again equal at 47 of our present days. The moon will then be about 340,000 miles away. This condition is not imminent, for Jeffreys calculates that it will arise in the year 50,000,000,000 A.D. Even then, the earth-moon system will not be stable, for solar tides will be acting, provided that the earth still has liquid oceans, and these tides will slowly bring the moon and earth closer together until at some time in the unimaginably distant future the moon will come within Roche's limit, be shattered, and will form a smaller, but denser, ring of fragments than that showpiece now circling Saturn.

If the increase in the length of the day had been going on in all past history at the rate determined by Jeffreys, it would have taken over 8×10^9 years to change the day from 4.8 hours to 24 hours. This is an impossibly long time. We must assume that the rate of change was more rapid in the past than at present, never less rapid, and also that the earth's beginning was about 4.5×10^9 years ago.

From theoretical considerations, Jeffreys (3) established that the loss of angular momentum of the earth was inversely proportional to the sixth power of the moon's distance. This principle allows a calculation of the moon's distance and the length of the day as functions of geologic time. The resulting equation[1] is

$$4.5a^{13/2} = T, \qquad (10\text{-}2)$$

where a is the moon's distance relative to its present distance and T is the time in billions of years since 4.5×10^9 years ago.

If the earth-moon system is essentially as old as the earth, then equation (10-2) probably represents the changes in the moon's distance during geologic time with fair accuracy. The moon, on any reasonable assumption, fled rapidly away from the earth in the early days. Even in a conservative view it reached half its present distance in about 5 per cent of geologic time; and, since the Cambrian period, which began the last ninth of geologic time 5×10^8 years ago, the moon has receded about 4,300 miles, the month has increased by only 18 hours, and the day has lengthened by about one-quarter of an hour, as Table 19 shows. Almost all the startling changes occurred in the first quarter of geologic time or early in the Archeozoic era.

These wild variations in the moon's distance have caused major changes on the earth, but, because the earth is so much more massive than its satellite, the effects on the moon were correspondingly greater. The moon was thus brought

---

[1] See Appendix 1, p. 433.

to present the same face toward the earth before the rotation of the latter had been appreciably affected by tidal friction.

The early astronomer, Cassini (1625–1712), developed three empirical laws (16) concerning the rotation of the moon:

1. The Moon rotates uniformly about an axis which is fixed with respect to the Moon itself. The period of this rotation is identical with the sidereal period of the Moon in its orbit, namely, 27.321661 days.

2. The pole of the lunar rotation z makes a constant angle (1°35′) with the pole of the ecliptic Z, which may here be regarded as a fixed point on the celestial sphere.

3. In consequence of the nearly uniform regression of the lunar node on the plane of the ecliptic and the nearly constant inclination of the lunar orbit (5°9′), the pole of the Moon's orbit P is known to describe a small circle about Z in a period of $18\frac{2}{3}$ years. The arc of a great circle zP contains also the pole Z. In other words, the planes of the lunar orbit and the lunar equator intersect on the ecliptic, the latter plane being intermediate between the two former.

TABLE 19
DISTANCE OF THE MOON DURING GEOLOGIC TIME
(Inverse Sixth-Power Law)

Distance (Miles)	Time (Unit = 10^6 Years)	Day (Hours)	Distance (Miles)	Time (Unit = 10^6 Years)	Day (Hours)
50,000	0.17	11.1	200,000	1,790	22.0
75,000	2.4	13.5	215,000	2,270	22.8
100,000	16	15.6	238,840	4,500	24.0
150,000	220	19.1			

These three observational sets may be interpreted on the basis of theory developed largely by Lagrange, Laplace, and Poisson. Summaries of their work may be found in standard textbooks (16, 17).

Cassini's first law would not be valid if the moon were homogeneous and spherical. It implies that the moon must possess unequal axes and that, if the moon is homogeneous, the axis pointing toward the earth is the largest.

The tidal bulge of the earth averages about 4 feet, very small in comparison with the equatorial bulge. The moon would possess three nearly equal axes at present if it were completely adjusted to the earth's tidal pull and its own centrifugal forces. In the case of a perfect adjustment, the moon would present an equipotential surface on which the effective gravity would everywhere be perpendicular. This is the type of surface which would be formed by a liquid.

Jeffreys (18) has solved the problem by assuming that the motion of the moon around the earth is one of steady revolution and that, as the moon always points the same face to the earth, the motion of each part of the moon is one of revolution with constant angular velocity about an axis perpendicular to the plane of the orbit and through the center of mass of the earth and moon together.

The effect of the total potential due to the earth and to the orbital motion is equivalent to a disturbing potential. He then assumed that the moon could adjust its shape to form an equipotential surface and found that the equation of the moon's surface so compensated for any distance from the earth would be

$$r = r'\left(1 + \frac{5}{12}\frac{M}{M'}\frac{r'^3}{c^3}\frac{7x^2 - 2y^2 - 5z^2}{r'^2}\right), \qquad (10\text{-}3)$$

where r = radius of the moon, r' = radius of mean equivalent lunar sphere, M = mass of earth, M' = mass of moon, and c = distance of moon in lunar radii. The semiaxes, x, y, and z, of the moon are, respectively,

$$r'\left(1 + \frac{35}{12}\frac{M}{M'}\frac{r'^3}{c^3}\right), \quad r'\left(1 - \frac{10}{12}\frac{M}{M'}\frac{r'^3}{c^3}\right), \quad r'\left(1 - \frac{25}{12}\frac{M}{M'}\frac{r'^3}{c^3}\right). \qquad (10\text{-}4)$$

These formulae give increments of $+125$, -36, and -89 feet, respectively. If the moon were completely adjusted to all gravitational potentials acting, the three axes would differ from the mean by these amounts at present. This is on the basis of a homogeneous moon. If the density varies acording to Wiechert's law (19) for the earth, the above increments would be multiplied by 0.9. Fortunately, the expressions are convenient for use in calculating the bulge existing on the moon at any past distance from the earth.

The principal moments of inertia of the moon should satisfy, to the first order,

$$\frac{C' - A'}{C'} = 5\frac{M}{M'}\frac{r'^3}{c^3} = 0.0000375, \qquad (10\text{-}5)$$

$$\frac{C' - B'}{C'} = \frac{5}{4}\frac{M}{M'}\frac{r'^3}{c^3} = 0.0000094, \qquad (10\text{-}6)$$

when $M/M' = 82$ and $c/r' = 221$. These values are in complete disagreement with the observed values. The observed value of $(C'-A')/C'$ is near 0.000629, which is about 17 times as large as theory would indicate. The observed values of

$$f = \frac{C' - B'}{C' - A'} \qquad (10\text{-}7)$$

are tabulated in Table 20. Hopmann (21) suggests that a mean value of $f = 0.721 \pm 0.012$ would be reasonably close to the correct figure. Hydrostatic theory calls for $f = 0.25$. The discrepancies found are so large that they cannot be ascribed to errors in the observations. They must have physical bases. The values quoted in equations (10-5) and (10-6) have been computed for the present separation of the earth and moon.

Laplace (29) was the first to notice this anomaly but was content to blame it on accidental distortions produced in the moon as it solidified. Jeffreys (3) realized that it could well be that the excessive bulge was a fossil tide, one formed when the moon was still plastic enough to adjust its figure to an equipotential surface. At some specific distance from the earth, the moon's layers solidified to the point where compensation no longer occurred; and, from that time on, as the moon receded, the primitive bulge remained as a fossil tide. The strength of the lunar rocks, if they at all resemble those found on earth, is amply high enough to allow the moon to maintain a permanent departure from hydrostatic equilibrium. This does not mean that the moon is solid all the way through. Because of the low gravitational pull, sufficient strength is available from the outer layers. The center of the moon may still be weak.

TABLE 20

DETERMINATIONS OF f

Observer	f	Observer	f
Stratton-Schlüter (20)	0.50 ± 0.03	Michailowski-Belkowitsch (25)	0.84 ± 0.080
Schlüter-Naumann (21)	$.71 \pm .030$	Banachiewicz-Jakowkin (21)	$.74 \pm .030$
Hartwig-Hayn (22)	$.73 \pm .070$	Jakowkin (26)	$.68 \pm .020$
Hayn (23)	$.77 \pm .040$	Hartwig-Koziel (27)	$.71 \pm .050$
Hartwig-Naumann (24)	0.71 ± 0.033	Weimer (28)	0.57 ± 0.04

With the observed fact that the $(C'-A')/C'$ ratio is 17 times the present theoretical equilibrium value, we may solve for the distance between the earth and moon, which would require the observed value of this ratio; it is found to be $17^{-1/3}a$ or $0.39a$ or 90,000 miles.

If the moon's shape is due to a frozen tidal bulge, the moon was once as close to the earth as 90,000 miles. The month would then have been 6.3 of our present days long. If, in fact, f were $\frac{1}{4}$ or near it, we would be justified in assuming that such was the case. As it is, f seems to be about 0.7–0.8 and theoretically is independent of the distance of the earth from the moon.

Jeffreys (30, 31) has also shown that if the moon's equator is approximately circular and the moon solidified considerably closer to the earth than it now is and if it were rotating freely at the time and not keeping one face always toward the earth, the ratios given in equations (10-5) and (10-6) would be considerably larger. Then, considering only the moon's rotation, the observed value of $(C'-A')/C' = 0.000629$ corresponds to a lunar day of 3.5 of our days. As Jeffreys points out, this is purely an *ad hoc* hypothesis with no independent confirmation. This requires that the polar axis be much shorter than either of the other two axes. As we shall see later, this is not the case.

The net result is that theory and observation simply indicate that the moon now is in far from a condition of hydrostatic equilibrium, and yet it is close enough to a former condition of hydrostatic equilibrium to warrant a good deal more study, for the correct interpretation of the puzzle may well yield the clue to the early history of the earth and moon system.

Kuiper (32), by a slight modification of terms, has tackled the problem in an attempt to determine the sequence of events. He postulates that the moon is a homogeneous triaxial ellipsoid and took the axis of rotation as unit length. The axis pointed toward the earth then becomes $1 + \beta$, and the third axis in the planes of the lunar equator and sky becomes $1 + \alpha$. Then, to a first-order approximation, the moment of inertia around the polar axis, C', is proportional to $1 + \alpha + \beta$. The moment around the axis pointing toward the earth, A', becomes $1 + \alpha$, and B' becomes $1 + \beta$. The ratios determining the motion of the ellipsoid then approximate

$$\frac{C' - A'}{B'} = \beta, \quad \frac{C' - B'}{A'} = \alpha, \quad \frac{B' - A'}{C'} = \gamma = \beta - \alpha. \quad (10\text{-}8)$$

Theory shows that it requires $C' > B' > A'$ or, more generally, $B' > A'$ and $(C' - B')(A' - C') < 0$. Otherwise, the physical librations of the moon either would not be stable or might show secular terms, which is contrary to observation.

Now, $\mu =$ the mean motion of the node of the lunar orbit on the ecliptic in terms of the mean motion of the moon itself, or $\mu = 0.004019$. The period of the node is 18.6 years.

The mean inclination of the lunar equator on the ecliptic, θ_0, has been found to be about $1°32'$, with values ranging from $1°31'10''$ determined by Koziel (27) to $1°33'50'' \pm 19''$ by Watts (33). The inclination of the moon's equator, θ_0, has been found (34) to depend on the inclination of the moon's orbital plane, i, and on β and μ as follows:

$$\theta_0 \cong \frac{3i\beta}{2\mu + \mu^2 - 3\beta}(1 - \tfrac{1}{2}\mu + \tfrac{5}{8}\mu^2) \cong i\left(\frac{2\mu}{3\beta} - 1\right)^{-1}. \quad (10\text{-}9)$$

Equation (10-9) may be used to determine β with a considerable degree of precision, unless β should be close to $\tfrac{2}{3}\mu = 0.00268$. This yields the value of $\beta = 0.00062$, with an uncertainty of about 1 per cent.

Much less is known about α or γ. The latter occurs as a factor in both a monthly and an annual term found in the theoretical expression for the physical libration in longitude; but the monthly term is too small to be observed, and the annual term is very poorly determined. Franz (35) has found $\gamma = 0.00032$, and Weimer (28) derived $\gamma = 0.00027 \pm 0.00003$. Kuiper suggests that the uncertainty may be greater than these figures. Apparently both α and γ are about $\tfrac{1}{2}\beta$ or larger.

Problem of the Moon's Motion and Shape

Kuiper (32) traced the early history of the moon by using the assumption that the calculated bulge is actually a fossil tidal bulge. The moon may then be traced back to its origin billions of years ago when it was very much closer to the earth. This dynamically determined bulge is approximately 2,300 feet above the mean sphere, and the semiaxis aligned with the earth is 3,500 feet greater than the average radius in the plane of the sky.

Plummer (36) shows that the theory of μ can be used to derive an explicit relationship between μ and the ratio a/a_0, where a and a_0 are the former and present mean distances of the moon from the earth. The result is

$$\mu = -0.00420 \left(\frac{a}{a_0}\right)^3 + 0.000117 \left(\frac{a}{a_0}\right)^{41/2} + \ldots \quad (10\text{-}10)$$

The minus sign of the larger term indicates that the motion of the nodes is a regression. As we move backward in time, μ decreases, with the dependence on a equally divided between the absolute motion of the node becoming slower and the mean motion of the moon becoming faster. As μ decreases rapidly with declining a, there must come a time when it will fall below the critical limit of $1.5\beta = 0.00093$.

The effect of the increasing departure from sphericity of the earth as we go backward in time, the moon approaching the earth and the earth rotating more rapidly, cannot significantly affect the motion of the nodes, μ, at the distances involved. The effect on μ from the oblateness of the earth is

$$\mu' = -\frac{3}{2}\frac{C-A}{Ma^2}, \quad (10\text{-}11)$$

where M is the mass of the earth and C and A are the moments of inertia of the earth. Inasmuch as this equation also shows a minus sign, the effects are additive to the solar effect on μ.

The present value of C is $0.334MR^2$ if R is the radius of the earth, while $(C-A)/C = 1/297$ and $a/R = 60$. The present value of μ' is found to be 5×10^{-7}, or 10^4 times smaller than the solar term. While the ratio μ/μ' changes more steeply than a^5 because the rotation period of the earth and therefore $(C-A)$ also depend on a, μ' will be small compared with μ unless a is very much less than $0.4a_0$. Therefore, it follows from equation (10-10) that μ will reach the critical limit 1.5β at $a = 0.608a_0$.

In Kuiper's attempt to see what happened as this limit was approached, he put $\mu = \frac{3}{2}(\beta + \varepsilon)$ in equation (10-9). Then

$$\theta_0 \cong \frac{i\beta}{\varepsilon}. \quad (10\text{-}12)$$

Unless i went to zero simultaneously with ε, the obliquity of the lunar equator must have increased indefinitely as the limit was approached.

The history of i is not well known. G. H. Darwin's (14, 37) work was carried out before the true nature of tidal friction was recognized, and his model calculations with bodily viscosity must be redone. They suggest that i has not changed drastically, except possibly at the very beginning of the moon's history.

There is no basis for assuming that i was zero at the beginning of the moon's existence. Kuiper was led to conclude that if the observed elongation of the moon, β, were the result of events at or immediately after the moon's birth and if the moon actually passed the critical point near $0.608a_0$, it toppled over on its center.

It is when we try to reconstruct the sequence of events that we run into trouble. If the moon were formed outside the critical distance, there is, of course, no question of its toppling over. There is also no question of the bulge being a fossil tidal bulge. It could not be one. The observed elongation must then be caused by irregularities in the distribution of materials in the moon, left as it accumulated. Conversely, if the moon were formed inside the critical limit, it could have been formed cold, as Urey (38) believes, and have remained cold throughout its history. In this case, the bulge represents accidental distortions.

The dynamically observed bulge corresponds in height to a tidal bulge at a distance of $0.39a_0$. This distance is less than the critical distance, $0.608a_0$, and hence a literal application of the existing theory would require the longest axis to move so as to become the axis of rotation, with the intermediate axis pointing toward the earth. This would then have happened because β would be greater than $\frac{2}{3}\mu$, which would require a negative i or θ, neither of which has physical meaning in this problem. The long axis would remain the axis of rotation until the critical distance was reached, at which time the moon would shift so as to point the long axis toward the earth, as at present. The picture is even more muddled if the moon were ever melted sufficiently to permit it to assume the shape, even approximately, of an equipotential surface.

It is generally accepted that the moon was accumulated from cold objects. If it melted, this was a subsequent development. If it melted, the effect of the tides would make the long axis point toward the earth. As long as the moon was molten or plastic, this shape would probably remain in this orientation; but if the moon ever cooled to the point where the adjustment in shape were not almost instantaneous, theory suggests that it would topple over until the long axis became the axis of rotation. Subsequent adjustments to the disturbing potential would re-establish the long axis in line with the earth, which again would be an unstable configuration. The term "the inconstant moon" would certainly have been applicable at that time.

If the theory is correct, there is no escape from this dilemma, provided that the moon liquefied closer to the earth than $0.608a_0$; and if the bulge is a fossil tidal bulge, it must have been formed and frozen inside this limit.

It will appear later that there are numerous observations pointing unambiguously to the fact that the moon was once at least partially melted and that the surface could and did adjust isostatically. If this is true, then the surface could also assume the distorted equipotential surface required by the proximity of the earth in the early days, and the bulge may well be a fossil tidal bulge.

It seems probable, at the present level of knowledge, that our theories concerning the rotational stability of satellites need thorough revision, with full account taken of tidal friction and possibly bodily friction. If this is completely done, it may well develop that we will not have to postulate the wild gyrations of a liquid or plastic moon, as theory seems to require.

On the basis of these equations, Kuiper (32) provisionally concluded that, regardless of where the moon solidified, β could not be interpreted as a fossil tidal bulge. Consistent with this conclusion, he suggested that the present values of α and β were likely to give some measure of the effects of the last major impacts received by the moon. They suggest differential deposits of new material in the three main directions of the order of 500 meters.

It is interesting to see just what physical differences correspond to the variance of f from the theoretical value of 0.25 if we accept the dynamically determined value of $f = 0.72$. The dynamically determined bulge is 3,500 feet above the average radius in the plane of the sky. The equatorial radius in the plane of the sky departs from the theoretical radius as determined from an equipotential surface by about 1,500 feet, just a little more than a quarter of a mile. This distance is many times less than the distortions of the surface produced by non-tidal forces. It is entirely possible that the theoretical value of $f = 0.25$ is more or less completely masked by the larger surface variations.

Jeffreys (13) approached the problem from a different direction. He showed that if both the earth and moon were originally fluid and rotated at about the same rate, the rotation of the moon would approach the rate of its revolution about 17,000 times as fast as that of the earth would. This conclusion assumes that the elasticity in the two bodies is not so great as to affect the order of magnitude of the height of the tides and that the phase lags of the tides were similar. The moon would then be brought always to present the same face toward the earth before the rotation of the earth had been greatly affected.

Because of the small size of the moon compared with the earth and the absence of a blanketing lunar atmosphere, the moon would have solidified somewhat earlier than the earth. From this point on, any tidal friction in the moon must have been in its bodily tide, for there can have been no shallow seas upon it. The elastic tide in the moon must be about one-fiftieth of the hydrostatic equilibrium tide. On both grounds, the ratio of 17,000 must have been greatly reduced after the moon's solidification, but the extent of the reduction is not known. There are no positive grounds for supposing any appreciable amount of bodily tidal friction in the earth, and, accordingly, there is no strong reason to assume it in the moon. It is sufficient to note that some mechanism in the past caused the moon always to point one face toward the earth. This mechanism must have been some form of tidal friction, and at least a limited amount of liquidity or plasticity in the moon seems to have been required.

Jeffreys then assumed that the moon was completely free from internal friction after solidification. He showed that the tidal friction within the earth's seas has acted through the period when the earth had oceans to drive the moon farther away from the earth. Now, if at any instant, the moon's rotation were exactly synchronized with its revolution around the earth, in the next instant the effects of tidal friction in the shallow terrestrial seas would have forced the moon very slightly farther away. The period of revolution would have grown longer, and the rotation rate would not have been affected. Even in the absence of any friction in the moon, the recession of the moon would have produced a perfectly imperceptible deviation of the axis of least moment from the line of centers.

It is clear that, regardless of its magnitude, this deviation has an important dynamical effect. The moon's longest axis pointing systematically to one side of the earth causes the couple on the moon produced by the earth's attraction to be, on an average, negative. It is this couple that reduces the moon's rate of rotation and keeps it equal to the period of revolution while the latter changes.

The question may be raised as to whether other complementary functions, such as the amplitude of the free libration in longitude, may have become great during the early history of the system. It is observed that this libration is now imperceptible. The present period of the free libration in longitude is about 40 months, and it remains, for all time, proportional to the period of revolution. Thus the change in the period of oscillation during a complete period is small compared with the period itself. In these circumstances, it is known that the amplitude is proportional to the square root of the current period. The amplitude is vanishingly small now, and if the moon had ever revolved in one of our present days, it would have been still smaller by a factor of about 5.

To explain the observed facts—that the moon's periods of rotation and revolution are equal and that its free libration in longitude is very small—we need not assume that tidal friction is still operating in its interior. It is sufficient that tidal friction should have been enough to produce these conditions before or soon after solidification, which is highly probable. Once produced, they would be permanently maintained by the earth's attraction on the moon's equatorial protuberance.

Several conclusions may be reached from a study of these paragraphs. The moon is not spherical but does possess an axis aligned with the earth, which is longer than the other two axes. This is a dynamically determined fact, not a physically observed condition.

Because the earth has oceans, friction develops in the shallow seas due to motion of water accelerated by the tidal pull of the moon. This friction acts as a brake on the rotation of the earth, reducing its angular momentum. This loss of angular momentum is compensated for by a reacting gain in angular momentum by the moon, which manifests itself in a slow increase in the radius of the moon's orbit.

Going back in time, we find that the moon was once closer to the earth than it is now and that the earth then rotated faster. This process may be carried out mathematically for a duration of 4.5 billion years. As the total amount of angular momentum in the system is constant, the month and the day would have been equal at about 4.8 hours each when the moon was at a distance of about 9,000 miles. Actually, the moon would have been destroyed if formed, or, more probably, it never would have coalesced at the small distance from the more massive earth. Stresses set up in a body such as the moon by the tremendous tidal pull of the earth would have ripped it asunder. The minimum distance at which the moon could have been formed is about 11,000 miles, and the corresponding month was then about 6.5 hours long. The mathematics of this journey backward in time do not imply that the moon was formed 11,000 miles from the earth but only that it might have been born this close.

Jeffreys' analysis of the stabilizing action of the earth on a lunar bulge would prevent the wild gyrations that Kuiper (32) has tentatively suggested based on Tisserand's (17) theory of the inclination of the moon's equator. There is no longer any reason to suppose that the very real bulge of the moon may not be a fossil tidal bulge. Other modes of origin are not excluded; but, at present, the frozen tidal-bulge hypothesis seems to be reasonable. If the f ratio were closer to $\frac{1}{4}$, then we would be on firm ground in suggesting that the bulge was tidally induced when the moon was close to the earth, but it is not too close to the theoretical value, and so some doubt must remain.

Two other alternative answers are given by Jeffreys' equations. Somehow the moon must have been brought into a position where the bulge became aligned permanently with the earth. If there were any tidal friction in the moon, this result would have occurred. Probably, but not certainly, this primitive tidal friction was engendered in the liquids of a partially melted moon. If there were no tidal friction in the youthful moon, then its rotation period must have been longer than the early month, so that the lunar day and month could have become equal as the moon receded from the earth. This hypothesis also requires a nice synchronization, so that the two periods became equal when the long axis was directed closely toward the earth. The velocity of recession of the moon was rather high at first, because tidal friction varies with the inverse sixth power the distance.

In any case, the existence of a large bulge in the figure of the moon is evidence of a considerable amount of strength in the outer layers unless the moon is completely compensated isostatically. As the moon seems to be the same age as the earth, about 4.5 billion years old, this bulge has existed essentially all this time. Regardless of whether it is an accidental distortion of the moon's shape or a fossil tidal bulge, this strength must have endured. It must have endured even when the tremendous events marking the terminal phases of the moon's formation were taking place. The true delineation of this bulge thus becomes of paramount importance, and the amount of time and energy required will have been well spent.

REFERENCES

1. FOTHERINGHAM, J. K. "Note on the Secular Accelerations of the Sun and Moon as Determined from the Ancient Lunar and Solar Eclipses, Occultations, and Equinox Observations," *M.N.*, **80,** 578, 1920.
2. ———. "A Solution of Ancient Eclipses of the Sun," *ibid.*, **81,** 104, 1920.
3. JEFFREYS, H. *The Earth*. Cambridge: Cambridge University Press, 1924.
4. SCHOCH, K. *D. Seculäre Accel. d. Mondes u.d. Sonne*. Berlin: Privately printed, 1926.
5. SITTER, W. DE. *Bull. Astr. Inst. Netherlands,* **4,** 21–38, 1927.
6. ———. *Ibid.*, pp. 57, 61.
7. JONES, SIR H. S. *M.N.*, **99,** 541–58, 1939.
8. TAYLOR, G. I. "Tidal Friction in the Irish Sea," *Phil. Trans., A,* **220,** 1, 1919.
9. HEISKANEN, W. "Uber den Einfluss der Gezeiten auf die säkulare Akzeleration des Mondes," *A.N.*, Vol. **214,** col. 81, 1921.
10. JEFFREYS, H. "Tidal Friction in Shallow Seas," *Phil. Trans., A,* **221,** 239, 1920.
11. MUNK, W., and MACDONALD, G. J. F. *The Rotation of the Earth*. New York: Cambridge University Press, 1960.
12. MACDONALD, G. J. F. "Interior of the Moon," *Science*, **133,** 1045, 1961.
13. JEFFREYS, H. *The Earth*, p. 237. 3d ed. Cambridge: Cambridge University Press, 1952.
14. DARWIN, G. H. *The Tides*. Boston and New York: Houghton Mifflin Co., 1898.
15. JEFFREYS, H. "The Resonance Theory of the Origin of the Moon, II," *M.N.*, **91,** 169, 1930.
16. PLUMMER, H. C. *An Introductory Treatise on Dynamical Astronomy*, chap. xxiii. Cambridge: Cambridge University Press, 1918.
17. TISSERAND, F. *Traité de mécanique céleste*, Vol. **2,** chap. xxviii. Paris: Gauthier-Villards et Fils, 1891.
18. JEFFREYS, H. Ref. 13, p. 158.
19. WIECHERT, E. "Ueber die Massenverteilung im Innern der Erde," *Nachr., Gesellsch. Wiss. Göttingen, math.-phys. Kl.*, p. 221, 1897.

20. STRATTON, F. J. M. *Mem. R. Astr. Soc.*, **59,** 257, 1909.
21. HOPMANN, J. "Selenodätische Untersuchungen #3," *Mitt. U.-Sternw.*, Vol. **6,** 1952.
22. HAYN, F. "Die Achsendrehung des Mondes," *A.N.*, **211,** 311, 1920.
23. ———. "Selenographische Koordinaten. III," *Sächsische Akad. d. Wiss., Leipzig,* 1907.
24. NAUMANN, H. "Selenographische Koordinaten. V," *Sächsische Akad. d. Wiss., Leipzig,* 1939.
25. BELKOWITSCH, I. W. *Engelhardt Obs. Bull., Kazan,* **10,** 28 1936.
26. JAKOWKIN and BELKOWITSCH, I. "Zur Frage nach der Bestimmung der Mondfigur vermittels Terminator Beobachtungen," *A.N.*, **256,** 305, 1935.
27. KOZIEL, K. "The Moon's Libration and Figure as Derived from Hartwig's Dorpat Heliometric Observations," *Acta Astr.*, Sér. a, Vol. **4,** Cracovie, 1948.
28. WEIMER, TH. *Recherches sélénographiques* (thesis). Paris: Gauthier-Villards, 1954.
29. LAPLACE, P. S. "Traite de mécanique céleste," *Œuvres*, Vols. **1–5.** Paris: Imprimerie Royale, 1843–46.
30. JEFFREYS, H. "On the Figures of the Earth and Moon," *M.N.*, **97,** 3, 1936.
31. ———. *M.N. Geophys. Suppl.*, **4,** 1–13, 1937.
32. KUIPER, G. P. "On the Origin of the Lunar Surface Features," *Proc. Nat. Acad. Sci.*, Vol. **40,** No. 12, 1954.
33. WATTS, C. B. "A New Method of Measuring the Inclination of the Moon's Equator," *A.J.*, Vol. **60,** No. 11, 1955.
34. TISSERAND, F. Ref. 17, eq. (61) (simplified).
35. FRANZ, J. "Die Konstanten der physischen Libration des Mondes," *Astr. Beob. U. Sternw. z. Königsberg*, Abt. 38.
36. PLUMMER, H. C. Ref. 16, p. 285.
37. DARWIN, G. H. *Scientific Papers,* Vol. **2.** Cambridge: Cambridge University Press, 1908.
38. UREY, H. C. *Geochim. et Cosmochim. Acta,* **1,** 207, 1951.

11

THE SHAPE
OF THE MOON
AND THE
NEW CONTOUR
MAP

That the moon has an equatorial bulge reasonably aligned with the earth is not subject to question. The exact nature of this bulge cannot be determined dynamically; only a blurred average bulge, which is a composite of all surface irregularities and density variations, can be so measured. Numerous astronomers have recognized the importance of a better delineation of the bulge and its relationship to other surface features.

Two methods of measurement have been attempted. Both these methods are simple in principle but extremely difficult to put into practice. In the first method, two photographic plates are taken at widely differing librations. This allows us to measure positions of lunar objects as photographed from two angles which may differ by perhaps 14°. If the moon were spherical, we could measure positions of such objects on one photograph; and, from the known distance and orientation of the earth and moon at the two times, we could predict exactly where these objects would appear on the second photograph. Actually, of course, the moon shows surface variations in height, relative to a sphere, of around 6 miles, with certain mountain peaks extending considerably higher above the low spots. As a result, the measures of the second photograph made on the same lunar objects will not agree with the predicted positions if the measurements are made carefully enough. The differences between the predicted and measured positions on the second plate are functions of the height of the object above or below the mean sphere. This method uses the principle of a mathematical stereopticon.

In the second method, certain observers have exploited the deformation of the terminator. It is obvious, even to the naked eye, that the terminator is never a smooth, curving line but is irregular. This is due to real variations in height of portions of the lunar surface. In low areas the terminator is reached sooner, relative to the sunlight side, than in the higher areas. Isolated high points may often be seen still bathed in sunlight at a considerable distance beyond the terminator. Theoretically, accurate measures of the position of the terminator at different times would yield a contour map of the moon of great accuracy. In practice, the exact measurement of the end of the sunlit region is fraught with many accidental and systematic errors. The terminator, in general, marks the region on the moon where the sun's rays are tangent to the surface. This line is thus very difficult to measure. No really satisfactory results have been obtained by this method.

The stereoscopic principle has been attempted several times. The first major attempt was by Franz (1), who in 1899 published the results of many years of work. The earlier measures of Franz of crater positions were all made at the telescope. Later measures were on photographic plates. Franz measured these minute shifts for 55 small craters spread widely over the lunar disk. From these widely scattered points, he derived a value for the x semi-axis of 1.00114 ± 0.00390. As 10^{-5} of the radius of the moon is 57.1 feet, the bulge is thus $6,500 \pm 22,000$ feet. The measured bulge is thus about twice the dynamically determined height, but the scatter of the points is excessive.

In 1904, Hayn (2) found the semiaxis which points toward the earth to be 1.0023, or 13,000 feet longer than the average radius in the plane of the sky. Hayn candidly admitted that the accuracy of the results was low. He worked only with the tiny but brilliant crater Mösting A, which is near the center of the disk.

Pickering (3) found 1.0013 ± 0.0012, but his measures were made on twenty points, all within a half-radius of the mean center of the disk, and, because of the small leverage which these restricted points supply, the probable error is large.

Saunder (4) found 1.00052 ± 0.00027 from thirty-eight points on four negatives. In this case, the chosen craters are essentially on the central meridian. This answer means a bulge of $3,000 \pm 1,550$ feet. However, if Saunder leaves out the values for Anaxagoras and Anaxagoras A, the two points farthest north, he finds 1.00123 ± 0.00029—very similar to the results of Franz and Pickering.

It should be pointed out that when the results of Franz and Saunder are plotted, the scatter of points in each case is large but quite comparable. Consequently, the two men have meant quite different things by their assignments of probable errors. The probable error of Franz's solution is given as fourteen times that by Saunder.

Saunder made a deliberate attempt to determine the eccentricity of the prime meridian only; but, except for his selection of points according to their longitude, neither Saunder nor Franz has tried to analyze the raw data according to the nature of the region near each point. For example, Saunder has included a few points which lie in the bottoms of deep craters. In neither list are the points segregated because of their upland or maria locations. Both lists have been published in full, fortunately, and consequently they may be so analyzed. The results are quite amazing and prove to have several very important consequences.

Before the measures of Saunder and Franz can be compared and combined, certain systematic differences must be eliminated. The heights of the 55 points of Franz were measured with respect to a mean sphere whose radius was 1.00056 times that used by Saunder. Hence 56 units must be subtracted from Franz's heights to put them on the same scale as that of Saunder. In addition, the heights measured by both men correspond to the elevation of the rims of small craters above or below the mean sphere. To reduce them to the true surface, an average of 35 units, or 2,000 feet, must be subtracted from all heights. Saunder reduced additional measures by Franz on 14 objects which were also in his own list. A total of 205 units must be subtracted from these heights to bring them to the same scale as the others. When these operations have been performed, the heights are strictly comparable, and we have 93 of them with which to deal. Of these, 6 have been rejected because they are so situated that their positions relative to the undisturbed surface cannot be determined. The remaining points are nearly equally distributed between maria and "continental" regions, 45 lying on the seas and 42 in the uplands.

These 87 points (5), nearly equally divided between continental and maria locations, have been reanalyzed according to their locations. The relationship between measured height as a function of distance from the center of the moon's visible disk for those points lying in the light upland areas and separately for those points lying in the darker maria show the same general trend as that derived from plots of Franz's blended measures. The scatter of the points around the mean curve is reduced by a factor of 2 for the points on the maria and by a factor of 3 for the upland points. In both cases, the scatter is much less than that found by Saunder.

DuFresne (6) has shown that statistically the larger variations in height found for the maria regions are real and correspond to a wider variation in

Shape of the Moon and the New Contour Map

height of portions of the dark areas relative to the mean sphere than is true for the brighter areas.

In 1949 (7), it was shown that the bulge of the moon as defined by the upland points was

$$0.00130 \pm 0.00012, \quad (11\text{-}1)$$

or

$$7{,}400 \pm 700 \text{ feet}. \quad (11\text{-}2)$$

The best fit to the maria points was

$$0.00117 \pm 0.00021, \quad (11\text{-}3)$$

or

$$6{,}700 \pm 1{,}200 \text{ feet}. \quad (11\text{-}4)$$

However, the maria averaged 100 units, or 5,700 feet, lower than the uplands near the center of the disk and 87 units, or 5,000 feet, lower at the limb. The average depression of the maria is thus very close to 1 mile.

The height of the bulge as measured on upland and maria is the same within 700 feet, or one part in ten, and hence it is quite probable that the difference is not real and that all parts of the lunar surface are affected by the same bulge, regardless of local irregularities. The most probable value for the lunar bulge thus is

$$0.00127 \pm 0.00010 \quad (11\text{-}5)$$

in linear measure; this means that the few measures of Franz and Saunder yield a bulge of

$$7{,}200 \pm 600 \text{ feet} \quad (11\text{-}6)$$

greater than the average radius in the plane of the sky.

This approximation to the bulge yields a value about twice that of the dynamically determined bulge but also shows clearly that the maria are great depressed areas. The effect of this real variation in the moon's surface contours must be taken into consideration in any future work on the rotation of the moon and its moments of inertia. It would seriously affect the value of f.

The foregoing conclusions are valid as far as they go, but they simply are not adequate to give us the data we need. Eighty-seven points distributed essentially at random over half the surface of the moon are equivalent to one point in every 80,000 square miles, an area roughly equal to one of the larger of the western states. It would be brash indeed to claim that a definitive contour map of the continental United States could be drawn from 48 height elevations, one in each state, particularly when the accuracy of the determination was low. Nevertheless, Franz (1) did attempt the solution from his 55 points. His map,

Figure 33, shows low areas in Mare Serenitatis, Mare Imbrium, and Oceanus Procellarum and a suggested low region near Mare Australe. The center of the disk appeared to average high.

Other astronomers, not satisfied with the early results, have tackled the problem. Ritter (8) used measures of the variations along the terminator and published a contour map (Fig. 34). His results are exceedingly difficult to reconcile with other determinations. Ritter's technique and results have been subjected to heavy criticism from Jakowkin and Belkowitsch (9) in the U.S.S.R. Mainka (10) also used the deformation of the terminator to determine the shape of the moon, while Weimer (11–13) used the stereoscopic method. Their results were reasonably consistent but gave a value for the bulge that is improbably large, about 0.004. Wirtz (14) found 0.0014.

In 1901, Franz (15) published a summary of positions of 150 moon craters. In 1952, Hopmann (16) at Vienna reworked Franz's data, giving new positions for each object in the list and new determinations of the height of the surface of the moon at each point. He did not publish a contour map for these new reductions.

In 1958, Schrutka (17) undertook a complete review of the original measures by Franz and, like Hopmann, derived a new set of positions for each crater and a new determination of the height above or below a mean sphere.

Later that year, Schrutka and Hopmann published a contour map of the moon (18) (Fig. 35), which shows certain similarities to Franz's contour map and also certain differences. In general, a bulge of the order of 9,500 feet was found, with the maria usually being considerably depressed. Schrutka and Hopmann, like Franz, drew their contour lines with complete disregard for obvious lunar features. Schrutka gave his determination of the probable error of a single height measure as $\pm 4{,}030$ feet (± 1.23 km). In either method—the stereoscopic or the deformation of the terminator—the measurements are extremely difficult. No one has ever succeeded in using the latter method satisfactorily.

In the stereoscopic method, the measures must be made to an accuracy approaching that required for stellar parallax work, and yet the objects to be measured are nowhere nearly stellar images. The little craters used generally range from 10 miles in diameter to 1 mile across, or approximately 10″ to 1″ of arc (geocentric). To complicate the problem, the craters are nearly circular but are illuminated by the sun in such fashion that the opposite sides of the crater often show widely differing brightnesses. The amount of shadow, both internal and external, is a function of the latitude of the crater and its distance from the terminator. On plates which are to be compared, the illumination is often from opposite directions. In addition, foreshortening changes the apparent shape of a crater.

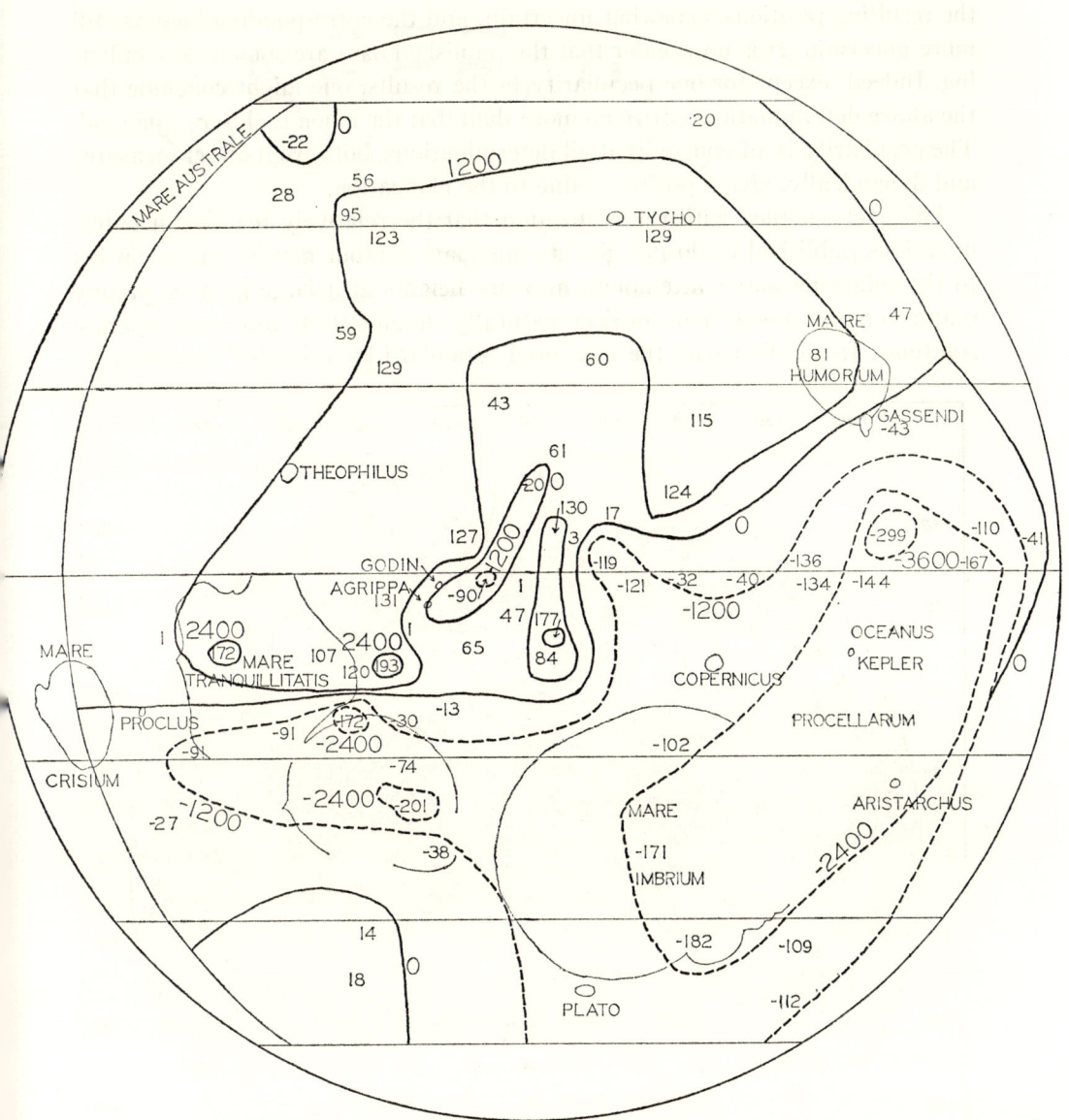

Fig. 33.—Contour map of the moon according to Franz. (Redrawn from Ref. 1.) Contour interval 1,200 meters. nit for individual height determinations 10^{-5} lunar radii.

All these factors combine to render the measures unusually difficult to make, the resulting positions somewhat uncertain, and the corresponding heights still more uncertain. It is no wonder that the published data are somewhat conflicting. Indeed, except for one peculiarity in the results, one might conclude that the above determinations prove no more than that the moon is closely spherical. The peculiarity is, of course, that all determinations, both from direct measures and dynamically, give a positive value to the elongation.

This fact, coupled with the realization that the relatively few height determinations published could not give an adequate contour map of the moon led to the following major attempt to measure heights at a large number of tiny craters on the moon. This project naturally divides itself into two separate solutions. In the first part the measured "standard co-ordinates" for each se-

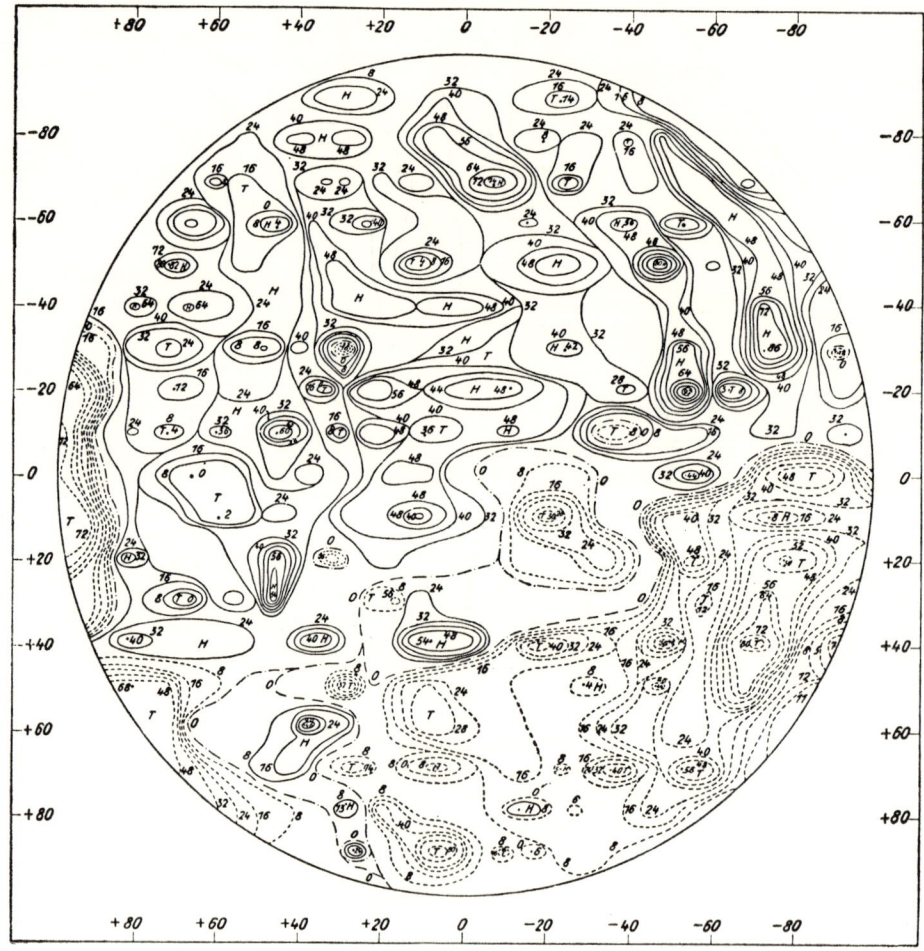

Fig. 34.—Contour map of the moon according to Ritter (8)

Shape of the Moon and the New Contour Map 219

lected point must be determined for each photographic plate. Then from these measured co-ordinates the final determinations of height can be made.

In all reductions of photographs, some means must be available to allow determination of plate constants. Unfortunately, the photographic plates that can be used do not possess auxiliary standard positions, such as background stars. The plates of the moon used by Watts at the United States Naval Observatory do have background stars from which plate constants can be determined and are excellent for his purpose in determining the contours near the lunar limb, but the smaller aperture of the telescope and the shorter focal length render these plates less attractive than those from certain other observatories for use in this stereoscopic problem.

Fig. 35.—Contour map of the moon according to Schrutka and Hopmann (18)

With his characteristic kindness, C. D. Shane, then director of the Lick Observatory, sent a complete file of photographs of the moon taken with the 36-inch telescope, and five photographs were selected from this file. Contact prints were then made on glass from the original negatives, and these five positives were measured.

There is really only one adequate set of standard positions of lunar-surface features. This is Franz's original set of 150 points (15). There are earlier measures. For example, Schmidt (19) gives a list of 157 formations whose positions have been determined as "points of the first order" by Lohrmann (20), Mädler (21), and Neison (22). To these we might add the position of Biot, found by Mädler; Horrocks, Halley, and Hipparchus L by Neison; with Triesnecker B, now Bruce, by Pritchard (4). This list includes practically all the fundamental points known at the turn of the century.

Mädler estimates the uncertainty attached to a single measure as 8″ or 9″ (geocentric), which would correspond to half a degree of selenographic longitude or latitude near the center of the disk and to a still greater amount as the limb is approached. The probable error derived from the measures themselves, after rejecting the unfavorable points, is stated by Neison (22) to be 7″.2. Rather less than half the whole number given by Schmidt depend on ten or more measures, whether by one or more observers.

Eleven points were measured five times, at least, by both Lohrmann and Mädler. The average difference in selenographic latitude is 17′, and in longitude 10′, corresponding, respectively, to 4″.4 and 2″.3 geocentric. Systematic errors, particularly in latitude, appear in the results.

When Saunder made his celebrated table of positions of lunar formations (23–26), he analyzed the available data and decided to use three fundamental points only. He took for a first point the position of Mösting A as determined by Franz from Schlüter's observations with the Königsberg heliometer. These were made under Bessel's direction for the determination of the moon's physical libration.

Pritchard obtained positions of Ptolemaeus A and Triesnecker B from measures on photographs. Saunder (4) himself made measures at the telescope resulting in a position for Ptolemaeus A. This was very close to the position determined by Pritchard, and a mean position was accepted. The third position chosen was Pritchard's for Triesnecker B.

With these three fundamental reference points, Saunder reduced his measures of numerous small formations and derived ξ and η positions for each photographic plate (4). Later he combined the data from these plates and, with the stereoscopic method, derived height determinations above and below a mean sphere for a series of formations along the central meridian of the moon.

As will be seen later, Saunder's height determinations are in fair agreement with those to be derived here, but his limited number of fundamental points suggests that we use his work only as a check. This leaves only Franz's 150 points, but even here the choice is complex. Hopmann's (16) 1952 re-reduction of Franz's points was made with improved libration constants. Schrutka's (17) 1958 work on these same points apparently represents a further improvement on Hopmann's work. Determinations of plate constants for my own measures, using Saunder's fundamental points, gave results very similar to those published by Schrutka and agreed less accurately with Hopmann. Consequently, it was decided to use the 150 points of Franz as reduced by Schrutka as fundamental points in the determination of plate constants. Any systematic errors inherent in the fundamental system of points will, of course, be reflected in the results;

TABLE 21*

DATA FOR LICK MOON PLATES

	Plate No.				
	1	2	3	4	5
Date (G.C.T.)....	1938 Aug. 20.533	1945 Oct. 26.574	1942 July 23.189	1938 May 7.176	1938 June 3.168
Phase..........	24.3 days	20.36	9.69	7.0	4.6
l..............	− 3°.91	+ 7°.41	− 4°.74	+ 4°.63	+ 4°.72
b..............	+ 3°.32	− 1°.36	− 5°.80	+ 7°.28	+ 7°.33
C_0............	− 5°.97	+ 8°.24	+ 13°.76	+ 17°.65	+ 16°.42
s'.............	988″.1	968″.1	985″.8	972″.1	984″.0
π_v............	− 49°.1	+ 25°.9	− 1°.3	+ 18°.5	+ 35°.2
z'.............	38°.6	17°.3	53°.6	36°.5	55°.0

* l and b are the topocentric librations, including physical; C_0 is the position angle of the axis, corrected for parallax; s' is the moon's semidiameter corrected for augmentation; π_v is the position angle of the vertex, measured from the moon's axis; and z' is the zenith distance corrected for parallax.

but the comparison with Saunder's results, which use only one of Franz's measures for a fundamental point, indicates that the errors in Schrutka's positions, on the average over the moon's disk, are far smaller than the absolute height variations, which are a principal aim of this project.

My measurements were made on the Gaertner parallax measuring instrument belonging to the Dearborn Observatory of Northwestern University. Inasmuch as the parallax machine was not in use, Dean Simeon E. Leland permitted me to use the instrument in a specially constructed room in my basement for a period of nearly a year. I am deeply grateful to Northwestern University and to Dean Leland for this exceptional favor.

The five Lick Observatory plates are listed in Table 21. The librations, refractions, and other constants were computed and furnished me by Dr. C. B. Watts, of the United States Naval Observatory. Dr. Watts took time from his busy schedule to confer with me and to write me concerning the progress of this work. His knowledge and interest are very much appreciated.

The method of measurement of each plate was as follows. The plate was mounted on the parallax measuring engine with the axis of the moon approximately vertical. With a fine needle, two tiny scratches were made in the emulsion, one on either side of the moon's image. By rotation of the plate, the two scratches were exactly aligned with the horizontal line which bisected the field of the projected image. Each mark must stay centered on the line as the carriage is moved in the x-direction.

Enlargements 30 inches in diameter on paper had previously been made of each of the five plates, and the specific formations to be measured had been identified and numbered. This eliminated any work directly on the glass positives and made for easy identification of objects.

Two measures were then made in the x-co-ordinate of each selected point. In almost all cases, the two measures agreed very closely. If agreement were not good, two additional measures were made and the average of all four measures accepted, unless it became obvious that a wrong identification had been made.

After this the plate was rotated 90° and the two needle scratches identified. Fine angular adjustments were made, so that the two scratches would remain centered on the vertical line in the field of the projected image. It was then assumed that measures could be made in the same sense as the first set except that the rotation of the plate now allowed the measures to be made for the y-direction.

Two more similar operations permitted measures to be made in the x-direction at 180° and y-direction at 270°. The measures in each set were averaged, and then the opposite sets were combined to give the final measured Cartesian co-ordinates of the individual lunar formations. The origin of these co-ordinates is thus an arbitrary point, and the center of the moon's disk is an unknown point.

To determine the co-ordinates of the center of the moon's disk, a new series of measures was made before the plate was removed from the measuring engine. The center of rotation of the photographic plate on the measuring engine did not coincide with the center of the moon's disk, but the co-ordinates of the center of rotation were exactly known. Measurements were made which yielded the distance from the center of rotation to a tangent to the moon's limb. Similar measures were made at intervals of 2° around the limb of the moon as seen from the center of rotation. In each direction, on approach to the terminator, measures were stopped before the deep shadows of limb craters distorted the position of the limb.

A series of radii had thus been obtained at uniform angular separations between the center of rotation, which was a fixed point, and tangents to the bright limb, which was essentially circular. Cartesian co-ordinates of these tangent points in a known system could then be derived. By least squares, a first

Shape of the Moon and the New Contour Map

approximation to the moon's radius on the plate and the co-ordinates of the center of the disk could be obtained.

With the known position of the moon's center in the original co-ordinate system and the radius, the measures of the positions of the selected lunar formations could be reduced to decimals of a lunar radius as projected onto the plane of the sky. The errors in this first approximation for each plate turned out to be of the order of 1 mile, or about $1''$ of arc, geocentric.

These values so obtained are the conically projected co-ordinates, $x + \delta x$ and $y + \delta y$. We need to have the orthogonally projected co-ordinates. In other words, we need to correct for the finite distance of the moon. The corrections may be written thus:

$$\delta x = (x + \delta x)[(z + \delta z)\sin s' + (r + \delta r)^2 \sin^2 s'], \tag{11-7}$$

$$\delta y = (y + \delta y)[(z + \delta z)\sin s' + (r + \delta r)^2 \sin^2 s']; \tag{11-8}$$

$x + \delta x$ and $y + \delta y$ are observed; $z + \delta z$ is found from

$$(x + \delta x)^2 + (y + \delta y)^2 + (z + \delta z)^2 = 1, \tag{11-9}$$

and s' is the moon's angular semidiameter, while

$$(r + \delta r)^2 = 1 - (x + \delta x)^2 - (y + \delta y)^2. \tag{11-10}$$

The terms in $\sin^2 s'$ are always less than 0.000025. The unit selected for this project is 10^{-5} lunar radii, or 57.1 feet.

With δx and δy computed for each point, the values of x and y—the orthogonally projected co-ordinates of each point—are known, and z may be computed on the assumption that all points lie on a sphere. The orientation of the moon on the photographic plate is not yet known.

Let x and y be the co-ordinates of any point corrected for finite distance and refraction, referred to the center and represented as fractions of the moon's radius; and z is the third co-ordinate of the same point formed from the equation

$$x^2 + y^2 + z^2 = 1, \tag{11-11}$$

and ξ, η, ζ the "standard" co-ordinates of the same point; θ the angle through which the co-ordinate axis of y must be turned to bring it into coincidence with the projection of the moon's central meridian, the positive direction of rotation being the same as for position angles. Then the usual formulae for transformation of co-ordinates give

$$\xi = (\cos l \cos \theta - \sin l \sin b \sin \theta)x - (\cos l \sin \theta + \sin l \sin b \cos \theta)y \\ + (\cos b \sin l)z, \tag{11-12}$$

and

$$\eta = (\cos b \sin \theta)x + (\cos b \cos \theta)y + (\sin b)z, \tag{11-13}$$

where l and b are the topocentric librations, including physical. These equations are in the form

$$\xi = Ax + By + Cz \qquad (11\text{-}14)$$

and

$$\eta = Dx + Ey + Fz, \qquad (11\text{-}15)$$

and the problem is to determine θ.

On each of the five plates, the measured x, y, z co-ordinates, corrected for everything except refraction, were substituted in equations (11-14) and (11-15) for three of Schrutka's points well separated on the disk. An approximate value of θ was then deduced for each plate.

With θ known with sufficient precision, the differential refraction corrections in each co-ordinate may be applied. The refraction corrections were based on tables in use at the United States Naval Observatory. The table is set up for the barometer at 30″ and the thermometer at 50° F. It is sufficiently accurate for the first approximation in this problem. At this stage we now have the first approximations to orthogonally projected co-ordinates for each point and each point corrected for refraction.

The further corrections to be applied to x and y are those for errors of center and radius, for deviations from standard differential refraction corrections, and for optical distortion. These corrections, along with the transformations necessary to deduce the values of the "standard" co-ordinates, ξ, η, ζ, are all linear, and, therefore, we have

$$\xi = Ax + By + Cz + D \qquad (11\text{-}16)$$

and

$$\eta = Ex + Fy + Gz + H, \qquad (11\text{-}17)$$

where $A, B, \ldots, H$ are constants for the plate. If these are known, the equations are sufficient to determine ξ, η, and ζ, which are connected by the relation

$$\xi^2 + \eta^2 + \zeta^2 = 1. \qquad (11\text{-}18)$$

In order to determine the constants, the accepted co-ordinates of the points which were common to the measures of Franz as reduced by Schrutka and to my list and which appeared on each individual plate were substituted in equations (11-16) and (11-17), and the eight constants for each plate were found by least-squares solutions. Because of the relatively small number of points common to Schrutka's list and mine, very small auxiliary corrections were later made to the plate constants for all plates except No. 3. The fundamental data needed in the reduction of the five plates are listed in Table 22.

The measured co-ordinates in the horizontal and vertical directions, given in millimeters, are listed in Table 43 (Appendix 2).

Shape of the Moon and the New Contour Map

Craters Nos. 697 and 698 were measured only on Plates 4 and 5 and were used only as fundamental points in the determinations of plate constants.

When the measured positions of these craters, peaks, and spots are corrected as indicated above and substituted in the proper equations, the "standard co-ordinates" ξ and η are found. These "standard co-ordinates" are listed by plates in columns 3–7 and 12–16 of Table 23 (Appendix 2); $\bar{\xi}$ and $\bar{\eta}$ are average co-ordinates derived from the individual plate measures.

The plates were selected so that Nos. 1 and 2 were well separated in angular orientation, and the librations in each co-ordinate were opposite in sign and

TABLE 22

FUNDAMENTAL PLATE AND REDUCTION CONSTANTS FOR LICK MOON PLATES

	Plate 1	Plate 2	Plate 3	Plate 4	Plate 5
Co-ordinate of moon's center (x)	68.47356 mm	79.20709 mm	113.99501 mm	130.33670 mm	111.41612 mm
Co-ordinate of moon's center (y)	95.86220 mm	93.86930 mm	90.08632 mm	93.85667 mm	90.00818 mm
Reciprocal of moon's radius	0.01188839	0.01212125	0.01188938	0.01205286	0.01193196
$\sin s'$	0.00479	0.00469	0.00478	0.00471	0.00477
$\sin^2 s'$	0.0000229	0.0000220	0.0000228	0.0000222	0.0000228
Refraction correction to x	$1.000426x + 0.000099y$	$1.000310x - 0.000015y$	$1.000310x + 0.000011y$	$1.000316x - 0.000049y$	$1.000516x - 0.000295y$
Refraction correction to y	$1.000395y + 0.000099x$	$1.000314y - 0.000015x$	$1.000875y + 0.000011x$	$1.000448y - 0.000049x$	$1.000726y - 0.000295x$
A	0.990522	0.948781	0.996156	0.996861	0.947609
B	−0.093807	0.276800	−0.003082	−0.050148	0.300110
C	−0.069636	0.128980	−0.083127	0.079947	0.080618
D	−0.00064	−0.00022	0.00084	−0.00011	0.00235
E	0.092806	−0.279699	−0.009597	0.037121	−0.311410
F	0.991217	0.958730	0.993790	0.992884	0.940921
G	0.058574	−0.024612	−0.102184	0.127660	0.126583
H	−0.00189	0.00071	0.00154	−0.00106	0.00109

reasonably similar in magnitude. The succeeding pairs of plates, 2–3, 3–4, and 3–5, were chosen because they were from widely different librations, thus giving differences in angular orientation so that the stereopticon principle could be applied. Because of these relative librations, $\bar{\xi}$ and $\bar{\eta}$, for points measured on Plates 1 and 2, 2 and 3, 3 and 4, and 3 and 5, are simply averages. For points measured on three plates, the average of the averages was computed. For example, if a point were found on Plates 2, 3, and 4, the average co-ordinates on 2 and 3 were averaged with the co-ordinates of the same objects on 3 and 4. This has the effect of giving double weight to the measures on the middle plate, but the result is that the $\bar{\xi}$ and the $\bar{\eta}$ co-ordinates are reasonably close to the standard co-ordinates of these points at zero librations.

If we assume only accidental errors to exist, both in Schrutka's co-ordinates and in mine, and that the errors are equally probable in each set, the probable error in each set is ±12.9 units or ±735 feet in the ξ co-ordinate and ±11.2 units or ±640 feet in the η co-ordinate. Similarly, the probable errors are found to be ±12.2 units in ξ and ±22.5 units in η when Saunder's measures are compared with mine and ±30.6 units in ξ and ±26.4 units in η when Franz's units are compared with mine.

Inasmuch as there are slight systematic variances between Saunder's, Franz's, and my reductions due to differences in libration constants, it is not strictly accurate to intercompare them in this fashion, but the low observed errors indicate that absolute positions for these lunar formations have been closely approximated. The positions of Gassendi ζ, Manilius, and Piton have been omitted in these calculations.

The next step is to determine the variations in altitude of each point relative to a spherical surface. The method used has been adapted from Saunder (4).

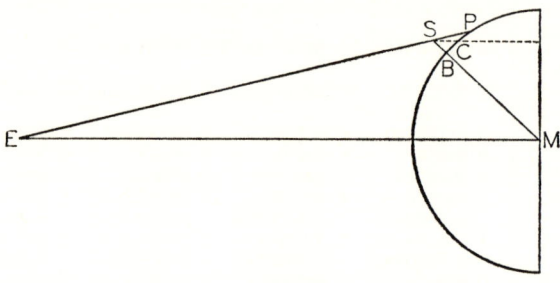

FIG. 36

Let M be the center of the moon (Fig. 36), E is the point of observation, S is a point whose altitude is to be determined. Let the radius MS cut the theoretical "mean sphere" whose surface nearly coincides with the actual surface of the moon in B. Let ES cut the same sphere in P. Let us assume that all the observed points are on the surface of a sphere. Suppose that this sphere is that whose radius is MB. Then the reduced co-ordinates, ξ and η, are those of P.

Now let M be taken as the origin, ME as the axis of z; x, y, z the co-ordinates of P, and $x + \delta x, y + \delta y, z + \delta z$ are the co-ordinates of B. Further, let $MB = 1$, $ME = d$, $BS = h$. The co-ordinates of S then are

$$(x + \delta x)(1 + h), \quad (y + \delta y)(1 + h), \quad (z + \delta z)(1 + h),$$

while the co-ordinates of E are 0, 0, d. Since the three points, E, S, P, are in a straight line, their projections on any one of the co-ordinate planes are in a straight line. This leads to

$$\begin{vmatrix} 0, & x, & (x + \delta x)(1 + h) \\ d, & z, & (z + \delta z)(1 + h) \\ 1, & 1, & 1 \end{vmatrix} = 0; \quad \begin{vmatrix} 0, & x, & (x + \delta x)(1 + h) \\ 0, & y, & (y + \delta y)(1 + h) \\ 1, & 1, & 1 \end{vmatrix} = 0, \quad (11\text{-}19)$$

Shape of the Moon and the New Contour Map

which reduce, respectively, to

$$x \delta z + (d - z) \delta x + \frac{hxd}{1+h} = 0, \quad (11\text{-}20)$$

$$x \delta y - y \delta x = 0. \quad (11\text{-}21)$$

Inasmuch as both B and P are on the sphere whose radius is unity, we find

$$x^2 + y^2 + z^2 = (x + \delta x)^2 + (y + \delta y)^2 + (z + \delta z)^2 = 1. \quad (11\text{-}22)$$

Now, if we put

$$h' = \frac{h}{1+h} \quad (11\text{-}23)$$

and neglect all squares of δx, δy, and δz and remember that

$$\frac{1}{d} = \sin s' \quad (11\text{-}24)$$

we derive

$$\delta x = -\frac{xz h'}{z - \sin s'}, \quad (11\text{-}25)$$

$$\delta y = -\frac{yz h'}{z - \sin s'}, \quad (11\text{-}26)$$

$$\delta z = \frac{h'(1 - z^2)}{z - \sin s'}. \quad (11\text{-}27)$$

Next, define ξ and η as the co-ordinates of P referred to the standard axes. Then $\xi + \delta \xi$ and $\eta + \delta \eta$ are those of B, and

$$\xi = Ax + By + Cz + D, \quad (11\text{-}28)$$

$$\eta = Ex + Fy + Gz + H, \quad (11\text{-}29)$$

where the direction cosines A, B, C, D, E, F, G, H have the values listed in Table 22. Consequently,

$$\delta \xi = Ax + By + Cz, \quad (11\text{-}30)$$

$$\delta \eta = Ex + Fy + Gz, \quad (11\text{-}31)$$

and

$$\delta \xi = \frac{C - \xi z}{z - \sin s'} h' = Ph', \quad (11\text{-}32)$$

$$\delta \eta = \frac{G - \eta z}{z - \sin s'} h' = Qh'. \quad (11\text{-}33)$$

Let ξ_1 and η_1 be the values of ξ and η given by any particular plate. P_1 and Q_1 are the corresponding values of P and Q, and $\bar{\xi}$ and $\bar{\eta}$ are the average co-ordinates from all plates on which the formation position was measured. The librations of this group of plates should each average approximately zero, so that $\bar{\xi}$ and $\bar{\eta}$ are close to the standard co-ordinates at zero librations.

Now $\bar{\xi} + \delta\bar{\xi}$ and $\bar{\eta} + \delta\bar{\eta}$ are the co-ordinates of B. These give us the following conditional equations from which we can find $\delta\bar{\xi}$ and $\delta\bar{\eta}$ and ultimately h:

$$\bar{\xi} + \delta\bar{\xi} = \xi_1 + P_1 h', \tag{11-34}$$

$$\bar{\eta} + \delta\bar{\eta} = \eta_1 + Q_1 h'. \tag{11-35}$$

Similar equations may be found for each of the other plates. We may rewrite equations (11-34) and (11-35) as follows:

$$\delta\bar{\xi} - P_1 h' + (\bar{\xi} - \xi_1) = 0, \tag{11-36}$$

$$\delta\bar{\eta} - Q_1 h' + (\bar{\eta} - \eta_1) = 0. \tag{11-37}$$

These equations may be solved by least squares, noting in particular that, by definition of $\bar{\xi}, \bar{\eta}$,

$$\Sigma(\bar{\xi} - \xi_1) = 0, \quad \Sigma(\bar{\eta} - \eta_1) = 0, \tag{11-38}$$

and therefore, if n is the number of plates, the normal equations become

$$n\delta\bar{\xi} - \Sigma P_1 h' = 0, \tag{11-39}$$

$$n\delta\bar{\eta} - \Sigma Q_1 h' = 0, \tag{11-40}$$

$$-\Sigma P_1 \delta\bar{\xi} - \Sigma Q_1 \delta\bar{\eta} + \Sigma(P_1^2 + Q_1^2)h' - \Sigma[P_1(\bar{\xi} - \xi_1) + Q_1(\bar{\eta} - \eta_1)] = 0. \tag{11-41}$$

These equations may be solved to yield

$$\delta\bar{\xi} = \frac{1}{n} \Sigma P_1 h' \text{ with weight } n, \tag{11-42}$$

$$\delta\bar{\eta} = \frac{1}{n} \Sigma Q_1 h' \text{ with weight } n, \tag{11-43}$$

$$h' = \frac{\Sigma[P_1(\bar{\xi} - \xi_1) + Q_1(\bar{\eta} - \eta_1)]}{\Sigma(P_1^2 + Q_1^2) - 1/n[(\Sigma P_1)^2 + (\Sigma Q_1)^2]}, \tag{11-44}$$

with the denominator being the weight of h'. Also

$$h = \frac{h'}{1 - h'}. \tag{11-45}$$

The weight of h' varies from a minimum of about 0.025 near the center of the moon's disk to greater than unity very near the limbs. The displacements of a crater due to differences in libration also increase proportionately as we move from the center toward the limb. Therefore, the accuracy of the determination of heights should be relatively insensible to the position of the object on the lunar disk, unless the observer experiences more trouble in measuring the position of formations in one part of the disk than in another.

Saunder felt that his measures became more accurate as the limb was approached, while my measures seem to be slightly less accurate in the outer half. In either case, the accuracies obtained are quite similar; the ranges appear to be no more than a ratio of 3 to 2. In this particular problem n was 2, as only successive pairs of plates were intercompared to yield height measures.

All in all, 733 lunar formations were measured. Of these, 37 were discarded, as it proved impossible to make satisfactory measures on a needed minimum of two plates in the following pairs: Plates 1 and 2, 2 and 3, 3 and 4, 3 and 5. Height determinations were made for 696 lunar formations. For nearly half this number, there were two determinations of height. This allowed a solution to be made for the probable error in a single height determination. It was found to be $\pm 39.7 \times 10^{-5}$ of the moon's radius, or $\pm 2{,}270$ feet, for the plate pairs 2, 3 and 3, 4. For the plate pairs 3, 4 and 3, 5 the probable error in a single height determination was found to be $\pm 48.8 \times 10^{-5}$ of the moon's radius, or $\pm 2{,}786$ feet. Each probable error is substantially less than that found by Schrutka ($\pm 4{,}030$ feet) or Saunder ($\pm 3{,}026$ feet).

The data are listed in Table 23 (Appendix 2). Columns 21–24 give the measured heights from the indicated pairs of plates. Column 25, $\bar{h}$, repeats the height measure from one of the preceding four columns if a single determination and gives a numerical average if there were two height determinations.

The very great majority of the formations measured were small craters. In a few cases they were isolated peaks or central peaks of craters. In a modest number of cases the markings were small white spots, which presumably were tiny ray formations around small craters below the limit of resolution of the photographic plates. The method of measuring was such that the height determination was that of the average crest of the crater rim, not the surface. In order to eliminate the effect of the varying sizes and varying rim heights, the following procedure was used.

The diameters of the craters were measured at least twice and the average values tabulated in column 26. Equation (7-6) gives a relationship between rim height and crater diameter for Class 1 craters, and most of the small craters measured were of Class 1. This curve was assumed to represent with reasonable accuracy these small craters, and the rim heights, column 27, were determined from it. The measured heights, $\bar{h}$, were then reduced by the rim height in each case to give the height, h, in column 28, for the lunar surface at the position of the formation relative to the mean sphere. Estimates of the heights of peaks were made from shadow lengths.

The height determinations may be compared with those of Schrutka, Franz, and Saunder in the next three columns.

Figure 37 shows four charts illustrating the accuracies of the height determinations. Figure 37, *A,* shows my measured heights as abscissae plotted against Franz's measures on the same points as ordinates. There is very little correlation. The 45° line represents perfect correlation; and, while there is a slight tendency to cluster around the line, it is very slight. The relationship between my measures and those of Saunder is shown in Figure 37, *B.* The correlation is better, but still not good. Figure 37, *C,* depicts the measures by Schrutka versus mine. The trend is clear statistically. The points Schrutka finds high are, on the average, high on my list, but still the scatter is uncomfortably large. Figure 37, *D,* shows my results compared with those of Hopmann. There clearly are large systematic differences.

Figure 38 is the plot of the results from my pairs of Plates 2 and 3 versus the

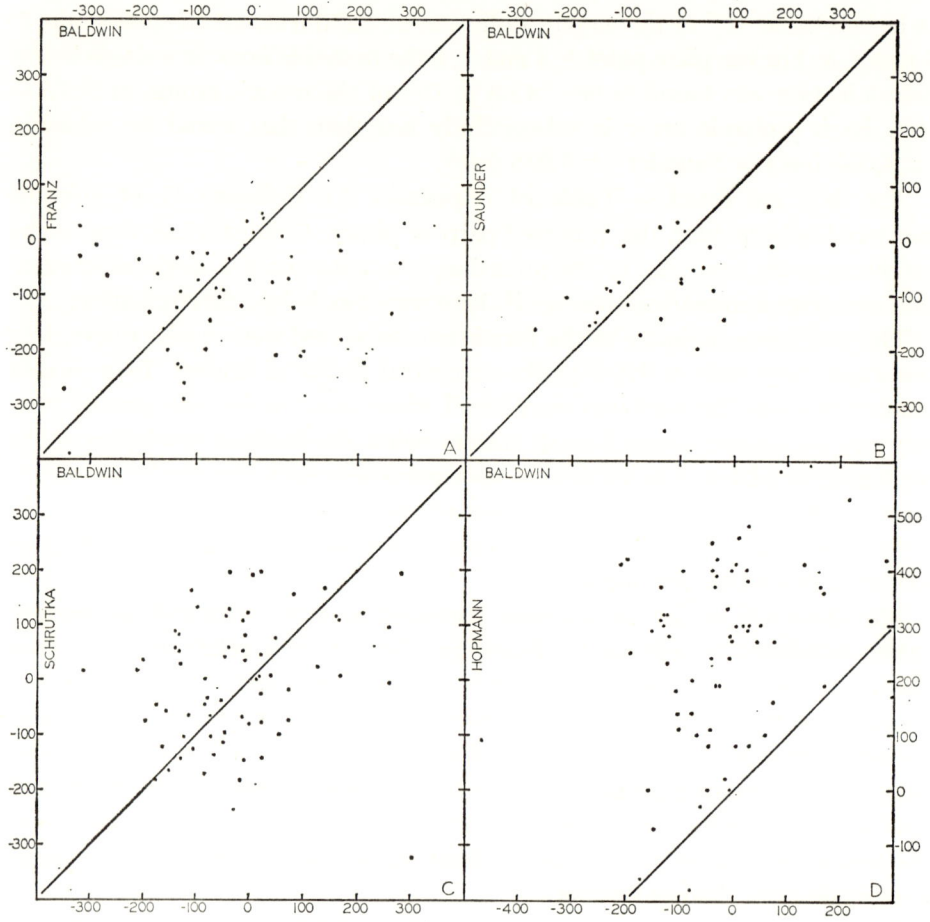

Fig. 37.—Correlations of individual height measures, relative to a sphere, of lunar features. (Baldwin versus Franz, Saunder, Hopmann, and Schrutka.)

heights derived for the same points on Plates 3 and 4. Here the scatter is markedly reduced, and there is no question of the trend. It was from these sets of data that the probable error of a given height determination of ±2,270 feet was established.

In passing, may I note that, even though this project has resulted in new accurate positions and heights for a large number of points and—as will be shown—they will yield a reasonably good contour map of the moon's surface, further improvement is still both possible and very desirable. In this respect,

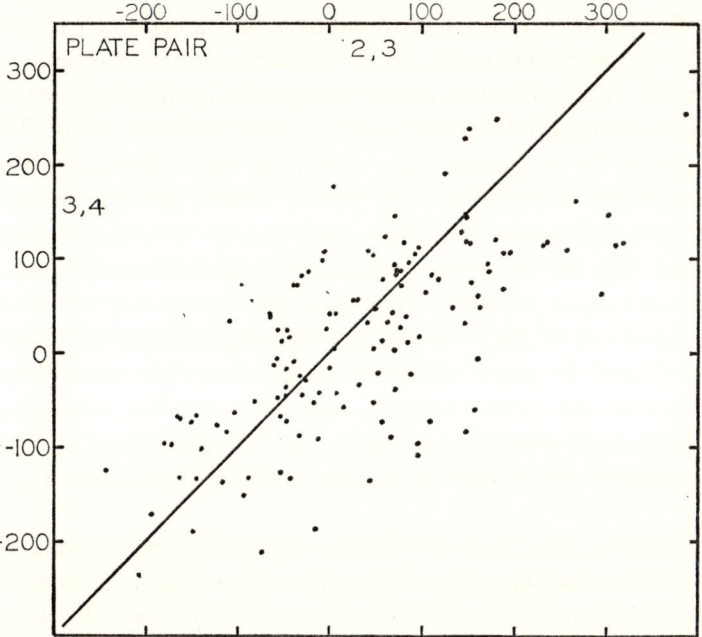

Fig. 38.—Correlations of individual height measures, relative to a sphere, of lunar features between plate pairs 2, 3 and 3, 4.

however, the measures have proved to be very difficult and time-consuming and the reduction of the measures to be extremely tedious. To improve on these results significantly would require larger-scale photographs, probably taken in infrared light with very large telescopes, to give clear pictures of very tiny craters. Improved measuring engines capable of handling the larger plates would be needed. Star images on the same plates to allow absolute plate constants to be found would be of great help. With all this technical advance, about twice the number of measured points would be required, say 1,500 minimum. Then the courageous man or woman who tackles this problem should plan to spend at least a year of full-time work, probably longer, on this job.

Many stages of the calculation can be done by IBM 650 computers, but large parts are not easily adapted to mechanical reductions. Portions of my calculations were run on the Northwestern University Dearborn Computing Center IBM 650 by Mr. Harry Rymer.

The formation of a contour map from these height determinations is straightforward. A dot was plotted on a 20-inch circle for each crater position. Beside each dot was written the corresponding height determination. Black was used for positive heights, red for the negative values. It was immediately apparent that the high and low areas on the moon were not distributed at random but that the red, or lower measures, were concentrated in the dark maria. This simply confirmed what Franz had found over half a century earlier.

Strip charts were then made, each covering a strip one-tenth of a lunar radius in width. Eighteen such charts were made by plotting measured height as a function of measured η in strips of $\xi = 0.9$–0.8, 0.8–0.7, etc. There are no measured points for ξ greater than ± 0.9. On each chart, the best line was drawn through the points. Similarly, twenty charts, plotting measured height versus ξ for one-tenth units of η, were made and the best lines drawn.

In the next step the two sets of charts were intercompared, and a new line was plotted on one set of charts. The new line represented the mean curve of height for the center of the strip and was halfway between the original line for this same strip and the points measured on the other right-angle strip charts for the same zone on the moon. Actually, there was excellent correlation between the north-south and the east-west charts. The resulting sets of strip charts gave a series of eighteen curves relating surface height as a function of η for steps of 0.1 in ξ.

At intervals of 0.1 in ξ from $\xi = 0.85$ to $\xi = -0.85$, points were plotted on a new chart at the values of η where h was 0, ± 50, ± 100, ..., ± 350 in units of 10^{-5} lunar radii. Connections of these points of equal altitude, using the strip charts as guides, were then made. The maria stood out obviously as low regions, and a fair map of the dark areas could have been made purely from the height determinations.

The nature of a contour map is such that lines cannot be drawn indiscriminately through the measured points. Changes of height are continuous, not discontinuous, and this forces certain relationships between contour lines and clarifies the positions of some lines which could not be explicitly placed from the measures. For example, there are slight but real differences in height on opposite sides of the Altai Mountains. It is clear that a single contour line cannot cross the high portions of this range. Therefore, the few points actually measured here force a skirting of the range for the $+50$ contour line, and sections of the entire Altai system through $360°$ are thus outlined, and its relationship to Mare Nectaris is emphasized. Certain regions on the moon show rapid changes in altitude, primarily in Mare Imbrium and Mare Serenitatis. The contour lines can be only schematic here, as they cannot be drawn closely enough spaced.

Several very interesting facts stand out. The great impact areas—Mare Humorum, Mare Nectaris, Mare Imbrium, Mare Crisium, and Mare Serenitatis—are all low lands. Southern Mare Foecunditatis is very deep. All of Mare Nubium and Oceanus Procellarum are deep, with the deepest point north of Mare Humorum. Strangely, a deep low zone is near Mare Australe, but, of all the low areas, this is the only one that is not filled with the dark material.

The deepest points are in Mare Imbrium, lower than −350 units. Three of the four highest regions are near the moon's center in the southwest quadrant. The other is east of Mare Humorum. These places show heights of more than +200 units. The total range of elevation is approximately 36,000 feet, or nearly 7 miles.

Based on this contour map, the region near the rim of the moon varies considerably but averages a little above 0 units high. Only in the southwest squadrant are there bright continental areas approaching the center. From any direction in this quadrant, the moon's surface systematically rises as the center is approached. This type of rise is characteristic of the bulge measured by so many others. In the southeast and northwest quadrants, there is real evidence that the surface rises from the limb toward the center, but the great dark areas intervene.

An even more striking bit of evidence is shown by the map. In every single case, the maria are variable in height. The shores are the highest parts, and they deepen toward the center. The impact maria are more symmetrical in all ways than the remaining dark areas, but, regardless of type, the dark matter fills the lowest parts of a larger system. The bright continental zones marginal to the maria, without exception, slope down toward the gray areas for about 100 miles, and the slope continues out onto the maria.

From this brief summary, we must conclude that the moon does possess a bulge generally aligned with the earth; the magnitude of the bulge cannot be precisely determined by inspection, but it appears to be substantial. The bulge is notably higher than the dynamically determined value, but this was a result that might have been forecast, because the dynamical bulge is an average figure which takes into account both the geometric shape of the entire moon and the variations in density. The maria are late-comers to the face of the moon and in their formation have significantly affected the bulge and general levels of the moon's surface.

Three of the available contour maps show many points of similarity and some differences. In general, they all show the bright areas to be uplands and the dark areas to be lowlands. A fourth, Ritter's map, does not agree well with the others.

Three maps indicate a low bright area south and east of Mare Australe. Ritter's map does not. Franz and I show the bright areas in the southeast quadrant to be high. Schrutka finds them low. In each case, the center of the disk rises as it is approached from the southwest quadrant.

Schrutka and I find Mare Foecunditatis low. Franz has no points in this region. Ritter finds it high.

The only major discrepancy is that Franz, from two points, and Ritter suggest that Mare Tranquillitatis is high, while Schrutka and I find it moderately high on the east and deeper in the center and west.

With allowances made for the large probable errors in heights and the few points on the maps of Franz and Schrutka, these maps are reasonably consistent with mine. Ritter finds Oceanus Procellarum and Mare Imbrium low, as do the others. He specifically finds all other places found to be low on the other maps to be high or average in height. He does agree with Watts that the west equatorial limb is low.

Two other projects should be mentioned at this point, although neither is yet complete. The Army Map Service of the Corps of Engineers, U.S. Army, Washington, D.C., is preparing a contour map of the moon where the contours are derived by a stereoscopic process. A preliminary copy of the western half of the moon has been issued, so that scientists may comment upon it and perhaps suggest improvements. This is a superb piece of work, and the printing of the map is beautifully done. The contours appear to be very accurate in local regions of the moon, but there is a question as to how accurate absolute contours will be from one section of the moon to another hundreds or thousands of miles apart. This project will undoubtedly have great value, particularly when instrumented vehicles, either with or without men, are landed on the moon; but it is probable that it will not add much to our knowledge of the gross contours and the nature of the bulge.

A similar project is being carried out by the U.S. Air Force Aeronautical Chart and Information Center of St. Louis, Missouri. This chart is being prepared with supervisory assistance from G. P. Kuiper, D. W. G. Arthur, and E. A. Whitaker. The positions of features on this chart have been determined through the use of selenographic control established primarily from the measures of Franz and Saunder.

The vertical datum is based on an assumed spherical figure and a lunar radius of 1,738 kilometers subsequently adjusted downward by 2.6 kilometers to minimize the extent of lunar surface of minus elevation values. Gradients of major surface undulations were established by interpolating Schrutka-Rechtenstamm's reductions of Franz's measurements of 150 moon craters. The probable error of comparative elevation values is estimated at 1,000 meters. The vertical datum so established is considered interim and will be refined as soon as an accurate figure of the moon is determined. The reproduction of this map is also excellent and quite comparable with that of the Army Map Service.

One last project whose conclusions are closely correlated with these contour data is the magnificent work of Watts at the Naval Observatory. For many years, Watts has been engaged in the detailed analysis of the variations in height in the regions of the limb. These measures can be made with accuracy and will ultimately result in a detailed contour map usable at all possible librations and covering a zone about 300 miles wide.

At my request, Dr. Watts drew up a detailed outline of the limb from two plates taken nearly at zero libration. The plates were taken at the United States Naval Observatory. The reductions on these plates approximate closely but are not exactly in the system of his final datum: Plate 112 of 1949 June 7.1 showed the west limb at $l = +0°.53$, $b = +0°.92$; Plate 227 of 1950 March 6.3 showed the east limb at $l = -0°.88$, $b = +0°.92$. Because of the foreshortening effect near the limb, a high spot, either in front of or behind the true limb, may dominate the contour for a considerable range of librations. These two plates were each taken near zero libration, but the apparent contour was not exactly that of the moon's surface at this libration because of this effect.

To eliminate the local irregularities from the systematic variations of the moon's surface at the limb, an approximate average height was taken from Watts's charts over 2° intervals. These variations were converted into units of 10^{-5} lunar radii and are tabulated around the limb on Figure 39. Through happenstance, the radius of the moon used by Watts agreed almost perfectly with that which I used, so his numerical data are consistent in absolute value with mine.

If we start at the south pole and move counterclockwise, we find that the polar limb values are missing, as on these plates the deep shadows affected the measures. A low area appears to agree well in position and magnitude with that inferred from the stereoscopic determinations. It circles Mare Australe, at least in the visible half, and shows twice on the limb measures. Immediately behind Mare Australe, the limb is high.

Directly west of Mare Foecunditatis, the limb is again high. When we reach Mare Smythii and Mare Marginis, it sinks markedly. Beside Mare Smythii is the lowest part of the limb. There are very few measures on the disk near this part of the limb, but the whole region, from about $\eta = -0.1$ to $\eta = +0.3$ and from Mare Foecunditatis and Mare Crisium to the limb, is full of dark zones and craters with dark interiors. Some observers have said that this region gives the impression of being marshy. It certainly is low.

Below Mare Marginis, the limb is systematically high for a considerable distance; and it is near here that the second largest bright continental, or upland, area is found. The few disk measures also show this area to be high.

236   *Shape of the Moon and the New Contour Map*

Watts finds the limb to be again low as it passes Mare Humboldtianum and then to rise to an average height, which it maintains past the pole and parallel to Mare Frigoris.

As Mare Frigoris blends into Sinus Roris and Oceanus Procellarum, the dark area approaches the limb. Coincidentally with this, the height of the surface along the limb becomes lower; and from the 20°–25° where Oceanus Procellarum essentially reaches the limb, the limb height measures become negative.

FIG. 39.—Contour map of the moon

As Oceanus Procellarum recedes from the limb, starting about $\eta = +0.4$, the rim heights rise again. A modest peak is shown in the region of the D'Alembert Mountains, which are mostly beyond the limb at mean libration, but, as the limb continues south, it drops below normal height for a considerable distance. There is no obvious reason why this dip should be where it is, but, judging by other similar zones, we might expect a considerable dark area just around the limb. Existing photographs of the back side of the moon do not show this section.

Wilkins and Moore (27) describe Mare Orientalis in this general region. It is apparently a large, shallow crater much like Bailly.

The limb near the Rook Mountains is high, and from the south the "marshy" area near Schickard and Wargentin is near another extensive low portion of the rim. The Doerfel Mountains appear very high, and, surprisingly, the rest of the limb to the south pole is low at this libration.

The correlations between the limb height measures by Watts and the stereoscopic height determinations on the disk are considered rather good. As a general thing, the limb is low when it lies near dark areas or "marshy" areas and is found to be high when a considerable expanse of bright continental land lies nearby.

From examination of the contour map (Fig. 39), it is clearly evident that the dark maria are, without exception, low. The ground slopes downward toward the maria for something of the order of 100 miles, except in the cases of the mountainous borders of the circular maria, and even here the dark centers are very low, and immediately outside the mountains are low annuli. It is gratifying to see that Watts's limb measures yield the identical relationship.

Because of these local distortions, sometimes amounting to miles, the plot of the contour of the limb, Figure 40, *A*, shows wide fluctuations which mask the normal limb contour. No systematic trend can be detected. The portions of the limb which have been affected by the maria and the limb mountains are indicated in Figure 40, *A*. In addition, both Watts's data and the contour map show that Mare Australe is circled by a major low ring. The intersections of this ring with the limb are also marked.

Mountain ranges clearly visible on the front side of the moon are always associated with large craters and circular maria, both of which are impact structures. It is not absolutely certain that this is the case with the limb mountains, but it is probably true there also. Only the Rook and the Doerfel Mountains are clearly visible at mean libration. The positions of these two ranges are also shown in Figure 40, *A*. All other limb ranges are nearly or completely hidden by the limb.

Inasmuch as both mountain ranges and maria are superficial structures which distort the limb locally but do not systematically modify it, it would

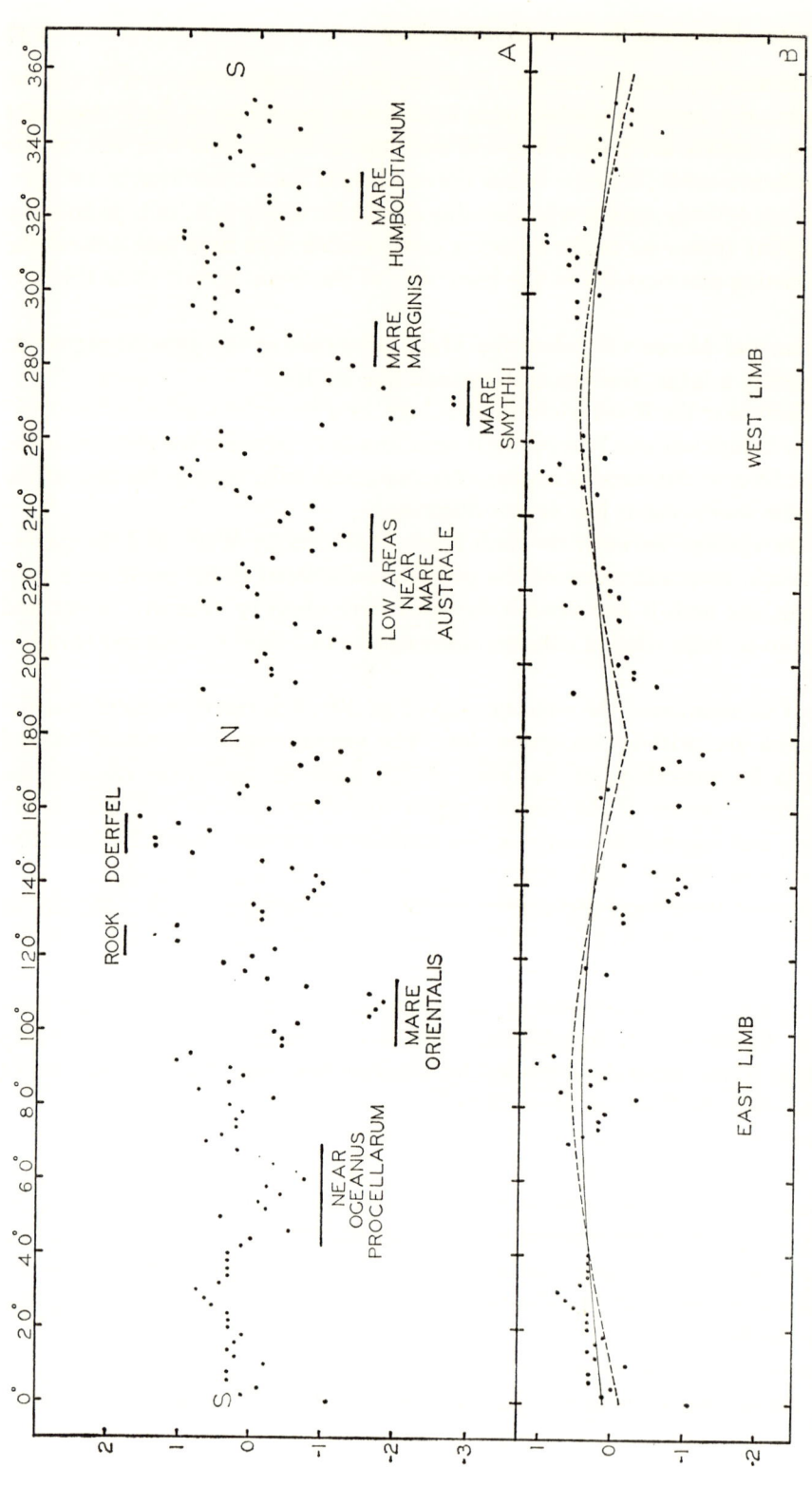

Fig. 40.—Limb-height measures by Watts, two-degree averages. *A*: all measures, with indicated high regions near mountain ranges and low regions near maria. *B*: those limb-height measures distant from mountain ranges and maria. *Dashed line*—limb contour from least-squares solution of Watts's data; *solid line*—limb contour from combined data (eq. [11-48]). *Horizontal scale*: degrees of arc around limb; *vertical scale*: unit is 100×10^{-5} lunar radii, or 5,710 feet.

Shape of the Moon and the New Contour Map

seem logical that the limb height measures near these structures should be eliminated and the bulge sought among the limb measures distant from the mountains and maria. Figure 40, *B*, shows these remaining limb height measures. They clearly indicate a smaller polar diameter than the equatorial diameter in the plane of the sky.

In the data of Watts there are 103 altitudes of points plotted in Figure 40, *B*, from those portions of the limb that are distant from maria and mountain ranges and should yield a contour of the undisturbed limb. A least-squares solution gave the polar axis as 0.99986 and the equatorial axis in the plane of the sky as 1.00058. This solution is plotted on Figure 40, *B*, as a dotted line. The mean height of the limb is $+22$ units.

We may apply the same principle to the measures of altitude on the face. Of the 696 points, more than half are in the maria or in the Apennine Mountains or so close to the shores of the maria that they lie in the downwarped areas. With these points eliminated, there are 218 points which should be representative of the continental regions. A least-squares solution yielded the three axes as follows:

$$\zeta = 1.00128, \quad \xi = 0.99992, \quad \eta = 1.00024. \qquad (11\text{-}46)$$

The average height of the limb is $+8$ units or 14 units (799 feet) lower than the radius used by Watts. The results clearly indicate a substantial bulge.

Jeffreys has shown that the departure of each axis from the mean sphere should be in the ratio of $+35, -10,$ and -25, regardless of the distance of the moon from the earth, provided that the form of the moon is that of a tidal bulge or a fossil tidal bulge. Applying this criterion to the results of the calculations on the bulge as defined by the continental points on the face, we find that the axes should be in the ratio of

$$\zeta = 1.00128, \quad \xi = 1.00025, \quad \eta = 0.99991. \qquad (11\text{-}47)$$

The measured axes in the plane of the sky differ from the theoretical axes by only 33 units, or 1,884 feet. Considering the small leverage afforded by the height measures scattered over the continental regions, this difference is not overly significant.

The Watts radius in the plane of the sky is 14 units greater than that derived from my measures. If these 14 units are subtracted from Watts's data, they should then be consistent with and on the same scale as my measures and may be combined to yield a triaxial spheroid. A least-squares solution from the 321 points gives

$$\zeta = 1.00135, \quad \xi = 1.00030, \quad \eta = 0.99995. \qquad (11\text{-}48)$$

In this case the radius in the plane of the sky is found to average 1.00012 rather than the expected 1.00008. The radius of the mean sphere is 1.000535.

The three components of a fossil tidal bulge scaled to equation (11-48) and proportional to Jeffrey's formulae (10-4) would be

$$\Delta \zeta = +81.5 \text{ units}, \qquad \Delta \xi = -23.3 \text{ units}, \qquad \Delta \eta = -58.2 \text{ units}. \qquad (11\text{-}49)$$

From these we derive the three axes as

$$\zeta = 1.001350, \qquad \xi = 1.000302, \qquad \eta = 0.999953, \qquad (11\text{-}50)$$

which are in perfect agreement with the three measured axes, (11-48). Actually the agreement is better than the precision of the solution would indicate, but nevertheless, if the original assumption be granted—namely, that the primal shape of the bulge may be determined from the limb height measures of Watts combined with my face height measures, both on the continental areas distant from mountains and maria—then we must conclude that the moon does still possess a fossil tidal bulge which shows three axes departing from a mean sphere by

$$\zeta = +4{,}655 \text{ feet}, \qquad \xi = -1{,}330 \text{ feet}, \qquad \eta = -3{,}325 \text{ feet}, \qquad (11\text{-}51)$$

and the axis aligned with the earth is 6,980 feet greater than the average axis in the plane of the sky. The solid line in Figure 40, B, is consistent with this solution but is raised 14 units to agree with the scale of Watts's data.

If the isostatic adjustments of the lava-covered areas have caused significant altitude changes in the continental regions beyond the marginal downwarpings, then the magnitude of the present bulge will less accurately represent the original bulge. The close agreement with the theoretical shape of the moon at a lesser distance than at present is evidence that the moon's bulge was controlled in size and shape by the tidal pull of the earth during the period when the maria were formed and the isostatic adjustments occurred and past the time when the moon stiffened so that it could no longer adjust its shape as it fled the earth.

A similar analysis may be made from the points lying within the maria, but the results must be regarded as questionable from several points of view. The maria exhibit wide variations in altitude, and these variations are systematic, in that the circular maria are deep basins and the irregular maria are shallower basins, but in each case the shore lines are higher than the centers. Clearly, the dark areas suffered vertical adjustments after they were formed.

Because of the peculiar nature of the circular maria, all points lying in them have been omitted from the following analysis. All other points on the maria, 198 in number, have been used to determine the three axes which they define. The results are $\zeta = 0.99938$, $\xi = 0.99871$, and $\eta = 0.99936$. There is thus a bulge of 3,826 feet for the axis aligned with the earth relative to the equatorial axis in the plane of the sky. This is somewhat less than the bulge as shown by the upland points but is larger than the dynamical bulge.

The η-axis is poorly determined, for there are only 23 points whose η co-

Shape of the Moon and the New Contour Map 241

ordinates are greater than 0.60 and but 12 points beyond $\eta = 0.75$. All these 23 points lie in the northern hemisphere.

The unequal distribution of the maria suggests that the intercomparison of the bulges as determined on the bright and dark regions should be between the ζ and ξ co-ordinates. These give 5,985 feet for the bright portions and 3,826 feet for the dark. Because of the real height variations within the maria, the agreement probably is as good as could be expected.

The maria are found to be depressed relative to the uplands by approximately 2 miles. This figure would have been increased if the data for the circular maria had been included. The derived bulge from the upland regions is

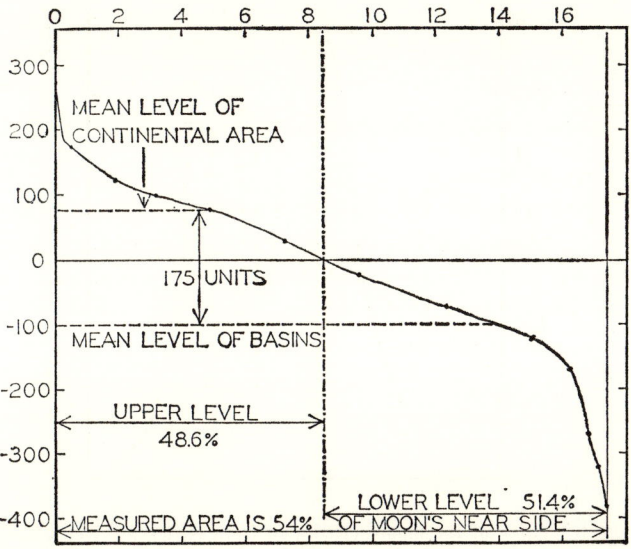

FIG. 41.—Hypsographic curve of the moon. (Redrawn from Hédervári, Ref. 28.) *Horizontal scale:* unit is 1,000 quadrate degrees; *vertical scale:* unit is 100×10^{-5} lunar radii, or 5,710 feet.

accepted as more closely representative of the form of the moon unmodified by isostatic adjustments. A height of 6,980 feet for the bulge aligned with the earth relative to the mean radius in the plane of the sky corresponds to a maximum separation of 69,000 miles between the earth and moon when the final change of form occurred.[1]

Hédervári (28), working from the preliminary announcement (29, 30) of the present contour map of the moon, has constructed hypsographic and hypsometric curves (Figs. 41 and 42). The extreme limb regions are not represented on the contour map. Hédervári's curves thus consider only the central 54 per cent of the moon's surface.

As is obvious from the differing appearance of the two hemispheres, the hypsographic curves for each differ. The upland areas of the southern hemi-

[1] See Appendix 3.

sphere greatly exceed the maria in area, while the reverse is true in the northern hemisphere. The hypsographic curves show two almost equal morphologic areas. This differs from the earlier charts (31-35), which gave a much larger continental area. Possibly the earlier charts included more limb areas, which would give this effect.

Hédervári's hypsometric curve shows a line with two maxima. The difference in level is about 175×10^{-5} of the radius of the moon, or perhaps 3,040 meters. On the earth the difference between the two main morphologic levels is about 4,800 meters.

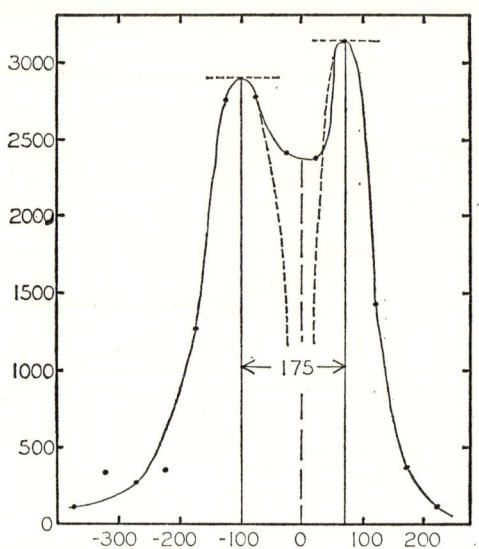

Fig. 42.—Hypsometric curve of the moon. (Redrawn from Hédervári, Ref. 28.) *Horizontal scale:* unit is 100×10^{-5} lunar radii, or 5,710 feet; *vertical scale:* unit is quadrate degrees.

Hédervári computes the thickness of the crust of the moon above the level corresponding to the Mohorovičić discontinuity as about 20 kilometers, as contrasted with Joksch's earlier estimate of 45 kilometers. The value for the thickness of the earth's continental crust is intermediate between these two figures.

This last conclusion should be treated with caution, because on the earth the continents apparently are derived by volcanic action from an older surface and materials below it and are thus younger than the basaltic layer which lies below the bottoms of the oceans and the continents. On the moon it is the darker and lower maria that are the younger features.

The equivalent of a hypsometric curve may be derived from Watts's limb data. There is a slight suggestion of the double hump for the west limb heights but none at all for the east limb. This is due to the fact that, in general, the maria do not extend past the limb.

Clearly, the moon is not now in isostatic equilibrium. The existence of a definite bulge, as determined dynamically, shows that. Two alternatives are left us. At some time in the past the moon was in isostatic equilibrium or at least close to it, or else it never was in balance.

If it never were in isostatic equilibrium, the moon must have had and now has substantial strength to enable it to resist for billions of years the pull of gravity. In this case, the shape of the moon cannot tell us much concerning the early history. Its shape must have been due to irregularities which developed during accretion. The broad, gentle slopes of the major undulations in the moon's surface belie this concept.

If it were ever in equilibrium, the moon must have been far closer to the earth than at present. In this case, the bulge is real. Its height would be a function of the distance between the earth and moon; and the corollary is that the moon at that time was capable of adjusting its shape to the tidal distortions produced by the earth. At the same time, the moon had a hard crust, for myriads of ancient craters were then formed. Later the lava flows of the maria developed and produced isostatic adjustments which greatly distorted the tidal bulge.

After the maria came, the moon became rigid while still so close to the earth that the modified bulge was still high. Then, as the earth and moon receded from each other, the strengths of the lunar rocks must have been enough to maintain a bulge of the order of 2 miles in height. Urey (36), from arguments based on Haskell's theory (37, 38) of viscosity of liquids, does not agree that the moon ever was weak enough to adjust isostatically.

It is my conclusion that the evidence clearly shows that the moon had to adjust isostatically under any circumstances. The effective lunar viscosity, particularly in the outer layers, is substantially greater than has been heretofore assumed. This is consistent with the observed fact that the lunar craters, which presumably are nearly as old as the moon, still exist, while they would quickly have disappeared on the earth because of isostatic adjustment according to the formulae of Haskell (49).

S. K. Runcorn (50) concludes that the observed bulge could be supported by convective currents which are driven by a temperature difference of 50° C. This, combined with the smaller size of the moon, results in a heat flow of one-fourth that of the earth per unit area. The velocity of flow of materials in the moon's mantle must then be one-fortieth that in the earth. The lunar viscosity will be two orders of magnitude higher than that in the earth's mantle, presumably the result of the lower pressures in the moon.

In this case, the bulge results from a second-order convection pattern; and the moon's surface, while not spherical, is an approximation to an equipotential surface disturbed by local irregularities. This offers a method of producing the density variations suggested by Urey, Elsasser, and Rochester (51) as an interpretation of the dynamically measured deviation of the moments of inertia of the moon from those of a sphere.

Runcorn's rising column of high-viscosity liquid pointing toward the earth would be warmer and hence less dense than other parts of the moon. This leads to a bulge aligned with the earth and a low ring at the limbs of the moon, where the denser currents descend after cooling.

Even though the moon's interior were in a state of convection and the surface in an equipotential condition, a dynamical bulge would appear, albeit lower than the physically measured bulge.

The assumption of a present-day convection in the moon is an interesting one and well worth pursuing. It leads to possible contradictions with the observed high rigidity of its outer layers.

Recent measures on orbital changes of man-made satellites of the earth tend to give corroborating evidence that the earth has greater rigidity than theory would permit. The interior of the earth is hot. The core, which is large, is liquid. The entire earth is still capable of adjusting isostatically to loads and relief of loads and to adjust to centrifugal forces. Yet, in spite of its effective plasticity, the earth has shown evidences of unexpected strength. It is possible that Runcorn's convection mechanism could produce distortions of the earth and possibly obviate the need to postulate great rigidity.

O'Keefe, Eckels, and Squires (39) from the orbit of the Vanguard satellite $1958\beta_2$ have determined that the earth has a slightly pear-shaped figure. This calculation indicates that the periodic changes in the orbital eccentricity of the satellite could be explained by the presence of a third zonal harmonic in the earth's gravitational field. In the present case, the third zonal harmonic modifies the geoid so that the stem of the pear is toward the north pole. The bulge of the pear is in the southern hemisphere. The amplitude of the harmonic is 0.0047 cm/sec^2 in the surface acceleration of gravity, or 15 meters of undulation in the geoid.

Vening Meinesz and Heiskanen (40) are quoted by O'Keefe (39):

> ... Assume that the earth's gravitational field is very nearly that of a fluid in equilibrium. They consider that the deviations from such an ellipsoid, in any given area, do not exceed about 30 milligal-megameter units—that is, they assume that one will not find deviations of more than 30 milligals over an area of 1000 kilometers on a side, or deviations of more than 3 milligals in an area 3000 kilometers on a side.

While (39):

> Our determination of the third-degree zonal harmonic shows that the hypothesis of Vening Meinesz and Heiskanen is not justified; for example, each of the polar areas has a value of about 120 milligal-megameters, and each of the equatorial belts a value more than twice as great.

The presence of a third harmonic of the amplitude [41] indicates a very substantial load on the surface of the earth. Following the arguments of Jeffreys, we may calculate the values of this load and the minimum stress required in the interior to support it. We find a crustal load of 2×10^7 dy/cm². We can choose between assuming that stresses of approximately this order of magnitude exist down to the core of the earth, or that stresses of about 4 times that amount exist in the uppermost 700 kilometers only. These stresses must be supported either by a mechanical strength larger than that usually assumed for the interior of the earth or by large-scale convection currents in the mantle.

Carey (42) has criticized the above conclusions on the basis that the same observations would be expected if the mean density of the mantle in the southern hemisphere is a little less than that in the northern hemisphere, but O'Keefe (43) has refuted this conception.

Cohen and Anderle (44) from observations of satellite Transit 1B (1960γ) have confirmed O'Keefe's conclusion that there exists a third-order harmonic in the earth's gravitational field. This confirmation is remarkable, in that the satellite orbit, the observational method, and the orbit-fitting procedures were all different from those used in the earlier study of Vanguard. King-Hele and Merson (45) had earlier suggested a similar north-south asymmetry in the earth's shape from observations of Sputnik 2. Kozai's calculations (46) give similar results.

A completely different distortion of the earth has been found recently by Izsak (47), of the Smithsonian Astrophysical Observatory. From accurate photographs of satellite positions (Vanguard II and III) he has determined that the equatorial cross-section of the earth is not exactly circular. The diameter passing through a point in the Atlantic just off Brazil, longitude 33°.15 west within a half-degree, is about 1,300 feet longer than the equatorial diameter at right angles to it. Reductions of data from the two satellites gave quite concordant results.

Since this observation a quiet argument has raged (48). Kaula, of NASA, from Minitrack observations of Vanguard I found that the equator was out of round only 10 per cent as much as Izsak determined. Then Kozai, of the Smithsonian Astrophysical Observatory, confirmed this result from observations of Vanguard I and II and Sputnik 1. More recently, Newton, of the Applied Physics Laboratory of Johns Hopkins University, raised the two-headed equatorial bulge back to 1,000 feet from reductions of Transit satellite observations. Regardless of the exact magnitude of the bulge, it exists, and the general size of it is closely determined.

These two sets of observations show that the earth, which is still capable of isostatic adjustments, does possess unexpected strength, and, by inference, the observed strength of the moon, with due account taken of its lower gravity, does not imply that the moon must be cold or cool in the interior.

REFERENCES

1. Franz, J. "Die Figur des Mondes," *Astr. Beobachtungen, Königsberg,* Vol. **38,** 1899.
2. Hayn, F. "Die Rotationselemente des Mondes und der Ort von Mösting A," *A.N.,* Vol. **165,** col. 305, 1904.
3. Pickering, W. H. *Harvard Ann.,* Vol. **51,** chap. iv, 1903.
4. Saunder, S. A. "First Attempt To Determine the Figure of the Moon," *M.N.,* **65,** 458, 1905.
5. Baldwin, R. B. *The Face of the Moon,* p. 187. Chicago: University of Chicago Press, 1949.
6. DuFresne, E. R. "Note on the High Relief of the Lunar Maria," *Ap. J.,* **124,** 638, 1956.
7. Baldwin, R. B. *The Face of the Moon,* p. 191.
8. Ritter, H. "Versuch einer Bestimmung von Schichtlinien auf dem Monde," *A.N.,* **252,** 157, 1934.
9. Jakowkin and Belkowitsch, J. "Zur Frage nach der Bestimmung der Mondfigur, vermittels Terminatorbeobachtungen," *A.N.,* **256,** 305, 1935.
10. Mainka, C. "Untersuchung über die Verlängerung des Mondes nach der Erde zu," *Mitt. Königliche U.-Sternw. Breslau,* **1,** 55–70, 1901.
11. Weimer, T. "Enregistrement de profils lunaires," *C.R. Inst. de France, Acad. Sci.,* **230,** 1834, 1950.
12. ———. "Atlas de profils lunaires," *Pub. Obs. de Paris,* 1952.
13. ———. "Recherches sélénographique; allongement de sélénoide, libration physique; profils lunaires," *Bull. Astr.,* **17,** 271, 1954.
14. Wirtz, C. "Zur Figur des Mondes," *Festschrift Heinrich Weber,* p. 480. Leipzig: Teubner, 1915.
15. Franz, J. "Ortsbestimmung von 150 Mondkratern," *Mitt. Sternw. Breslau,* Vol. **1,** 1901.
16. Hopmann, J. "Selenodätische Untersuchungen," *Mitt. Sternw. Wien,* **6,** 13, 1952.
17. Schrutka-Rechtenstamm, G. "Neureduktion der 150 Mondpunkte der Breslauer Messungen von J. Franz," *Mitt. Sternw. Wien,* **9,** 251, 1958.
18. Schrutka-Rechtenstamm, G., and Hopmann, J. "Die Figur des Mondes," *Sitzb. d. Öst. Akad. Wiss. Math.-naturw. Kl.,* Abt. II, p. 167. Vols. **8–10,** 1958.
19. Schmidt, J. F. J. *Die Charte der Gebirge des Mondes.* Berlin: Dietrich Reimer, 1878.
20. Lohrmann, W. G. From unfinished map. Published by Schmidt, 1878.
21. Beer, W., and Mädler, J. H. *Der Mond.* Berlin: Simon Schropp, 1837.
22. Neison, Edmund (pseud.), Neville, E. N. *The Moon and the Condition and Configuration of Its Surface,* p. 576. London: Longmans, Green & Co., 1876.
23. Saunder, S. A. "The Determination of Selenographic Positions and the Measurement of Lunar Photographs, First Paper," *M.N.,* **60,** 174, 1900.
24. ———. "The Determination of Selenographic Positions and the Measurement of Lunar Photographs, Second Paper: Determination of a First Group of Standard

Points by Measures Made at the Telescope and on Photographs," *ibid.*, **62,** 41, 1901.
25. ———— "The Determination of Selenographic Positions and the Measurement of Lunar Photographs, Third Paper: Results of the Measurement of Four Paris Negatives," *Mem. R. Astr. Soc.*, **57,** 1, 1908.
26. ————. "The Determination of Selenographic Positions and the Measurement of Lunar Photographs, Fifth Paper: Results of the Measurements of Two Yerkes Negatives," *ibid.*, **60,** 1, 1915.
27. WILKINS, H. P., and MOORE, P. *The Moon, a Complete Description of the Surface of the Moon, Containing the 300-Inch Wilkins Lunar Map.* New York: Macmillan Co., 1955.
28. HÉDERVÁRI, P. Prepublication information. Letter of June 14, 1961.
29. BALDWIN, R. B. "A Lunar Contour Map," *Sky and Telescope,* **21,** 84, 1961.
30. ————. Translation of "A Lunar Contour Map," *Kagaku Yomiuri,* **7,** 75, 1961. Tokyo: Press Yomiuri.
31. JOKSCH, H. C. *Zs. f. Geophys.*, **25,** 3, 1957.
32. BROCKHAUS, K., and JOKSCH, H. C. *Zs. f. Geophys.*, Vol. **26,** 1960.
33. ENGEL, K. H. *J. ILS*, 1959.
34. HÉDERVÁRI, P. *J. ILS*, 1959, 1960.
35. ————. *Beitr. z. Geophys.*, 1961.
36. UREY, H. C. "Origin and History of the Moon," prepublication copy, 1961.
37. HASKELL, N. A. *Physics,* **6,** 265, 1935.
38. ————. *Ibid.*, **7,** 56, 1936.
39. O'KEEFE, J. A., ECKELS, A., and SQUIRES, R. K. "Vanguard Measurements Give Pear-shaped Component of Earth's Figure," *Science,* **129,** 565, 1959.
40. HEISKANEN, W. A., and MEINESZ, F. A. V. *The Earth and Its Gravity Field,* p. 72. New York: McGraw-Hill Book Co., Inc., 1958.
41. O'KEEFE, J. A., and BATCHLOR, C. D. *A.J.*, **62,** 183, 1957.
42. CAREY, S. W. "North-South Asymmetry of the Earth's Figure," *Science,* **130,** 978, 1959.
43. O'KEEFE, J. A. "North-South Asymmetry of the Earth's Figure," *Science,* **130,** 979, 1959.
44. COHEN, A. J., and ANDERLE, R. J. "Verification of Earth's 'Pear Shape' Gravitational Harmonic," *Science,* **132,** 807, 1960.
45. KING-HELE, D. G., and MERSON, R. H. *J. Brit. Interplanetary Soc.*, **16,** 471, 1958.
46. KOZAI, Y. *Smithsonian. Ap. Obs., Spec. Rept.*, No. 22, 1959.
47. IZSAK, I. G. "Ellipticity of the Earth's Equator," note in *Sky and Telescope,* **21,** 275, 1961.
48. Note in "Science and the Citizen," *Sci. American,* **205,** 82, 1961.
49. HASKELL, N. A. "The Viscosity of the Asthenosphere," *Amer. J. Sci.*, **33,** 22, 1937.
50. RUNCORN, S. K. "Convection in the Moon," *Nature,* **195,** 1150, 1962.
51. UREY, H. C., ELSASSER, W. M., and ROCHESTER, M. G. "Note on the Internal Structure of the Moon," *Ap. J.*, **129,** 842, 1959.

12

NATURE OF THE LUNAR-SURFACE MATERIALS AS DETERMINED BY REFLECTED LIGHT

The nature of the lunar-surface materials may be studied by numerous methods. Reasonably consistent results concerning their physical nature may be obtained, but essentially nothing is known of their chemical nature.

Almost one hundred years ago, Zöllner (1) determined from the measured amounts of light received from the moon at different angles, i.e., different portions of the lunar surface at a given phase, that the average slope of the materials composing the moon's crust is 52° from the horizontal in the brighter or continental regions. Zöllner worked in the visible spectrum not far from 5,000 A. This average angle of 52° is not an angle of repose. On earth, the angle of repose for most substances is between 30° and 40°. For crushed limestone, iron, and copper ores and similar materials, it is about 37°. For Lake Michigan sands, it is about 32°. With a lesser gravity, as would obtain on the moon, the angle of repose would be larger but not greatly so.

The measured angles of the great slopes on the moon are all relatively small. Ashbrook has shown that even the face of the Straight Wall has a slope of about 41° ± 3° (2). Consequently, the average value of 52°, coupled with the small wave lengths at which the observations were made, indicates that the lunar surface in the bright regions is composed of a rough, loose surface of cracked and shattered stone and smaller particles down to and including a fine dust.

Fesenkov, Parenago, and Staude (3) have concluded from the high directivity of the back-scattering of light from the lunar surface that it is composed of comparatively large grains, loosely bound together and partially shadowing each other.

The lesser variations of reflected light with phase angle for the maria and the darker color suggest that the sharp angularity of the uplands is not repeated, at least on as wild a scale. A surface of steep, rough, and vesicular lava and some dust is indicated.

The brightness of the moon changes uniformly with phase angle near full moon (4). The average value over the same phase angles for thirty-four asteroids is 0.030 mag. per degree. For the moon, it is 0.028; and for Mercury it is 0.032. Mars, Venus, and the earth, which have atmospheres, have similar values around 0.015. Gehrels (5) found 0.03 mag. per degree for 20 Massalia for phases between 7° and 20°, but at phase angles smaller than 7° the factor was larger.

Any material which reflects and scatters light will polarize it in a fashion and to a degree that will yield information on the nature of the material. Arago (6) first discovered in 1811 that the moon's light was polarized. Secchi (7, 8) showed that the polarization effects were not produced in a lunar atmosphere but were from the scattering of light from the moon's hard crust.

In 1873 Petrushevskiy (9) gave a sound theoretical foundation to polarization studies of the moon. In the same year, Rosse (10, 11) made the first measures of the degree of polarization as a function of lunar phase. He showed that the maximum polarization was attained at the quadratures.

In 1890, Landerer (12–14) tried to determine the composition of the lunar rocks in the maria from polarization measures. He found the angle of complete polarization to be about 33°, and from the computation of the refractive indices for the maria he concluded that the maria were composed of vitroporphyry.

Lyot at the French Observatory at Meudon has determined the variation in the fraction of the moon's light, which is polarized as a function of the angle of vision (15, 16). A characteristic curve is obtained similar to that given in the laboratory by a surface covered by volcanic ash or dust.

The proportion of the light which is polarized does not suggest a lava surface for the brighter regions. It varies inversely with the reflectivity of the portions of the moon which have been studied, as Secchi showed (7) in 1859. The maria give consistently higher values than do the upland areas.

The plane polarization of light varies markedly with lunar phase angle. For the maria, there is a maximum polarization at phase angles 100°–110° and at

280°–290°. The peak values range between 10 and 16 per cent for different maria. The polarization is zero at 22°–23° and at 327°–328°. At 0°, full moon, there is a slight negative polarization of around 1 per cent. The upland areas show identical variations but with approximately half the amplitude.

Markov (17, 95) showed that dark-bottomed craters, such as Schickard or Grimaldi, exhibit less polarization than do maria of the same albedo.

In 1923 Barabashov (18) showed that the maxima of polarization were so broad that no positive identification with terrestrial rocks could be made.

Laboratory work shows that dark, opaque rocks and other substances polarize light more or less completely at certain phase angles. Light-colored rocks and materials into which light can penetrate, even for short distances, and be reflected, polarize light relatively little.

Lyot has found the polarization of the moon, Mercury, and the asteroid Vesta to be essentially the same. Since, in all probability, we cannot assume that Vesta is covered with volcanic ash, it seems likely that there are considerable quantities of meteoritic dust and associated fragmental materials on this tiny body.

The low percentage of plane polarization observed in the moon is evidence that in neither the uplands nor the dark lowlands do we have tremendous exposures of bare rock. A modest layer of dust, ash, or sandlike material must cover most of the moon.

In 1927, Wright (19) found that exposed surfaces of basalt, obsidian, or any other basic rock yielded a degree of polarization in reflected light considerably above the low values observed for the moon. His results are reasonably consistent with the presence on the moon of substances like pumice or powdered rock. He suggested that the powders could have been formed from granite or sandstone but not from basalt.

Sinton (20) in 1956 concluded that the surface dust might have been formed from basalt on the basis of a mass-absorption coefficient of the surface material for a wave length of approximately 1.5 mm, which was determined from observations with optical techniques.

Lyot did not discuss the variations in polarization with color. Umov (21, 22) and Toporets (23) have found that when polarized light is reflected from dull-colored surfaces, there is a selective depolarization which is a minimum for those colors that are absorbed most strongly by the surface. For the moon, the greatest polarization must occur in the blue end of the spectrum.

Dollfus and Cailleux (24) have thoroughly studied the relationship between the form of the polarization curve and the physical and chemical structure of various materials. They concluded that (*a*) vitreous sands do not give negative polarization; (*b*) quartz grains produce a very small negative polarization; (*c*) powdered substances which absorb light weakly produce a weaker negative polarization and smaller values of the inversion angle, α, than the moon; (*d*) opaque granulated substances (emery) are more similar to the moon in terms of these two characteristics of the polarization curves; and, (*e*) finally, powdered and opaque substances of the type of volcanic ashes give, according to Dollfus, a negative polarization down to —3 per cent at $\alpha \approx 5°$ and a transition of the polarization through zero at $\alpha \approx 15°$. Dollfus and Cailleux interpreted these factors as implying that the moon was covered with volcanic ash.

Dollfus (25) showed that quartz sand having grains with crystalline faces had polarization characteristics quite different from the moon. In general, his results indicate a tendency for the negative polarization to increase as the albedos and the dimensions of the particles decrease; also, the angle at which the polarization changes from $+$ to $-$ increases as the grain size becomes smaller. He concluded that the moon's top surface was highly pulverized.

TABLE 24
THE ALBEDOS OF SURFACES OF VARIOUS MATERIALS (30, 97)

Material	Albedo	Material	Albedo
Snow	0.9 –0.5	Sandstone	0.22
Clouds	.9 – .6	Clay, schist	.25
Limestone	.56	Granite	.24
Sand	.34– .29	Basalt	.14
Ice	.37	Stone meteorites	.18
Water	.45–0.03	Fusion crust of meteorites	.05
Trachyte lava	.10	Chernozem (black soil)	0.05–0.07
Basalt lava	0.06		

Extensive studies by Fesenkov and his students (26) in 1935–41 and by Dzhapiashvili (27) in 1957 have related polarization changes with phase angle for numerous areas on the lunar surface as small as 25 km in diameter. They have verified and amplified the work of Lyot and Dollfus. They, too, concluded that the moon must have a fine-grained surface layer. Gehrels' new measures (28, 29, 96) of polarization at McDonald Observatory suggest surface particle sizes smaller than 0.3 μ (1.2 $\times$ 10^{-5} inch) in diameter.

The albedo of any substance depends on the direction from which it is viewed and the direction from which it is illuminated. Table 24 lists the albedos of surfaces of various materials when the surface is perpendicular to both the line of sight and the illumination. The values given are only approximate, as rather wide ranges in reflecting power may be observed for each subject material.

The integrated albedo of the entire hemisphere of the moon is approximately 0.073 (31). The photographic albedo is about 0.054 (32), while the albedo in the infrared, as determined with a thermocouple, is perhaps 0.10. Many studies have been made concerning variations in brightness over the lunar disk. This variation is found to be up to 3.5 times, or 1.4 mag. In 1919 Götz (33) published the absolute photographic brightnesses of 55 features at full moon and their albedos.

In 1921, Rosenberg (34) obtained the visual brightnesses of the same features and made a comparison with the absolute photographic brightness. This gave him a first approximation to the colors of lunar formations. Similar, but more comprehensive, work has been done by Markov (35-37).

In 1893–95 Wislizenus (38) made the first studies of 20 selected features on the moon and their changes in brightness with changes in phase angle. This was done by means of a stellar photometer and by comparing the disappearance of an artificial star on the background of one lunar feature or another. His results were not very accurate, but they led to work in 1915 by Wirtz (39) and in 1918 by Barabashov (40), who demonstrated clearly that the maximum brightness of any region occurred approximately at full moon. Barabashov explained this result as evidence that the lunar surface is extremely porous. In 1923, Markov extended the work of Barabashov and demonstrated its essential correctness for the entire lunar surface.

Sharonov (41-43), working in Leningrad in 1928 and in Yerevan in 1935, obtained coefficients of brightness of 56 lunar features at different phases with a visual surface photometer. He showed that, among similarly located features, the light-colored formations such as rays and the insides of certain craters gave the sharpest curves, while the darker spots gave much flatter curves. He continued this work in 1938 at Tashkent.

As a result of all work from 1818 to 1946, coupled with the extensive earlier work, Barabashov and Chekirda (44) have concluded that (1) the brightness of all features increases extremely rapidly near full moon; (2) the brightness of different parts of the lunar surface—seas, mountain regions, and craters—changes approximately according to the same relation; (3) the maximum brightness of the light rays lasts longer than that of the bright regions adjoining them; (4) at the northern and southern limbs, the change in brightness is extremely great for an insignificant change in the angle of incidence, i; and (5) for features with an identical albedo, the brightness is identical at full moon. This makes it possible to intercompare the albedos of different parts of the lunar surface from their brightnesses at full moon.

Barabashov and Chekirda (45) attempted to intercompare the reflecting power of different terrestrial rocks, both in their natural state and in a pulverized state, with the observations of reflection of light from the moon and Mars. They found that the best agreement with the curves of the change of brightness of the lunar seas was given by artificially created surfaces with sharp irregularities. They suggest that the surface of the moon is extremely porous and that it is covered with fragments of volcanic rocks or numerous intersecting clefts. The clefts were presumed to be the result of extensive cracking caused by temperature fluctuations. They also concluded that the light-colored rays could not be powder-like substances, since powder or dust with fine grains more nearly approximates Lambert's law than the sharp peaks observed near full moon.

Wirtz (39) first and then Fedorets (46) established the fact that the peaks of the reflection for various objects did not always occur at full moon but that certain features gave curves reaching maximum somewhat before or after full moon. The earlier observations indicated a tendency for a formation in the eastern hemisphere to attain maximum brightness before full moon and after full moon for objects in the western hemisphere. All later observations have indicated that formations reached maximum brightness at or after full moon, regardless of location. In 1946, Minnaert made a series of calibrated photographs with the 40-inch refractor at Yerkes Observatory.

Recently van Diggelen (47), of Utrecht Observatory, investigated photometric properties of the floors of thirty-eight craters on these photographs. Lunation curves relating reflected light intensity to phase angle were accurately derived by combining his own work with that of several other observers, all reduced to the same photometric system. In general, the results confirmed those of Barabashov and Fedorets. In all cases, the maxima of these lunation curves occurred close to phase angle 0°—at full moon, irrespective of the crater's location on the lunar disk.

However, several craters were found with maximum intensity occurring slightly after zero phase. Tycho and Aristarchus have 10° lags, Proclus has a 5° lag, for examples. Nearly all the craters showing this anomaly have bright rays visible at full moon and presumably are of relatively recent formation. Older features, including Alphonsus and Grimaldi, habitually have maxima precisely at full moon. Grimaldi was one of the formations found by Wirtz to have a maximum before full moon.

The shapes of the curves are functions of the selenographic longitudes of the craters, as is necessary, but the forms are independent of latitude. This was shown in 1949 by Tschunko (48). The heights are dependent to a large extent on the albedos.

Van Diggelen's measures yielded Bond albedos of 0.04–0.09. He tried to find theoretical interpretations of his observed results. He showed that the Lambert and Lommel-Seeliger laws were not even close to the observed curves. More complicated phase functions, such as the empirical ones of Öpik (49) in 1924 and Fesenkov (50) in 1928, also do not fit. It is not possible to obtain a satisfactory agreement by using the scattering functions derived by Chandrasekhar for planetary atmospheres (51).

The probable answer was given many years ago by Russell, Dugan, and Stewart (52). They stated:

The half moon, though apparently of half the area of the full moon, is only one ninth as bright. Part of this difference arises from the fact that in the region near the terminator of the half moon the sun's rays strike the surface very obliquely, and therefore illuminate it feebly; but most of it must be due to the rough character of the lunar surface, which causes it to be more or less darkened, except at the full, by the shadows cast by its own irregularities. The shadows of the mountains which are visible with the telescope are probably of less importance than those of innumerable small irregularities, perhaps no bigger than bowlders or even pebbles. A homely illustration of the same principle is that a broken road of rough but white snow appears darker than the surrounding smooth snow if one looks toward the sun, and brighter if one looks the other way.

Barabashov in 1924 (40, 53) and Bennett (54) in 1938 proposed that much of the moon's surface was covered by random hemispherical pits. The attempt failed to fit the observed lunation curves in the steep portions near maximum. Bennett then postulated deeper pits in the form of half-ellipsoids and found a better fit. Van Diggelen further assumed the pits to be separated by ash-covered level spaces. This proved to be even better but still not perfect.

Because of the difficulty in fitting theory to observation, van Diggelen carried out a series of laboratory experiments in which he sought to find terrestrial materials which would duplicate the observed lunation curves. He tried volcanic ash, which is similar in reflecting power to crater floors; small glass beads, analogous to tektites, which may occur on the moon; and metal plates covered with small pits or elevations. Strangely enough, the best match was obtained with a spongy material having numerous fine clefts, the lichen known as *Cladonia rangiferina*.

Van Diggelen, of course, made no assumption that there were living lichens on the moon's surface, but he did conclude that the moon's surface was not merely a layer of dust, as has been suggested, but had an irregular, spongy character with many randomly placed, small-scale depressions and elevations.

No one has yet explained why the relatively young craters with bright rays show the slightly displaced maxima of their lunation curves. Craters both east and west of the meridian show the same lag.

Riyvés (55) made a very artificial assumption and found a result which rather closely approximated the variations of brightness with phase. He postulated a cylinder covered with grooves which had vertical walls and flat bottoms. There are supposed to be smooth strips between the grooves. Riyvés found that the walls of the grooves would have little effect when the phase angle exceeded 110°. For larger phase angles the phase curve is that of a Lambert sphere, the radiation coming from the regions between the grooves.

When the model was applied to the moon, Riyvés found that Lambert's law corresponded to the contribution from the smooth areas at smaller phase angles. If this is subtracted from the observed lunar radiation, what is left is from the grooved area. He found satisfactory correlation if two-thirds of the moon were covered with grooves whose depths were a little greater than their widths. Although this is purely an *ad hoc* model, it suggests the direction in which the ultimate solution will be found.

It is essentially certain now that the effects of myriads of tiny shadows on the airless moon are more important in determining the brightness variations with phase than is the physical and chemical nature of the lunar rocks.

TABLE 25
THE ALBEDOS OF CERTAIN LUNAR FORMATIONS (56)

Formation	Co-ordinates		Albedo
	β	λ	
Sinus Medii	+ 7°	− 8°	0.054
Mare Nubium	−23	−14	.062
Mare Serenitatis	+28	+15	.070
Mare Nectaris	−15	+33	.080
Mare Nectaris	− 7	+26	.089
Palus Somnii	+13	+43	.095
Upland Area	0	+70	.100
Ptolemaeus	−44	−53	.108
Aristoteles	+50	+17	.110
Copernicus	+10	−20	.120
Tycho	−43	−12	.137
Tycho, ray	−23	+25	.163
Aristarchus	+23	−47	0.176

Clearly there are wide variations in the brightness of different portions of the moon's surface (see Table 25), but these variations are all limited to relatively dark materials. Most of the materials such as found on the earth—snow, sand, sandstone, and granite—are excluded because they reflect far too much light. The materials most clearly resembling the light regions on the moon seem to be comparable with stony meteorites in reflecting power. For the regions of medium reflectivity, the trachytic lavas, certain dark tuffs, volcanic ash and slag; for the darkest regions, basaltic lavas, dark volcanic ash, and slag; and the fusion crusts of meteorites give comparable reflectivities. On the whole, the moon is almost as dark as if it were made of coal. The mere fact of similar reflectivity does not, of course, indicate the type of rocks existing on the moon, for there are many materials with almost identical albedos.

Pettit and Nicholson (57) in 1930 showed that the limb of the moon is actually brighter than would be expected from the trends in other parts of the surface. Markov (58) had found similar results two years before. The limbs are actually about 60 per cent brighter than the adjacent maria. Fielder felt that this effect had two causes—the screening of the lower and darker region by brighter mountains and the fact that the rays are often interrupted by elevations and are apparently crowded together when the limb is viewed (59). Any luminescence of the rays would amplify the effect. Sytinskaya (56) proposed that the lunar surface was covered with a dark scoria originating from bedrock through high temperatures due to meteoritic impact explosions.

A closer limitation of the problem of the nature of the moon's surface can be made by studying the colors and distributions of colors of different regions on the moon. There are no tremendous color differences on any part of the moon, but there are clear indications that certain regions reflect specific wave lengths in a different fashion than do others; and, consequently, we may say that certain places are redder or greener or bluer than others.

The study of this part of the problem has been the province of many authors. It has been known for many years that the light of the entire moon is slightly yellower than the sun. From observations in 1948–54, Sharonov (60, 61) determined the color excess, D, to be $0^m.332$. The preliminary work of Rosenberg (62) and Götz (63) showed that there actually were color excesses in various parts of the lunar surface relative to the average, the largest of which was about $0^m.17$. The accuracy of these measures was low, but they do indicate real differences in the colors of the surfaces.

In 1908–10, Wilsing and Scheiner (64-66), at Potsdam, made the first series of spectrophotometric observations of the moon with a visual spectrophotometer. While these observations were relatively inaccurate, they did show that the lighter regions of the moon were redder than the darker regions. They suggested that the distribution of energy in the spectrum and the albedo of the light spot in the northwestern part of Mare Serenitatis and of the light region between Proclus and Macrobius corresponded fairly well with that of volcanic ash, while the dark region in Mare Imbrium was similar to a coarse-grained basalt. The bottom of Copernicus looked like a dark volcanic rock, porphyrite.

In 1911 Meithe and Seegert (67) photographed the moon through two filters —a red filter with a pass band of 6,000–7,000 A and an ultraviolet filter of 3,300–3,600 A. By superimposing these photographs, they obtained an approximation to a two-colored photograph of the moon. The places which reflected ultraviolet rays strongly appear on this picture to have a green color, while the places reflecting red rays strongly are represented by orange. Parts of the moon reflecting red and ultraviolet rays identically remain gray and pink. These were subsequently confirmed in the 1926 survey made at Lick Observatory by Wright (68).

In 1952 Barabashov began a systematic program of photographing the moon on color film at the prime focus of a 270-mm reflector (69). On the color prints obtained from the color negatives, greenish, reddish, bluish, rust-color, violet, and other hues may be clearly seen at different places on the lunar surface. His results confirmed and extended the other results of Meithe and Seegert.

Barabashov found that Mare Crisium is greenish in its southern part and rust-colored in its northern part. In Mare Foecunditatis the colors are inverted. Mare Tranquillitatis shows a blue tint. Mare Serenitatis is reddish in the middle, but on the southern and northern shores there are greenish bands extending into the sea adjacent to the shores. Mare Imbrium is quite spotted with reddish, greenish, and rust-colored regions. The sizes of these spots vary. The smallest spots which he could distinguish are not over 22 km in diameter. In general, Oceanus Procellarum is greenish, only its southern end appears to have a brown tint. Mare Humorum is reddish, just as is Mare Nubium. Green spots are often seen in the latter.

Firsoff (70) has pointed out that color photography is not an impartial test. Any dominant color tends to smother all the rest. Overexposure gives a red bias; underexposure exaggerates the shorter wave lengths.

The region of Tycho has a greater reddish hue; some of its rays also are reddish. Mare Frigoris is reddish. In the center of the lunar disk and up toward the southern limb, there are reddish regions, while the mountain regions adjacent to Mare Tranquillitatis are greenish. The bottom of Plato has a reddish cast.

South of Mare Imbrium there are five very intense green spots, while Wood's region near Aristarchus is quite reddish. Barabashov noted that many parts of the lunar maria have a mosaic structure, whereby small regions of extremely different coloring are closely adjacent to one another.

Barabashov and Chekirda (71, 72) undertook a special study to determine the quantitative differences in color on the surface of the moon. Their purpose was to establish definitely the intensities of the colors and to compare them with the colors of terrestrial rocks. They used a 200-mm refractor with a lunar-solar camera. The photographs were taken through filters having pass bands with the following maxima: infrared, 8,400 A; red, 6,500; green, 5,020; blue, 4,150; ultraviolet, 3,650. Seventy-two regions were studied, and they were relatively evenly distributed over the disk. They obtained the following basic results:

1. The maximum deviations of the brightness of the 72 measured regions from a point selected for reference (No. 33) and having selenographic coordinates $\varphi = -29°20'$; $\lambda = -64°00'$ are: for the infra-red filter, from $+0^m185$ to -0^m115, i.e., 0^m300 in all; for the red filter, from $+0^m157$ to -0^m109, i.e., 0^m266 in all; for the blue filter, from $+0^m243$ to -0^m102, i.e., 0^m345 in all; and, finally, for the ultraviolet filter, from $+0^m192$ to -0^m159, i.e., 0^m351 in all.

Thus, the deviations are greatest for the ultra-violet filter and least for the red filter.

2. For craters with systems of light-colored rays (Tycho, Copernicus, Kepler, Aristarchus), the fluctuations in brightness in the different rays are considerably less than for the entire group of measured formations: in the infra-red part of the spectrum, the average deviations are $0^m.157$, in the red $0^m.108$, in the blue $0^m.128$, and in the ultra-violet $0^m.088$. Here, the fluctuations are greatest for the infra-red rays and least for the ultra-violet rays.

3. For mountainous regions, these fluctuations are greater than for the craters with ray systems. However, they do not exceed $0^m.287$. Just as for the craters with ray systems, the fluctuations are greatest in the red rays and least in the ultra-violet rays.

4. For maria and bays, the fluctuations are maximum in the infra-red and ultra-violet rays—$0^m.247$ and $0^m.229$—and minimum in the red rays $0^m.147$.

5. For the seas with irregular outlines, we have the smallest amplitude of fluctuations: $0^m.074$ in infra-red rays; $0^m.026$ in red rays; $0^m.038$ in blue rays; $0^m.079$ in ultra-violet rays. However, just as for the other seas, this amplitude is maximum in the ultra-violet and infra-red regions of the spectrum, and minimum in the red region of the spectrum.

6. On the basis of these measurements, we conclude that:
   a) Craters with ray systems are, generally speaking, reddish;
   b) Seas and the bottoms of craters, and also the light bands about Tycho are reddish;
   c) Certain mountainous regions, especially in the southern hemisphere of the moon, are also reddish;
   d) Certain mountainous regions and irregularly outlined seas are greenish on the whole;
   e) Certain light rays are also reddish.

All of this can be clearly seen on the color photographs.

The results obtained make it possible to conclude that the reddish hue, as is clearly seen on the color photographs, is especially characteristic of the Mare Serenitatis and of the mountainous regions in the southern part of the lunar disc. In the Mare Imbrium, as in the Oceanus Procellarum, reddish and greenish hues occur. In the center of the disc, there exist fairly large spots of an intensive greenish color. A number of bright mountainous regions, especially on the northwestern limb of the disc of the moon, are greenish.

To the north of the Mare Humorum and the Mare Nubium, reddish hues predominate. Apparently, many ancient formations have a reddish hue, while the younger formations have a greenish hue.

Similar work was done in 1954 by Yezerskiy and Fedorets (73), and a program is now being conducted at the Kharkov Astronomical Observatory.

These studies show that the ring about the crater Tycho and the light substances lining the craters of Copernicus and Kepler are reddish. The interior of

Copernicus is less red than that of Tycho, while the interior of Kepler is still more pale. While there is a similarity in the changes of color in these systems, there are important differences apparently connected with the nature of the material from which the crater was formed. On the average, the rays of Tycho are as red as the crater interior up to a distance of 7.7 times the crater diameter. Thereafter, this redness decreases rapidly.

The rays of Copernicus are not so red as those of Tycho; and, at a distance of about 1 diameter from Copernicus, a blue color begins to predominate. The whitish color of the rays of Kepler begins to change toward a blue color at the edge of the crater and continues to have a blue tint for well over 100 kilome-

TABLE 26

COMPARISON OF LUNAR REGIONS WITH TERRESTRIAL ROCKS STUDIED

Wave Length (A)	Moon, Reddest Regions	Red Quartz Porphyry	Moon, Red Regions	Quartzose Sandstone	Moon, Greenest Regions	Volcanic Ash
8,400	-0^m722	-0^m642	-0^m884	-0^m879	-0^m761	-0^m727
6,500	$-.502$	$-.435$	$-.674$	$-.557$	$-.453$	$-.473$
5,020	.000	.000	.000	.000	.000	.000
4,150	$+.390$	$+.438$	$+.277$	$+.245$	$+.301$	$+.251$
3,650	$+0.157$	$+0.143$	$+0.439$	$+0.346$	$+0.475$	$+0.461$

Wave Length (A)	Moon, Bluest Regions	Iron Quartzite	Moon, Brown Regions	Quartzose Sandstone
8,400	-0^m831	-0^m641	-0^m910	-0^m879
6,500	$-.557$	$-.531$	$-.612$	$-.557$
5,020	.000	.000	.000	.000
4,150	$+.138$	$+.119$	$+.194$	$+.245$
3,650	$+0.200$	$+0.212$	$+0.398$	$+0.346$

ters. Firsoff (70) states that most of the bright crater rims, floors, and rays are brightest in green.

Barabashov and Chekirda (74) photographed, with the same equipment and plates, various terrestrial rocks. This made it possible to determine directly the distribution of energy in the spectrum of reflected light from the terrestrial rocks. The rocks included chiefly those selected by Khabakov, the existence of which on the moon was most likely in his opinion, and also rocks selected by the authors and distinguishable by their noticeable color. In all, there were 49 selected examples of terrestrial rocks, and, of these, there were a few which showed colors similar to the various colored regions of the lunar surface.

Barabashov and Chekirda point out that undoubtedly other rocks would be found with similar color distributions. The conclusions given in Table 26 are only indicative and not final.

Various attempts have been made to apply spectrographic methods to this problem by Teyfel (75, 76) and others. Their results generally confirm the color photographic work. While the results of these extensive series of observations are interesting and informative, they are still quite frustrating. They do not yet permit us to pinpoint the exact nature of the lunar materials. Exact color definitions have not yet been determined.

Relatively few rocks have a distribution of reflectivities as determined spectrographically such that they may be similar to lunar materials. Two such rocks are tuff from Armenia and volcanic ash. Tuff resembles certain mountain regions, and volcanic ash resembles the region of Sinus Medii.

The spectrophotometric measures of Wilsing and Scheiner of lunar features and terrestrial rocks, the photographic measures of lunar features and rocks through five filters by Barabashov and Chekirda with rocks measured on the spectroelectrophotometer show that the following rocks best match the moon in terms of two characteristics—albedo and spectral distribution: according to Wilsing and Scheiner, volcanic ash, coarse-grained basalt (dolerite, porphyrite, liparitic pitchstone, and Mandelstein basalt); according to Barabashov and Chekirda, volcanic ash, red quartz porphyry, iron quartzite, tuff (Armenia), and volcanic ash (Armenia).

Inasmuch as many areas of the lunar surface differ appreciably in both albedo and color, the materials making them up must also differ. The lunar surface is non-homogeneous. This is beautifully brought to our attention by the detailed drawings of the floors of Plato, Grimaldi, Eratosthenes, and Copernicus by Firsoff (77). Plato shows nine distinct color patches made up of at least four different colors—dull rusty red, smoky brownish-gray, brimstone yellow, and a dirty green. Grimaldi shows ten areas of at least three colors. Copernicus and Eratosthenes are quite variegated. The once simple structure of the lunar surface is slowly being shown to be of extreme complexity.

Minnaert (78) believes that the photometric properties of the lunar surface indicate a porous condition, while the polarimetric measures call for a thin dust layer covering all details.

Ever since the eclipse of October 4, 1884, when Lowe (79) remarked that it was a particularly "dark eclipse," especially the total phase as compared with previous ones, astronomers have tried to determine whether or not the moon's rocks fluoresce. In the same year Burder (80) suggested that the surface of the moon might be in some degree self-luminous.

In 1932, Link (81) outlined a theory of the intensity of various parts of an eclipsed moon. This theory discounted any luminescent properties of the moon's rocks. Later (82) he found that refraction of the sun's rays in the earth's atmosphere could not explain the observed excesses of illumination from a quantitative point of view. He then proposed that the moon was in some degree self-luminous, noting that the solar corona would be less eclipsed than the disk of the sun and that the corona was the source of ultraviolet radiation which might excite minerals on the moon's surface.

In 1950, Link (83-85) summarized numerous observations which suggested, from observed excesses of illumination in the penumbra, that as much as 10 per cent of the light of the full moon might have been due to luminescence.

Dubois (86) has suggested that both ultraviolet solar radiation and corpuscular radiation from the sun might produce characteristic emissions from lunar rocks.

Because of a suggestion by Link, Kosyrev (87) and Dubois (88, 89) tried to observe residual intensity in the Fraunhofer lines of the lunar spectrum. Both men concluded that luminescence of this type existed. Kosyrev worked with the H and K lines of ionized calcium.

Dubois (88) claimed to have detected fluorescence in the following sunlit regions of the colors indicated:

Red	Oceanus Procellarum
Red and yellow	Regiomontanus (floor)
Red and green	Mare Humorum
Red, yellow, and blue	Mare Tranquillitatis, Sinus Medii, Mare Serenitatis, Tycho
Red, yellow, blue, and violet	Mare Imbrium
Yellow and green	Mare Crisium
Yellow and blue	Mare Nubium
Green	Southwest limb
No fluorescence detected	Copernicus

In a later paper, Dubois (89) summarized the works of many men and showed that 46 out of 86 regions gave a definite luminescence. In general, the dark areas showed more luminescence than did the continents. Perhaps this is due to the greater ease of detecting luminescence against a dark surface.

Various terrestrial materials are known to fluoresce. A green luminescence on the moon could conceivably be associated with willemites, cadmium, or beryllium silicates or phenacites.

In 1959, Dubois (90) noted that, in general, there were two types of luminescence bands. The wider bands (of widths about 200 A) were usually associated with lunar regions emitting a green luminescence. He explained these as identical with bands of willemites and silicates of the alkaline earths with other less active minerals. He suggested that narrower bands which could not be identified with those of any specific mineral might be explained by the superposition of absorption and emission bands.

Kosyrev's study (87) of the H and K lines of calcium found a few places where luminescence existed. His spectrograms of the maria, continents, Wood's Spot, and the craters Plato, Schickard, and Copernicus gave negative results. The ray system of the Aristarchus-Herodotus area did show a definite amount of luminescence, which was greatest after full moon. He indicated that the luminescent intensity at full moon sometimes reached a maximum of 13 per cent of the intensity of the rays at the same wave length. Kosyrev concluded that there was a luminescent substance situated at the bottoms of small depressions connected with the rays of Aristarchus and that this substance did not contain even very tiny amounts of iron because iron was "a very active extinguisher of luminescence." Kosyrev also suggested that the substance, in order to exhibit luminescence, must have a very low temperature, probably below 0° C, and that, in order to have such a low temperature when the sun was high, it must have a very high albedo—about 0.3–0.4, which is quite comparable with that of white sand. He suggested a variety of quartz.

Platt (91) has taken the suggestion that the interstellar dust particles presumably responsible for the reddening of distant stars may consist largely of random aggregates of unsaturated and free-radical molecular species rich in carbon, nitrogen, and oxygen (92, 93), and he postulated that the moon's surface might be covered by such materials. As an alternative, the surface layers may have been extensively damaged by radiation. Such particles are of the order of 0.5 μ in diameter and would strongly absorb light in the visible region of the spectrum (94). The prediction of surface disorder and opacity, possibly accompanied by porosity in layers of recombined dust, would help account for two perennially striking properties of the moon's surface—the generally low albedo, about 7 per cent, and the low thermal conductivity.

Because surfaces of such a chemically unstable character could also be highly reactive and extensive exothermic reactions might easily be triggered, Platt was led to make his now famous comment, "The first man who plants a rubber boot on a lunar surface may be in for an unpleasant surprise."

References

1. ZÖLLNER, J. K. F. *Photometrische Untersuchungen mit besonderer Rücksicht auf die physische Beschaffenheit der Himmelskörper.* Leipzig: Englemann, 1865.
2. ASHBROOK, J. "The Lunar Straight Wall," *Pub. A.S.P.*, **72**, 55, 1960.
3. FESENKOV, V. G., PARENAGO, P., and STAUDE, N. *Pub. Astr. Inst. Russ.*, **4**, 1, 1928.
4. WATSON, F. G. *Between the Planets.* Philadelphia: Blakiston Co., 1941.
5. GEHRELS, T. "Photometric Studies of Asteroids. V. The Light-Curve and Phase Function of 20 Massalia," *Ap. J.*, **123**, 331, 1956.
6. ARAGO, F. *Œuvres.* 1811.
7. SECCHI, A. Letter from the Rev. Father Secchi to Admiral Manners, *M.N.*, **19**, 289, 1859.
8. ———. Letter from the Rev. Father Secchi to the Astronomer Royal, *ibid.*, **20**, 186, 1860.

9. PETRUSHEVSKIY, F. F. "Plan fizicheskogo issledovaniya poverkhnosti Luny" ("Plan for the Physical Study of the Surface of the Moon"), *Prot. Russk. Fig.-khim o-va*, Vol. **9**, 219, 1873.
10. PARSONS, W. (the third EARL OF ROSSE). "Preliminary Note on Some Measurements of the Polarization of the Light Coming from the Moon and from the Planet Venus," *Sci. Proc. R. Soc. Dublin*, **1**, 19, 1878.
11. ———. "Temperature of Moon's Surface," *Nature*, **16**, 438, 1877.
12. LANDERER, J. J. "Sur l'angle de polarisation des roches ignées et sur les premières deductions sélénologique qui s'y rapportent," *C.R.*, **111**, 210, 1890.
13. ———. "Sur l'angle de polarisation de la lune," *ibid.*, **109**, 360, 1889.
14. ———. "Sur la polarisation de la lumière lunaire," *ibid.*, **150**, 1164, 1910.
15. LYOT, B. "Étude des surfaces planètes par la polarisation," *C.R.*, **177**, 1015, 1923.
16. ———. "Polarisation de la lune et des planètes Mars et Mercure," *ibid.*, **178**, 1796, 1924.
17. MARKOV, A. V. "Rezul'taty opytnykh issledovaniy polyarizatsii detaley lunnoy poverkhnosti" ("The Results of the Experimental Studies of the Polarization of Features of the Lunar Surface"), *Izv., Glav. Astr. Obs.*, **20** (No. 158), 138, 1958.
18. BARABASHOV, N. "Polarimetrische Beobachtungen an der Mondoberfläche und am Gesteinen," *A.N.*, **229**, 14, 1926.
19. WRIGHT, F. E. "Polarization of Light Reflected from Rough Surfaces, with Special Reference to the Light Reflected by the Moon," *Proc. Nat. Acad. Sci.*, **13**, 535, 1927.
20. SINTON, W. M. "Observation of a Lunar Eclipse at 1.5 mm," *Ap. J.*, **123**, 325, 1956.
21. UMOV, N.A. "Eine spectropolariskopische Methode zur Erforschung der Lichtabsorption und der Natur der Farbstoffe," *Phys. Zs.*, **6**, 674, 1905.
22. ———. "Chromatische Depolarisation durch Lichtzerstreuung," *ibid.*, **13**, 962, 1912.
23. TOPORETS, A. S. "Ob effekte Umova" ("The Umov Effect"), *Zhur. eksp. i teor. fiz.*, **20**, 396, 1950.
24. DOLLFUS, A., and CAILLEUX, A. "Étude polarimetrique de la lumière renvoyée par quelques sables et limons," *C.R.*, **230**, 1411, 1950.
25. DOLLFUS, A. "Polarisation de la lumière renvoyée par les corps solides et les nuages naturels," *Ann. d' ap.*, **19**, 83, 1956.
26. FESENKOV, V. G. "Détermination de la polarisation de la couronne solaire," *Astr. Zhur.*, **12**, 109, 1935.
27. DZHAPIASHVILI, V. P. "Issledovaniye polyarizatsionnykh svoystv obrazovaniy lunnoy poverkhnosti po elektrofotometricheskim izmereniyam" ("Study of the Polarization Properties of Formations on the Lunar Surface by Means of Electrophotometric Measurements"), *Byull. Abastumanskoy Obs.*, No. 21, 1957.
28. GEHRELS, T. *A.J.*, **64**, 332, 1959.
29. ———. *Lowell Obs. Bull.*, **4**, 300, 1960.
30. Quoted from page 132 of an English translation (unpublished) of A. V. MARKOV, *The Moon*, 1961.

31. Russell, H. N. "The Albedo of the Planets and Their Satellites," *Ap.J.*, **43,** 173, 1916.
32. Rougier, G. "Photometric Comparison of the Moon and Sun, Photoelectric Albedo of the Moon," *C.R.*, **202,** 463, 1936.
33. Götz, F. W. P., "Helligkeitsverhältnisse und Albedo von 55 ausgewählten Stellen der Mondoberfläche bei mittlerem Vollmond," *Veröff. Sternw. Öst. Tübingen*, Vol. **1,** No. 2, Part 1, 1919.
34. Rosenberg, H. "Photometrische Messungen der Mondoberfläche und das Flächenphotometer der Sternwarte Österberg," *A.N.*, **214,** 137, 1921.
35. Markov, A. V. "Fotograficheskaya yarkost' i otrazhatel'naya sposobnost' detaley lunnoy poverkhnosti" ("Photographic Brightness and Reflecting Power of Features of the Lunar Surface"), *Astr. Zhur.*, **4,** 60, 1927.
36. ———. "Raspredeleniye yarkosti po disku Luny y polnoluniye" ("The Distribution of Brightness over the Disk of the Moon at Full Moon"), *ibid.*, **25,** 172, 1948.
37. ———. "Otrazhatel'naya sposobnost' i nokazateli tsveta Zemli i detaley Luny po elektricheskim promeram" ("The Reflecting Power and Color Indices of the Earth and of Features of the Moon from Electrical Measurements"), *Byull. Abastumanskoy Astr. Obs.*, No. 11, p. 107, 1950.
38. Wislizenus, W. F. "Selenophotometrische Beobachtungen," *A.N.*, **201,** 289, 1915.
39. Wirtz, C. "W. F. Wislizenus' Selenophotometrische Beobachtungen," *A.N.*, 201, 289, 1915.
40. Barabashov, N. "Über die Reflexion des Lichtes an der Mondoberfläche und an porösen Flächen," *A.N.*, **221,** 289, 1924.
41. Sharonov, V. V. "Opredeleniye absolyutnoy otrezhatel'noy sposobnosti poverkhnosti Luny i planet" ("Determination of the Absolute Reflecting Power of the Surfaces of the Moon and Planets"), *Trudy Astr. Obs., Leningradskogo U.*, **6,** 33, 1936.
42. ———. "Opyt opredeleniye absolyutnykh znacheniy koeffitsiyenta yarkosti lunnoy poverkhnosti" ("An Attempt To Determine the Absolute Values of the Brightness Factor of the Lunar Surface"), *Uchenyye zapiski Leningradskogo U.*, **31,** 28, 1939.
43. Sharonov, V. V., and Sytinskaya, N. N. "Issledovaniya otrazhatel'noy sposobnosti lunnoy poverkhnosti" ("Studies of the Reflecting Power of the Lunar Surface"), *Zapiski Leningrad. Ges. U.*, **4** (No. 153), 114, 1952.
44. Barabashov, N. P., and Chekirda, A. T. "Fotograficheskaya fotometriya lunnoy poverkhnosti" ("Photographic Photometry of the Lunar Surface"), *Pub. Khar'kovskoy Astr. Obs.*, No. 8, p. 29, 1948.
45. ———. "Ob otrazhenii sveta ot poverkhnosti Luny i Marsa" ("The Reflection of Light from the Surface of the Moon and Mars"), *Astr. Zhur.*, **22,** 11, 1945.
46. Fedorets, V. A. "Fotograficheskaya fotometriya lunnoy poverkhnosti" ("Photographic Photometry of the Lunar Surface"), *Uchenyye zapiski Khar'kovskogo Ges. U.*, **42,** 49, 1952.

47. DIGGELEN, J. VAN. "Photometric Properties of Lunar Crater Floors," *Recherches Astr. Obs. Utrecht*, **14,** Part 2, 1, 1946.
48. TSCHUNKO, H. F. A. "Theoretical Photometry of the Moon," *Zs. f. Astr.*, **26,** 279, 1949.
49. ÖPIK, E. "Photometric Measures on the Moon and the Earthshine," *Tartu Obs. Pub.*, **26,** 3, 1924.
50. FESENKOV, V. G., and PARENAGO, P. "Photometry of the Moon," *Russ. Inst. Ap.*, **4,** 1, 1928.
51. CHANDRASEKHAR, S. A Series of Papers on "Radiative Equilibrium." A comprehensive account was given by Chandrasekhar in *Bull. Amer. Math. Soc.*, **53,** 41, 1947.
52. RUSSELL, H. N., DUGAN, R. S., and STEWART, J. Q. *Astronomy*, **1,** 173. Boston: Ginn & Co., 1926.
53. BARABASHOV, N. "Études spectrophotométriques de la surface lunaire," *Akad. Nauk SSSR, Astr. Zhur.*, **1** (Nos. 3–4), 44, 1924.
54. BENNETT, A. L. "A Photovisual Investigation of the Brightnesses of 59 Areas on the Moon," *Ap.J.*, **88,** 1, 1938.
55. RIYVÉS, B. G. *Pub. Astr. Obs. Tartu*, **32,** 129, 1952.
56. SYTINSKAYA, N. N. "Svodnyy Katalog absolyutnykh znacheniy vizual'noy otrazhatel'noy sposobnosti 104 lunnykh ob'yektov" ("Summary Catalogue of the Absolute Values of the Visual Reflecting Power of 104 Lunar Features"), *Astr. Zhur.*, **30,** 295, 1953.
57. PETTIT, E., and NICHOLSON, S. B. "Lunar Radiation and Temperatures," *Ap.J.*, **71,** 102, 1930; also *Mt. W. Contr.*, No. 392.
58. MARKOV, A. "The Absolute Photographic Brightness of Details of the Lunar Surface," *A.N.*, **231,** 57, 1928.
59. FIELDER, G. *Structure of the Moon's Surface*, p. 49. London: Pergamon Press, 1961.
60. SHARANOV, V. V. "Svodka sravneniy tsveta Luny i Solntsa" ("Summary of Comparisons of the Color of the Moon and the Sun"), *Astr. Tsirk. Akad. Nauk SSSR*, No. 157, p. 19, 1955.
61. ———. "Issledovaniya po kolorimetrii Luny. II. Novaya obrabotka potsdamskoy spektrofotometrii" ("Studies on the Colorimetry of the Moon. II. New Processing of the Potsdam Spectrophotometry"), *Vest. Leningradskogo U.*, No. 1, p. 155, 1956.
62. ROSENBERG, H. "Photometrische Messungen der Mondoberfläche und das Flächenphotometer der Sternwarte Österberg," *A.N.*, **214,** 5121, 1921.
63. GÖTZ, F. W. P. "Photographische Photometrie der Mondoberfläche veröff," *Sternw. Öst. Tübingen*, Vol. **1,** No. 2, 1919.
64. WILSING, I., and SCHEINER, I. "Vergleichende spektralphotometrische Beobachtungen am Monde und an Gesteinen nebst Albedobestimmungen an letzeren," *Pub. Ap. Obs. Potsdam*, Vol. **20,** No. 61, 1909.
65. ———. *Ibid.*, Vol. **24,** No. 74, 1909.
66. ———. *Ibid.*, No. 77, 1921.

67. MEITHE, A., and SEEGERT, B. "Über qualitative Verscheidenheiten des von den einzelnen Teilen der Mondoberfläche reflektierten Lichter," *A.N.*, **188,** 9, 239, 1911.
68. WRIGHT, W. H. "The Moon as Photographed by Light of Different Colors," *Pub. A.S.P.*, **41,** 125, 1929.
69. BARABASHOV, N. P. "Issledovaniye fizicheskikh usloviy na Lune i planetakh" ("Study of the Physical Conditions on the Moon and on Planets"), *Izd-vo Khar'kovskogo Ges. U.*, 1952.
70. FIRSOFF, V. A. "Color on the Moon," *Sky and Telescope*, **17,** 328, 1958.
71. BARABASHOV, N. P. "O tsvetnykh kontrastakh no poverkhnosti Luny" ("Color Contrasts on the Surface of the Moon"), *Tsirk. Khar'kovskoy Astr. Obs.*, No. 2, p. 3, 1953.
72. BARABASHOV, N. P., and CHEKIRDA, A. T. "O tsvetnykh kontrastakh lunnoy poverkhnosti" ("Color Contrasts on the Lunar Surface"), *Trudy Astr. Obs., Khar'kovskogo U.*, Vol. **3** (No. 11), 18, 1955.
73. YEZERSKIY, V. I., and FEDORETS, V. A. "Opyt fotograficheskoy spektrofotometrii lunnoy poverkhnosti" ("An Attempt at Photographic Spectrophotometry of the Lunar Surface"), *Astr. Tsirk. Akad. Nauk SSSR*. No. 159, p. 18, 1955.
74. BARABASHOV, N. P., and CHEKIRDA, A. T. "O tsvete svetlykh luchey kraterov Tikho, Kopernik i Kepler" ("The Color of the Light-colored Rays of Tycho, Copernicus, and Kepler"), *Tsirk. Khar'kovskoy Astr. Obs.*, No. 13, p. 3, 1955.
75. TEYFEL, V. G., "O tsvetovykh kontrastakh no lunnoy poverkhnosti y vidimoy oblasti spektra" ("Color Contrast on the Lunar Surface in the Visible Part of the Spectrum"), *Astr. Tsirk. Akad. Nauk SSSR*, No. 179, p. 8, 1957.
76. ———. "On the Difference in the Spectral Properties of Areas of the Lunar Surface," *Russ. Astr. J.*, **36,** 1041, 1959.
77. FIRSOFF, V. A. "Color on the Moon," *Sky and Telescope*, **17,** 329, 1958.
78. MINNAERT, M. "Photometry of the Moon," in *Planets and Satellites*, ed. G. P. KUIPER and B. M. MIDDLEHURST, p. 213. ("The Solar System," Vol. III.) Chicago: University of Chicago Press, 1961.
79. LOWE, E. J. "The Recent Eclipse of the Moon," *Nature*, **30,** 590, 1884.
80. BURDER, G. F. "The Recent Eclipse of the Moon," *Nature*, **30,** 590, 1884.
81. LINK, F. "Théorie photométrique des éclipses de la lune," *Bull. Astr.*, **8** (Ser. 2), 77, 1932.
82. ———. "Sur la luminescence de la lune," *C.R.*, **223,** 976, 1946.
83. ———. "Le Rôle de l'ozone atmosphérique dans les éclipses de la lune," *Ann. d'ap.*, **9,** 227, 1946.
84. ———. "Photométrie photographique de l'éclipse de la lune du 18 décembre, 1945," *C.R.*, **223,** 718, 1946.
85. ———. "Problèmes relatifs au rayonnement de l'atmosphère solaire," *Trans. I.A.U.*, **7,** 135, 1950.
86. DUBOIS, J. "Peut-on observer sur la lune des phénomènes de luminescence?" *Astronomie*, p. 225, June, 1956.

87. Kosyrev, N. A. "Luminescence of the Lunar Surface and the Intensity of the Corpuscular Radiation from the Sun," *Pub. Crimean Ap. Obs.*, **16,** 148, 1956.
88. Dubois, J. "Peut-on observer sur la lune des phénomènes de luminescence?" *Astronomie,* p. 297, July–August, 1956.
89. ———. "Sur l'existence de la luminescence lunaire. Résultats obtenus," *C.R. Soc. Franç. de Phys.*, Séance du 25 October, 1956.
90. ———. "Contribution à l'étude de la luminescence lunaire," *Rozpravy Československé Akad. Ved.*, **69,** Part 6, 1, 1959.
91. Platt, J. R. "On the Nature and Color of the Moon's Surface," *Science,* **127,** 1502, 1958.
92. Donn, B. *Mém. Soc. R. Soc. Liège.* Sér. 4, **15,** 571, 1954.
93. Donn, B., and Urey, H. C. "On the Mechanism of Comet Outbursts and the Chemical Composition of Comets," *Ap.J.*, **123,** 339, 1956.
94. Platt, J. R. "On the Optical Properties of Interstellar Dust," *Ap.J.*, **123,** 486, 1956.
95. Markov, A. V. *Pulkovo Bull.*, No. 158, 1958.
96. Gehrels, T. "Polarization of Light of Moon and Planets," paper presented at the one hundred-third meeting of A.A.S., August 30–September 2, 1959, Toronto, Canada.
97. *The Moon: A Russian View,* ed. A. V. Markov, p. 124. Chicago: University of Chicago Press, 1962.

13

NATURE OF THE LUNAR-SURFACE MATERIALS AS DETERMINED BY HEAT MEASURES AT INFRARED AND RADIO FREQUENCIES

An effort parallel with the measurements made on reflection effects is the determination of temperature variations on the moon from the amount of heat it radiates. Although this problem is a complex one, the method has been used with great success. There are two thermal cycles which can be studied by optical or radio techniques. The first is the lunar month, where the point in question is subjected to sunlight for 2 weeks and then darkness for 2 weeks. The other is the short period of a lunar eclipse. Thermal effects at the long cycle should be from farther below the moon's surface than at the shorter cycle.

Lord Rosse, with his great reflector, made the first serious effort in this direction (1). He found that the lunar equator, 3 days after full moon, showed a range of temperatures from 473° to 348° K. These values are systematically higher than later determinations of the temperature, and so it is probable that he did not succeed in completely separating the reflected and the radiant light.

Langley (2, 3) later denied these results, finding that the surface temperature rarely rose above freezing, while it often dropped to 73° K. Peal (4) and later Fauth (5) claimed that this was so because there was no lunar atmosphere and hence there could be no storing-up of heat. It would be reflected immediately or re-radiated into space. That this point of view is erroneous is clear, for when the black-body laws are applied, it is found that a perfect black body at the moon's mean distance from the sun would have an average temperature of 277° K.

Very (6, 7), at Allegheny, found that the temperature of the moon rose continuously from sunrise to sometime in the afternoon. At an altitude of the sun of 15°, the temperature was 273° K; and when it was vertical, the still rising temperature was at the boiling point, 373° K. Working with a vacuum thermocouple on the 100-inch telescope, Pettit and Nicholson (8) closely checked Very's first two values but not the third. They measured the heat of the subsolar point on the full moon to be 407° K. From this point, the temperature dropped as the limbs were approached. At 0.5 radius, the temperature was measured to be 395° K. At 0.75 radius, it was 375° K. At 0.9 radius it was 350° K, while close to the limb it had dropped to about 340° K. When the temperature of the subsolar point was measured at the quarter phases, a surprise resulted, for it was only 354° K.

This difference must be accounted for by the great roughness of the surface. One must be careful in assigning a unique temperature to a particular place on an airless body. The measured value will vary in a manner depending on the orientation of the surface irregularities toward both the sun and the earth and on the percentage of the sunlit surface that is hidden from a terrestrial observer (9).

Average temperatures as determined at full moon are higher than the boiling point of water, 373° K, over one-eighth of the entire lunar surface, an area 1,600 miles in diameter. Yet, even under a midday sun, a piece of lunar material placed in a shadow would drop nearly 200° C in temperature in a relatively short time.

Using the only early determination of the temperature at lunar midnight, that of 120° ± 5° K by Pettit and Nicholson (8), Wesselink (10) found a reasonable consistency with this datum in a calculation of the variations in temperature at the central point of the disk through a lunation from its midday temperature of 374° K, on the assumption that the surface was homogeneous and composed of dust *in vacuo*. Jaeger (11–13) tried a similar calculation on the assumption of a dust layer over a rock layer, which is a good thermal conductor. Without a complete curve of the temperatures throughout a lunar night, the two models cannot be distinguished.

Sinton (14, 15), in 1960, determined variations in temperature over the lunar disk at ten different phases by infrared methods at 8.8 μ. In general, his work corroborated the earlier work. In particular, he established that the darker maria became hotter than the brighter areas by perhaps 10° K. Scans through Copernicus showed an interior cooler than normal by about 10° K and a warm sunlit rim. He did establish that the midnight temperature was about 125° K, as did Pettit and Nicholson, although he was unable to make a thermal contour map of the night side.

Numerous measures of temperatures have been made at longer wave lengths. Much of this information has been summarized by Garstang (16). Coates (17), following the work by Gibson at 8.6 mm (18, 19), chose a wave length of 4.3 mm. He constructed a heat map of the moon at three different phases and stated:

> The following conclusions may be drawn from the examinations of these lunar maps. In general, at the level of 4.3-mm emission, the maria heat up more rapidly and also cool off more rapidly than the mountain regions. Mare Imbrium appears to be an exception to this, remaining cooler than its surroundings throughout the lunar cycle. Since it is apparent that the different regions on the moon have different characteristics at millimeter wave lengths, it is important that the previous measurements at longer wave lengths be repeated with higher resolution, in order to separate the characteristics of the different regions.

Piddington and Minnett (20) worked at a wave length of 1.25 cm and determined the effective temperature as a function of phase during three lunations. The temperatures were 239° K averaged over the entire disk and 249° K for the center. These values varied about ±40° and ±52° K, respectively, during a lunation. Maximum temperatures came 3 days after full. These results do not agree with Pettit and Nicholson's measures made in the near infrared, as was shown by Troitsky (21).

Reduction of observations such as these are complicated by the partial transparency of rock materials to these wave lengths. This causes a phase lag in the observed radiation relative to the lunar phase angle during a lunation. Hence the temperatures, as predicted here, theoretically depend on a parameter involving the coefficient of absorption, as well as σ in equation (13-1) for a homogeneous lunar crust. Because the whole moon was observed in transit, there are certain geometric variations which are troublesome.

Piddington and Minnett find that their results can be correlated with a thin dust layer less than 1 cm thick over a layer of greater thermal conductivity. Jaeger (11, 22) found that they are compatible with either such a layered structure or a homogeneous dust surface.

Troitsky and Zelinskaya (24), at the longer wave length of 3.2 cm, found a temperature of 170° K. There was no certain variation with phase. The radiation received did not vary more than ±7 per cent during a lunation. Measurements were made on 70 days in 1952.

Akabane (25), in Japan, working at 10 cm, found temperatures over the disk to range from 240° to 390° K, with a variation during the month of ±50°. The maximum temperature came a few days after full. This is the only "long"-wave study that shows such a variation; it should be repeated.

Denisse and LeRoux (26) at 33 cm and Seeger, Westerhout, and Conway (27), working at 75 cm, found no variations in temperature during a lunation. Other similar studies are by Dicke and Beringer (28) at 1.25 cm; Zelinskaya, Troitsky, and Fedoseyev (29) at 1.63 cm; Amenitskii, Noskova, and Salomonovich (30) at 8 mm; Grebenkemper (31) at 2.2 cm; and Westerhout (32) at 21.6 cm.

In studies at microwave lengths, the range of the variation in temperature decreases with increasing wave length. Solids, such as rocks, which are poor electrical conductors, are partially transparent to microwave radiation, even though they are opaque to visible or infrared radiation. It thus follows that microwave thermal radiation from the moon does not originate at the surface but in a region below the surface. The temperatures which we observe are an average of the real physical temperatures of the layers traversed. Hence the larger the wave length, the greater the penetration and the less the variation, because most of the changes in temperature are surface effects.

Pawsey and Bracewell (33) find that the temperature depends on two parameters, $(K\rho c)^{-1/2}$ and $C = \alpha(KP/\rho c)^{1/2}$, where P is the length of the lunar month and α is a constant connected with the depth to which microwave radiation penetrates the lunar material. A large value of α (and C) implies little penetration, small α means great penetration; $1/\alpha$ equals the average depth of emission of observed radiation.

Gibson's (18) work at 8.6 mm during a lunation is in excellent agreement with Jaeger's theoretical work with $\alpha = 0.28$ and radiation coming from 3 cm deep, on the average. Piddington and Minnett (20), working at 1.25 cm, find a smaller α. Troitsky and Zelinskaya (24) at 3.2 cm find $\alpha = 0.1$ and the average depth to be 10 cm. These preliminary determinations of the penetration of the order of three times the wave length are consistent with a lunar surface containing some metallic oxides. Sinton's (34) measures at 1.5 mm yield $\alpha = 6$, or a mean penetration of 1.7 mm. In the laboratory he found $\alpha = 4$ for basalt and for a meteorite a value of roughly 20.

The June 14, 1927, eclipse was studied by Pettit and Nicholson (8) in an attempt to find temperature variations. Although the point crossed by the greatest chord of the earth's shadow was only 48 seconds of arc from the south limb, they were successful in securing measured temperature changes throughout the eclipse. The temperature was 342° K before the eclipse. It dropped immediately at the beginning of the partial phase, which lasted 1 hour, and was 210° K at the beginning of totality. The rapid drop continued for another 20 minutes, reaching 170° K. Thereafter, for the remainder of totality—2 hours and 20 minutes—the drop was very slow. The minimum at the end of totality was 152° K. For 20 minutes during the first part of the partial phase, the temperature did not change greatly, rising to only 160° K, but thereafter it rose rapidly, attaining 330° K in the next half-hour.

Epstein (35) showed that these data imply that the lunar-surface materials make an exceedingly good insulating layer, but, because of a faulty approximation, he inferred a misleadingly large coefficient of the thermal conductivity. He assumed that the rate of thermal radiation was proportional to the fourth power of the initial temperature, which is incorrect in view of the large range of temperatures observed by Pettit and Nicholson. He concluded that the most probable substance was pumice. There also seems to be some question of units in Epstein's solution.

Pettit (36, 37) made measures on the October 27, 1939, eclipse, using the atmospheric transmission window between 8 and 14 μ. He found results similar to those in the 1927 eclipse, although the temperature curve was shifted upward because the observed area was nearer to the center of the disk. The temperature dropped from about 372° K at first contact to 207° K at second contact. The umbral portion of the eclipse found the temperature slowly falling to about 175° K. The receiver used here had a much larger projected area than did the one used in 1927. The 1939 eclipse was centered primarily on the light uplands, but about 25 per cent of the area was filled by Mare Vaporum and a tiny bit by Sinus Aestuum.

Wesselink (38), Lettau (39, 40), and Jaeger (11) correctly interpreted the 1939 results of Pettit as showing that the fall in temperature observed was consistent with a thin homogeneous lunar surface composed of a powder or dust *in vacuo*, possibly covering a second layer of different characteristics. The basis for this conclusion was work published in 1910 (41) and 1911 (42) by Smoluchowski on eleven powders of grain size smaller than 0.1 mm at gas pressures as low as 0.06 mm of mercury. Kannuluik and Martin (43) have confirmed Smoluchowski's work in general, although they derived a somewhat higher value for the thermal conductivity of powders in vacuum.

If the crust of the moon is homogeneous, the solutions of the equation of heat conduction, for a particular eclipse on the model considered, form a one-parameter family in the thermal parameters when these are assumed to be independent of temperature. This parameter is

$$\sigma = K\rho c, \qquad (13\text{-}1)$$

where K is the coefficient of thermal conductivity, ρ is the density, and c is the specific heat capacity. Wesselink showed that the heat capacity, ρc, per unit volume did not vary widely for common minerals. Therefore, the procedure amounts to a method of deriving a value of K. It is found to be very low, possibly lower than all materials save only Smoluchowski's powders *in vacuo*.

If K is small, the lunar surface is a poor conductor of heat. Absorbed solar radiation will not be conducted far into the moon. It will affect only a thin layer, and the temperature will vary widely and show rapid changes. If K had proved to be large, the heat would have been conducted to greater depths, and the temperature variations at the surface would have been much smaller.

Wesselink showed that, at lunar midday, only about 1 per cent of the heat of the sun was conducted through the surface layers. The rest was either reflected or absorbed and immediately re-radiated by the warm surface. He also showed that thermal effects during an eclipse were negligible below 7 mm and that effects during a lunation were also inappreciable below this level, provided that the dust were at least this thick. If there were a thin layer of dust overlying solid rock, the temperature variations during a lunation could penetrate 2–3 meters. Wesselink (38) may be quoted as saying:

The conclusion from all this work is that the surface of the moon is covered with dust but that no information is forthcoming on the depth of the dust layer. The extremely low values of the thermal conductivity provide yet another confirmation of the absence of any appreciable lunar atmosphere. The dusty nature of the surface is in agreement with conclusions from studies on the variations with phase of reflected moonlight.

Gilvarry (44) analyzed Pettit's results and concluded that if the dark areas had been composed of bare rock, the effect would have been easily observed. He found that essentially the entire area under observation must have been dust-covered. He rejects the lava hypothesis and provisionally accepts Gold's dust hypothesis. His reasoning would seem to be that if the dark areas were lava, then dust would cover them and eliminate the color differences, and hence we do not need to call on lava to account for the color difference.

Gilvarry's calculations illustrate the fact that, in the penumbral phase of the eclipse, the temperature fall agreed well with the predicted curve from a dust surface but that, in the umbral portion, a better fit was attained by assuming that the dust was only 2 mm thick in the upland portions studied and that the Mare Vaporum area was thicker dust. His curves all agreed within about 10° K, while Pettit's data for the 1927 and 1939 eclipses had uncertainties of $\pm 6°$ and $\pm 3°$ K, respectively.

The data do indicate that the surface may be dust-covered but are not sufficiently accurate yet to allow a specification of the thickness of the dust. As Gilvarry points out, there is a possibility of an even better fit if sufficient solutions of the heat-conductivity equation were available to carry out properly the necessary optimization over two parameters. For a rigorous determination of the optimum fit on this model, one should probably take into account the dependence of the radiative component of the thermal conductivity of dust on the third power of the absolute temperature. Jaeger and Harper (45) found that this variation alone was not sufficient to reconcile the small differences between theory and observation for the case of homogeneous dust.

Gilvarry (46) remarks:

> It has been established to this point that the eclipse observations of Pettit and the theoretical analysis exclude the possibility of exposed lava, specifically on Mare Vaporum and, by implication, on any mare. This deduction makes it impossible to salvage the hypothesis of the presence of lava without violating the requirements of Ockham's razor. To be specific, one cannot logically explain the contradiction of the lava hypothesis with Pettit's results by postulating obscurant dust on the lava of Mare Vaporum, for in that case the *dust* would have to be dark and smooth, and the additional assumption would vitiate the prime evidence of the appearance of the mare surfaces for the identification as lava in the first instance. Any such resolution of the difficulty would involve *ad hoc* assumption of some combination of the three processes of dust creation, transport and denigration which Gold invokes to explain the nature of the maria without the intervention of lava at all. Thus the main conclusion of this paper is that a logical argument for inferring the presence of lava on the lunar surface no longer exists, unless the reasoning is divorced from the purely superficial appearance presented by the surface of the maria.

Gilvarry's main point in the article is correct that there is no direct measurement of radiation from the moon which would require the dark areas to be lava.

It is implicit in Gilvarry's argument that a transfer of dust occurred from the uplands to the lowlands and that the thick deposits of dust changed color by some process when they settled down in the maria. The argument against the lava nature of the maria fails if the dust were produced *in situ* in uplands and in lowlands and if the color is closely related to the color of the material from which it came.

In Sinton's (34) measures by optical techniques of the temperature variations during the January 19, 1954, eclipse, at 1.5-mm wave length, he found that the temperatures fell from 300° to 170° K at the end of totality. There was a substantial lag relative to the incident solar radiation. The low temperature did not return to normal until sometime after the moon had left the penumbra. The maximum temperature was lower and the minimum temperature higher than Pettit and Nicholson had found at earlier eclipses.

Gibson (47, 48) reported that measurements at 8.6 mm showed no evidence of temperature variations during two total lunar eclipses. In his detailed observations of the total lunar eclipse of March 13, 1960, at a wave length of 8.6 mm, Gibson found a maximum possible temperature variation near the center of the disk of 2° and a probable variation of 1° or less. The attenuation constant was determined to be 0.02–0.03/cm.

Other radio wave-length eclipse studies are by Mitchell and Whitehurst (49) at 7.5 mm on November 18, 1956; Kajdanovsky, Turusbekov, and Khaikin (50) at 3.2 cm on January 29, 1953; and Mezger and Strassl (51) at 21 cm on June 13, 1957. Castelli, Ferioli, and Aarons (52) observed the March 13, 1960, eclipse at 10- and 23-cm wave lengths and found no brightness temperature changes exceeding 2.5 per cent. A portion of the September 5, 1960, eclipse was observed at 8.6 mm by Tyler and Copeland (53), who found a decrease in brightness temperature of 20°.

Only the 1.5-mm observation disclosed a significant variation in brightness amounting to a reduction of about 40 per cent. Because of instrumental limitations, the 7.5-mm observation could not reveal changes of less than 10 per cent or 20° in temperature.

Gibson (48), in discussing the results of the March, 1960, eclipse, points out that the attenuation constant, α, was found to be 0.02–0.03/cm, as contrasted with his previous value of 0.28/cm obtained for the same wave length from measurements through a lunation instead of an eclipse (18). In the latter he found new moon temperatures of 145° and 225° K at full moon. This conflict rules out the possibility that the surface of the moon is strictly homogeneous to a depth of as much as 3 cm, with the thermal conductivity consistent with the infrared observations. He stated that it could be shown that the 8.6-mm eclipse and lunation values for α were not appreciably drawn together by assuming major values for thermal conductivity in a homogeneous surface material.

Gibson obtained agreement between infrared and the 8.6-mm results on the assumption that the lunar surface was stratified, with the top stratum thin and essentially transparent at the radio wave length. By combining various surface parameters as tabulated by Jaeger (12), he derived, for an emissivity of 0.85, a temperature range of 238°–149° K, which corresponds fairly well with Gibson's (17, 47) values for the 8.6-mm wave length of 225° and 145° K, or 229° and 165° K, as given by Salomonovich (54) for the 8-mm wave length. The parameters which gave the best fit corresponded to a depth of 4.5 mm for the top layer, which is sufficient to be compatible with the eclipse interpretations and is also adequate to yield the observed midnight infrared surface temperature of 120° K, as pointed out by Jaeger. The thermal conductivity of the second layer is about 16 times that of the top layer.

In order to explain the temperature variations as determined over a much wider range of wave lengths, Gibson found it necessary to postulate either that the electric conductivity of the substratum varied faster with frequency than theory would suggest or that the observed results might be explained by a third stratum in the lunar surface which lay well below the depth of penetration for wave lengths of 1 cm, but which influenced the thermal emission for wave lengths of about 3 cm.

A reduced brightness variation during the month could then be attributed to increased thermal conductivity in the third layer, which would increase the temperature gradient through the second layer and thereby reduce the mean temperature fluctuation at the depths most important in 3-cm emission. Alternatively, it may be that if the third layer has reduced electrical conductivity compared with the intermediate layer, the depth from which 3-cm wave lengths may be received may be increased, and the reduced temperature fluctuation at the greater depths would account for the reduced brightness variation actually observed. The properties of increased thermal conductivity and decreased electrical conductivity compared with the intermediate layer are suggestive of a more compacted material, possibly fused solid, whose attenuation may be more like that of sand than the intermediate layer. Gibson concludes that the experimental results may be considered to indicate that the lunar surface is stratified, with a shallow top layer somewhat like terrestrial sand covering a porous substratum of high electric conductivity, which in turn may cover a third layer somewhat similar to terrestrial rock. Gibson (55) reported:

> As to the question of where on the moon the radio beam was centered, this may be considered to be about a fifth of the lunar radius west of the center of the lunar disk. It was approximately this point in the west-to-east scans that maximum radio brightness was observed, the deviation of this position from the center of the disk being due to the lag of the surface heating with respect to lunar phase. However, the area encompassed by the radio beam is quite large, even to the half-power points some twelve minutes in diameter apart, and most of the lunar face affects the observed radio brightness.

The 12-minute-diameter area was centered near the crater Godin, and possibly as much as two-thirds of the area covered was of maria type.

Barabashov and Chekirda (56) recently reviewed optical evidence relating to the lunar surface, and they concluded that neither fine powder nor a fused material was likely but rather that a coarsely fragmented tuffaceous rock, together with volcanic ash in some places, seemed probable. Their estimate of the size of the fragments was 3–10 mm, and they considered it unlikely that the fragmented or pulverized top layers could be more than several centimeters deep.

Saari and Shorthill (57) reported infrared observations of the craters Tycho, Copernicus, Aristarchus, and Proclus during the September 5, 1960, eclipse. The first three remained warmer by about 40° than their environs during the eclipse. Sinton (58) found the same effect at Tycho, and he interpreted these results as indicating that the crater floor was composed of rock with a thin film of dust 0.3 mm thick, while the neighborhood of the crater would be covered by much thicker dust. This is consistent with Sinton's recent observation that Copernicus is normally slightly cooler than its surroundings outside eclipse but in the daylight. Shorthill, Borough, and Conley (59) found similar enhanced radiation during a lunar eclipse at Aristarchus, Copernicus, and Tycho, but not at Eratosthenes or other craters near Tycho. The first three are relatively new craters with rays; the others are older and rayless.

Piddington and Minnett (20) combined their own radio observations with those of Pettit and Nicholson (19) in the infrared, both during a lunar eclipse. They derived

$$d = 610 K(K'\rho'c')^{-1/2}, \qquad (13\text{-}2)$$

where d is the thickness and K is the thermal conductivity of a thin layer of material lying above the substratum with constants K', ρ', and c'.

Jaeger and Harper later examined the observations of both pairs of men and determined that if $(K\rho c)^{-1/2} = 1,030$ c.g.s., they could nearly, but not quite, match the eclipse data of 1939. They suggested that the solid subsurface was not quite homogeneous.

It is possible that K depends on temperature. Also Muncey (60) in 1958 had noted that if K and c are proportional to the absolute temperature, then $(K\rho c)^{-1/2}$ could be of the order of 200 or 300 at 300° K, rather than about 1,000 c.g.s. These constants must be thoroughly studied.

On the assumption that all constants were correct, Jaeger and Harper (45) varied K', ρ', c', and d and found that the best fit to the eclipse curve was given by $d \sim 2$ mm and $(K'\rho'c')^{-1/2} \sim 100$. They concluded that the substratum corresponded to "terrestrial values for substances such as pumice or gravel." From a comparison with a curve for $(K\rho c)^{-1/2} = 20$, valid for bare rock, they feel that not more than 5 per cent of the observed area could have been exposed rock.

Fremlin (61) assumed a rather high heat flux and a very low value of the thermal conductivity. Jaeger (22) questioned these assumptions, and this led to a reply by Fremlin (62), in which he concluded that if there were patches of bare rock amid larger areas of dust, $(K\rho c)^{-1/2}$ could be 20 for the rock and as large as 2,700 for the dust. With these figures, he closely matched Pettit's observed eclipse curve when 4.8 per cent of the surface was exposed rock. Fremlin also stated that, by varying the albedos of different kinds of rock, there would be no difficulty in using a value for K for dust which was even lower than that employed in his previous paper.

Fremlin's conclusions may ultimately turn out to be correct; they certainly are more logical than to conceive of the moon as a surface of uniform layers. It must be remembered, though, that at present there is no observational backing for his very low values of K.

MacDonald (63) has studied the heat balance of the moon and finds that the energy radiated from the surface due to the internal original or radioactive heat is in the range of 4–16 ergs/cm² sec. For comparison, the value for the earth is about 50 ergs/cm² sec. This radiation is very tiny compared with the incident light and heat from the sun. It could maintain the moon's surface at temperatures in the range of 16°–23° K. Clearly, the temperatures of the moon's outer layers are in a cyclically balanced state, with the sun furnishing almost all the heat.

Sinton (64) finds that his measures of the temperature of the lunar subsolar point indicate a total heat flux of 1.85 cal cm^{-2} min^{-1}. The expected value is 1.75 cal cm^{-2} min^{-1} as derived from the solar constant. He worked at 8.8 μ. Sinton suggested that the extra 0.10 cal is probably produced by corpuscular radiation received from the sun.

Biermann (65, 66), in his analysis of the accelerations produced in comet tails, assumed that 10^{11}–10^{13} electrons and protons with velocities between 1,000 and 3,000 km/sec are incident per cm² near the earth. The required 0.10 could be produced by 2×10^{12} protons per cm² per second of velocity 2,000 km/sec.

A reciprocal approach is to project radar waves from the earth to the moon in extremely short pulses and to observe the reflected waves. These experiments have been conducted in the wave-length range of 10 cm to 3 meters (67–77). The first radar echoes from the moon were obtained in 1946 by the United States Army Signal Corps. In general, experiments of this type show that the moon is a remarkably good reflector for the 10-cm waves and reflects the longer wave lengths in progressively poorer fashion.

In visible light, the moon normally is brightest at the center of the disk, neglecting local variations, and shades off gradually toward the limb, approximately according to the Lommel-Seeliger law. At microwave lengths, the brightness of the moon falls off according to Lambert's law, which is a cosine function. At these frequencies, the lunar disk fades more rapidly away from the center than it does in optical frequencies. If the moon were a polished sphere, it would appear to the eye as a faint disk with a tremendously bright solar image reflected from the appropriate position.

At microwave frequencies, a similar phenomenon is manifest. Inasmuch as the radar waves are sent from the earth, the bright specular reflection appears near the center of the disk. According to Pettengill's observations (78) at 70 cm (440 megacycles), the specular reflection is about fifty times as bright as the more diffuse Lambert scattering-law reflection from nearby (see Fig. 43). The original discovery of this effect was by Trexler (73).

Lunar-Surface Materials by Heat Measures

Accurate measures of the time taken by a short pulse to go to the moon and back demonstrate that the specular reflection comes from the nearest region of the moon. The duration of the sharp reflected peak is only a few hundred microseconds. It was therefore concluded that it was produced by the reflection from an area near the center of the moon having relatively gentle slopes. Trexler noted that 50 per cent of the power came from the first 5 miles

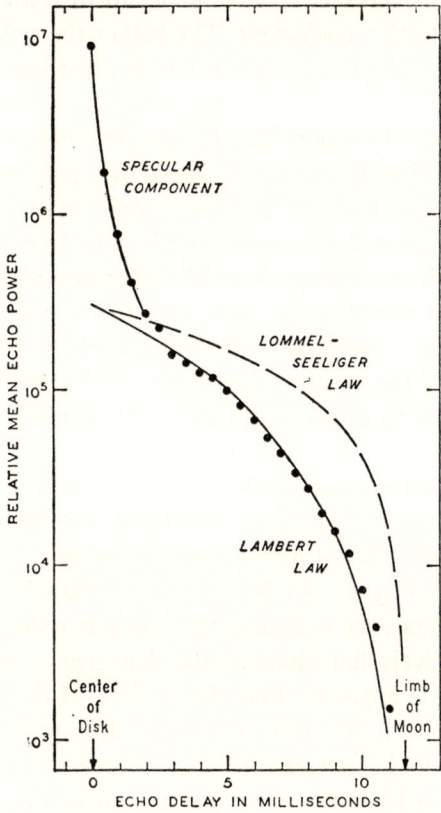

Fig. 43.—Radar brightness of the moon's disk at various distances from its center at 440 megacycles per second. (According to Pettengill, Ref. 78.) (Courtesy Lincoln Laboratory of Massachusetts Institute of Technology.)

of depth, or a circle of only 105 miles in radius. This is about one-tenth the radius of the moon. Lovell (79) states that the main scattering takes place within an area whose radius is about one-third that of the moon.

Senior and Siegel (77) discussed the theory of radar scattering by the moon, pinpointing the mathematical theory behind both the specular and the diffuse reflection. They note that from 20 to 30 areas are present in the specular reflection but do not identify them with known features on the moon. They find that the data suggest that, whereas the specular areas nearest the center of the disk are portions of spherical surfaces, the areas farther back are possibly crater-like.

The longer the pulse length, the more there are of these reflectors that can be covered by a single pulse at a given instant of time, and the higher the return that one obtains. According to Siegel (80), if the reflection coefficient were as low as 10^{-4} power, then the amplitude of each return would agree with a perfectly smooth homogeneous sphere made up of material constants which yielded the above reflection coefficient. The higher the reflection coefficient, the smaller is the percentage of the surface which contributes to the specular reflection.

Victor and Stevens (81) report that the planet Venus appears to be a much better radio reflector than the moon. Relative to a polished, conducting sphere of the same size, Venus appears to have a reflectivity of 10–15 per cent at 2,388 megacycles per second, or a wave length of 12.56 cm. Similar experiments on the moon with their continuous-wave bistatic radar yielded reflectivity numbers of 2 per cent. It would seem, then, that the specular reflection from the moon is composed of a series of reflections, each of which is produced by multiple reflections of the radar pulses from small portions of the local area. Only those areas properly aligned with the earth would contribute to the specular reflection.

The radar signal analysis has not been accomplished for pulse lengths below 2 microseconds, and, as a result, it is very hard to determine the effect of distortions on the pulse that is not due to many reflectors. All the principal reflectors seem to be covered by a pulse length of 300–600 microseconds.

As it actually happens, the region of the specular reflection is at least partly occupied by the relatively flat Sinus Medii. The region does have in it numerous small craters. It is apparent that the surface of the moon in this general area is reasonably smooth between the craters. It is probable that the average slope of the craters and irregularities in the area are sufficient that energy scattered according to Lambert's law will return to the earth as a very tiny fraction of the specular reflection. There is a parallel here to receiving light from a mirror which has a series of holes punched in it. The effect of the holes is lost in the glare from the between-holes surface.

This interpretation is consistent with saying that if no crater exists in a given area and a portion of this area is exactly perpendicular to the radar pulse, then a specular reflection will occur. If a crater exists in the area, it will scatter the energy incident on it, and the specular reflection will be somewhat reduced, but its character will not be changed. Tiny craters which have been filled with dust will act like undisturbed smooth ground.

There is no inconsistency between the optical and the microwave appearances of the moon. They refer to the same body. The optical measures show that the moon has many irregularities; the radar reflections show that there are some reasonably flat areas between the irregularities. We do not know what percentage of the total area is flat enough to produce a specular reflection, but it probably is small. Yaplee, Bruton, Craig, and Roman (68) have found such near-specular reflections on earth above dry sandy soil at a 10-cm wave length.

Senior and Siegel find that the lunar material has a very low relative permittivity, approximately 1.1, but suggest that this is not necessarily inconsistent with a solid lunar material and that selected powders might conceivably fill this bill.

All observations are consistent with a thin dust layer on the moon. Numerous particles of the size of sand and gravel are not excluded. Whipple (82) has modified the dust hypothesis slightly:

From observations of the moon made optically in the infrared by radio and by radar it is concluded that the moon's surface consists of a thin insulating layer overlying denser material, producing a surface that is a very poor reflector at all frequencies so far observed. At dimensions comparable to the wavelength of light, the surface is almost rough, but relatively smooth at wavelengths and dimensions comparable to 10 cm. The moon's surface is subjected to bombardment by meteoritic material and corpuscular radiations (protons) and to a slow rain of dust. Molecules, atoms, and ions are a consequence of the meteoric bombardment. It is concluded that these processes coupled with solar radiation will tend to cement the dust into a weak, semiporous matrix and that no appreciable quantity of loose dust will be found on the moon's surface.

Sytinskaya (83–86) thinks that, based on photometric, colorimetric, and polarimetric investigations, it must be concluded that the action of micrometeoritic impacts would be sufficient to convert the lunar rocks into a slaglike composition.

Urey (87) has correctly inferred that any deep areas of dust will be reconsolidated into a rocklike mass at not too great a distance below the surface.

Gold (88, 89) has suggested that dust is formed in the uplands and transported to the lowlands. This will be discussed later, but there are reasons for thinking that this process has not operated extensively on the moon. If such a massive transportation of dust has not occurred, then the known dust must have been hurled into place more or less at random, by meteoritic collisions, or it was produced *in situ*. If the dust were produced locally, how deeply are the surface layers pulverized?

During 1959, at a Colloquium on Lunar Matters at the Vought Astronautics Division of Chance-Vought Aircraft, Inc., in Dallas, Texas, the subject of the thickness of the dust layers on the moon was thoroughly discussed. The opinion was expressed there that the dust was essentially produced *in situ* and that the mechanism was that of micrometeoritic impact and emanations from the sun, both corpuscular and radiant. Some dust must come from faraway meteoritic impacts, but, at the present rate of infalls, this is relatively unimportant. Almost all the material striking the moon is in the form of micrometeorites. These tiny objects cannot penetrate the lunar surface by much more than a millimeter. Large objects, such as those which produce the "shooting stars," will penetrate slightly farther, but the production of dust is a self-limiting phenomenon. A dust layer will prevent more dust from being formed beneath it. If a very large meteorite strikes, it will blast away the dust and surface materials, some to remote parts of the moon and some to leave the moon. In this case, the formation of a thin layer of local dust will start all over.

Radiation from the sun possesses small penetrating power. It can only produce surface dust. Cosmic rays, of course, will penetrate farther; but, even here, the depth is measured in feet, and if the dust is formed, it will be shallow. Whipple (82) and Öpik (23) later reached similar conclusions.

An interesting point is that the eclipse measures show that the areas of exposed rocks must be less than 5 per cent of the area studied, whether the region is light or dark. The remainder of the areas where the temperatures have been measured must be dust-covered or else covered by some other form of insulating material which has K as low as dust in a vacuum.

The observed fact that the exposed lunar surface is usually quite rough in the upland areas or in craters suggests questioning the blind adherence to the doctrine of universal dust. May not the action of the sun, cosmic rays, and even meteoritic materials lead to a modification of the lunar rocks into a very porous form which would, in a vacuum, possess a value of the thermal conductivity equal to that observed? Dust, by itself, should fall away from steep slopes, and Zöllner has shown that the average small-scale slope of the lunar surface materials is 52°. A mechanism seems required that will produce from the lunar materials a very highly insulating material and leave it where it was formed. This might be a process requiring millennia or more to complete.

It is noted that, even under terrestrial air, there are many dozens of commercial insulating materials which have very low thermal conductivities, of the order of 10^{-2} that of granite or basalt. This value approaches rather closely that possessed by the lunar-surface layers.

References

1. Rosse, Earl of. "On the Radiation of Heat from the Moon, the Law of Its Absorption by Our Atmosphere, and of Its Variation in Amount with Her Phases," *Phil. Trans. R. Soc., London*, **163**, 587, 1873.
2. Langley, S. P. "On the Temperature of the Surface of the Moon," *Mem. Nat. Acad. Sci.*, **3**, 13, 1885.

3. ———. "The Temperature of the Moon," *ibid.*, **4,** 107, 1889.
4. PEAL, S. *A Short Abstract of the Theory of Lunar Surfacing by Glaciation,* p. 10. Calcutta: Band & Mookeyee, 1886.
5. FAUTH, P. *Mondesschicksal; wie er ward und untergeht; eine glazialkosmogonische Studie.* Leipzig: R. Voigtländer, 1925.
6. VERY, F. W. "The Temperature of the Moon," *Ap. J.*, **24,** 351, 1906.
7. ———. "Temperature Assigned by Langley to the Moon," *Science,* **37,** 949, 1913.
8. PETTIT, E., and NICHOLSON, S. B. "Lunar Radiation and Temperatures," *Ap. J.*, **71,** 102, 1930.
9. PETTIT, E. "Lunar Radiation as Related to Phase," *Ap. J.*, **81,** 17, 1935.
10. WESSELINK, A. J. "Heat Conductivity and Nature of the Lunar Surface Material," *Bull. Astr. Inst. Netherlands,* **10,** 351, 1948.
11. JAEGER, J. C. "Conduction of Heat in a Solid with Periodic Boundary Conditions, with an Application to the Surface Temperature of the Moon," *Proc. Cambridge Phil. Soc.*, **2,** 355, 1953.
12. ———. "The Surface Temperature of the Moon," *Australian J. Phys.*, **6,** 10, 1953.
13. ———. "Sub-Surface Temperatures on the Moon," *Nature,* **183,** 1316, 1959.
14. SINTON, W. M. "Heat Maps of the Moon," *Sky and Telescope,* **19,** 348, 1960.
15. GEOFFRION, A. R., KORNER, M., and SINTON, W. M. "Isothermal Contours of the Moon," *Lowell Obs. Bull.*, No. 106, 1960.
16. GARSTANG, R. H. "The Surface Temperature of the Moon," *J. Brit. Astr. Assoc.*, **68,** 155, 1958.
17. COATES, R. J. "Lunar Brightness Variations with Phase at 4.3-mm Wave Length," *Ap. J.*, **133,** 723, 1961.
18. GIBSON, J. E. "Thermal Radiation of the Moon at 0.86 cm. Wave Length," *U.S. Naval Res. Lab. Rept. 4894, ASTIA,* AD-143, p. 403, 1957.
19. ———. "Lunar Thermal Radiation at 35 KMC," *Proc. Inst. Radio Eng.*, **46,** 280, 1958.
20. PIDDINGTON, J. H., and MINNETT, H. C. "Microwave Thermal Radiation from the Moon," *Australian J. Sci. Res. (Ser. A)*, **2,** 63, 1949.
21. TROITSKY, V. S. "K teorii radioizlucheniya luny" ("On the Theory of Radio Emission of the Moon"), *Akad. Nauk SSSR, Astr. Zhur.*, **31,** 511, 1954.
22. JAEGER, J. C. "Sub-Surface Temperatures on the Moon," *Nature,* **183,** 4671, 1316, 1959.
23. ÖPIK, E. J. "Surface Properties of the Moon," Prepublication copy, 1961.
24. TROITSKY, V. S., and ZELINSKAYA, M. R. "Determination of Certain Properties of the Surface Layers of the Moon from Its Radio-Emission at a Wavelength of 3.2 cm," *Russ. Astr. J.*, **32,** 550, 1955.
25. AKABANE, K. "Lunar Radiation at 3000 Mc/s," *Proc. Japan. Acad.*, **31,** 161, 1955.
26. Unpublished. Quoted by SEEGER, WESTERHOUT, and CONWAY, Ref. 27.

27. Seeger, C. L., Westerhout, G., and Conway, R. G. "Observation of Discrete Sources, the Coma Cluster, the Moon, and the Andromeda Nebula at a Wavelength of 75 cm," *Ap. J.*, **126,** 585, 1957.
28. Dicke, R. H., and Beringer, R. "Microwave Radiation from the Sun and Moon Measured with a Microwave Radiometer," *Ap. J.*, **103,** 375, 1946.
29. Zelinskaya, M. R., Troitsky, V. S., and Fedoseyev, L. I. "Radio Emission of the Moon on 1.63 cm," *Russ. Astr. J.*, **36,** 643, 1959.
30. Amenitskii, N. A., Noskova, R. I., and Salomonovich, A. E. "The Radio-Image of the Moon at 8 mm," *Russ. Astr. J.*, **37,** 185, 1960.
31. Grebenkemper, C. J. "Lunar Radiation at a Wavelength of 2.2. cm," *U.S. Naval Res. Lab. Rept. 5151*, p. 1, 1958.
32. Quoted by Seeger, Westerhout, and Conway, Ref. 27.
33. Pawsey, J. L., and Bracewell, R. N. *Radio Astronomy*, chap. viii. Oxford: Clarendon Press, 1955.
34. Sinton, W. M. "Observations of Solar and Lunar Radiation at 1.5 mm," *J. Opt. Soc. America*, **45,** 975, 1955.
35. Epstein, P. S. "What Is the Moon Made Of?" *Phys. Rev.* (Ser. 2), **33,** 269, 1929.
36. Pettit, E. "Radiation Measurements on the Eclipsed Moon," *Contr. Mt. W. Obs.*, No. 627, 1940.
37. ———. *Ap. J.*, **91,** 408, 1940.
38. Wesselink, A. J. "Heat Conductivity and the Nature of the Lunar Surface Material," *Bull. Astr. Inst. Netherlands*, **10,** 351, 1948.
39. Lettau, H. "On the Heat Budget of the Moon and the Surface Temperature Variations during a Lunar Eclipse," *Geofis. pura e appl.* (Milan), **19,** 1, 1951.
40. ———. "Was Sagen die Temperaturen des Mondes über seine Oberfläche aus?" *Die Umschau über die Fortschritte in Wissenschaft und Technik* (Frankfurt am Main), **52,** 417, 1952.
41. Smoluchowski, M. *Bull. Acad. Sci. Cracovie (A)*, Vol. **129,** 1910.
42. ———. *Ibid.*, p. 548, 1911; also *Internat. Crit. Tables*, **2,** 315.
43. Kannuluik, W. G., and Martin, L. H. "Conduction of Heat in Powders," *Proc. R. Soc. London, A,* **141,** 144, 1933.
44. Gilvarry, J. J. "The Nature of the Lunar Maria," *Ap. J.*, **127,** 751, 1958.
45. Jaeger, J. C., and Harper, A. F. A. "Nature of the Surface of the Moon," *Nature*, **166,** 1026, 1950.
46. Ref. 44, p. 760.
47. Gibson, J. E. *Proc. I.R.E.*, **46,** 280, 1960.
48. ———. "Lunar Surface Characteristics Indicated by the March, 1960, Eclipse and Other Observations," *Ap. J.*, **133,** 1072, 1961.
49. Mitchell, F. H., and Whitehurst, R. N. *A Radio Study of the Sun and Moon at Millimeter Wavelengths*, University of Alabama Phys. Dept. Rept. under OOR Contract, 1958.
50. Kajdanovsky, N. L., Turusbekov, M. T., and Khaikin, S. E. *Pub. 5th Conf. on Cosmogony (Moscow)*, p. 347, 1956.

51. Mezger, P. G., and Strassl, H. *Planetary Space Sci.*, **1**, 213, 1959.
52. Castelli, J. B., Ferioli, C. P., and Aarons, J. *A.J.*, **65**, 485, 1960.
53. Tyler, W. C., and Copeland, J. Paper presented at 107th meeting A.A.S., December, 1960.
54. Salomonovich, A. E. *Astr. Zhur.*, **35**, 129, 1958.
55. Gibson, J. E. Personal communication, June 26, 1961.
56. Barabashov, N. P., and Chekirda, A. T. *Astr. Zhur.*, **36**, 851, 1959.
57. Saari, J. M., and Shorthill, R. W. Paper presented at 107th meeting A.A.S., December, 1960.
58. Sinton, W. M. "Eclipse Temperatures of the Lunar Crater Tycho," *Lowell Obs. Bull.*, No. 108, 1960.
59. Shorthill, R. W., Borough, H. C., and Conley, J. M. "Enhanced Lunar Thermal Radiation during a Lunar Eclipse," *Pub. A.S.P.*, **72**, 429, 481, 1960.
60. Muncey, R. W. "Calculations of Lunar Temperatures," *Nature*, **181**, 1458, 1958.
61. Fremlin, J. H. "Volcanoes on the Moon," *Nature*, **183**, 239, 1959.
62. ———. "Sub-Surface Temperatures on the Moon," *ibid.*, p. 1317.
63. MacDonald, G. J. F. "Interior of the Moon," *Science*, **133**, 1045, 1961.
64. Sinton, W. M. "Is the Moon Heated in Part by Solar Corpuscular Radiation?" *Pub. A.S.P.*, **72**, 362, 1960.
65. Biermann, L. *Zs. f. Ap.*, **29**, 274, 1951.
66. ———. *Mém. Soc. R. Sci. Liège*, Ser. 4, **13**, 251, 1953 (*Liège Inst. d'Ap. Repr. No. 352*).
67. Hey, and Hughes, V. A. Personal communication to K. M. Siegel, April 28, 1958.
68. Yaplee, B. S., Bruton, R. H., Craig, K. J., and Roman, N. G. "Radar Echoes from the Moon at a Wavelength of 10 cm.," *Proc. I.R.E.*, **46**, 293, 1958.
69. Aarons, J., Peters, H., and Whitney, H. Personal communication to K. M. Siegel, October 29–30, 1959.
70. Blevis, B. C. Personal communications to K. M. Siegel, October 8, 1957, and June 3, 1958.
71. Fricker, S. J., Ingalls, R. P., Mason, W. C., and Stone, M. L. "UHF Moon Reflections," paper presented at U.R.S.I.-I.R.E. meeting, Washington, D.C., 1958.
72. Leadabrand. Personal communication to K. M. Siegel, October 30, 1959.
73. Trexler, J. H. "Lunar Radio Echoes," *Proc. I.R.E.*, **46**, 286, 1958.
74. Webb, H. D. Personal communication to K. M. Siegel, October 30, 1959.
75. Evans, J. V. "The Scattering of Radio Waves by the Moon," *Proc. Phys. Soc.*, B, **70**, 1105, 1957.
76. Browne, I. C., Evans, J. V., Hargreaves, J. K., and Murray, W. A. S. "Radio Echoes from the Moon," *Proc. Phys. Soc.*, B, **69**, 901, 1956.
77. Senior, T. B. A., and Siegel, K. M. "A Theory of Radar Scattering by the Moon," Prepublication copy. Summarizes data from Refs. 61–70.
78. Green, P. E., Jr., and Pettengill, G. H. "Exploring the Solar System by Radar," *Sky and Telescope*, **20**, 9, 1960.

79. Lovell, A. C. B. Report on British Work carried out in Radio Astronomy since 1954, presented at XII General Assembly at U.R.S.I., Boulder, Colorado, 1957.
80. Siegel, K. M. Personal communication.
81. Victor, W. K., and Stevens, R. "Exploration of Venus by Radar," *Science,* **134,** 46, 1961.
82. Whipple, F. L. "On the Lunar Dust Layer," *Smithsonian Inst. Ap. Obs.,* Cambridge, Mass., April 29, 1958. *Data Pub.* (Washington, D.C.), 2 ASTRO-12 (abstr.).
83. Sytinskaya, N. N. "The Nature of the Moon's Surface (Using Photometric and Colorimetric Observations)," *Priroda,* **41,** 93, 1952 (in Russian).
84. ———. "Les Explosions des météorites, origine de la modification de la surface de la lune," *Akad. Nauk SSSR Voprosy Kosmogonii,* **5,** 13, 1957 (French summary).
85. ———. "Origin and Nature of Lunar Surface Crust Based on Data of Comparative Study of Brightness," *Leningrad U. Uchënyye Zapinski,* No. 190, 1957. (In Russian. English translation on file U.S. Geol. Surv., Military Geology Branch.)
86. ———. "New Data of Meteor-Slag Theory of the Formation of Outer Layers of the Lunar Surface," *Akad. Nauk SSSR, A.J.,* **36,** 315, 1959 (English summary).
87. Urey, H. C. "The Origin of the Moon's Surface Features," *Sky and Telescope,* **15,** 2, 1956.
88. Gold, T. "The Lunar Surface," *M.N.,* **115,** 585, 1955.
89. ———. "Dust on the Moon," in *Vistas in Astronautics: Second Annual Astronautics Symposium,* Vol. **2.** New York: Pergamon Press, 1959.

14

TEKTITES

In 1952 Nininger (1) suggested that tektites might be portions of the lunar surface that had been blasted off the moon by meteoritic impacts. O'Keefe (2) in 1958 revised and generalized the support for the argument that the materials for the moon's surface could, in fact, be similar to those of tektites found on the earth. Varsavsky (3) and Gold (4) have accepted this hypothesis. The tektites are peculiar glassy objects, possibly of extra-terrestrial origin. They are found in various parts of the world, but their distribution is non-random.

Senior and Siegel (5) have determined electromagnetic constants from radar data for the surface materials of the key scattering centers of the moon. Olte and Siegel (6) have determined them for tektites in the laboratory. The two results do not agree, and Olte and Siegel conclude from this either that the lunar surface is not made of material such as the tektites they measured or that, if tektite material is found on the moon, it is found in a crushed form with a very low packing factor. Alternatively, the material of the moon's surface may be converted into tektitic material by meteoritic impact, or material ejected from the moon by this process is converted into tektites in the passage through the earth's atmosphere.

Conversely, Kohman's observations of radioactive Al^{26} and Be^{10} formed by cosmic-ray bombardment showed that the level of radioactivity was considerably above any conceivable levels of production by cosmic or terrestrial radiations at or beneath the earth's surface. This finding, if correct, virtually eliminates not only the earth but also the moon as the source of tektites. Kohman suggests that the tektites arrived at the earth's surface as a loose cluster of glassy objects from outside the solar system (7). Anders (8) has recently shown that the tektites do not contain Al^{26}. If confirmed, this would eliminate the outer-space origin of these bodies.

On February 9, 1913, a great cluster of meteors appeared over southern Saskatchewan, passed over northern Michigan, Ontario, and disappeared out to sea near New York City. O'Keefe (9) has called these the "Cyrillids" after St. Cyril's Day, February 9. Several scientists (10–18) have contended that the objects almost certainly were small earth satellites moving in an orbit of low eccentricity. LaPaz (19), Fenner (20), and O'Keefe (2, 21, 22) have suggested a connection between the Cyrillid shower and tektites. Apparently, no tektites were found resulting from this shower, but the principle seems established that such orbits could lead to localized distributions of meteoritic or tektitic remnants.

Conder (23) in 1934 and Spencer (24) in 1937 were probably the first to advocate fusion of terrestrial sediments by meteoritic impact. Chemical and magnetic measurements indicate that tektites were either melted at temperatures in excess of 2,500° C or heated at this temperature for periods in excess of 20 minutes (25).

Urey (26) has proposed that the tektites are by-products of a direct collision between the earth and the nucleus of a comet. The explosion following the collision would melt sedimentary rocks and scatter them over a wide area. The idea was expressed only as a possibility, and the landing places of the comet or comets were not identified.

Baker (27) questioned Urey's cometary-impact hypothesis, saying that a few primary forms developed in an extra-terrestrial environment and were subsequently modified by aerodynamic phenomena during high-speed earthward flights through the earth's atmosphere. Urey (28, 29) vigorously defended his hypothesis. Chapman's (30) calculations on possible velocities of entry into the terrestrial atmosphere indicate that the australites could have come from the moon.

Barnes (31) has made a life-study of tektites and has concluded that they arose from sedimentary rocks in the earth's crust which had been very suddenly heated by meteoritic impact. Cohen's (34) work shows that tektites could probably have been produced also from such rocks as quartz and porphyry. Smith and Hey (35) have found fragments of quartz in tektites.

The measures of radioactive decay of potassium 40 to argon 40 made independently by Zahringer in Heidelberg and Reynolds in California yield ages of different clusters of tektites ranging from 45 million years for the Texas bediasites to as little as 5 thousand years for the Australian australites.

Barnes tentatively identifies the 20-million-year-old moldavites with the nearby impact crater of similar age, the Rieskessel. The Ashanti Crater in Ghana may be the source of the tektites found on the Ivory Coast.

Other clusters cannot yet be identified with impact craters. The australites, in particular, are puzzling, for they show two heating cycles, including one due to rapid passage through the air. Chapman (43) indicated that the second cycle of heating occurred as the tektites struck the earth's atmosphere at about 11 or 12 km/sec as they descended from the moon.

Barnes suggests that the australites and the Darwin glass arose from a large meteoritic collision on the nearest portion of the Antarctic continent, Wilkes Land. This must have been a tremendous collision, and the crater should still exist.

Schmidt (41), in response to Barnes's suggestion, noted that the 1959–60 U.S. Antarctic over-snow traverse in Victoria Land–Wilkes Land found no gravity or magnetic data supporting the meteoritic impact postulated to have occurred in that area. Schmidt noted that the traverse might not have crossed the proper area, but he did comment that the thick ice and snow cover of this part of Antarctica, which amounted to 8,000–9,000 feet, could have diminished, if not prevented, any large-scale tektite shower if the meteorite did fall in that area.

Barnes (42) answered Schmidt and agreed that the thick ice cover diminished the possibility that the impact crater could be in that portion of Antarctica. However, he felt that, in view of the rayed distribution of ejecta from impacts and the spotty distribution of australites in Australia, the impact site might be anywhere in Antarctica where the ice was thin or absent 5,000 years ago. In such a case, it would be necessary to assume either that the rays of ejected tektites missed southern Africa or South America or that the tektites are present in the southern parts of these continents and have not been found.

In view of the fact that much, although not all, of the Antarctic land mass is covered by deep ice, the argument of Schmidt is a very strong one. In order for a crater to be formed by impact in ice 8,000–9,000 feet thick and to reach the rock layers beneath the ice and to produce tektites from these rock layers, the crater would have had to be, at a minimum, in the 26–30-mile diameter range. This is an impact almost comparable with that of the Vredefort structure, and it does not seem possible that such a crater could have been formed in Antarctica without its remnants being obvious to the pilots of the innumerable planes which have flown over the area.

Recently (44) it has been suggested from the potassium-argon ages that there are only three families of tektites. Comparison of chemical analyses (45) of 66 specimens of tektites shows that their compositions are similar to, but not identical with, those of various sedimentary rocks and certain igneous rocks.

Reid and Cohen (36) have recently identified the presence of coesite in the Darwin glass found in Western Tasmania. This discovery of 100 ppm of coesite shows that Darwin glass is an impactite produced from terrestrial materials containing SiO_2 under the high pressures and temperatures generated by meteoritic impact. The fact that the Darwin glass is found resting directly on diverse bedrock types precludes any hypothesis of volcanic or fulguritic origin. Darwin glass and Libyan glass are objects similar to tektites but perhaps are not true tektites. They certainly are of terrestrial origin.

Barnes (31) visited the site of the Wabar craters in Arabia in 1960 and found numerous bits of ordinary impact glass, as Philby (37) had described, but many of them were in the shapes of dumbbells, spheres, teardrops, etc., all of the splash forms that are observed in tektites. A similar observation was made by Alderman (38) many years ago at the Henbury Main Crater in Australia.

It has very recently been established that at least one type of tektites—the philippinites—contain tiny spherules of nickel-iron and kamacite (39). These particles closely resemble the similar metallic inclusions in the silica glass of the Wabar craters (37, 40).

The study of these strange glasses has now progressed to the point where it is clear that they originated in meteoritic (or cometary) impacts and that their compositions are similar to, but not necessarily identical with, terrestrial sediments and some igneous rocks. The probability is high that at least some of them were formed on and from the earth, but even this is not definite. Further work is still being done and may yet demonstrate that the tektites are of lunar origin. It is very difficult to account for the double cycle of heating shown by the australites on the basis of a terrestrial origin.

In a very thorough study Pinson (46) showed that the Rb/Sr ratios in tektites were within the range found in terrestrial shales and granites but were considerably less variable. It was also shown that the variations in the Sr^{87}/Sr^{86} ratio in tektites are quite different from what one would expect to find, if they are the results of random fusion of terrestrial materials, particularly acid rocks.

Pinson also stated that tektites could not have been derived from any chondrites which have been measured but that they could have been derived from achondrites approximately 500–1,000 million years ago or from the upper mantle approximately 350–500 million years ago.

Similarly, his work on the major element compositions of tektites does not support the theory of formation by random fusion of terrestrial materials whether by impact of meteorites, asteroids, comets, or lightning. The sympathetic variation of the Sr^{87}/Sr^{86} ratios with the Rb/Sr ratios observed in tektites suggests that they differentiated from a common or similar source material approximately 175 million years ago, but their K-Ar ages indicate that they were last heated less than 30 million years ago. These two "ages," together with their similar composition and widespread distribution, cannot be explained by any proposed theory of terrestrial origin.

Thus Pinson was forced tentatively to accept an extra-terrestrial origin for the formation of tektites, fully aware that an extra-terrestrial origin raises many new questions. He says:

The Moon is often suggested as the parent body. If the lunar surface is approximately 4.5 b.y. old the observed Sr^{87}/Sr^{86} ratios in tektites must have been generated by a material having a low Rb/Sr ratio, such as basalt. But how were tektites produced from a basaltic lunar surface? Another possibility is that the surface of the Moon has undergone recent differentiation, an occurrence that is contrary to most theories of lunar history. Perhaps, new studies of the Moon will be able to answer some of these questions, but at present it appears that all suggested theories of the origin of tektites contain difficulties.

The subject is a very important one and merits a great deal of further study.

REFERENCES

1. NININGER, H. H. *Out of the Sky*, p. 302. Denver, Colo.: Denver University Press, 1952.
2. O'KEEFE, J. A. *Nature*, **181,** 172, 1958.
3. VARSAVSKY, C. M. *Smithsonian Inst. Ap. Obs., Tech. Rept.*, No. 4, 1957.
4. O'KEEFE, J. A., VARSAVSKY, C. M., and GOLD, T. *Nature*, **181,** 172, 1958.
5. SENIOR, T. B. A., and SIEGEL, K. M. *J. Res. Nat. Bur. Stand.-D. Radio Prop.*, **64D,** 217, 1960.
6. OLTE, A., and SIEGEL, K. M. "Distinction between the Electromagnetic Constants of Tektites and Libyan Desert Glass and Their Effect on Lunar Surface Theory," *Ap. J.*, **133,** 2, 706, 1961.
7. KOHMAN, T. P. *Nature*, July 26, 1958.
8. ANDERS, E. *Geochim. et Cosmochim. Acta*, **19,** 53, 1960.
9. O'KEEFE, J. A. "Tektites as Natural Earth Satellites," *Science*, **133,** 562, 1961.
10. CHANT, C. A. *J.R.A.S. Canada*, **7,** 145, 1913.
11. BURNS, G. J. *J. Brit. Astr. Assoc.*, **24,** 111, 1913.
12. DENNING, W. F. *J.R.A.S. Canada*, **7,** 404, 1913.
13. DAVIDSON, M. *J. Brit. Astr. Assoc.*, **24,** 148, 1913.
14. PICKERING, W. H. *Pop. Astr.*, **30,** 632, 1922.
15. FISHER, W. *Pop. Astr.*, **36,** 398, 1928.
16. LAPAZ, L. *Meteoritics*, **1,** 402, 1956.
17. O'KEEFE, J. A. *J.R.A.S. Canada*, **53,** 59, 1959.
18. MEBANE, A. D. *Meteoritics*, **1,** 405, 1956.
19. LAPAZ, L. *Pop. Astr.*, **46,** 227, 1938.
20. FENNER, C. *Proc. R. Soc. Sci. Australia*, **62,** 208, 1938.
21. O'KEEFE, J. A. *NASA Technical Note D-490*, 1960.
22. ———. *Space Research*, ed. H. K. BIJL, p. 1080. Amsterdam: North-Holland Publishing Co., 1960.
23. CONDER. *Industrial and Mining Standard of Australia*, **89,** 329, 1934.
24. SPENCER, L. J. "The Tektite Problem," *Mineralog. Mag.*, **24,** 503, 1937.

25. FRIEDMAN, I., THORPE, A., and SENFTLE, F. E. "Comparison of the Chemical Composition and Magnetic Properties of Tektites and Glasses Formed by Fusion of Terrestrial Rocks," *Nature,* **187,** 1089, 1960.
26. UREY, H. C. *Nature,* March 16, 1957.
27. BAKER, G. *Nature,* **185,** 291, 1960.
28. UREY, H. C. "Origin of Tektites," *Nature,* **187,** 855, 1960.
29. ———. "Origin of Tektites," *ibid.,* **181,** 1458, 1958.
30. CHAPMAN, D. R. "Recent Re-entry Research and the Cosmic Origin of Tektites," *Nature,* **188,** 353, 1960.
31. BARNES, V. E. "Tektites," *Sci. American,* **205,** 58, 1961.
32. ———. *Geo. Times,* **1,** 16, 1957.
33. ———. "Origin of Tektites," *Nature,* **181,** 1457, 1958.
34. COHEN, A. J. "Trace Element Relationships and Terrestrial Origin of Tektites," *Nature,* **188,** 653, 1960.
35. SMITH, C., and HEY, J. S. *Bull. Inst. Franç. Afrique Noire,* **14,** 762, 1952.
36. REID, A. M., and COHEN, A. J. "Discovery of Coesite in Darwin Glass," paper presented at first western meeting, American Geophysical Union, Los Angeles, California, December, 1961.
37. PHILBY, H. ST. J. "Rub'al Khali: An Account of Exploration in the Great South Desert of Arabia," *Geog. J.,* **81,** 1, 1933.
38. ALDERMAN, A. R. "Meteorite Craters at Henbury, Central Australia," *Mineralog. Mag.,* **23,** 19, 1932.
39. CHAO, E. C. T., ADLER, I., DWORNIK, E. J., and LITTLER, J. "Metallic Spherules in Tektites from Isabela, Philippine Islands," *Science,* **135,** 97, 1962.
40. SPENCER, L. J. "Meteoritic Craters as Topographic Features on the Earth's Surface," *Geog. J.,* **81,** 227, 1933.
41. SCHMIDT, R. A. Letter in *Sci. American,* **206,** 12, 1962.
42. BARNES, V. E. Letter answering Schmidt, *ibid.*
43. CHAPMAN, D. R. "Origin of Tektites," *Lunar and Planetary Exploration Colloquium,* II, **4,** 37, 1961.
44. ANDERS, E. *Lunar and Planetary Exploration Colloquium,* p. 44.
45. LOWMAN, P. D., JR. *Lunar and Planetary Exploration Colloquium,* p. 40.
46. PINSON, W. H., JR. "Chemical and Physical Studies of Tektites," Final Report prepared for Geophysics Research Directorate, Air Force Cambridge Research Laboratories, Office of Aerospace Research, United States Air Force, Bedford, Massachusetts, February 15, 1962.

15

ANALYSES OF EARLIER THEORIES OF THE MOON'S HISTORY

It is difficult to say when the moon was formed, for the process is continuing even today. As an arbitrary definition, the terminal phases of the moon's formation came when the moon reached approximately its present size. The latest evidence is that rocks in the crust of the earth are up to 3 billion years old, that the earth as a whole is 4.5–4.6 billion years old, and that meteorites and other objects in the solar system took their present forms about the same time in the past (1).

Reynolds (2) recently found a high concentration of Xe^{129} in the chondritic meteorite which fell in Richardton, N.D., in 1919. This isotope is a decay product of I^{129}, and he derived a time of formation of the meteorite of 3.5×10^8 years after element formation. Katcoff, Schaeffer, and Hastings (3), basing their calculations on terrestrial xenon, found 2.7×10^8 years for the formation of the earth. Edwards and Borst (4) similarly found 3.5×10^8 years for the formation of the moon. Borst (5) states that the coincidence of these three ages gives strong support to the hypothesis that the planets were formed in a relatively short period.

The terminal period of the formation of the moon's crust, as contrasted with the main building period, was not extremely short. It must have stretched over a considerable period of time. Certain evidences of change in the processes occurring in crater formation and subsequent changes in crater shapes have been noted.

The studies made by Arthur (6), Shoemaker (7), and others have indicated rather clearly that the craters possess a random distribution in the bright areas of the moon and a different, but also random, distribution in the dark areas. The dark matter has covered large portions of the lunar crust, obliterating a great many craters. The postmare craters are relatively few in number and are superimposed at random over the dark areas and hence may be presumed to be similarly distributed in the bright areas over the earlier formations. Close inspection suggests that this is true. The dark material of the maria thus gives us a basic relative dating criterion, dividing two stages in the moon's history at some ancient, but undetermined, date in the past.

Urey (8) has given cogent reasons for believing that the moon and earth were built by accretion from cold planetesimals which had formed earlier from the primordial materials of the proto-earth. Öpik (9, 10), Urey (11), and Alter (12) have suggested that the moon is a captured sister planet and not the earth's twin, but this should not change the mode of origin, because presumably a process of this nature which was valid for the earth would be valid for any other terrestrial planet in the solar system.

Although there is considerable disagreement concerning the stages in the formation of the solar system before the cool planetesimals were formed, it is rather generally agreed that these small bodies did form and then were brought together to form the planets and the moon. There is a wide diversity of opinion concerning the history of the body of the moon after it reached its present size.

Obviously, we do not know what was the moon's condition when it was half-grown, but we do see portions of its surface which are very old. These portions are thoroughly peppered with craters. These craters of all sizes grade smoothly into new-appearing later craters, and most of them undoubtedly were formed by impacts of bodies from space. The data of several preceding chapters define rather closely the nature of the bodies which could produce these craters, and they are found to be consistent with the sizes of large meteorites up to the small asteroids. These are the sizes of objects which we would expect to be produced by Urey's model of the formation of cold planetesimals. So far, there is no conflict. It is in the next stage of the moon's development that the many theories differ.

Urey's publications (8, 11, 13, 14) say that the moon was formed cold and remained cold or at least solid throughout its entire history. It has been gradually warming, perhaps because of radioactivity in its rocks, but never has become hot enough to liquefy the lunar rocks. The dark areas he usually has felt to be lava, but at times he seems to lean toward Gold's deep-dust hypothesis in the sense that he has suggested that the dark areas are dust and gravel ejected from the Imbrium area by the great impact explosion.

To explain the dark areas as lava flows, Urey has postulated that the impacts of large, low-velocity planetesimals led to liquefaction of these objects rather than true explosions, and this lava remained for the most part in the great impact craters, although in the case of the largest, Mare Imbrium, a great deal of overflow occurred. The lava is thus of external origin and was deposited at the same time as the great circular maria were formed.

Kuiper's model (15) of the formation of the solar system also postulates that the moon was formed by planetesimals and was originally cold but that radioactive heat, starting at least 5,000,000,000 years ago, gradually warmed the moon until the impacts of the great planetesimals triggered a release of the subsurface lavas.

Urey (13) has severely criticized this sequence of events. Baldwin (16) has pointed out that a considerable period of time elapsed between the impacts which formed Mare Imbrium and other similar structures and the rise of the lava. The lava was postulated as coming from below. Gold (17) disinterred T. J. J. See's (18) old hypothesis that the dark areas were not lava, but were thick layers of dust. It is certain that these four hypotheses are mutually exclusive.

Let us examine each hypothesis, starting with Gold's dust, for this is certainly the most unusual approach. Gold's (17) solution to this problem rests on two assumptions:

Recent discussions of the lunar surface have mostly favoured the interpretation of meteoritic impacts for the generation of the craters, and of lava flows for the formation of the flat surfaces. There are two reasons for departing from the lava hypothesis: firstly, because there are some severe obstacles to that interpretation connected with the distribution of subsequent impacts on the plains and the absence of a satisfactory time sequence interpretation; and secondly, because it may be unnecessary to invoke both processes if the impacts alone seem sufficient together with other effects which are known to be present or which must be inferred in any case. The omission of the lava flow hypothesis is an attractive step also, because of the absence or great rarety of lunar features that resemble the formations associated with terrestrial volcanic activity, although a far greater intensity of such activity has had to be assumed.

Gold's reasons for his hypothesis do not stand up under analysis.

It is quite apparent that the numbers of craters formed before the great dark areas developed are many times greater per unit area than the numbers of the postmare craters. This observation simply places the maria as having been formed relatively, but not necessarily absolutely, late in the major sequence of events. It tells us that the crater-forming impacts had not ceased, although the rate had markedly lessened. An alternative solution is that there were two sources of crater-forming bodies. The premare craters could then have been formed from one source. The maria were formed in some manner after this source had been almost or completely depleted, and later the present supply of planetesimals or meteorites from the asteroid belt became available to form lunar craters. The choice between these two ways in which the impact craters might have been formed is not germane to the testing of Gold's hypothesis. In either case it is necessary only to show, first, that the postmare craters are distributed at random on the maria and, second, that the numbers of these craters as functions of their sizes are similar for the various maria.

TABLE 27

SIZE-FREQUENCY DISTRIBUTION OF PRIMARY IMPACT CRATERS ON LUNAR MARIA*

	Approximate Square Kilometers	Crater Diameter in Miles								Craters per 10^5 Km^2
		1	2	4	8	16	32	64	128	
Mare Imbrium............	864,000	199	117	37	10	5	1	0		42.7
Lacus Somniorum.........	64,500 ⎫	103	68	41	15	5	2	0		46.5
Mare Frigoris............	439,000 ⎭									
Mare Serenitatis..........	318,000	88	41	7	1	1	0	0		43.4
Mare Foecunditatis.......	311,000	56	34	28	6	3	1	1		41.5
Mare Tranquillitatis......	402,000	89	57	39	11	6	0	1		50.5
Palus Epidemiarum.......	288,000 ⎫									
Mare Nubium............	261,000 ⎬	111	64	27	11	0	1	0		32.6
Mare Humorum..........	107,000 ⎭									
Mare Nectaris............	96,400	26	16	2	1	0	0	0		46.7
Mare Crisium............	165,000	{ 20 (39)	10 (15)	6 (6)	4 (4)	0 (0)	0 (0)	0 (0)		24.2 (38.8)†
Total................	3,315,900	{ 692 (711)	407 (412) }	187	59	20	5	2		{ 41.4 (42.1)†

* Modified from Shoemaker and Hackman (7).
† From author's count.

These two studies have already been made. Arthur, in the reference quoted earlier, has shown beyond doubt that the craters within the borders of Mare Imbrium are distributed at random. This confirms their non-lunar origin as opposed to a crater-forming process operating from within. Shoemaker and Hackman (7) have made counts of postmare craters on almost all the major lunar dark areas and find that they are distributed with amazingly similar frequencies (see Table 27). With the possible exception of the dark matter in Mare Crisium, all other areas show the same frequency pattern. The great dark areas, judging from this test, were all formed at essentially the same time. Even Mare Crisium, which shows a slight deficit of craters under 4 miles in diameter, departs from the norm by less than a factor of 2. The craters of Mare Crisium, coming so close to the limb as it does, are exceedingly difficult to see on many photographs. Close examinations of several such pictures yielded counts similar to that of Shoemaker and Hackman, but a study under a 6-power glass of a contact-glass positive from the original negative of a Lick Observatory photograph of June 2, 1938, when the moon was at age 4.6 days, disclosed 54 craters under the 4-mile-diameter limit. These pits in general were hard to see but are considered real in every case. With this change, Shoemaker and Hackman's value of 24.2 postmare craters per 10^5 km^2 becomes 38.8 such craters. This value is close to the average of the other dark areas and indicates that the dark part of Mare Crisium was formed at about the same time as the others.

These observations eliminate Gold's first assumption. The observations lead to a consistent sequence of events on a relative time scale, and there are no difficulties concerning the distribution of the postmare craters.

As for Gold's second reason—namely, that impacts alone, or together with other effects known or inferred to be present, are sufficient to account for the dark areas—we can subject this to check also.

There is not the slightest question but that the moon is covered by a shallow layer of dust. This is not Gold's point. His suggestion is that various erosional forces have acted on the entire surface to produce dust. He then proposes a mechanism whereby the dust can be transported over the lunar surface, and, finally, he postulates that the dark areas are the places where the dust has accumulated and that the dust there has tremendous depth.

All authorities agree that craters are of widely differing ages. It has often been assumed that the older a crater is, the more chance there has been for another impact to deform the older pit. The more deformed pits are statistically more shallow and their rims average lower than do the Class 1, or new-appearing, craters.

Gold assumes that these variations in dimensions are due exclusively or dominantly to erosion and then looks for the eroded material, or dust, and finds it in the dark regions. He states (17):

> If erosion is assumed to be the cause of the rounding of the features of old craters, as discussed earlier, then it is necessary to find the eroded material. This cannot reasonably be expected to have all left the Moon entirely. If an estimate is made of the amounts missing from the rims of all the many overlapping old craters, then this cannot come to less than the equivalent of a 300 foot depth if it were distributed over the entire surface, and probably a great deal more. Though on the lava flow hypothesis it could be supposed that the eroded material was so distributed that it was all drowned, the question would remain as to the whereabouts of the material eroded after the supposed lava flooding. This is still a substantial depth, at any rate deeper by a large factor than is necessary to make it optically opaque. Even if the erosion process were not to take the erosion product any further than is strictly required by the change of shapes, there would still have to be large areas of the flat plains covered by that mate-

rial. This is clearest in the cases of flat-bottomed craters whose erosion on the inner crater walls must be assumed to have covered over at least part of the flat bottom, yet the colour there even close to the sharp edge is the same as that of the flat surfaces elsewhere. A more diffuse distribution of the eroded material would only strengthen this argument, so it must be concluded that the darker shade is that of eroded material even if it came from lighter rock.

Certain comments are pertinent. In essentially all rocky materials, including those which have been considered as possibly present on the moon, fragmented material is the same color as the original, but powdered rock such as would be produced by meteoritic impact is considerably lighter than the rock from which it was formed. The rays from Tycho in the bright uplands and Copernicus in the darker areas are all considerably lighter than the parent rock. This is opposite to the Gold dust theory. There the dust is the dark material, and the original rock is light.

Gold also does not distinguish between the colors of the bottoms of the great walled plains. Clavius does not show a different color on its interior plain from the walls, but Schickard, for example, shows two very dark interior areas. Most of the great craters in the upland regions do not possess dark centers, although some do, particularly the ones in the lower areas near the maria. The existence of the great flat floors in these old craters cannot be attributed to erosion and infilling from the walls.

This conclusion is buttressed by the discussion in Chapter 9, where it was shown that the maximum possible loss of materials from the crater rims was from 200 to 400 per cent too small, at a minimum, to account for the flat bottoms of the old and large craters. From both points of view it must be concluded that the tremendous production of dust on the moon expected by Gold simply never occurred.

Some amount of dust must be expected on every part of the moon. With the known rates of infalls of meteorites, poorly defined though they are, this is unavoidable. All measures of the heat absorbed and emitted by the moon confirm it, but they merely show that a thin surface layer is present. Dust produced *in situ* by other than impacts might well be the color of the adjacent rock.

The fact that rays exist around many of the postmare craters is not consistent with a large and continuing dust production. There are many hundreds of large and small ray craters scattered essentially at random over the moon's disk. There is no reason to assume that they are all very recent; they form too large a percentage of the postmare craters. If a ray crater is produced on the dark material and the dark material is composed of dust, why are the rays lighter than the dust?

Petterson and Rotschi (19) have estimated from the nickel content of deep ocean deposits that the quantity of meteoritic material currently being deposited on the earth is of the order of 10^6 tons per year.

Gold (17) concludes from this:

Therefore an opaque blanket would be spread over the entire surface, and renewed once every million years. (This consideration would be invalidated if in each little impact an amount of vapour was freed and allowed to escape from the Moon entirely, of a mass of the same order as or exceeding that of the micro-meteorite; or if a fraction escaped the period would be lengthened. The material making up the rays of the craters would have to be at least some centimeters thick if it is not to be covered over in too short a time to account for the absence of similar craters on the Earth. Until the rays are submerged they would have to be able to shed dust just like the highlands.) A process of migration of dust from high ground to the lowlands can again be invoked to account for denudation of the high ground and for that reason for a differentiation in colour. Again the actual surface in the flat regions would have to be dust, and nothing is gained by supposing lava to have been darker than rock.

By contrast, Hawkins (20) finds the rate of infall of meteoritic material at present to be 10^3 times less than Petterson and Rotschi. This leads to a negligible dust layer.

From satellite measurements LaGow and Alexander (21), Dubin (22), and Nazarova (23, 24) have found that the amount of meteoritic material entering the earth's atmosphere each day is of the order of 10^9 gm (4×10^5 tons per year). It is not yet clear how much of this material actually enters the atmosphere and how much is trapped in the higher satellite regions.

During major meteoritic showers the rate of influx of particles increases by factors of from 10^1 to 10^4. The last-named figure is unusual. Whipple (25–27) showed that the number of particles falls off with increasing height above the earth by a factor of 10^2–10^4 toward the density of the cloud which produces the zodiacal light.

The above data correspond to a covering of dust on the earth of 1 cm in thickness each 40,000,000 years if we assume an average density of such particles to be 3 and also that the above material does enter the earth's atmosphere as postulated.

The falloff in particle density with increasing height above the earth leads to the conclusion that the moon would accumulate a layer 1 cm thick in 4,000,000,000 years or more. Where the exact rate lies is still not certain, but the evidence of the ray craters suggests that the moon does not accumulate a great deal of dust from space.

The number of large ray craters, including Copernicus, Kepler, Aristarchus, Tycho, and the numerous small craters with rays or halos must have been accumulated over many hundreds of millions of years. All observations of meteoritic impacts as a function of crater diameter demand that the majority of the ray craters cannot have been produced recently. The rays are often, but not always, present at craters where ages are measured in hundreds of millions of years and absent at craters whose ages are necessarily of the order of several billion years. From all points of view, then, erosion must occur on the moon, but it is of minor importance in explaining the present appearance of the ancient craters and the thickness of the maria.

Gold has reached an interesting conclusion from his premise that the maria are dust. The area of the bright highlands is greater than that of the dark lowlands. More than half the dust must have been produced in the bright areas and then transported by some mechanism to the region of the maria. He then searched for a mechanism which would allow the eroded material—fine dust— to "flow" like a fluid over tiny gradients.

A requirement is for the transport of dust particles to the minimum height that they can reach to occur at a rapid rate compared with the other processes of the formation or freeing of dust at the surface or its acquisition from the outside. He states (17):

The requirement is for the average speed of flow not to diminish appreciably with the angle of the slope until a very small angle is reached. It should not be a flow restricted by the equivalent of viscosity in a liquid, for this would leave slight gradients persisting for long and remove steep gradients very quickly. The type of flow required would be one where at any instant a thin layer on the surface behaves like a non-viscous liquid whilst the remainder underneath remains stiff. This would then allow steep slopes to persist but would assure that the deposits of dust possess a flat surface. Such a "fluidization" of a surface layer would have to be a process resulting from an external energy source, providing an agitation for the dust particles on or near the surface.

Gold mentions several mechanisms which might "fluidize" the dust. Among these are the effects of an evaporation and condensation cycle of a suitable vapor, the effects of micrometeorites, and the effect of electrostatic forces. In most of his writings he seems to lean most heavily on the last-named. Such forces will arise as a consequence of the photoemission of electrons from the surface due to the ultraviolet light of the sun, or they could arise in some larger electrical process connected with solar events of the type that cause aurorae and magnetic storms on the earth.

The process of photoemission in the absence of an atmosphere, and especially of moisture, will result in an erratic distribution of charge on the irradiated surface. The surface should be a good insulator. "While the average positive charge will be inadequate to lift particles off, it is not clear that the chance distributions in small localities could not do it. . . . If particles were frequently dislodged by electrostatic forces, then again a net flow would result on the surface in accordance with the requirement" (17). It is by some such process, a form of Goldian movement, that the maria are supposed to have been formed. Warner (28, 29) has given Gold's hypothesis some backing.

If Gold is correct, then we have a choice. Either almost all the dust was produced in the past and now has been transported onto the maria, or the dust changes color when it crosses the boundaries of the maria. Gold postulates a mechanism whereby the dust darkens with age. Polarization, radio, and thermal measurements are unanimous in suggesting the presence of a thin layer of dust in the bright uplands.

The new contour map of the moon (30) shows that at least half the bright upland area drains not toward the maria but toward the limb, and yet the limb is not dark.

The magnificent crater, Theophilus, is a postmare crater. Under certain angles of illumination, remnant rays may be seen spread on Mare Nectaris and Mare Tranquillitatis. Presumably the faintness of the rays means that Theophilus is quite old, approaching the maria in age.

The contour of this pit is accurately known from the measures made by McMath, Petrie, and Sawyer (31) on McMath-Hulbert motion pictures (Fig. 32, p. 192). The bottom is everywhere curved, and the inside of this crater shows a far greater slope than is usually to be found on the lunar surface. At the crater origin, the materials of the walls must have been highly fragmented or pulverized, so that the formation of dust would be greater than would be normal on the lunar surface outside craters. Since Theophilus was formed, there has been no appreciable production or transportation of dust in a place where conditions should have been highly favorable to Gold's mechanism if it existed as a major force.

The great rille systems, in general, mark the edges of the maria. Migrating dust, to reach the central regions of the maria, would have to cross these great trenches. Certainly, the dust could not climb back out, once it had fallen in. The actual volume of the depressions of the rilles is negligible compared with the volume of the materials forming the maria. It must be very much less than 1 per cent. Because the rilles are not filled up, they must have been produced by a still active moon in the last few million years, or else the erosional forces have effectively ceased, or the Goldian movement has stopped during the period since the rilles were formed.

Conversely, there is a great deal of evidence for a lava nature for the maria, only two bits of which will be mentioned here. There is specific evidence of melting and erosions by very hot lavas. East of Alpetragius is a slightly smaller ring, and only the highest portions of that ring project above the dark area. These projections are low and generally rounded. The following craters, to name but a few, show melting and erosive action on the seaward side: Fracastorius, Letronne, Doppelmayer, and Posidonius. A great many other similar cases could be pointed out near the shores of any mare. For example, immediately north of Pitatus is a half-ring a few miles in diameter. The north half of this ring is entirely gone as if melted away, but the southern half is complete. The area of Palisa, both to the south and to the east, shows a muted and blurred outline as though it had been invaded by the lava flows.

No distinction in principle can be made among Arzachel, where a moderate crater fill is evident, Ptolemaeus, where the crater is filled apparently to ground level, and Wargentin, which is filled to the brim. Lava forced from below can easily account for these examples. It would be unusual to find a mechanism which would allow hundreds of cubic miles of dust to climb a wall some thousands of feet high just to fill one crater.

Only one conclusion can be drawn: the dark areas are not dust and never have been dust. They are great lava flows and have a thin wash of dust over the surface. This dust has had three sources—micrometeoritic material from space, ejected rock flour from distant craters, and dust generated from the local dark solidified lavas. The last is clearly the predominant type of dust. Were it not so, we would not be able to distinguish the maria by color. The wide variations in color which appear on the maria in large and small areas give added evidence that no mass migration of dust has occurred. Gold's hypothesis does not satisfy the observations.

Gilvarry in 1957 (47) suggested that Gold's considerations implied that the moon was formed cold and is now cold. Its strongly triaxial shape reflects the strengths of the lunar rocks, and the moon never was in isostatic equilibrium. Because on the earth an area of high isostatic gravity anomaly is often an area of high earthquake activity of the tectonic type, he postulated a high level of seismic activity of the tectonic type on the moon. He thought that this activity represented a major erosive agent capable of reducing surface rock to rubble and possibly, with the co-operation of other agents, to fine particles and dust. The time scale would be short compared with the lunar age.

A lunar seismic wave exhibiting the maximum amplitudes of ground acceleration (500 cm/sec^2) and horizontal displacement (60 cm) observed for an earthquake (48) could impart a horizontal velocity of approximately 200 cm/sec to superficial dust. The maximum height of the trajectory of a dust particle from a perfectly elastic collision would be about 100 cm. Gilvarry pointed out that these were extreme assumptions, but he claimed that the mechanism could cause a levitation of the dust sufficient to permit it to flow down low gradients.

The same arguments may be used against this hypothesis as against Gold's mechanism. The observations indicate no large flow of dust on the moon, and, by inference, the amount of seismic activity there is limited. This is not to be taken as meaning that the moon is completely quiescent, but probably the seismic activity is relatively minor at present, although numerous microseisms may well occur.

Gilvarry in 1960 (49) also revived the hypothesis of Beard (50) and Davis (51) that the moon was once covered with water, although Gilvarry limits the oceans to the regions of the maria. The dark color of the lava was attributed to carbon from primitive life left in the rocks as the seas finally dried up. The Swann bands of C_2 observed by Kosyrev (52) in the efflux from the crater Alphonsus were supposed to be from this carbon as heated by a meteoritic impact. The particular tektites which seem to be formed from sedimentary rocks could have come from the moon's sedimentary rocks. The argument stands or falls on the presence of liquid water on the moon at some time in the past. In a subsequent chapter it will be argued that the evidence precludes the existence of lunar oceans at any time.

The case is quite different with the theories of Urey (8, 11, 13, 14). He has been responsible for many of the advances in our understanding of the early history of the development of the planets and the moon. He has taken the primitive planetesimal hypothesis of Chamberlain and Moulton (32) and developed it into a clear, logical theory which fits a great many of the observed facts. His positive contributions include explanations of the observed amounts of the various elements in the planets and the moon, the origin of the planetesimals, and the building-up of the planets and moon from cold fragments. In this last he independently reached the same conclusions as did Gilbert (33) before the turn of the century.

In a recent paper Urey (20) discussed the duration of the intense bombardment processes on the moon and concluded that the period was very short and that it occurred before the earth was completely assembled. In this solution the moon was a sister planet captured during a passage close to the earth by the dissipation of relative energy in tidal effects or in collisions with residual objects which had not yet coalesced with the earth. The massive collisions were the causes of the maria.

Urey's model for the moon specifies that it was built by accretion from cold particles which had approximately the chemical composition of chondritic meteorites. The sequence of events before this is as follows (34):

(1) A disc nebula at temperatures so low that hydrogen is partially condensed to the solid state existed, though no process for its formation is discussed. (2) This broke up into masses of lunar size together with the quota of cosmic gases. (3) These objects collided later to produce very fine solid material, and part of this material and the gases were lost to space owing to particle radiation from the sun. (4) Subsequently the residue accumulated into the planets and the immediate parents of the meteorites.

Fish, Goles, and Anders (35) have criticized certain of Urey's conclusions concerning the early history of the moon and similar bodies. It is beyond the scope of this study to discuss the very early stages of the solar system. It is perhaps sufficient to say that the formation of the moon is well explained by the accretion of cold particles of a considerable range of sizes and that Urey's theory provided those particles. From the random collisions of these tiny objects, larger centers would be formed and gradually the moon nucleus would grow. At first, its temperature would be low. Its composition would be essentially uniform.

It is at this point that Urey's ideas and mine part company. He feels that the moon never became melted throughout and that this conclusion is demanded by the theory of conduction of heat through a body as large as the moon. If the moon were ever liquid, the inner regions would still be liquid or essentially at the melting point, and therefore isostatic adjustments would quickly eliminate any major surface irregularities. The new contour map of the moon in this book, the excellent work by Watts (36) on the variation in height of the limb, and the existence of a dynamical bulge all point out that the moon is not spherical and that surface irregularities within some hundreds of miles often amount to several miles. If this is so and the hot moon could not cool down, then such variations in height could not exist, but they do exist and therefore the moon is not hot. If it is not hot now, it never could have been hot. So runs Urey's argument.

Urey has not minimized the hot-moon theory without thoroughly considering it.[1] He has analyzed possible methods by which the moon could be heated. Among these are short-lived and long-lived radioactive elements, gravitational energy, and chemical energies. He has argued against the hot-moon idea by what seems to be three assumptions. He has assumed that a solid crustal layer could not exist over a liquid and that the surface irregularities and the bulge could not be supported by a moon with a weak center. He has rejected the earlier views that the lavas came from below and has substituted the idea that the lavas were actually lavas but that they resulted from the melting of the low-velocity objects which struck the moon to form the circular maria.

Let us examine the last process first. In the laboratory we have projected missiles of various compositions, including stone, against many targets (37–40). Heat is produced, and dust and explosive energy, but never an appreciable amount of liquid. Poulter (41) has accelerated two slabs of anorthosite rock against each other at a relative velocity of about 2.4 km/sec and found that they were reduced to a moderately fine powder, but not melted. All the material could not have been melted at this velocity of impact.

---

[1] At the Lunar Exploration Conference held at the Virginia Polytechnic Institute, Blacksburg, Virginia, August 12–18, 1962, Urey took the position that, while the moon probably had been cold or cool throughout its history, it might have been hot or even melted.

At terrestrial meteoritic craters, most of which were produced by low-velocity impacts of 1–15 miles per second, there is little evidence of extensive melting. Even at the Wabar craters there is only a relatively little silica glass, a far cry from fluid lava. Consequently, there is no reason to postulate that the impacts on the moon produced the observed hundreds of thousands of cubic miles of lava. The smaller, but still violent, impacts which formed the normal craters did not liquefy rocks. The only possibility is that the very low velocity of impact and the large sizes of the objects which formed Mare Imbrium and others like it could have created conditions which would have melted the missiles and some of the lunar landscape rather than vaporize them. But, alas, even this possibility is denied. Without exception, it can be shown that the great circular maria were formed dry and that the lava came considerably later.

In the Mare Imbrium structure there are three large craters which definitely were formed after the great crater and before the lavas came. They are Sinus Iridum, Archimedes, and Plato. Sinus Iridum is not, as Urey feels, the place where the Imbrium planetesimal first struck. It is an independent crater, very similar to Clavius, though much younger. Sinus Iridum was formed after Imbrium. It developed on the wall of the larger object and eliminated the wall. The north half of Sinus Iridum is beautifully developed in the high rim area of Mare Imbrium. The southern half of the crater is buried in the impact area of the mare. The lost rainbow rim is not lost; it is merely misplaced, for it exists to this day under the lavas. Its outline is clearly shown for most of its length by wrinkles in the dark floor of Mare Imbrium. Plato was also formed on the northern wall of an existing Mare Imbrium, but it does not overlap the edge. It has formed a projecting rim into the sea. Archimedes was formed on the shelf area outside the impact region of the larger sea.

All three of these great craters are now flooded with frozen lava. There is no sign on the dark floor of Mare Imbrium that explosions produced them. These signs are buried. None of them could have been in its present position and come through the explosion which produced Imbrium and still escaped unscathed. They clearly are post-Imbrium and prelava.

Five other similar but smaller craters can be detected. Wallace, on the shelf in the south, is a mere partial ring projecting above the lavas. Just north of Aristillus is a drowned ring, and Cassini is similarly placed in time. Sinus Gay Lussac is a flooded pit similar to the smaller Archimedes M.

It would be stretching the long arm of coincidence too far to postulate that these eight objects were all formed in the short period when the lava sheet was still liquid—a time to be measured in years at the most—when no others as large as four of the eight were formed on Mare Imbrium in the billions of years since.

Other possible craters of this type may exist. There is an oval series of ridges and markings in western Sinus Iridum. "Ancient Newton" south of Plato may be an old crater. Firsoff (53) suggests that several others are prelava craters, including one which claims Piton as a remnant of its rim.

The existence of additional old craters would only strengthen the argument of this section, but, although the markings as outlined by Firsoff do exist, their identification as ancient craters which have been buried by the lavas is not established.

The much smaller Mare Humorum also shows the same time sequence. First, the great dry crater was formed. Then came several smaller craters, including Doppelmayer, Lee and the larger crater it overlaps, Vitello, Hippalus, Gassendi, and some smaller craters on the east shore. Then the lavas came and drowned and melted and eroded the newer pits. All are now in a very sad state of repair.

Mare Serenitatis is adjacent to Mare Imbrium. It is older than Imbrium because the flying fragments from the larger explosion have nearly destroyed the Haemus Mountains bordering Serenitatis and forced an apparent alignment of them toward the impact area of Mare Imbrium. Some of the Haemus grooves disappear beneath the dark layers of Serenitatis; yet nowhere on the latter surface do we find any evidence of action from the Imbrium explosion.

Along the western shore the two prime examples of craters older than the lava but younger than the formation of Serenitatis are Le Monnier and Posidonius. At least three other nearby craters are similar.

Still farther to the west lies Mare Crisium. The craters Yerkes, Lick, Cooke, and several others on the floor all postdate the Crisium impact and predate the lavas. On the rim of Mare Crisium, we find Eimmart and possibly Condorcet.

There is one, and there are suggestions of other such datable craters in the limb sea—Mare Humboldtianum.

Mare Nectaris completes the roster of major, easily visible circular maria. Fracastorius is the prime example of the craters of this type, but Bohnenberger to the west and Beaumont to the east are of similar type. In the list in Table 28, there are 55 craters of this special type. It must be remembered that it does not take 55 craters of this type to prove that the circular maria were formed dry and that the lavas came later. It takes only one unambiguous case, and we have many.

Fielder (42) has constructed a histogram based on Young's crater counts (43). It gives the numbers of craters in 3-km-diameter steps from 15 to 135 km. There are 1,281 of these pits in the part of the disk lying within 70° of the center, and it includes named craters outside these limits. Cumulative totals of craters smaller than 135 km in diameter were developed and are listed in Table 29. Similar cumulative totals were found for the postcircular maria–prelava craters, omitting Sinus Iridum, which is outside Young's limits. At each diameter interval, the ratio of the two cumulative totals was calculated. The data are consistent for craters larger than 42 km. The average ratio was found to be 19 to 1, in the sense that, in all size ranges above 42 km, the Young list contained roughly 19 times as many craters as there are of our special type.

For craters smaller than 42 km, the numbers in the circular maria projecting above the lavas become steadily less, relative to Young's data. This is probably

TABLE 28

POSTDRY CIRCULAR MARIA, PRELAVA CRATERS

In Mare Crisium		In Mare Humorum		In Mare Nectaris	
Formation	Diameter (Miles)	Formation	Diameter (Miles)	Formation	Diameter (Miles)
NE. of Cape Agarum	30	Gassendi	68.6	Fracastorius	74.8
Lick	21	NE. of Doppelmayer	49	Beaumont	32.4
Yerkes	19	Near Lee	45	Between Daguerre and Mädler	32
F	16	Doppelmayer	40	Daguerre	27
Cooke	14	Hippalus	38	Bohnenberger	22
Cooke B	14	W. of Agatharchides	35	N. of Bohnenberger	18
N. of Yerkes	12	Lee	28	SE. of Isidorus H	14
S. of Yerkes	12	Vitello	27.8	4 Craters N. of Bohnenberger	14
N. of 30-mile crater	8	Mersenius D	21		13
N. of Yerkes	7	Puiseux	17		13
		Loewy	13		6
In Mare Imbrium		Between Doppelmayer and Mersenius D	13	E. of first 14-mile crater Between Daguerre and Bohnenberger	6
Formation	Diameter (Miles)	Mare Humorum H	9		5
		Mare Humorum C	8	Isidorus H	5
Sinus Iridum	165	Mare Humorum E	7		
Plato	62.6	B west of Gassendi	7	**In Mare Serenitatis**	
Archimedes	50.0	SW. of Mersenius D	4		
Cassini	35.2			Formation	Diameter (Miles)
Sinus Gay Lussac	22	**In Mare Humboldtianum**			
N. of Aristillus	18			Posidonius	62.9
Wallace	14	Formation	Diameter (Miles)	Le Monnier	45
Archimedes M	7			Littrow	22
		On far NW	24	Littrow A	16
				Le Monnier Z	6

due to the smaller craters' being completely covered by the lava in the thicker parts of the flow and appearing only where the lava is relatively thin.

Fielder states that Young's data should be multiplied by 3 to be representative of the entire lunar surface. Therefore, $19 \times 3 = 57$, or the special-class craters are 57 times more rare than the total numbers of craters in these size ranges.

But the area covered by the circular maria totals approximately 1,620,000 km², while the area of the visible half of the moon is 19,000,000 km². The ratio is 11.7 to 1. Therefore, the number of craters produced between the formation of the dry circular maria and the coming of the lava relative to the number of all lunar craters of comparable sizes is in the ratio of $11.7/57 = 0.2$, or the observed numbers of all craters per unit area is 5 times higher than those within the circular maria. On the assumption that this rate was constant from the time of the earliest observable crater to the coming of the lava, the period of time between the formation of the dry circular maria and the coming of the lava was

TABLE 29
CUMULATIVE CRATER COUNTS

Diameter Range (Km)	No. of Craters (Young's Data)	Cumulative Total Craters	No. of Craters (from Table 28)	Cumulative Total Craters	Ratio of Cumulative Totals
132–35	3	3	. . .	. . .	. . .
129–32	4	7	. . .	. . .	. . .
126–29	2	9	. . .	. . .	. . .
123–26	6	15	. . .	. . .	. . .
120–23	4	19	1	1	19
117–20	4	23	. . .	1	23
114–17	4	27	. . .	1	27
111–14	2	29	. . .	1	29
108–11	6	35	1	2	17.5
105–8	1	36	. . .	2	18
102–5	6	42	. . .	2	21
99–102	4	46	2	4	11.5
96–99	7	53	. . .	4	13.2
93–96	7	60	. . .	4	15
90–93	3	63	. . .	4	15.8
87–90	6	69	. . .	4	17.2
84–87	12	81	. . .	4	20.2
81–84	9	90	. . .	4	22.5
78–81	8	98	2	6	16.3
75–78	16	114	. . .	6	19
72–75	8	122	2	8	15.2
69–72	18	140	. . .	8	17.5
66–69	15	155	. . .	8	19.4
63–66	14	169	1	9	18.8
60–63	18	187	1	10	18.7
57–60	21	208	. . .	10	20.8
54–57	28	236	2	12	19.7
51–54	23	259	2	14	18.5
48–51	20	279	1	15	18.6
45–48	36	315	. . .	15	21.0
42–45	42	357	3	18	19.8
39–42	65	422	. . .	18	23.4
36–39	59	481	1	19	25.3
33–36	62	543	5	24	22.6
30–33	82	625	1	25	25.0
27–30	95	720	3	28	25.7
24–27	106	826	2	30	27.5
21–24	137	963	5	35	27.5
18–21	153	1116	6	41	27.2
15–18	165	1281	. . .	41	31.2
12–15	. . .	. . .	3	44	. . .
9–12	. . .	. . .	7	51	. . .
6–9	. . .	. . .	3	54	. . .

at least one-fifth as long as the total period, on the average. Mare Humorum, Mare Nectaris, and possibly Mare Serenitatis came early in this period, Mare Crisium somewhat later, while Mare Imbrium developed, geologically speaking, shortly before the lavas came.

The lava flows divide the crater-forming period into two distinct times. In the earlier period, the rate of infall was heavy. After the lavas, the rate of crater formation was very slow. Because of this, the postmare craters have been neglected in this computation.

Presumably, the rate of infall declined continually from the origin of the moon to the present rate, and therefore the period of time during which the great circular maria were dry may have been substantially longer than was estimated. In fact, the ratio must be considerably underestimated, for there are few datable craters observed in the central impact areas of any of the great circular maria except Mare Nectaris and Mare Crisium. Several craters of this type project partially into the impact area of Mare Humorum. We cannot give an absolute dating to the maria by this method, but all evidence suggests that the lavas came so late in the moon's history that there is no longer a need for us to postulate a sudden cessation of infalls at the time the dark areas were formed. This process gives only a relative dating of events.

It is known from the crater counts of Shoemaker and Hackman (7) that all the dark maria are essentially of the same age. The great circular maria craters themselves were undoubtedly formed at rather widely differing times. Humorum and Nectaris seem to be the oldest. Imbrium is definitely the youngest. Based on the assumption of a continuous decline in the rate of infalls to the moon, it is suggested that the lava flows may be hundreds of millions of years, or even billions of years, younger than the earliest observable surface markings. This rough dating of the lava flows is in better agreement with reasonable thermal histories of the moon such as have been developed by MacDonald (44), Baldwin (46), and as expressed in this volume.

While the method is rough, it is conclusive. Urey's hypothesis that the lavas were formed by the impacts is not sustained. The lavas came from the body of the moon considerably after the great circular maria craters were produced and came nearly at the end of the major period of impacts of meteorites with the moon's surface. Urey (45) suggests that the duration of intense bombardment of the moon was very short, possibly only a few thousands of years. Such a short period does not seem reasonable, based on the discussion of this chapter and that of Chapter 9 concerning progressive changes in the forms of the older, premare craters.

Inasmuch as Gold's dust hypothesis must be discarded and Urey's melting of the planetesimals on impact does not fit the observations, we must reach one of two conclusions. Either the dark areas are the result of some completely unknown or unrecognized process, or else the moon's body did produce the outpourings of lava. If the latter choice is made—and frankly I see no other realistic choice—then we must admit that the observations demand that the moon, at least once, became hot enough to melt at least a portion of its body.

All theories of the history and structure of the moon must then be in harmony with these observations. Those which do not permit the existence of liquid lavas in the moon well after the formation of the moon's crust must be discarded. The theories of the heat balance and cooling of the moon need revision.

In 1954, Kuiper (15) presented his views on the early history of the moon and on its structure. While he has gone into far greater detail, the essence of his argument is that the moon had once been melted completely by long-lived radioactive heating, that the premare craters were produced by left-over fragments from the formation of the earth and moon, and that the maria were outpourings of lava triggered by the impacts which produced the great circular maria.

The idea of the melting of the moon was not original with Kuiper. It was clearly anticipated in *The Face of the Moon* in 1949, and, indeed, most earlier authors other than Gilbert (33) assumed it. Urey (8) and Gilbert have shown that the moon collected from cold objects. Urey has given theoretical reasons why the moon could never have melted completely. Kuiper has pushed the origin of the moon back to possibly 6×10^9 years, in order to permit the melting to be produced from the increased amounts of thorium, potassium, and the two uraniums which could have been present then.

Both ideas cannot be right. Either the moon melted, or it did not. In a later chapter, additional evidence will be cited to indicate that the moon did melt, but Kuiper's interpretation of the subsequent events is not completely in accord with the observations.

Both Urey and Kuiper have associated the great lava flows with the collisions which produced the circular maria. Urey found the lava to be from the body of the planetoid. Kuiper specified that the moon was liquid below a thin crust, perhaps 16 km thick, composed of uncompacted accreted material. Even at a depth of 16 km, the crustal material should soon have been sintered and compacted. Urey has correctly shown that this thin crust could not be stable over liquid rock of lesser density.

The same arguments as those just used against Urey's hypothesis are valid against Kuiper's. The crust of the moon was hard and thick when the giant circular maria were formed. These tremendous craters were formed dry. There was no extensive lava produced from the colliding body. There was no sudden release of liquid rock from below. The magmas finally appeared in these areas, but they came much later. Urey (13) concludes one of his papers with the following:

> It seems to me that Kuiper's theory is not internally consistent. Perhaps a theory of the moon which assumes that it was completely molten in its early history can be devised, even though he has not done so. Such a theory would probably best assume that the irregular shape of the moon is due to a frozen tidal wave and that the thermal conductivity of lunar material is considerably higher than current estimates for terrestrial materials. Then it should be assumed that the irregular shape and the surface features were formed very late in its history when it had acquired a very rigid structure.
>
> In a recent publication I made an attempt to explore possible radioactive heating effects during the early history of the solar system. Very complicated and puzzling histories must be assumed. Though the process of the moon's origin was undoubtedly more intricate than anyone has the courage to imagine, it is my belief today, as it has been for some five years, that very probably the moon was accumulated at low temperatures from a primitive dust cloud of solar composition with the iron in oxidized states and that the concentrations of radioactive substances within the moon are sufficiently low that melting has never occurred. In fact, I believe that present temperatures are so low that the interior of the moon has a high strength and that such low temperatures require the moon to have been formed at low temperatures and never to have been melted at any time.

I agree with Urey that Kuiper's theory is not internally consistent. It is also probably true that the moon was formed from cold objects, but it is also an observed fact that the dark areas cannot be explained as dust and are only to be considered as lava. If this is so, the lavas came from the body of the moon, which must have been at least partially melted. If the body of the moon became partially melted, it was and still is hot at some unknown distance below the surface. If the moon was and is hot, we must account for the observed fact that there are real variations in height of the surface. A very real strength of the outer layers is indicated.

The bringing-together of these apparently inconsistent observations and theories is one of the goals of the remaining chapters.

References

1. REYNOLDS, J. H. "The Age of the Elements in the Solar System," *Sci. American,* **203,** 171, 1960.
2. ———. *Phys. Rev. Letters,* **4,** 8, 1960.
3. KATCOFF, S., SCHAEFFER, O. A., and HASTINGS, J. M. *Phys. Rev.,* **82,** 688, 1951.
4. EDWARDS, W. F., and BORST, L. B. *Science,* **127,** 325, 1958.
5. BORST, L. B. "Time of Planet Formation," *Science,* **131,** 566, 1960.

6. Arthur, D. W. G. "The Distribution of Lunar Craters," *J. Brit. Astr. Assoc.,* **64,** 127, 1954.
7. Shoemaker, E. M., and Hackman, R. J. "Interplanetary Correlation of Geologic Time," paper presented at the seventh annual meeting of the American Astronautical Society, Dallas, Texas, January, 1961.
8. Urey, H. C. *The Planets.* New Haven: Yale University Press, 1952.
9. Öpik, E. J. "Tidal Deformation and the Origin of the Moon," Prepublication copy, 1961.
10. ———. "Surface Properties of the Moon," Prepublication copy, 1961.
11. Urey, H. C. Prepublication copy of revision of *The Planets,* chap. 2, "The Moon." 1960.
12. Alter, D. "The Evolution of the Moon," *Pub. A.S.P.,* **73,** 5, 1961.
13. Urey, H. C. "Some Criticism of 'On the Origin of the Lunar Surface Features' by G. P. Kuiper," *Proc. Nat. Acad. Sci.,* **41,** 423, 1955.
14. ———. *Ibid.,* **42,** 889, 1956.
15. Kuiper, G. P. "On the Origin of the Lunar Surface Features," *Proc. Nat. Acad. Sci.,* **40,** 1096, 1954.
16. Baldwin, R. B. *The Face of the Moon,* p. 211. Chicago: University of Chicago Press, 1949.
17. Gold, T. "The Lunar Surface," *M.N.,* **115,** 585, 1955.
18. See, T. J. J. "Origin of the So-called Craters on the Moon by the Impact of Satellites, and the Relations of These Satellite Indentations to the Obliquities of the Planet," *Pop. Astr.,* **18,** 173, 137, 1910; also in *Pub. A.S.P.,* **22,** 13, 1910.
19. Petterson, H., and Rotschi, H. *Geochim. et Cosmochim. Acta,* **2,** 81, 1952.
20. Hawkins, G. S. "Asteroidal Fragments," *A.J.,* **65,** 318, 1960.
21. LaGow, H. E., and Alexander, W. M. "Recent Direct Measurements by Satellites of Cosmic Dust in the Vicinity of the Earth," *NASA Technical Note D-488.* 1960.
22. Dubin, M. In Ref. 25, 1961.
23. Nazarova, T. N. "Results of Exploring Meteoric Matter with Instrumentation of Sputnik III and Space Probes," paper presented at the eleventh International Astronautical Congress, 1960.
24. Komissarov, O. D., Nazarova, T. N., Neugodov, L. N., Poloskov, S. M., and Rusakov, L. Z. "Investigations of Micro-Meteorites with the Aid of Rockets and Satellites," *J. Amer. Rocket Soc.,* **29,** 742, 1959.
25. Whipple, F. L. "Dust and Meteorites," in "Space Flight Report to the Nation," American Rocket Society, New York Coliseum, October 9–15, 1961.
26. ———. "Particulate Contents of Space," presented on October 24, 1960, at the third Symposium on the Medical and Biological Aspects of the Energies of Space, at the U.S. School of Aviation Medicine, Brooks Air Force Base, Texas.
27. ———. "The Dust Cloud about the Earth," *Nature,* **189,** 127, 1961.
28. Warner, B. "The Lunar Maria," *Planetary and Space Sci.,* **5,** 283, 1961.
29. ———. "Accretion and Erosion on the Surface of the Moon," *ibid.,* p. 321, 1961.
30. Baldwin, R. B. "A Lunar Contour Map," *Sky and Telescope,* **21,** 84, 1961.

31. McMath, R. R., Petrie, R. M., and Sawyer, H. E. "Relative Lunar Heights and Topography by Means of the Motion Picture Negative," *Pub. Obs. U. Mich.*, **6,** 67, 1937.
32. Moulton, F. R. *An Introduction to Astronomy.* New York: Macmillan Co., 1905.
33. Gilbert, G. K. "The Moon's Face," *Bull. Phil. Soc. Washington,* **12,** 241, 1893.
34. Urey, H. C. Paper presented at Leningrad Symposium on the Moon, December, 1960.
35. Fish, R. A., Goles, G. G., and Anders, E. "The Record in the Meteorites. III. On the Development of Meteorites in Asteroidal Bodies," *Ap. J.,* **132,** 243, 1960.
36. Watts, C. B. Unpublished material, U.S. Naval Observatory.
37. De Boer, K., *A.N.,* **222,** 199, 1924; **223,** 177, 1924; **230,** 217, 1926.
38. ———. *Sirius,* **55,** 6, 1922, and **56,** 61, 1923.
39. Partridge, W. S., and Van Fleet, H. B. "Similarities between Lunar and High-Velocity Impact Craters," *Ap. J.,* **128,** 416, 1958.
40. Charters, A. C. "High Speed Impact," *Sci. American,* **203,** 128, 1960.
41. Poulter, T. C. Private communication to H. C. Urey.
42. Fielder, G. *Structure of the Moon's Surface,* p. 219. London: Pergamon Press, 1961.
43. Young, J. "Statistical Investigation of the Diameters and Distribution of Lunar Craters," *J. Brit. Astr. Assoc.,* **50,** 309, 1940.
44. MacDonald, G. J. F. "Interior of the Moon," *Science,* **133,** 1045, 1961.
45. Urey, H. C. "The Duration of Intense Bombardment Processes on the Moon," *Ap. J.,* **132,** 502, 1960.
46. Baldwin, R. B. *The Face of the Moon.* Chicago: University of Chicago Press, 1949.
47. Gilvarry, J. J. "Nature of the Lunar Surface," *Nature,* **180,** 911, 1957.
48. Heck, N. H. *Earthquakes,* p. 208. Princeton: Princeton University Press, 1936.
49. Gilvarry, J. J. "Origin and Nature of Lunar Surface Features," *Nature,* **188,** 886, 1960.
50. Beard, D. P. "The Impact Origin of the Moon's Craters," *Pop. Astr.,* **25,** 167, 1917.
51. Davis, E. G. "Origin of the Moon's Craters," *Monthly Evening Sky Map,* **40,** 10, 1946.
52. Kosyrev, N. A. "Observations of a Volcanic Process on the Moon," *Sky and Telescope,* **18,** 184, 1959.
53. Firsoff, V. A. *Surface of the Moon.* London: Hutchinson, 1961.

16

THE CIRCULAR MARIA AND RELATED STRUCTURES

The impact of a planetoid capable of producing one of the great dry, circular lunar maria is such a stupendous event that it is difficult to bridge the gap between it and the lesser explosions which formed the normal craters. If extrapolations based on crater diameter may be made, equation (8-12) (p. 176) suggests that the kinetic energies of the objects which produced these round structures were as listed in Table 30, and, depending on the velocities, the planetoid diameters may be calculated. None of these great structures is limited by its apparent boundary. The effects of the explosion are evident for hundreds of miles beyond.

Mare Nectaris is said to be 180 miles in diameter. This refers only to the dark area. Actually, Mare Nectaris (Pl. XX) is the basin of a great crater about 279 miles in diameter. It is surrounded by uplands forming the remnant of the original crater wall. On the west the Pyrenees are found, and on the east the ridge running south from Theophilus and Catherina. Fracastorius has destroyed the southern rim.

Wilkins and Moore say: "The entire plain appears to be concave, and the ridges (parallel to the edges) mark the successive levels" (1). The contour map

PLATE XX.—Mare Nectaris and the Altai Ring scarp. The Rheita Valley and several other valleys appear south of Mare Nectaris. (Lick Observatory, moon age 17.8 days, October 22, 1937.)

PLATE XXI.—Mare Crisium area. (Lick Observatory, moon age 4.6 days, June 2, 1938.)

PLATE XXII.—Mare Humorum area. (Mount Wilson.)

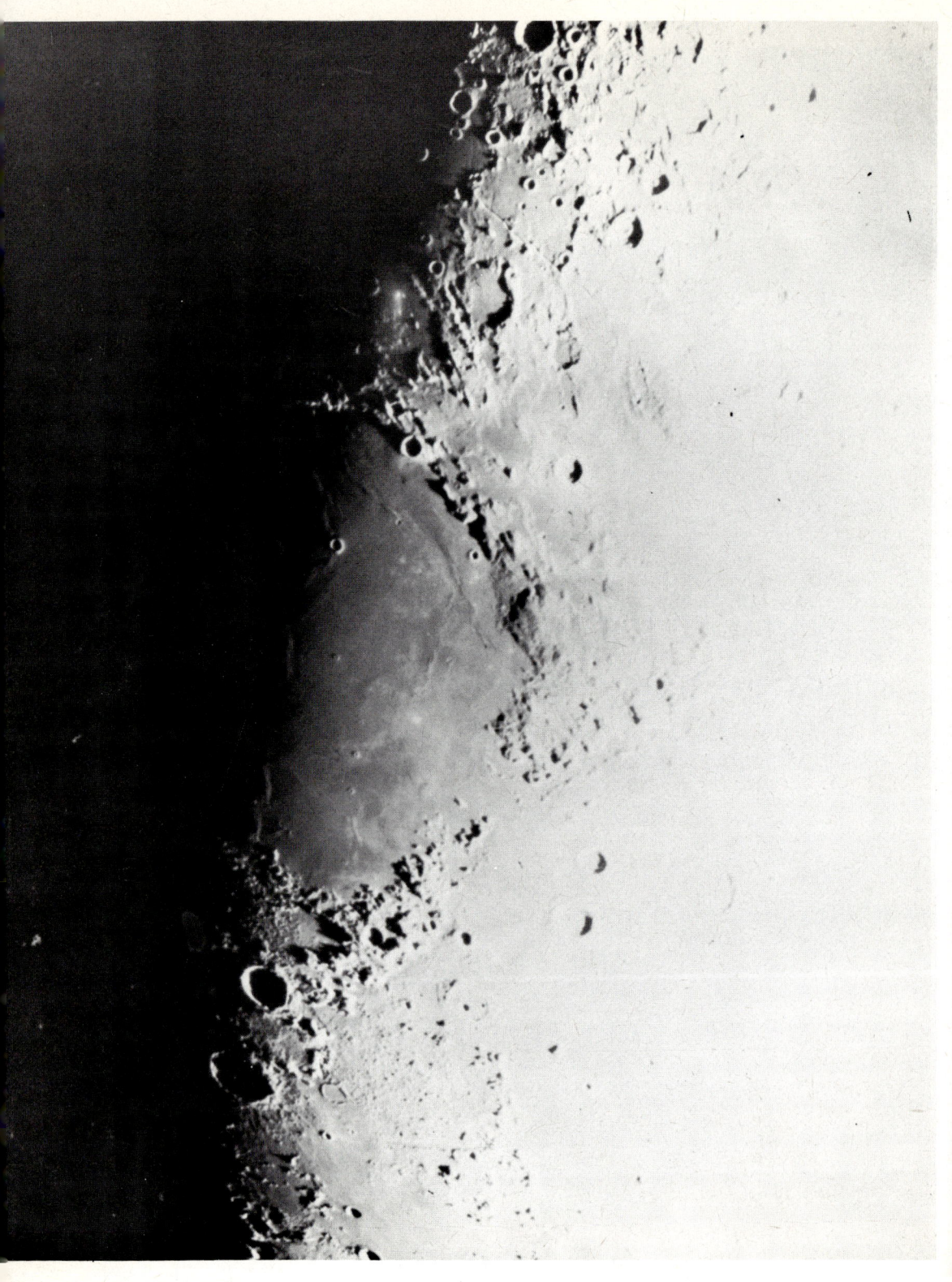

PLATE XXIII.—Mare Serenitatis, the Haemus Mountains, and markings radial to Mare Imbrium. (Mount Wilson photograph No. 393. Courtesy Dinsmore Alter, July 7, 1958.)

PLATE XXIV.—Mare Imbrium and its radial valleys. (Lick Observatory, moon age 20.40 days, September 6, 1936.)

PLATE XXV.—Ptolemaeus region. (Mount Wilson.)

PLATE XXVI.—Ricochet groove produced by British 4".5 AA shell. The "crater" was 12 feet long, 3 feet wide, and 1 foot deep. It had a raised rim, and most of the ejected material went sideways.

(p. 236) shows this effect clearly. There is a shallow shelf area along the shores of Mare Nectaris where the lavas are not thick.

Surrounding Mare Nectaris on all sides is a broad, flat plain liberally sprinkled with new and old craters. It is bounded throughout almost 360° by a relatively sharp scarp. The highest portion of the scarp, that in the southeast, is called the Altai Mountains and averages 4,000–6,000 feet high, with occasional peaks going up considerably higher. One west of Polybius has been determined to be 13,000 feet high.

Starting at the west wall of Piccolomini, the Altai Mountains extend in an arc concentric with Mare Nectaris. It is usually considered to end west of Catherina, but this long arc marks only the highest portion. The Altai scarp continues along the same circle through Kant and extends to just south of

TABLE 30

Possible Sizes of Planetoids Which Produced the Circular Maria

Mare	Crater Diameter (Miles)	D	E	Missile Diameter (Miles) $\rho = 3 \mathrm{Gm/Cm^3}$	
				Impact Velocity (2 m.p.s.)	10 m.p.s.
Imbrium..........	421	6.347	25.125	118	40
Serenitatis........	367 ?	6.287	24.942	103	35
Nectaris..........	279	6.168	24.580	78	27
Humorum.........	287	6.181	24.620	80	27
Crisium...........	330	6.241	24.802	92	31

Moltke. At this point there is a 70-mile gap which is lava-filled. After the gap it reappears, rather low, and extends to the west a little south of Censorinus.

The weakened scarp again disappears below the lava, this time that of Mare Foecunditatis, but emerges with renewed strength west of Magelhaens and proceeds in an irregular fashion through Monge and Monge B. Probably an ancient crater, which includes Cook within its borders, has deformed the earlier scarp. Beyond this point the ridge becomes more impressive and sweeps southward and eastward in a great arc, nearly touching Neander and completing the circle at Piccolomini. The latter is a more recent crater which has destroyed a portion of the great ring.

Mare Nectaris is not placed fortuitously in the center of the Altai scarp. They each are parts of a much larger and very ancient structure. The contour map shows that the region between the mountainous borders of the mare and the Altai ring is depressed by an average amount of roughly $\frac{1}{2}$–$\frac{3}{4}$ mile. The Altai circle is sharp on the inner wall and averages about 15 miles wide.

Beyond the Altai range are several of the most puzzling features on the moon. They are great valleys, all of which are radial to Mare Nectaris but not necessarily to the center of it. The most outstanding of them is the Rheita valley, which is about 115 miles long and has a maximum width of 15 miles. The valley is remarkably uniform in width and depth. It commences about 15 miles toward Mare Nectaris from the south wall of Rheita. The great crater is a later formation and encroaches on the valley. Midway along the valley is a modest, unnamed crater which extends completely across the depression. In the valley the crater walls are lower than at the outside parts of the crater. At the south end of the Rheita valley, another valley is found, making an angle of perhaps 30° with the Rheita valley. Although astronomers often include this as part of the former and mention the "bend" in the Rheita valley, the two are really parts of separate and distinct systems.

Parallel to the Rheita valley and about 20 miles away is another, but weaker, depression. Alter (2) showed that the area between the two valleys was also depressed. Extending from the Altai scarp into Brenner is another long, but narrower, valley. A shorter one bumps into the east side of Janssen, and at this point meets a wider and longer valley of another system. A similar valley is tangent to the west wall of Riccius.

To the west is the fantastic valley structure which starts at the Altai ring north of Reichenbach A and B and stretches interminably to the west through Snellius and far beyond. Snellius and several smaller craters are later formations than the valley.

These valleys and a few other fainter ones are all found in the south and southwest sides of the Altai Mountains. In general, they are close to the places where the mountainous ring is highest.

Mare Nectaris is old. With the possible exception of Mare Humorum, it is the oldest of the great circular maria. It was formed in a region of the moon thickly covered with craters. Possibly Cyrillus and Catherina are still older. They have been horribly deformed, far more than most craters. The shock wave could have so destroyed them.

The object which caused this great structure probably fell nearly vertically, because the crater is nearly symmetrically placed within the mountainous ring. Probably its path was slightly toward the southwest, as the axis of symmetry lies in this direction. The great crater, 279 miles in diameter, does possess a rim, but it is relatively low.

At smaller craters it has been observed that, as the crater diameter becomes larger, the crater depth also becomes greater, and the rim height slowly increases. At the same time, the effect of the curvature of the moon's surface becomes more and more important. It has been shown that the bottom of a crater is always lower than the chord whose length is equal to the crater diameter, until the crater diameter reaches about 200 miles. At larger craters the geometry would seem to demand that they appear more or less as flat spots on the moon. The rims shade gradually into the crater bottom rather than abruptly, as at the smaller craters. Mare Nectaris possesses just such an appearance.

The Altai ring is interpreted to be a shock wave generated by the impact and sent racing outward until it froze into position. The diameter of the Altai ring, which is tolerably circular, is 562 miles as measured along a radius from the center of the moon's disk. The Nectaris crater is 279 miles. The ratio is 2.01 to 1.

On the assumptions that the Altai scarp is 562 miles in inner diameter, 15 miles wide, and has an average height of 2,000 feet and that the moon's surface density is 2 gm/cm^3, we may calculate the energy required to pick this mass up against the lunar gravity and hurl it skyward like a projectile so that it would rise 2,000 feet. It turns out to be about five orders of magnitude less than was required to produce Mare Nectaris (Table 30). The difference in magnitude of these two numbers is quite encouraging.

A most interesting plate was taken at Lick Observatory on 1938 June $3^d04^h02^m$ with the moon at phase 4.58 days. It is reproduced in Alter's *Introduction to the Moon* (2) as Plate 40. It clearly shows that Mare Nectaris is surrounded by three ring anticlines. The inner is the crater rim itself and the outer is the Altai ring. Midway between the two is the hitherto unrecognized second anticline. It appears well defined for nearly 180°, and it is known to connect Cyrillus and Catherina, although this region is in shadow on this photograph.

The radial valleys have been suggested to be grooves cut by flying fragments from the impact (3). This seemed logical, because they were radial to the impact point and generally possessed slightly raised edges, particularly the Rheita valley. This suggestion has raised a storm of protest. Fielder (4) says:

> The Rheita valley might give to the casual observer the impression of being a long, shallow trough with a sharp bend in it, becoming narrower and deeper after the bed [*sic*]. In actual fact it is in the main a complex chain of craters. Other valleys of this type are common and are usually called crater chains.

Alter (5) says:

> Plates VI and VII, made near sunrise and sunset demonstrate that the "Valley" is merely a deep trench along the western part of a depression which has not been recorded previously, and which is wide enough that it would be permissible to consider it as a shallow mountain walled plain. Plate VIII, made at phase 4.1 days, with grazing sunlight, adds the fact that the valley itself is composed of seven confluent craters, of varying sizes, whose mutual walls have mostly collapsed. Except for these craters, no sign of any sort of explosion, either internal or impact, is observable.

If the Rheita valley were alone on the moon, one might possibly accept Fielder's interpretation that it was merely an alignment of contiguous craters, but it is not alone. It is an integral, though major, part of a great system of similar structures, all of which are aligned with Mare Nectaris. Whatever the origin of these valleys, they must have been formed as part of the great event which produced the complete structure.

This leaves only two alternatives. If the Mare Nectaris structure is older than the craters which are now found in the Rheita valley region, then the valley could have been formed by the impact of the great masses ejected from the explosion area. It is not a valid argument to say that there is no known theory of grazing impacts that will explain these valleys. Didion's experiments yielded equations which are valid at lower velocities than seem required here. At above 3,500 feet per second we are moving in unknown territory. In spite of objections, the grazing-impact theory is still a possibility. It is noted that there is no tremendous series of grazing impacts around Mare Nectaris, as there is around the eccentric Mare Imbrium.

The other choice is that the valleys represent subsidence phenomena. At the impact of the great planetoid, the area surrounding Mare Nectaris must have vibrated and twisted violently. The Altai ring is clear evidence of the magnitude of some of these forces. It is entirely possible that in a similar manner the lunar crust may have cracked on several radii from the impact and that, associated with the radial faulting, there occurred massive subsidence. Crater features would have been carried down with the surface and so still be visible crossing the valleys. Because of the observed number of craters crossing the Rheita valley, it seems more probable that the structure is a subsidence valley.

In either case the formation of these valleys was a by-product of the great impact and the transference of planetoid momentum into the motion of the lunar crust.

The Mare Crisium object came from slightly west of south. In the direction opposite to its approach the crater margin is highest and widest and roughest. As at Mare Nectaris, the shore areas of the lava center are shallow and several postimpact, prelava craters are visible (Pl. XXI).

Surrounding the great crater is a depressed shelf area, particularly evident in the northern half. It is bounded on the outer edge by a scarp, similar to the Altai ring but less prominent. There is some evidence that this ring anticline is considerably wider than the Altai ring, perhaps 90 miles wide at the most.

The diameter of the Crisium pit in the north-south direction is 286 miles. The radius of the anticline in the north direction is 303 miles. The relative sizes are thus in the ratio of 2.12 to 1.

Mare Crisium is not circular. The foreshortening due to its near limb posi-

Circular Maria and Related Structures

tion makes the north-south axis seem longer, but actually it is considerably less than the east-west axis, which is about 367 miles wide.

Part of this excess width is due to the lava's breaking through the west wall and the flooding of a considerable area of shallow shelf. In the opposite direction, not far from Proclus, there is a corresponding gap in the mountainous border of the mare. This pair of gaps lies essentially at right angles to the postulated direction of approach of the planetoid, and the major development of the Crisium structure is to the north of this line between the gaps.

There is no great valley system around Crisium, as there is at Nectaris. There is nothing to correspond to the Rheita valley, but there are numerous evidences of ejectamenta, particularly to the north. Many of the elongated markings lie inside the shock-wave scarp. They are particularly well developed between Macrobius and Cleomedes, but they are by no means limited to that location. Markings aligned with Mare Crisium may be found in most directions but are few in number in the south. In the north, some of the radial grooves lie beyond the shock scarp. Mare Crisium is also a depressed area. The lava fill shows clear evidence of subsidence. The center has dropped the most.

Mare Humorum is very similar to the two preceding objects. It is a lava-filled sunken basin. Its mountainous borders give it a diameter of 287 miles. It is nearly centrally placed in an incomplete ring anticline 594 miles in diameter. The ratio of sizes here is 2.07 to 1. Inside the ring is a depressed ring syncline, parts of which are lava-filled (Pl. XXII).

The Humorum planetoid came from almost due north, but the angle of fall was not far from perpendicular, because the structure is nearly symmetrical. The southern part is slightly better developed. It is in this direction that most of the elongated markings are found. The width of the shock scarp is greater than at Mare Nectaris, but less than at Mare Crisium. There is no great valley system such as that sported by Mare Nectaris.

Mare Humboldtianum is so close to the limb that it can be described only in generalities. It is mountain-bordered, about 185 miles in diameter, lava-filled, although perhaps not completely so. There is some evidence of a ring scarp surrounding, with about twice the crater diameter. No certain radial valleys have been found. It is a normal mare of this type, just a little smaller than usual.

Mare Serenitatis is the most difficult of the ancient maria to delineate, particularly in the northwest. This mare is usually considered to be the area of the dark lava flow, but this is only a portion of the structure (Pl. XXIII).

Mare Serenitatis is ancient of days. It may well prove to be the oldest of all these vast objects. It certainly was old when Mare Imbrium was born. It may prove to be an old structure partially overlying a still older object.

The southwest rim is marked by the Haemus Mountains, from where they emerge from the hinterlands of the Apennines to the place where they disappear at Point Acherusia. This was once a proud and high range. It has been acted upon by outside forces in numerous manners. The whole range has been battered by debris flying from the Imbrian collision zone. The normal pattern of the mountain range has been forcibly altered to align with Imbrium. Much of the Haemus Mountains has been lowered in the region edging Mare Vaporum by the same force that depressed Mare Vaporum. The entire border of Mare Serenitatis is now less prominent than it once was.

Beyond the gap to Mare Tranquillitatis comes Mount Argaeus, which marks the resumption of the Serenitatis rim. It continues as the Taurus Mountains toward Franklin. Then the rim becomes discontinuous before it reaches the crater. Although a high ridge extends toward Franklin, it is probable that the Serenitatis rim curves in toward Maury and then on to Mason, where it joins the Caucasus beyond Eudoxus. This rim thus outlines a tremendous oval considerably elongated in a northwest-southeast direction. It is here suggested that this northern and western extension of Mare Serenitatis may mark a still older formation which has been nearly obliterated.

Is there a shock scarp around this great crater? Starting near Agrippa and running west exactly parallel to the Ariadaeus Rille is a ridge which could be so interpreted. If this is so, it appears to tie into a series of drowned craters in a line across Mare Tranquillitatis. These craters probably indicate a shallow bottom to the sea under the lava. The ridge continues parallel to the Taurus range and is well shown as a ridge convex toward Mare Crisium between Macrobius and the Crisium shock scarp. It is well shown on a Lick Observatory photograph taken October 22, 1937, with the moon at age 17.8 days. It does not appear clearly on a Lick Observatory plate of June 2, 1938, with the moon at age 4.2 days. The ring may possibly be traced north of Atlas and disappears near Aristoteles.

The whole Serenitatis structure is so old and so distorted that our picture of it is extremely unsatisfactory. It certainly is an ancient crater of the type under discussion, but the fine details evident at the others are too blurred to permit exact definition.

As a suggestion, it is inferred that the planetoid came from the northwest at a very low angle and spent most of its force in the southeastern end, which is now lava-filled. It created the Haemus Mountains as the highest part of the rim, sent a shock wave out to the region of Agrippa, and, as usual, the region between the rim and the shock scarp is depressed. The shock scarp is thus roughly twice the crater radius from the center of the structure.

There are no radial valleys or grooves aligned with Mare Serenitatis, but north of Hipparchus are several ridges which point toward Serenitatis. At best it is an unhappy picture we draw of this object.

Circular Maria and Related Structures

Mare Imbrium offers the greatest contrast. It is clear and crisp. It is the youngest of the major circular maria. Unlike the similar but smaller structures, the shelf area between the crater pit proper and the shock-wave ring is almost completely covered with lava (Pl. XXIV). In addition, Mare Imbrium is by far the most eccentric structure. Between Mare Imbrium and Mare Serenitatis is a gap approximately 40 miles wide. South of this gap is the moon's greatest and wildest mountain range, the Apennines. In the hinterlands of the Apennines is a great, jagged slope of debris stretching about 400 miles along the range and up to 150 miles from the scarp. Shortly before Eratosthenes is reached, the mountain range suddenly becomes much narrower. Beyond Eratosthenes it disappears and reappears north of Copernicus as the Carpathians, a mountain range of peculiar mien. Its outlines are muted, and the ridges and valleys have been forced into alignment with the impact area in the great sea. The Carpathians fade away to the east and disappear beneath the lavas.

After a gap, isolated peaks appear inside the Carpathian sweep near to Euler. Again there is a gap, and the next reappearance is the Harbinger mountain range near Aristarchus. Beyond the scattered Harbingers, the mountains reappear out of the lavas and approach Sinus Iridum from the east. The Jura Mountains are simply the high border of the Bay of Rainbows. Since Iridum was formed after Mare Imbrium and has seriously distorted the rim, the mountains are highest at this point. Beyond Sinus Iridum, the rim of Imbrium fines out to nearly nothing east of Plato, but beyond Plato the blocky Alps arise.

Between the Alps and the Caucasus Mountains is another gap symmetrically placed relative to the one at the east end of the Carpathians. The Caucasus is aligned with the west end of the Apennines.

The construction of the Imbrian shock-wave mountain range is thus divided into two parts. The first is an almost parabolic arc from the Caucasus to the end of the Carpathians. The second is a lower and less rugged area marking the northern boundary of the great sea. It has a smaller radius, and its ends lie inside the parabolic ranges, with open gaps at each side.

The interpretation of the structure is that the colliding planetesimal came into the lunar surface at a low angle from the north-northeast and at a low velocity. Both low angle of approach and low velocity are required to produce the extreme case of bilateral symmetry here observed and a marked lack of radial symmetry.

The object struck close to the present position of crater B, not far south of Helicon. The crater produced by the impact measures about 421 miles in diameter, but it has been nearly buried by the later lava. Only a few isolated peaks remain to mark the moon's greatest crater. Moving clockwise (on a photograph) the rim is delineated by the Straight Range, the Teneriffe Mountains,

Pico, β, Spitzbergen, Archimedes ζ, Lambert γ, La Hire, La Hire α, and Caroline Herschel ζ. These few peaks outline a great circle approximately in the focus of the parabolic shock front.

The sequence of events would seem to be as follows. The impact occurred when the 100-mile-wide planetoid struck. An explosion started immediately and lasted for at least half a minute as the object plowed into the ground and the shock waves raced in both directions from the impact area. The first shock front was broad and diffuse. It spread out in all directions and traveled up to the outer edge of Mare Vaporum in the south and Mare Frigoris in the north. It then froze in position as a broad, low ring anticline. Presumably it is present in the region of Mare Serenitatis and Oceanus Procellarum but is lost beneath the lavas.

Following the ring anticline is a broad ring syncline. It is now filled with the lavas of Mare Vaporum, Sinus Aestuum, Sinus Medii, and Oceanus Procellarum. In the north the ring syncline is marked by Sinus Roris and by Mare Frigoris, although on the extreme west a part of the depression may be due to a similar structure around Mare Serenitatis. The primary shock wave is the second one which became frozen into place as the mountainous arcs surrounding Mare Imbrium.

The central area, including the impact crater itself, had absorbed a tremendous amount of momentum, and there had been a great rebound. How long this rebound remained fixed as a structural dome is not certain, but there is definite evidence of its slumping. All along the Apennine front there are great masses which have slid down the scarp and out onto the plain. In the region west of Mount Huyghens this slumping has carried thousands of cubic miles of rock into Mare Imbrium for 20–30 miles. Much of the material in the highlands near Archimedes and in the Apennine range appears to have been emplaced violently, but whether from the body of the moon or from the planetoid, as Urey suggests, cannot be determined.

In order to produce the lack of radial symmetry observed, the impact velocity must have been very low, perhaps about 2 miles per second. At impact the shock wave would race on ahead of the fragments spewed from the collision area. The mountainous border would thus already be in place when the fragments arrived. The Apennines show a definite preferred orientation toward the impact point. The Carpathians in particular show the sculpturing effects of moving masses, as does the Caucasus. The mountains on the northern rim are not greatly marked.

In the region to the west the outstanding feature is the Alpine valley. It is an essentially perfect cut through the Alps, starting as a fine feature, widening to the middle, and becoming narrower at the western end. It is the most prominent feature of a whole family of such objects. Between the Alpine valley and the Caucasus range are at least eleven fainter but similar valleys. They are almost, but not quite, parallel and have a radiant at the impact point. A Lick photograph of June 11, 1938, with the moon at 10.35 days, shows this group well.

Beyond Mare Frigoris, particularly northwest of Plato, are numerous valleys with raised sides. One north of Archytas is very much like the Alpine valley.

Kuiper (6) has felt that the Alpine valley is a radial crack in the lunar surface produced by the impact. The ground separated and, for some reason, did not come completely together again, probably because some of the ground dropped and wedged the crack. This may be the correct interpretation, but at the present state of knowledge the suggestion is no more probable than that of a grazing impact of material ejected from the collision zone.

It is to the south and southwest of Mare Imbrium that the greatest clustering of valleys radial to the central crater is found. The entire western end of Mare Vaporum is heavily scored. The Haemus Mountains have been almost obliterated. Countless valleys, great and small, have been ripped through this range. Behind the mountains, in the lowlands, the valleys are found in similar numbers. In the lowest parts they are now lava-filled. Several dip down and are lost under Mare Tranquillitatis. East of Manilius and in Sinus Aestuum, where the lavas are thicker, i.e., the prelava surface was lower, the valleys are not found, but wherever the original surface projects, the valleys are found. In the Bode-Ukert highland the area is stippled with these radial valleys. They clearly have picked the highest spots for their appearance. The walls of Pallas have been struck repeatedly. The high regions near Ukert are furrowed. The lower places have mostly been missed.

This pattern exists wherever the valleys radial to Mare Imbrium are found. It can only mean that the valleys were dug by missiles from the collision zone. These missiles would naturally have a preference for the higher ramparts, coming as they did so close to the ground on their trajectories.

Beyond the lava-filled Sinus Medii a tremendous development of the radial valleys starts. The entire central area of the moon has been raked. Great gashes are found in the Ptolemaeus-Hipparchus area. In general, these furrows are long and narrow. They average perhaps 15 miles in length and 2–3 miles wide. They have raised rims. They are not found on the bottoms of the great lava-filled craters in the area.

Alter (5) has made an attempt to explain a peculiar feature in the Ptolemaeus region as due to subsidence:

> There are numerous narrow valleys in this area that have been interpreted by some able students as gashes cut by fragments of an asteroid, assumed to have struck first in the Mare Imbrium area. The principal one of these markings is shown as Plate X [Pl. XXV]. It is roughly 150 miles long and ends in the eastern wall of Alphonsus.

Any satisfactory hypothesis to account for these peculiar features must not neglect the following data:

1. The "gashes" are parallel to each other.
2. They are parallel to the east and west walls of Ptolemaeus and to the great rill on its eastern floor.
3. They are parallel to the great ridge on the floor of Alphonsus. Also parallel to them on this same floor is a conspicuous rill and most of a still more prominent one at the extreme west.
4. They are parallel to rills of craterlets on the northern floor of Ptolemaeus.
5. They are roughly parallel to the Straight Wall, which forms part of the western shore of Mare Nubium.

The reader should study very closely the details of the picture shown here. Surely the photograph demonstrates that this longest example cannot be the result of any cut.

These valleylike features are approximately perpendicular to the slope gradient between the Maria Nubium and Nectaris. The best guess today is that they were opened along faults at the times when sinkings produced these seas.

His points may be answered as follows:

1. The "gashes" are not parallel to each other. They form parts of a great fan of similar markings open through most of a full circle, except to the lava cover east of Mare Imbrium. All these markings radiate from the impact area in Mare Imbrium and hence are reasonably parallel only to nearby valleys.
2. The valleys are parallel to the east and west wall of Ptolemaeus in the same sense. In addition, the walls of Ptolemaeus are deeply marked by similar valleys. Any such valley occurring on the east or west wall of this crater will tend to force an alignment in the directions radial to Mare Imbrium. This is particularly evident on the east wall. Missiles which have struck the north and south walls simply leave gaps in the rim. They cannot force such an alignment there. Wilkins and Moore speak of the great rille in the crater floor as "one of which develops into a distinct cleft which after following the contour of the wall, enters a pass in the northeast rampart" (7).
3. The great ridge on the floor of Alphonsus is parallel to nearby valleys, which may well have had a similar origin. The supposedly parallel rille on the west is quite irregular and is nearly in the form of a semicircle.
4. I cannot identify these objects from the description. There are certain markings in this area radial to Herschel, a very much later formation.
5. The valleys are nearly parallel to the Straight Wall, but the latter deviates from parallelism in the wrong direction. If its axis is projected northward, it is found to miss the impact area of Mare Imbrium by a small amount.

It is interesting to trace the development and cause of the 150-mile-long valley which Alter (5) says cannot be due to missile grooving. As we come from Mare Imbrium, the valley begins southwest of Lalande under a small, perfect crater lettered "C." The first few miles of the valley are through a lowland, and the valley walls are low and indistinct. It quickly penetrates a ridge, and the walls become correspondingly higher. Beyond the ridge is a gap, and the valley reappears beyond the gap, offset half its width to the west. In similar fashion the valley extends southward. At times it doubles. Wherever it goes through an upland, the valley walls are prominent. As it passes through lowlands, the valleys almost or completely disappear. Near Ptolemaeus the multiple nature of the structure is apparent. The southern end stops abruptly in the wall of Alphonsus.

This long, narrow marking is not one structure. It is the result of at least ten, probably more, impacts by missiles from Mare Imbrium, which were ejected on almost identical paths. It is not a subsidence effect. Gilbert in 1893 showed that the valleys and other structures were radial to Mare Imbrium and correctly stated that a huge explosion in Mare Imbrium had ejected materials in all directions and that "the lunar furrows were really formed by the forceful movement of a hard body . . . controlled by its own inertia" (8).

The valleys in this general area were discussed by Steavenson (9), who stressed that they must have had a common origin not associated with the origin of the craters. He suggested a grazing-impact hypothesis of a swarm of meteorites. Davidson (10) reached similar conclusions. Darney (11) in 1933 pointed out that the valleys were radial to Mare Imbrium. Neither Darney's work nor Gilbert's received the attention that should have been awarded to them. Their work was forgotten, and the subject was raised independently only in 1942 (12, 13).

In past studies it has generally been assumed that the objects which produced these radial valleys were solids. My earlier thoughts were that the valleys were produced by flying materials from the lunar crust. Kuiper (6) in 1954 and Fielder (14) in 1955 gave this view some backing. Urey (15) in 1952 suggested that the valleys were produced by the impacts of masses of nickel-iron which had been imbedded in the Imbrium planetesimal and had broken loose at the impact.

Kuiper (6) has found blocks at the south ends of certain of the valleys which were slightly smaller than the width of the valleys. He suggests that they were the remnants of the missiles. However, the blocks are not always in the far end of the valleys, and it is doubtful whether they are parts of the missiles themselves. There is relatively little information on the penetrating power of projectiles and essentially none concerning grazing impacts.

My own observations of artillery shells during World War II and certain experiments on the sandy beaches of Lake Michigan, using various types of bullets, guns, and muzzle velocities, show that a dud shell striking at an angle of fall of less than 10° will often ricochet and will then produce a long, narrow groove. If a bullet hits a flat, sandy area, it will also produce a groove if the angle of fall is less than 1° for a soft bullet and less than 3° for a steel-jacketed bullet. In either case, if the bullet suffers a significant loss of velocity at the impact, it will produce a round, shallow crater. The higher the velocity, the more chance there is of producing a groove by a ricochet. The velocities were in the range of 900–3,000 feet per second.

Plate XXVI shows a groove of this general type produced by a British AA 4″.5 shell which struck at about 10° and burst in a low-order explosion.

Didion at Metz in 1834–35 conducted experiments on the penetrating power of projectiles into different materials. This work was published in 1838 (16). His equations and constants, as developed by Fielder (17), are as follows:

The equation of motion of a projectile of mass m and radius r, moving with velocity v through a resistive medium may be put in the form

$$m\left(-v\frac{dv}{ds}\right) = \pi r^2 (\alpha + \beta v^2) \qquad [16\text{-}1]$$

where α and β are constants specific to the medium. Let V be the velocity of the projectile at the commencement of valley-formation, and v its velocity at the end of valley-formation. Then

$$m\int_v^V \frac{v\,dv}{\alpha + \beta v^2} = \pi r^2 \int_0^s ds \qquad [16\text{-}2]$$

Hence,

$$s = \frac{2r\rho}{3\beta} 2.3026 \log_{10}\left(\frac{\alpha + \beta V^2}{\alpha + \beta v^2}\right) \qquad [16\text{-}3]$$

where ρ is the density of the projectile.

One may write

$$s = \frac{C r \rho}{\beta} \log_{10}\left(\frac{\alpha + \beta V^2}{\alpha + \beta v^2}\right) \qquad [16\text{-}4]$$

where C is a constant which does not depend on the medium which is being penetrated. The constant for a particular medium may be evaluated experimentally in the following manner. From equation [16-4], if two projectiles commence penetration at different initial velocities and penetrate to distances s and s' respectively,

$$\frac{s}{s'} = \frac{\log[1 + (\beta/\alpha) V^2]}{\log[1 + (\beta/\alpha) V'^2]} \qquad [16\text{-}5]$$

The value of β/α which is found from this equation may then be used in

$$\beta = \frac{C r \rho}{s} \log_{10}\left(1 + \frac{\beta}{\alpha} V^2\right) \qquad [16\text{-}6]$$

which follows from equation [16-4], in order to evaluate β. Hence, knowing the ratio β/α, α may be found. The constant C may then be found from equation [16-6]. Davidson [10] found $C \simeq 157$, and this value will be adopted here.

Now equation [16-4], which was developed by Didion and which, in principle, may be applied to an investigation of the formation of any lunar grooves that may have been gouged out by the ploughing action of projectiles, is of little use for application to specific cases, owing to the fact that so many of the quantities in it are uncertain. An attempt was therefore made to minimize such uncertainties.

For all the grooves in the Ptolemaeus region, described above, the initial velocity, V, can be considered to have the same value. Thus, the ratio of the energy of a [spherical] projectile to its projected area would be proportional to its radius. This means that, under the limitations of this simple consideration, the wider grooves would be expected to be the longer. The evidence indicates that the final velocity v must be taken as zero. Hence it is seen from equation [16-4] that the ratio s/r would be expected to be constant. Direct comparisons on the photographs produced the following rough results for this ratio:

[TABLE 31]

DATA ON VALLEYS RADIAL TO MARE IMBRIUM

Valley	a	b	c	d	e	f	Average
$s/2r$	8	....	5 or 10	9	6	6	8
s/b	4	....	3 or 6	4.5	2	2.7	4
$b/2r$	2	....	1.7	2	3	2.2	2

The average value of s/r was found to be about 16 and this was used in conjunction with $v = 0$ and $V = 1.68$ km sec^{-1} (circular velocity). In Table [31], b represents the breadth of a valley, which has been taken to be the distance between the highest parts of its bordering walls. It is seen that the ratio of the breadth of a valley to the breadth of the associated block is roughly constant and equal to two.

Now the constants α and β are specific to the Moon's crustal materials and are not known. Didion [16] gave $\alpha = 4.35 \times 10^5$ and $\beta = 88$ for sand mixed with gravel. For concrete, Baldwin gave $\alpha = 4.34 \times 10^7$ and $\beta = 40.5$ and said [3] that the lunar materials would offer a resistance "many times less than solid rock, although perhaps higher than would be developed by sand and gravel." Urey [15] remarked that "... moist materials offer much less resistance to projectiles and hence substantially higher values than the minimum and somewhat less than the maximum are likely to apply to the Moon's surface." Now α is the less critical of the two constants, and this was therefore chosen to be 10^7. The more critical constant β was calculated for a projectile of density 3 g cm^{-3}, by writing equation [16-4] in the form

$$\frac{3C\rho}{2\beta} \log_{10}\left(\frac{\alpha + \beta V^2}{\alpha + \beta v^2}\right) = \frac{s}{r} \qquad [16\text{-}7]$$

In equation [16-7], α and β have been replaced by $\tfrac{2}{3}\alpha$ and $\tfrac{2}{3}\beta$ respectively, in accordance with Didion's experimental evidence for elongated projectiles. Using the values given above, it was found that $\beta \sim 50$, which is certainly very reasonable. The density ρ might of course have been nearer 8 g cm^{-3}, as Urey [15] believes, but it should be noticed that a measured length s has been inserted in equation [16-7] whereas, in actual fact, there would have been spaces where resistive forces were considerably smaller than those assumed in the derivation of the formula, and hence s should have been reduced by an unknown factor. The value of ρ can be considered to incorporate this factor.

It was shown by Didion that equation [16-1] held for velocities ranging from 0.09 to 0.54 km sec^{-1}. The analysis which has been pursued above incorporated velocities as large as 1.68 km sec^{-1}: at such velocities, there might be reason to doubt that equation [16-1] would hold equally well, but, nevertheless, it is reasonable to suppose that it would, at least, provide a first order result, and the value of β which has emerged from the analysis is of the order of the value required to satisfy the conditions which very probably apply to the lunar rocks.

Dynamically, then, assuming that substantial fracturing of the missiles would not ensue soon after first contact—and this assumption has not been proved—it would seem as if the process envisaged was wholly plausible.

Sterne (18) showed that steel balls 17/32 inch in diameter, striking at 4,500 feet per second, would penetrate into aluminum for seven times their diameter. For material one-third the density, the penetration would be three times as great. The back-pressure at impact was calculated by Urey (15) to be 50,000 kg cm^{-2}, and the steel balls were only slightly deformed.

Similar experiments using basalt and Paricutin lava balls of the same size showed that they broke at velocities of impact between 0.25 and 0.27 km sec^{-1} and 0.15 and 0.16 km sec^{-1}, respectively. The former broke when the back-pressures were between 1,600 and 1,900 kg cm^{-2}, and the latter when they were between 600 and 660 kg cm^{-2}.

If the Imbrian missiles had been of Urey's nickel-iron, there is an outside chance that they would have survived the impact, but I would feel much surer if they had been of tungsten carbide. If the missiles had been of lunar material or similar material from the planetoid, they would have shattered on impact.

The grooving experiments on Lake Michigan sand showed that the material ejected from the furrows went directly sideways. Part of it would be available to form the raised rims so often noticed.

A most interesting letter was received from Dr. John Stanley, chairman of the Department of Zoölogy at McGill University, Montreal, Quebec, Canada, and with his permission is quoted here almost in full (19). His observations may well be pertinent to the problem of the formation of grooves:

When I was in England during the war, I had occasion to examine from a scientific viewpoint a number of bomb and V-2 craters. One in particular, made by a 1500 kilo bomb would have interested you. The bomb detonated on the water surface of an Emergency water supply tank standing in the middle of a broad street, paved with end-grain wood block. The actual point of maximum detonation velocity was a few inches below the road surface, the bomb travelling the 4 feet of water thickness while the fuse got going. Thus fragments were thrown out about $1\frac{1}{2}$ to 2 inches below the road surface. These fragments, moving at velocities of some thousands of feet per second (probably) exerted a planer-like action on the pavement, and produced a most remarkable system of "rays" extending out some 20 feet, in the form of incised vee-shaped grooves in the pavement. It is interesting to note that at these high speeds, there does not seem to be any sort of "hydroplane effect" tending to lift the moving fragment. I found some fragments at the ends of rays, and it was obvious from the less cleanly cut grain in the wood, that they had slowed down in the last few feet. The same bomb produced some short rays in a stone building, about two feet long and an inch deep. One could also observe differences in the pavement rays, due to small differences in pavement level, and in one ray, there was an interruption where it crossed a shallow burned patch from an old incendiary. Incidentally, I was about 500 feet from this bomb, and actually watched it strike, and your description of a striking meteor is exactly correct.

I have often suspected such an effect at shell bursts, but normally a shell bursts above ground or below ground, so the effect is difficult to detect. If this principle holds at the grazing impacts such as produced the lunar valleys radial to Mare Imbrium, it would certainly help to explain the great lengths of the valleys.

In 1957 a paper was presented at the third annual Contour Machining Conference in Los Angeles (20). While extrapolating from the machining of various metals to grooving the lunar surface is a far step, perhaps an impossible one, the results of certain theories developed by Dr. Salomon and Dr. Von Karman and experimental work by Lockheed Aircraft Corporation scientists may conceivably illustrate certain principles which may show how missiles can penetrate the lunar crust.

It is well known that, for each metal and each type of tool, there is a low-velocity regime in which excellent cutting results are obtained. At higher cutting speeds, measured in surface feet per minute, the tools begin to fail. Beyond what has been called the "valley of death," where cutting tools fail rapidly under high rates of speed, there is a cutting velocity where this does not occur. This is a major effect.

As a result of study of these theories, Lockheed initiated a series of tests intended to get some rough idea as to what happens to the workpiece and tools at very high velocities. The first problem was one of attempting to achieve the desired velocities, and, because there are a number of gun enthusiasts at Lockheed who are constantly trying to obtain flatter trajectories with their rifle loads, it was decided to use the same techniques to supply some answers quickly and at a moderate cost.

A 30-06 caliber Mauser, having a smooth barrel bored out to accept a 0.300-inch-diameter slug, was used. The slugs were 2 inches long and were made of

AISI 4340 steel, heat-treated to 280,000 psi, u.t.s. On the end of the barrel was mounted a tool holder. Cartridge cases were loaded with various grades and quantities of rifle powders.

First firings were made across single-point tools ground to a 90° included angle and in various alloys and carbides. Initial firings were chronographed, and muzzle velocities averaging 2,200 feet per second were recorded, or 132,000 s.f.m. A second series of tests was run with a flat-nose tool, and with these the surface feet per minute was increased to 162,000. In both series of tests, a definite cutting action was observed without tool failure, except for a carbide tip in the first test.

The cuts were smooth, the second group of slugs measuring 20 micro-inches. An increase in hardness of one to four points Rockwell was measured to a maximum of 0.010 inch below the cut surface, but otherwise no effects on structure or strength were measurable. To date, no chips have been observed. The heat of cutting all went into the chip, which was instantly oxidized. All materials tested, whether ferrous or non-ferrous, showed a similar effect.

This, of course, does not prove that such an effect would occur at impacts on the lunar surfaces, but at least it is an interesting approach. The cutting tool must be harder than the target material.

If, on the contrary, it is finally shown that the back-pressure at the impact would shatter the missile as it shatters the lunar surface, then we have two choices. Either the shattered missile material continued on and formed each valley, or a new type of missile must be found.

It is possible that the impact of the Imbrium planetoid generated jets much like those from shaped charges. On the airless moon, these jets, ejected as they are almost on mathematical lines, might continue on a very low trajectory, an elliptical orbit, until striking. Their actions might be sufficient to gouge out the observed thousands of valleys.

Regardless of the exact mechanisms, there can be no doubt that some type of missile was radiated from the Imbrium explosion in all directions and in great numbers. These missiles produced the myriads of the Imbrium valleys.

The Alpine valley and the Mare Nectaris valleys, which are substantially larger, may prove to be subsidence structures; but until it can be shown how subsidence valleys can have raised rims, such as one finds at the Alpine valley and the Rheita valley, the impact theory will be a possibility.

It has generally been thought that the missiles came from the lunar crust as it was expelled from the region of the impact or from the smashed material of the planetoid. Urey suggested that the missiles were nickel-iron bodies imbedded in the planetoid.

An alternative is that the missiles came from the back side of the planetoid and were spalled off as the shock wave reached the back and was reflected.

They would then correspond to the meteoritic masses found near the Arizona Meteorite Crater without evidence of their being heated.

These objects would be spalled off while the planetoid still extended above ground. Therefore, the fact that the shock wave would race ahead of the missile would not cause the mountainous border of Mare Imbrium to act as a barrier, as Fielder (21) and Firsoff (22) have considered.

Sinus Iridum shows occasional radial markings, but they are poorly defined. The Jura Mountains seems to be wrinkled parallel to the north shore of the bay. This may be a result of horizontal compression from the explosion. The object which produced Sinus Iridum struck before Mare Imbrium had become lava-filled. It struck just about at the position of the original wall of Mare Imbrium or even a trace below the wall. The resultant crater must have been quite distorted, with the southern edge low in the pit and the northern ramparts high on the wall. The height of the Imbrium wall would cause the explosion to act in that direction as though it were rather deeply buried. Compression of the rocks would be expected under these circumstances.

The basic pattern is repeated, with minor variations, at all the circular maria. There is a vast central crater with a relatively inconspicuous rim. The central pit is lava-filled. Great quantities of momentum were transmitted to the crust, for great shock waves had been produced. Each of these shock waves moved outward to produce a ring about twice the diameter of the crater. These shock waves were noticed at a few of the better-displayed lunar craters and at certain terrestrial meteoritic craters and at all cryptovolcanic structures.

In all cases, some ejection of material occurred. At Mare Imbrium tremendous amounts were broadcast. The planetoid here came in at a low angle and low velocity. At Mare Nectaris, little material was ejected unless the valley system really is a series of gouges and not subsidence effects. The Nectarian object came from nearly the vertical; and, judging from the width of the shock ring, it was much smaller and struck at a higher velocity than did the Imbrium planetoid. Between the craters and the shock anticline is a depressed ring or syncline which may or may not be lava-covered. In the largest example, the ring anticline is surrounded by a well-developed syncline, lava-filled, and a suggestion of an outer ring anticline. Associated with each are rille families, but these will be discussed in a subsequent chapter.

In all cases, the great craters were formed dry, and the lavas came considerably later. The only exception to this is the tremendous and very ancient crater remnant east of Mare Nectaris and high on the bulge, which appears always to have been dry.

The shock-wave rings are high and prominent at the circular maria but are relatively insignificant at the normal craters. The rings appear faintly only at sunrise or sunset and have been detected around Aristillus, Bailly, Bullialdus, and Atlas. A possible shock ring may be seen in the usual place on photographs of Copernicus when it is very near to the terminator. Northeast of Clavius is a raised, scarplike ridge at approximately the usual proportionate distance.

Inasmuch as the shock scarps surrounding so many impact structures average in size close to twice the crater diameter, the scaling equations relating the shock-ring diameter to the initial energy released would have an exponent similar to that for the crater itself, or in the range of about 1/3.6 to 1/3.4.

References

1. WILKINS, H. P., and MOORE, P. *The Moon,* p. 158. London: Faber & Faber, Ltd., 1955.
2. ALTER, D. *Introduction to the Moon.* Los Angeles, Calif.: Griffith Observatory, 1958.
3. BALDWIN, R. B. *The Face of the Moon.* Chicago: University of Chicago Press, 1949.
4. FIELDER, G. *Structure of the Moon's Surface,* p. 162. London: Pergamon Press, 1961.
5. ALTER, D. "Peculiar Features of the Lunar Surface," *Pub. A.S.P.,* **70,** 416, 489, 1958.
6. KUIPER, G. P. "On the Origin of the Lunar Surface Features," *Proc. Nat. Acad. Sci.,* **40,** 1096, 1954.
7. Ref. 1, p. 62.
8. GILBERT, G. K. "The Moon's Face," *Bull. Phil. Soc. Washington,* **12,** 241, 1893.
9. STEAVENSON, W. H. "The Lunar Furrows," *J. Brit. Astr. Assoc.,* **29,** 165, 1919.
10. DAVIDSON, M. "The Lunar Furrows," *J. Brit. Astr. Assoc.,* **29,** 194, 1919.
11. DARNEY, M. "Le System Imbrien," *Bull. Soc. Astr. France,* **47,** 452, 1933.
12. BALDWIN, R. B. "The Meteoritic Origin of the Lunar Craters," *Pop. Astr.,* **50,** 365, 1942.
13. ———. "The Meteoritic Origin of Lunar Structures," *ibid.,* **51,** 117, 1943.
14. FIELDER, G. "A Study of the Valley System Radial to Mare Imbrium," *J. Brit. Astr. Assoc.,* **66,** 26, 1955.
15. UREY, H. C. *The Planets.* New Haven: Yale University Press, 1952.
16. DIDION, I. *Traité de balistique.* Paris, 1838; 2d ed., 1860.
17. Ref. 4, p. 169.
18. Private communication to H. C. UREY, quoted in Ref. 15, p. 45.
19. STANLEY, J. Personal communication, October 2, 1949.
20. PETERSEN, A. H. "How Will We Shape the New Materials?" *Machine and Tool Blue Book.* Wheaton, Ill.: Hitchcock, December, 1957.
21. Ref. 4, p. 174.
22. FIRSOFF, V. A. *Strange World of the Moon.* London: Hutchinson, 1959.

17

THE LAVA FLOWS

So far, there have been only two major theories advanced to explain the dark-gray areas of the moon. For generations, scientists had assumed that they were lava. Then See (1) and later Gold (2) suggested that they were deep pools of dust. It is here contended that the evidence precludes the existence of the dust pools. Therefore, the alternative would seem to be that they are composed of some kind of lava from the body of the moon.

It is possible that the moon never possessed even a moderately dense atmosphere. It is too small a body. If the moon ever went through a massive degassing period, its atmosphere could have been several orders of magnitude denser than at present; but, even so, it probably was always a good approximation to a vacuum.

Under these conditions, the forces of erosion were always of minor importance; and it is probable that in the surface of the moon we see a history of 4.5 billion years superimposed, the later on the earlier. Portions of the uplands should be part of the moon's primal crust under a thin coating of dust.

It is now believed that both the earth and the moon are about 4.5 billion years old. No portions of the rocky layers of the earth have been discovered which are much over half the earth's age. No portion of its original crust has ever been found and probably never will be. The earth is so active, geologically and chemically, that the original crustal materials, if still *in situ*, have been metamorphosed beyond recognition. The moon clearly has been less active.

333

Probably portions of its original crust are visible. Its rocks will have undergone much less change.

The history of the moon is the history of the earth, except that the moon is so much smaller that its geologic and chemical processes have been slowed down by perhaps several orders of magnitude relative to the earth. We must attempt to infer the early history of the earth from its present condition. The magnitude of this task is evident when it is considered that, in spite of centuries of study, the details of the primitive earth are still matters of controversy on almost every point.

It is usually accepted that at present a generalized cross-section across a stable continental margin would show that the Mohorovičić discontinuity is about 35 km deep under the continents but that near the seashore it starts to rise and is only about 5 km down under the ocean.

Below the Mohorovičić discontinuity lie ultrabasic mantle rocks. Above this lies a layer of basic rock, about 5 km thick. It may be a basalt or a hydrated serpentine or some similar material. Some authorities state that it is roughly the same thickness under both land and sea. Others deny its existence under the continents. Under the sediments of the continents and above the basic layer are nearly 30 km of acid crustal rocks. Below the surface the knowledge of the rock character is determined largely from earthquake-wave velocities, gravimetric measures, heat flow, and theory.

The earth at present shows two great features—the continents and the ocean basins. The steep continental slopes form natural boundaries between them. At present the boundary is about 200 meters below sea level. On the average, the ocean basins are depressed nearly 5,000 meters relative to the continents.

Seismic work within the last ten years (3–6) has shown that the velocities measured beneath continents range from 6.0 km/sec (characteristic of acid rocks, about 65 per cent SiO_2) at the surface to 6.5 km/sec (probably due to basic rocks, about 50 per cent SiO_2) at 35 km. Below the Mohorovičić discontinuity, the velocity suddenly jumps to 8.2 km/sec (probably due to ultrabasic rocks, 40 per cent SiO_2). The lowest velocities are missing under the oceans, and the higher-velocity layers are correspondingly closer to the ocean bottoms.

On April 12, 1961, scientists aboard the "Cuss I," the National Academy of Science's drilling vessel, reported that they had penetrated into the floor of the Pacific Ocean off Guadalupe Island beneath the sedimentary layer and had recovered a 10-foot core. The material proved to be basalt, as had been foretold by seismic work. The drill penetrated 10 feet into this layer after going through nearly 600 feet of unconsolidated sediments plus 2.2 miles of ocean. The basalt may have been from an ancient flow from the volcanic Guadalupe Islands.

These figures, combined with gravity observations, indicate that the conti-

Lava Flows 335

nents are high because they are light. The continents and ocean basins are in approximate isostatic equilibrium. Wilson (7) says:

> Each continent may be divided into structures of three types—old stable shields, mountain ranges, and continental shelves—and it is becoming apparent that there is a mountain-building process which converts shelves into mountains, while mountains, by processes of erosion, are in time converted into provinces of shields. There seems to be a process of growth by which continents expand and encroach upon ocean basins.
>
> There is a suggestion that ocean floors represent those parts of the original crust of the earth which have been least altered and that they owe their general level to that cause.

The continents, then, have developed subsequently from the lower layers of basic and ultrabasic rock.

On the continents are great regions where lavas have been spilled out into laminated sheets. Similarly, there are great dark areas on the moon which are presumed to be of lava of unknown composition but which may be compared with the terrestrial examples.

The great lunar lava flows present many peculiarities which almost defy interpretation. Clearly, they are not one flow, but many. They are generally dark in color, but this is a relative term. The moon is everywhere a very poor reflector. The darker color of the maria suggests that they are composed of more basic materials than the uplands.

The projections of crater rims above the lavas give evidence of the total thickness of these layers. The southern part of Mare Nubium and the shores of Oceanus Procellarum are shallow; that is, they are formed of lava of the order of 3,000 feet thick. The central basin of Oceanus Procellarum and the central regions of the circular maria have much thicker lava covers.

The maria at present are, without exception, depressed regions of the lunar crust. They average nearly 2 miles lower than the bright areas. The ground normally slopes down toward the seas for a distance of approximately 100 miles. This slope continues out onto the lava-filled areas, which are deepest in their central regions. Each of the circular maria exhibits a series of steps concentric with the borders. These steps show them to be vast, shallow saucers. The surfaces of the dark areas are nowhere flat but exhibit sunken regions and often broad undulations.

Inasmuch as there does not seem to be any possibility of a non-lava origin for these markings, this implies that the lava flows underwent compaction and shrinkage or withdrawal of the still liquid magmas back within the crust. If the moon is in partial isostatic balance, the lavas are denser than the brighter upland materials.

The new contour map and DuFresne (8), from earlier measures, have established that the average relief in the maria is greater than in the upland areas. There are great ridges and ridge systems in all lava zones. Often they are associated with the over-all structure, outlining sublava features which may be inferred by other evidence to be present.

Associated with the dark areas are systems of rilles which form great families. The most prominent characteristic of the rilles is their preference for the marginal zones of the lava flows.

The lunar lavas are often of widely differing degrees of gray. Some flows are almost as bright as the uplands. Others are as black as coal. Often these contrasting shades are side by side. Other colors are known. Contrasting degrees of brightness often appear on what seem to be single flows.

There are no visible terminal ridges at such boundaries which might mark the end of a flow. Hence it is imperative that the lava flows were of extremely fluid lavas. The more fluid a lava, the more basic is its composition. The more basic its composition, the higher is the necessary melting temperature.

How are these great lava flows on the moon to be interpreted? What are the mechanisms which led to such great outpourings? The most obvious thought is to seek their origin in terms of flows from volcanoes, but, in spite of diligent search, there is no counterpart of a large terrestrial volcanic cone anywhere on the moon. Very tiny cinder cones may be present and be invisible because of their small size, but there is nothing like Fujiyama on the visible face. There are certain objects which resemble great shield volcanoes. They are called "domes" and are always found on the lava flows themselves. They cannot be the sources of the lava. Bullard points out that on the earth:

At the bottom of the crust the temperature cannot be much above 450° C and is certainly far below the melting point and below the temperature of the lava emerging from volcanoes. There must therefore be some mechanism for providing heat to volcanoes. This mechanism must be local and intermittent in its action; volcanoes are uncommon objects on the earth's surface; they occur only in special places and remain active in a given area for only a fraction of geological time. It may be shown [9] that chemical action is an inadequate source for the energy; it is also unlikely that the heat comes from an exceptional accumulation of radioactive material. There is no generally accepted view as to where it does come from, but there seems good reason to believe that it is derived from the dissipation of mechanical energy associated with earthquakes.

When an earthquake takes place, energy that has been stored as elastic energy is suddenly released, and part of it is radiated as elastic waves. This energy is gradually dissipated throughout the earth and converted into heat. The energy is only 1.1 ergs/cm^3 year and can have no important thermal effects. The remainder of the elastic energy that gave rise to the earthquake is dissipated as heat near the focus by friction at fault planes and by plastic distortion and fracturing of rock. No detailed calculations have been made, but it is likely that this energy is at least as great as that ap-

Lava Flows

pearing as seismic waves. This heat is produced in a very small fraction of the earth's volume and is probably sufficient to produce melting on a large scale [10].

This view ascribes earthquakes and volcanic heat to the mechanical energy associated with the distortion of the crust. It explains very naturally the general association of volcanoes and earthquakes without requiring particular earthquakes to be associated with particular eruptions. On this view it would be expected that volcanic activity would develop later than earthquakes in an area and would continue after the earthquakes had ceased. It is difficult to say whether this is so, but it is possible that Italy, Sicily, and the African Rift Valleys provide examples, as they have active volcanoes and rather minor seismic activity which may be the relic of greater activity in the past. The theory could not account for the volcanoes of the Pacific, as there are very few earthquakes there. This is not a serious objection, as we have other reasons for expecting high temperatures under the oceans [11].

Under the oceans there is some reason to suppose that the radioactivity is distributed through a greater depth than it is beneath the continents. This raises a dilemma which was first pointed out by Dr. R. Revelle (unpublished). If the radioactivity is distributed through too great a depth, the temperature near the base of the radioactive layer will rise above the melting point, which is inconsistent with the propagation of S waves. On the other hand, if the layer is made too thin, the rocks beneath the oceans will have to be assumed to contain an improbably large amount of radioactive material. Temperatures have been calculated for layers 200 and 100 km thick. . . . It appears that the melting point would be exceeded if the thickness were greater than about 150 km. The heat generation needed to give a heat flow of 1.2×10^{-6} cal/cm² sec would be 5.3×10^{-14} cal/cm³ sec for a 200-km layer, and 9.2×10^{-14} for a 100-km layer, after allowance has been made for original heat and for that from 4 km of basalt. These values are so much greater than the observed average of 0.4×10^{-14} cal/cm³ sec for ultrabasic rocks as to raise a real difficulty. It would be of great interest to have more measurements of the radioactivity of basalts and ultrabasic rocks from the oceans and also to examine the amplitudes of S at distances between 10° and 20° for oceanic and continental paths separately. . . . It is possible that the difficulty may be avoided by the transport of heat by convection in the mantle either continuously or intermittently . . . but it would then be necessary to explain why the material beneath the oceans did not become mixed with that beneath the continents. It must be admitted that the relation between the oceanic and continental heat flows raises difficulties to which no satisfactory solution is known [12].

Since the moon does not show normal volcanoes, it may be presumed to be, at the most, only moderately active seismically. The great lava flows on earth or moon, then, did not arise volcanically. Indeed, their vast extent is evidence against this mode of origin.

On the earth are several tremendous sheets of lava. The Columbia River Plateau in Washington, with extensions into adjacent portions of Idaho and

Oregon, covers 200,000 square miles. In places, the cuts of the deep canyons of the Columbia River, Snake River, and other tributaries show that the lava pile is up to 4,000 feet thick.

The Deccan trap of western India now extends over a compact area of 200,000 square miles, but erosion has been at work ever since it was formed 40–50 million years ago and has detached many areas as outliers. Originally, the extent of the Deccan trap as a continuous plateau was nearly 500,000 square miles. In places, the thickness of the piled-up lava sheets attains 10,000 feet (13). These regions are quite comparable with the huge flows on the moon.

The great lava flows of this type are apparently quite similar to the flows on the moon, but the lunar flows are ancient, as shown by the superimposed craters, while the described terrestrial sheets are "modern." Geologists have discovered that similar tremendous sheets of lava have appeared at intervals on the earth since the beginning of geologic time (13). The existence of the earthly flows demonstrates the principle that such lavas can be extruded upward through a cool crust and spread widely over the ground. On the moon without an atmosphere, the cooling of the lavas would be slower.

In all these tremendous terrestrial lava-covered areas, the same sequences are found. The sheets are not single but are composed of hundreds or thousands of flows. The typical sizes of flows range from a few acres to perhaps 3,000 square miles. The flows are from 4 feet to 500 feet thick. They average about 35 feet in thickness for the Snake River and southeastern Oregon plateaus; the Columbia River Plateau and the Deccan trap each average about 85 feet per flow.

The lavas arose from great cracks in the crust, usually near the edges of the flows. These openings occur as swarms of closely spaced cracks. They range from 5 to 150 feet wide and average about 30 feet. In length, the spread is from 100 feet to perhaps 15 miles.

The lava of the usual flow is a tholeiitic basalt and olivine basalt, although occasional rhyolite or more acid flows are known. The gross lava densities range from 2.8 to 2.98, averaging about 2.95. The powder density is a little over 3.0.

The extruded lavas must have been hotter than 1,100° C but were not over about 1,400° C, because they did not melt the rocks beneath the flow or on the edges of the dikes.

No good data on the viscosities of the flow lavas are available, but the Hawaiian lavas erupt at about 1,100°–1,150° C. They have viscosities of about 10^3 poises near the vents, increasing to more than 10^7 at a distance. In general, at a given temperature, the viscosity increases as the acid nature increases.

Many of the lava-flow top surfaces are reasonably smooth and of the pahoehoe type, but large areas are known with very rough and vesicular aa lava surfaces marking the escape of gases.

The individual flows solidified in a characteristic fashion, from both the top and the bottom. The central portion remained fluid the longest. It was very common to have the lava flow out from under the top layer, allowing the top to subside. The solidified top of a flow often broke into blocks, but evidence of the sinking of these blocks into the fluid is rare. It is known in deep lava pools within craters. There are no evidences of major systems of cracks developing on the solidified flows, but they normally show numerous minor joints, and columnar jointing is common.

The photographs of the moon and particularly the ones of greater contrast show that the dark areas are not of uniform brightness. Several great sheets are apparent on Mare Imbrium, including the darkest, which fills Sinus Iridum and fingers out into the sea to the south and west.

In the Mare Vaporum and Sinus Aestuum regions are numerous smaller dark flows. The number could be indefinitely expanded over the entire dark surface.

Each of the dark regions also shows variations in color. Often different colors are concentrated in small areas.

It is quite probable that the lunar lavas were extruded much as were the terrestrial sheets in numerous smaller flows.

There are many evidences that the seaward walls of old craters on the edges of the maria have been destroyed. It is possible they have simply been overwhelmed by the lavas or conceivably that they have been melted and eroded. If the latter proves to be the case, the lunar flows probably were much thicker than their terrestrial counterparts.

As these lavas reached the surface, they were exposed to a near-vacuum, and hence it seems probable that the outer layers of the flows are highly vesicular, made so by the rapid escape of gases.

The lava flows have been discussed here as if they all occurred almost simultaneously. Actually, the lava flows appeared over an extended period of time, judging by crater counts. There is about the same number of small craters per unit area in Ptolemaeus, for example, as there is in nearby Flammarion. Each of these areas is lighter in color than an irregular flow ending just north of Flammarion. This flow may be traced beyond the Triesnecker area. The crater density south of the Hyginus Rille and up to near Flammarion is somewhat less than on the older flow. The Mare Vaporum lavas are still younger. A tongue extends along the south side of the east branch of the Hyginus Rille. The surface is obviously changed, and the crater frequency here and to the north is lower than near Triesnecker. The larger craters on the lavas are of meteoritic origin, but many of the smaller ones, particularly on the older flows, may be of internal origin. All this defines a series of lava flows over an extended period of time but limited to the time after the dry Mare Imbrium was formed, for the Imbrium radial markings do not streak the floor of Ptolemaeus.

REFERENCES

1. SEE, T. J. J. "Origin of the So-called Craters on the Moon by the Impact of Satellites, and the Relations of These Satellite Indentations to the Obliquities of the Planet," *Pop. Astr.*, **18,** 137, 1910.
2. GOLD, T. "The Lunar Surface," *M.N.*, **115,** 585, 1955.
3. TATEL, H. E., TUVE, M. A., and ADAMS, L. H. *Internat. Assoc. Seism.*, *XI Conf.*, *Résumés*, p. 27, 1951.
4. ———. *Proc. Amer. Phil. Soc.*, **97,** 658, 1953.
5. WILLMORE, P. L., HALES, A. L., and GANE, P. G. *Bull. Seism. Soc. America*, **42,** 53, 1952.
6. HODGSON, J. H. *Pub. Dom. Obs., Ottawa*, **16,** 113, 1953.
7. WILSON, J. T. "The Development and Structure of the Crust," in *The Earth as a Planet*, ed. G. P. KUIPER, p. 143. Chicago: University of Chicago Press, 1954.
8. DUFRESNE, E. R. "Note on the High Relief of the Lunar Maria," *Ap. J.*, **124,** 3, 638, 1956.
9. GRATTON, L. C. *Amer. J. Sci.*, **243A,** 135, 1945.
10. BULLARD, E. "The Interior of the Earth," in *The Earth as a Planet*, ed. G. P. KUIPER, p. 120. Chicago: University of Chicago Press, 1954.
11. ———. *Ibid.*, p. 121.
12. ———. *Ibid.*, p. 117.
13. LONGWELL, C. R., KNOPF, A., and FLINT, R. F. *Outlines of Physical Geology*, p. 227. In *Outlines of Geology*. 2d ed. London: John Wiley & Sons, 1945.

18

THE ATMOSPHERE OF THE MOON

When it comes to the atmosphere of the moon, there is not much to talk about. It is practically non-existent. Yet, even so, this trace which must exist—this vacuum better than we can produce on earth—could tell us many things about the history of the moon, were we only able to analyze it. Most of what we think we know about it is sheer speculation. All observations, except two, designed to detect a lunar atmosphere have failed to do so.

Early astronomers thought that they had detected twilight effects. Often amateurs claimed to have seen meteors streak through the lunar air. It is now known that such observations represent wishful thinking, for the lunar atmosphere is many times more rare than the atmosphere of the earth at any height where meteors could appear (1, 2).

Fesenkov (3) made one of the most serious efforts to detect lunar air by its polarizing effects. He failed, but he succeeded in proving that the upper limit of density is far lower than had previously been suggested. Struve (4) in a review of Fesenkov said:

Fessenkoff examined with a piece of polaroid filter the faintly luminous area near the center of the moon, on the dark side of the terminator, at first and last quarters. The surface brightness of this area is considerable, and the test for polarization by rotating the polaroid presents no great difficulty. Fessenkoff found no change in the surface brightness as the polaroid was turned, and he concludes—presumably on the basis of laboratory tests—that the ratio of the brightness at radial and at tangential orientations of the axis of the polaroid cannot be in excess of $n = 1.04$.

If the diffuse light which Fessenkoff observed had been produced in its entirety by "twilight" in the lunar atmosphere, the polarization should have been complete, because the phase angle at first and at last quarters is 90°. But the light is mostly caused by scattering in the earth's atmosphere—that is, by an ordinary lunar halo, and perhaps by a small amount of the earth-lit surface of the moon. This background illumination should be almost completely unpolarized. If we designate this background surface brightness as b and the hypothetical polarized light of the lunar twilight as c, we have

$$n = \frac{\frac{1}{2}b + c}{\frac{1}{2}b} \quad \text{or} \quad \frac{c}{b} = \frac{1}{2}(n-1). \qquad [18\text{-}1]$$

If L is the amount of solar radiation received by a unit of surface on the moon oriented at right angles to the radiation, ρ is the density of the lunar atmosphere, and μ is the coefficient of scattering, then the surface brightness of an element of the moon located on the dark side of the terminator is

$$\mu L \rho d h. \qquad [18\text{-}2]$$

Integrating this over the entire thickness of the lunar atmosphere and designating by m the mass of a vertical column of unit cross-section, we find that

$$\mu L m = \tfrac{1}{2}(n-1)b. \qquad [18\text{-}3]$$

The quantity b was determined by Fessenkoff in the following manner. He measured the surface brightness of the sky in the vicinity of the sun and found it to be twice that of a standard plane white surface illuminated by the sun. He next computed the ratio between the surface brightness of the solar halo and that produced by a source whose stellar magnitude is 14.17 mag. fainter than the sun but which has a similar distribution of light over its surface. He then computed the difference arising from the facts that the lunar observations were made near the terminator and the moon was at first or at last quarter. Since the surface brightness of the solar halo in terms of the brightness of the standard surface depends upon the scattering power of air and upon the mass of the terrestrial atmosphere, he obtains an expression for b which involves this mass, M. It is assumed that the coefficients of scattering, per unit mass, in the atmosphere of the moon, μ, and in the atmosphere of the earth, μ_1, are the same. The final result is an expression for the ratio of the two masses,

$$\frac{m}{M} = 0.196 \times 10^{-4}(n-1). \qquad [18\text{-}4]$$

From the observations, $n - 1 < 0.04$. Hence

$$\frac{m}{M} < 10^{-6}. \qquad [18\text{-}5]$$

From equation [18-4], we find a maximum numerical limit of 0.784×10^{-6}; and, because the surface gravity of the moon is only one-sixth that of the earth, the lunar atmosphere will extend correspondingly higher. The density of the moon's atmosphere relative to that of the earth is directly proportional to the ratio of the air masses times the ratio of the surface gravities. Therefore, as an upper limit, we have

$$\frac{\rho_m}{\rho_e} = 1.3 \times 10^{-7}. \qquad [18\text{-}6]$$

This result has been subjected to strong criticism by Lyot and Dollfus (5), who pointed out that the strongly polarized ashen light was some 100 times brighter than that possible for a lunar atmosphere.

Dollfus (6), in 1953, established that the upper limit of density of the lunar atmosphere is 10^{-9} that of the terrestrial atmosphere. In 1956, he (7) was able to extend this limit to 10^{-10} atmosphere by polarimetric and photometric observations with the coronagraph.

Aurorae in the earth's air are usually found at heights between 53 miles and some height over 180 miles. In spite of having been searched for many times, aurorae have never been observed in the lunar atmosphere (8). From this we may assume that neither oxygen nor nitrogen is present in the lunar atmosphere to partial densities comparable with those which are observed in the earth's atmosphere in the above height range where strong aurorae are observed. These densities on earth (9) are from 2.5×10^{-5} to about 8×10^{-14} times standard air density. Possibly, faint aurorae might escape observation, and thus the lunar atmospheric density from this test could be of the order of 10^{-15} atmosphere.

On occasion, the moon passes in front of the Crab Nebula. This glowing mass is known to be the remnant shell of an ancient supernova observed by the Chinese in A.D. 1054. It still emits light and radio waves by a synchrotron mechanism.

It was reasoned that if there were a lunar atmosphere, it would be extremely tenuous. Because of its low density, exposed as it is to various solar radiations, it should be highly ionized. If the radio emission from the Crab Nebula passes through this ionized lunar atmosphere, the latter should modify the radio waves in a measurable fashion, and the degree of the change is a measure of the density of the moon's atmosphere. In 1955, such an occultation took place, and Costain, Elsmore, and Whitfield observed it. A preliminary report was issued in 1955 (10) and a more complete one in 1956 (11).

The distribution of radio brightness is dependent on wavelength and, at such an occultation, has a shape which is best explained by refraction at the day side of the moon's disk. Elsmore's value for the electron density is 1,000 electrons per cubic centimeter. He attributed the distortion in the radio waves to a lunar atmosphere of about 2×10^{-13} times the density of the standard terrestrial atmosphere at sea level and in no circumstances larger than 6×10^{-13} (12).

It has long been known that the small mass of the moon will not permit it to retain gases of low molecular weight. Kuiper (13) suggests that, because of the low velocity of escape, only 2.37 km/sec, molecules of weight less than 60 will escape during the hot lunar day. Kuiper searched for SO_2 lines in the 3,000–3,180 A range, but did not find them.

Edwards and Borst (14) have evaluated several sources of the rare gases krypton and xenon, which are thought to be too heavy to escape. These sources are as follows: (1) spontaneous fission of uranium-238, (2) fission of uranium-235 by thermal neutrons from cosmic rays, (3) thermal fission of uranium-235 from the reaction

$$O^{18}(\alpha, n)Ne^{21}, \qquad [18\text{-}7]$$

(4) xenon production from iodine-129 produced during the formation of the elements, and (5) primeval gases trapped in rock.

From their discussion, it would appear that only three mechanisms are adequate to account for the radio observations of density. These are (1) spontaneous fission of uranium-238, (2) xenon-129 from iodine-129, and (3) primeval gases. The spontaneous fission of uranium-238 certainly occurs, but, unless there is a marked concentration of the element in the near surface layers, most of the resulting krypton will be trapped in lunar rock. The mechanism involving residual primeval gases is based on analogy with the earth and is quantitatively too large. If a correction factor is applied to this mechanism, we could accept the data from meteorites, which give values much too small.

The probable existence of iodine-129 in the moon is based on the theories of element formation. Edwards and Borst (14) say:

> If we accept this assumption, it is then possible to estimate the interval between element formation and the formation of the satellite in its present state. The interval of 4×10^8 years would not appear to be contradicted by other observations.

Their three alternatives, if mutually exclusive, lead to explicit predictions of the composition of the lunar atmosphere. Edwards and Borst give the isotopic compositions, but, in general, if iodine-129 alone is responsible, there will be only xenon-129; if spontaneous fission, there will be 5 per cent krypton and 95 per cent xenon; if primeval gases, 93 per cent krypton and 7 per cent xenon.

Other materials possibly present in a lunar atmosphere include argon, which is a by-product of the radioactive decay of potassium, and the usual volcanic gases, such as SO_2, CO_2, and H_2O. Vestine (15) has made estimates of the rates of emission of these gases based on terrestrial data. He finds $J \simeq 5 \times 10^5/cm^2$ sec for argon and $J \simeq 10^{10}/cm^2$ sec for the volcanic gases.

With these rates of emission and the assumption that the rate of escape of the gases from the moon is determined by the maximum temperature found by thermocouple measurements, 370° K, Herring and Licht (16) find that the moon should have an atmosphere of 10^{-4} atmosphere of argon, 4×10^{-8} atmosphere of H_2O, and more than 1 atmosphere of SO_2 and CO_2. Vestine has pointed out that the effects of photodissociation will markedly reduce the latter values and also that the age of the moon, 4.5×10^9 years, does not allow enough time for SO_2 and CO_2 to build up to equilibrium.

Herring and Licht show that the ejection rates of krypton and xenon as derived by Edwards and Borst will lead to a negligible contribution of these gases to the lunar atmosphere, as compared with the above gases. Öpik (17) has shown that the low ionization potentials of krypton and xenon will cause them to escape the moon.

Since the lunar atmosphere is far less dense than the above calculations would indicate, Herring and Licht argue that some mechanism must operate to reduce the amount of such materials in the neighborhood of the moon. They suggest that the solar wind could do so. It is usually assumed that the solar wind is composed of protons with an abundance of $10^3/cm^3$ at a velocity of 10^8 cm/sec. If such a high-velocity proton were involved in an elastic collision with an atom, the proton, on the average, would transfer 1 kev of kinetic energy, which would permit the atom to escape the moon. On the assumption that every atom so struck would leave the moon, Herring and Licht find that the maximum density of argon would be 10^{-15} atmosphere, and, still neglecting photodissociation, it would be 10^{-11} atmosphere for the volcanic gases.

The moon does not possess a strong magnetic field. Clement (18) has estimated it to be less than 200 gammas, and the instruments in the Russian Lunik rocket which crashed onto the moon suggested that the field is less than 50 gammas. Such a weak magnetic field would not be able to shield the lunar atmosphere from the solar wind.

Recently Herring and Licht (19, 20) reported that both hydrogen and argon would be produced in a lunar atmosphere by bombardment of the moon's surface by the protons of the solar wind. They calculated the density as about 40,000 atoms for each square centimeter of the lunar surface, but, unlike a normal atmosphere, the density decreases very rapidly away from the surface.

It is also true that helium is produced by the action of cosmic rays (21, 22), but, because of its light mass, it would quickly depart the moon. If helium is produced by cosmic rays, then every other nuclide below iron should also be produced. Those of gaseous nature would also leave the neighborhood of the moon quickly.

Firsoff (23) attempted to calculate the maximum density of a lunar atmosphere on the assumption that it was condensed from the interplanetary medium. He found 10^7 particles per cubic centimeter. Brandt (24) objected to this value because Firsoff had used a mean molecular weight of 25 and a temperature of 250° K and a cutoff at 1,000-km height.

With temperatures of the interplanetary medium between 10,000° and 100,000° K (25) and a mean molecular weight of $\frac{1}{2}$, Brandt finds that the density of a lunar atmosphere cannot be more than a few per cent higher than that of the interplanetary medium. Elsasser's (26) value for the latter is $\approx 10^3$ particles per cubic centimeter. Therefore, Brandt finds that the maximum density at the surface must be between 10^3 and 10^6 particles per cubic centimeter and probably much closer to the 10^3 figure.

Firsoff (27) countered Brandt's argument by specifying that comparatively large amounts of gas may be held by persorption in the porous or pulverous materials, or both, of the lunar surface, which is in the condition of "permafrost" at a depth of the order of 1 meter.

Some of these gases should be liberated by the rising temperature due to sunlight, forming a low "skin," which is resorbed in the cold of night. This mechanism would permit a reasonably high density during the daytime, yet it would be undetectable by Dollfus' method on the night side of the terminator.

Firsoff also points out that if the lunar atmosphere is composed primarily of argon, which has a high ionization potential, it may not be highly ionized to any extent in depth. This might lead to an upward revision of the density as determined by Costain, Elsmore, and Whitfield from the radio observations of the occulted Crab Nebula. Firsoff further cites the 1956 observation of the occultation of the discrete radio source associated with Kepler's nova by Rishbeth and Little (28). They found that there was a positive response when the source was still 3' behind the lunar disk. This may indicate a refraction far above that found in the occultation of the Crab Nebula.

Brandt (29) reviewed Firsoff's arguments again and declined to change his original position. He further pointed out that Rishbeth and Little had stated that the visible remnant of Kepler's nova lay 3' inside the limb but that the best available position would put the radio source much nearer the limb than the optical remnant.

Öpik (30) has considered the effect on the escape rate when the thickness of an atmosphere becomes less than one mean free path. Öpik and Singer (31) have discussed the case of a lunar atmosphere formed by gravitational accretion of interplanetary gas. They find a result which lies between the estimates of Firsoff and Brandt.

The preceding arguments basically have had to do with the present state of the moon's atmosphere or an earlier one quite like it. It is generally considered (32–34) that the atmosphere of the earth was effectively missing "in the beginning" and that the present atmosphere and hydrosphere were gradually extruded from the earth and then evolved into their present forms.

Starting with Stoney (35), many scientists have treated the thermal escape of a planetary atmosphere (36–38). Spitzer (39) presented a model which was rather close to physical reality. In this paper he discussed the escape of gases from the earth and Mars, showing that, on the basis of thermal motions alone, hydrogen would escape from both earth and Mars quickly at any reasonable temperature, while helium should quickly leave Mars, but that it required a very high temperature of the upper atmosphere of the earth to permit it to escape. Heavier elements would not be appreciably lost by the earth in any reasonable time scale, but, unless the atmosphere of Mars were cool, many of them would escape. Use of his formulae for the moon would indicate a quick escape of a lunar atmosphere, probably about as rapidly as it was formed.

Gilvarry (40–43) has attempted to modify Spitzer's work by certain changes in assumptions and has concluded that if the moon released its water vapors very early and very quickly, they would form a series of deep oceans and an atmosphere composed almost entirely of water vapor. Under such assumptions and with his modifications of Spitzer's theory, the oceans and the atmosphere would last for perhaps 3,000,000,000 years. The atmosphere would be continuously replenished from the oceans.

Although his mathematical assumptions may prove to be correct, dissociation of the water vapor by ultraviolet light from the sun has not been considered, and this should speed up the process very substantially.

Pickering (44) killed the idea of the past existence of large bodies of water on the moon when he said:

> The lunar atmosphere, on account of the gravitative constant, can never have been very dense, like our own, and the rapid evaporation from extensive lunar oceans under low pressure and exposed to the tropical rays of the sun would have produced deeply eroded valleys and extensive river systems, which are conspicuous on the moon only by their absence.

It is highly probable that water vapor and other gases did appear from the body of the moon, but all observational evidence is that they escaped as rapidly as they appeared and that the atmosphere of the moon was always very rare.

In a recent lecture,[1] von Braun stated that a test of a Saturn first-stage rocket had carried about 100 tons of water to a height of about 90 miles, where it was dumped. The atmospheric density there was of the order of 10^{-8} that at sea level, still many orders of magnitude greater than could have been expected on the moon at any time in its history. The water practically exploded, for it expanded at about 2 km/sec.

The net result is that the authorities are arguing over the composition and proper order of magnitude of density of a lunar atmosphere which is so rare that it is beyond our present ability to duplicate it in the laboratory.

The moon's atmosphere is so tenuous that it can have no appreciable effect

---

[1] Public lecture by Wernher von Braun at Lunar Exploration Conference, Virginia Polytechnic Institute, Blacksburg, Virginia, August 12–18, 1962.

on the surface conditions. This has undoubtedly held true throughout the time span covered by the observable features of the moon's surface.

The importance of the lunar atmosphere for us still lies in the future. It lies in the possibility that it will ultimately tell us more about the moon's history.

References

1. Fielder, G. *Structure of the Moon's Surface*. London: Pergamon Press, 1961.
2. Vaucouleurs, G. de. "Recherche de météores lunaires," *Astronomie*, p. 267, 1947.
3. Fesenkov, V. G. "On the Mass of the Moon's Atmosphere," *Astr. J. Soviet Union*, **20,** 1, 1943.
4. Struve, O. "An Upper Limit for the Mass of the Lunar Atmosphere," Review of Reference 3, *Ap. J.*, **100,** 104, 1944.
5. Lyot, B., and Dollfus, A. "Recherche d'une atmosphère au voisinage de la lune," *C.R.*, **229,** 1277, 1949.
6. Dollfus, A. *C.R.*, **234,** 2046, 1952.
7. ———. *Ann. d'ap.*, **19,** 71, 1956.
8. Khan, M. A. R. "A Decisive Test for the Presence of Even a Highly Rarefied Lunar Atmosphere," *Pop. Astr.*, **54,** 312, 1946.
9. Harris, I., and Jastrow, R. "Upper Atmosphere Densities from Minitrack Observations on Sputnik I," *Science*, **127,** 471, 1958.
10. Elsmore, B., and Whitfield, G. R. *Nature*, **176,** 457, 1955.
11. Costain, C. H., Elsmore, B., and Whitfield, G. R. *M.N.*, **116,** 380, 1956.
12. Elsmore, B. *Phil. Mag.*, **2,** 1040, 1957.
13. Kuiper, G. P. *The Atmospheres of the Earth and Planets*, chap. 12. Chicago: University of Chicago Press, 1949.
14. Edwards, W. F., and Borst, L. B. "Possible Sources of a Lunar Atmosphere," *Science*, **127,** 325, 1958.
15. Vestine, E. H. *RAND Corp. Research Mem. RM-2106*. 1958.
16. Herring, J. R., and Licht, A. L. "Effect of the Solar Wind on the Lunar Atmosphere," *Science*, **130,** 266, 1959.
17. Öpik, E. J. "The Density of the Lunar Atmosphere," *Irish Astr. J.*, **4,** 186, 1957.
18. Clement, G. R. *RAND Corp. Paper P-833*, 1956 revised.
19. Herring, J. R., and Licht, A. L. Paper presented before Symposium at National Aeronautics and Space Administration, Goddard Space Center, 1959.
20. ———. "Moon's Atmosphere," *Science News Letter*, **77,** 4, 51, 1960.
21. Bauer, C. A. *Phys. Rev.*, **72,** 354, 1948.
22. Huntley, H. E. *Nature*, **161,** 356, 1948.
23. Firsoff, V. A. *Science*, **130,** 1337, 1959.
24. Brandt, J. C. "Density of the Lunar Atmosphere," *Science*, **131,** 1606, 1960.
25. Chamberlain, J. W. *Ap. J.*, **131,** 47, 1960.
26. Elsasser, H. *Mitt. Astr. Gesellsch.*, **2,** 61, 1957.
27. Firsoff, V. A. "Density of the Lunar Atmosphere," *Science*, **131,** 1669, 1960.
28. Rishbeth, H., and Little, A. G. *Observatory*, **77,** 897, 1957.

29. Brandt, J. C. *Science,* **131,** 1671, 1960.
30. Öpik, E. J. *Irish Astr. J.,* **4,** 6, 1957.
31. Öpik, E. J., and Singer, S. F. "Density of the Lunar Atmosphere," *Science,* **133,** 1419, 1961.
32. Rubey, W. W. *Crust of the Earth,* ed. A. Poldervaart, p. 631. New York: Geol. Soc. America, 1955.
33. Brown, H. In *The Atmospheres of the Earth and Planets,* ed. G. P. Kuiper, p. 258. 2d ed. Chicago: University of Chicago Press, 1952.
34. Suess, H. *J. Geol.,* **57,** 600, 1949.
35. Stoney, G. J. *Ap. J.,* **11,** 251, 357, 1897.
36. Jeans, J. H. *The Dynamical Theory of Gases,* p. 342. 3d ed. Cambridge: Cambridge University Press, 1921.
37. Jones, J. E. *Trans. Cambridge Phil. Soc.,* **22,** 535, 1923.
38. Milne, E. A. *Trans. Cambridge Phil. Soc.,* **22,** 483, 1923.
39. Spitzer, L. In *The Atmospheres of the Earth and Planets,* ed. G. P. Kuiper, p. 211. 2d ed. Chicago: University of Chicago Press, 1952.
40. Gilvarry, J. J. "Physical Parameters of the Atmospheric Escape Layer," *Nature,* **188,** 804, 1960.
41. ———. "Origin and Nature of Lunar Surface Features," *ibid.,* p. 886.
42. ———. "Escape of Planetary Atmospheres. I. Escape Layer," *Physics of Fluids,* **4,** 1, 1961.
43. ———. "Escape of Planetary Atmospheres. II. Lifetimes of Minor Constituents," *ibid.,* **4,** 1, 1961.
44. Pickering, W. H. *The Moon.* New York: Doubleday, 1903.

19

THE LUNAR RAYS

Scattered widely over the lunar surface, including the back side, are hundreds of bright, fresh-appearing craters which show exterior appendages. These latter take three forms. Immediately surrounding the pit is a bright nimbus, which may or may not have a dark central halo just outside the pit. Beyond the nimbus extend rays, lighter in tone than the normal surface. Some small craters show only the nimbus.

The rays are generally, but not always, arcs of great circles. Usually they radiate from the area of the central crater, but numerous examples are known where the ray, projected backward, would miss the crater. Occasionally the rays take the form of loops or transverse segments. The rays diverge from the craters but usually remain roughly constant in width. They cross all types of surfaces, but where they cross elevations, they are often seen to be prominent on the side of the crest toward the crater, and "shadow zones" may sometimes be found in the opposite direction (1).

The ray craters are scattered essentially at random over the lunar disk. There are approximately equal numbers of them in each quadrant (2). This observation identifies them as being almost entirely or entirely postmare in date. This is confirmed when we notice that all ray craters are sharp and crisp in appearance. They are Class 1 craters, and it is also known that the Class 1 craters are distributed essentially at random over the bright and dark areas alike (3).

The ray craters are of all sizes. Among the large ones we find Langrenus,

Lunar Rays

Theophilus, Copernicus, and Tycho. The smaller ones are legion. They range down to craters too small to be resolved. A great many of the baby ray formations are known. Examination shows them to be quite like the larger forms in all but size.

Pease, using the 100-inch reflector, noted that practically every white spot showing on the Mount Wilson photographs stands out distinctly as a crater pit with its rim projecting above the surface; the bright rays are seen as the illuminated sides of low mounds, which always cast shadows in the same direction, as do the neighboring craters.

There must be something about the moon which causes astronomers and others to suffer severe attacks of imagination. The complicated theories of the origin of the rays offer good evidence of this effect. Over fifty years ago Tomkins (4–8) first proposed the saline efflorescence theory of the origin of the lunar rays. He was also the author of the laccolithic theory of the origin of the craters.

It seems that in northern India lies the Salt Range. This is a volcanic region. From this area run three streamers where the upward movement of water under a hot sun has led to evaporation and the deposition of salts. The first of these streamers runs south to the sea; one goes west over the Frontier Province and probably across Iran as far as Lake Van, while the third follows the watershed of the Ganges and Jumna rivers for over 800 miles. Similar markings are known in certain Asian deserts and in the Libyan Desert. While light-colored streaks have been formed on earth by this process, there is nothing to connect them with the lunar rays. The salt zones are not appendages of any crater, volcanic or otherwise.

Fauth (9) suggested that the rays were ice crystals blown through holes in the crater walls. In 1885, Nasmyth and Carpenter (10) felt that the shattering of the lunar globe around the focal points (craters) was the origin of the rays. They postulated that the force was a volcanic one and that the pressure was internal. Emission of lavas from the cracks was supposed to produce the actual rays.

Much later Darney (11) analyzed various theories of ray formation and suggested that the rays might perhaps be fractures in the lunar crust. In 1954, Alter (12, 13) adopted this point of view. In a long and thorough paper he defended this hypothesis vigorously. In his view, the cracks were induced in the lunar crust by the impacts of the great meteorites which produced the ray craters. He concludes (14):

If rays were due to clouds of dust puffed out from craters, they would: (a) be almost radial; (b) not form the patterns found in the Copernican area; (c) not follow the pattern of the Tychonian-Furnerian area; (d) not show the great common pattern of the Copernicus-Kepler-Olbers-Seleucus-Aristarchus triangle.

The most probable conclusions appear to be:

1. Short rays are associated directly with craterlike formations which brighten under a high sun.
2. Long rays are complex and often can be observed progressing from one crater to another.
3. The rays lie along cracks which themselves are not observable. The visible rays are either dust from gases that have escaped from the cracks, or a staining of the rocks by such gases.

In addition, Alter makes a great point of an observation that a long ray from Tycho seems to stop as it reaches a ray from Stevinus; and similarly a ray from Olbers, which passes along the northern boundary of the "disturbed area" northeast of Aristarchus, was instrumental in stopping that disturbance. Alter's observations are correct as far as they go, but his conclusions do not seem warranted. There are numerous cases of rays crossing other rays from different systems. There is no real indication of one ray system adversely affecting another as it probably would do if the globe were shattered by the impacts.

At the larger craters, the rays extend for hundreds of miles. At the Vredefort structure there are numerous radial cracks which resulted from the explosion. These cracks are of the order of 10 miles long (15). It would appear that there is no great tendency at meteoritic impacts to split the crust radially to the extent required by Alter's ray mechanism. The case is different at the tremendous impacts which led to the formation of the circular maria, but in these cases no rays have been found, although certain radial valleys have been identified which may be due to subsidence, presumably along radial fractures.

As a general thing, the rays become visible about 12 hours after sunrise and brighten for 1 or 2 days. They remain visible, slowly becoming more conspicuous until full moon. They disappear about 12 hours before sunset. Their visibility is primarily a function of the angle between the line of sight from the earth and the direction of light from the sun, not of the angle of illumination, as at full moon the rays abound all over the disk.

MacDonald (16) pointed out that the rays were visible both when the observer's line of sight and the sun's rays were parallel, at full moon and when they were perpendicular, at the quarter-phase. Pickering (17) noted that the rays became invisible only when the sun was within 10° of the local lunar horizon and that "it is only the angle of incidence [of light] which is of consequence." Lenham (18) reached a similar conclusion.

This conclusion is somewhat different from that reached just above. The distinction is that in the first case we are dealing with relative changes in the brightnesses of rays and ground and in the second case we are dealing with the absolute reflectivity of the rays as a function of the elevations of the sun.

Pickering's conclusion also is not universally true. The Keplerian rays can be seen easily right up to the terminator when only the top of Kepler's rim is illuminated. Similarly, the rays extending westward from Olbers may be seen clearly when Olbers is on the terminator.

Numerous attempts have been made to explain these observations. Buell and Stewart (19) ran laboratory experiments by preparing a surface of small particles of basalt with pulverized basalt simulating the rays. They concluded that the rays were streaks of dark, normal material mixed with powder. Naturally, the pulverized fragments have settled to the bottom of any interstices in the lunar surface.

Fielder (20) has questioned the validity of this solution, because Buell and Stewart did not test for limb effects; and it would appear that reproducible effects would not have been obtained. Fielder has also attempted to explain the visibility patterns of the rays on the basis of a model of the moon coated uniformly with hemispherical protuberances and a model in which the surface is coated with hemispherical concavities. Neither model gave a true representation of the rays, but a superposition of the two gave fair concordance. The second of the two Fielder models is similar to those proposed by Schoenberg (21, 22), Barabashov (23), Bennett (24), and others to account for variations in surface brightness of the moon as a function of phase.

Actually, some model must be devised so that the brighter material of the lunar rays remains visible to a terrestrial observer, regardless of its position on the moon, because rays may be seen clear out to the limb. Only a percentage of the ray material can be above the surface because the rays tend to disappear near the terminator when the sun's beams approach the local horizontal. It is reasonable to think that the lunar surface is rough, irregular, and vesicular. Models based on uniformity and symmetry will never fit the observations. It is clear that some of the bright ray material must lie on the surface, while the remainder is depressed somewhat below.

Back in 1907, Pickering (17) suggested two possible geometric solutions. He felt that the rays were due to a highly reflecting material lying in the bottom of cracks in the darker surface or that they were due to bright pellets on a darker surface. A combination of these two ideas gives the best fit to the observed brightness variations.

Recently, O'Keefe (25) proposed that the rays were composed of small transparent spheres, presumably of a glassy nature. Van Diggelen (26) demonstrated by laboratory measurements on such a model that glass beads do not give a very satisfactory agreement.

In 1906, Tomkins (27), who was not an advocate of the impact theory of the origin of craters, stated that many of the rays had minute craterlets along them. In 1949 (28), I amplified this observation to show that the rays contained dark as well as light matter. It was noted that, along many of the longer rays from Tycho and, in particular, Copernicus, there is a definite preference of small craters for the regions covered by the rays. Many of these "on-the-ray" craters are elongated and point roughly to the main crater. This is best shown southwest of Copernicus. Schwinner (29) also proposed a subsurface origin for rays, while Spurr (30) advocated a volcanic-ash theory.

Based on the information gained from the terrestrial explosion and meteoritic craters, it has been found that the materials ejected from a crater pit are largely pushed into place but that a small percentage of the mass is ejected great distances. At the Rieskessel, masses have been found at least 44 miles

TABLE 32

RAY CRATERS

Crater Name	Crater Diameter (Miles)	Diameter of Ray Pattern (Miles)	Crater Name	Crater Diameter (Miles)	Diameter of Ray Pattern (Miles)
Agrippa	28.5	170	Lalande	15.2	200
Alfraganus	12.8	215	Lansberg A	5.6	27
Anaxagoras	31.9	600	Langrenus	82.2	950
S. of Apollonius	3	43	Littrow B	2.0	16
Aristarchus	24.7	270	Lubiniesky C	10.0	43
Aristillus	35.1	400	Mädler	17.9	170
Autolycus	24.4	270	Manilius	24.0	200
Bode	11.3	85	Marco Polo B	5.0	32
Byrgius A	10.3	260	Menelaus	16.8	270
Campanus A	6.8	32	Mersenius C	9.5	49
Censorinus	4.2	38	Mösting A	8.2	32
Cleomedes A	8.1	54	Mösting C	2.0	11
Copernicus	56.7	750	Olbers	41.8	500
Crüger Y	2.7	16	Pickering	7.6	100
Darney	9.1	70	Proclus	19.1	400
Dionysius	11.0	86	Pytheas	11.8	32
Euclides	6.9	70	Stevinus A	15.6	400
Eudoxus A	8.7	65	Strabo	34.2	400
Euler	16.6	130	Taquet	4.6	16
Furnerius A	12.5	240	Taruntius	35.3	200
Gambart A	7.2	49	Theophilus	63.8	675
Geminus A	10.8	175	Timaeus	20.3	215
Godin	22.3	235	Timocharis	21.9	86
Hind C	10.8	86	Triesnecker	16.5	194
Kepler	20.2	400	Tycho	54.0	1900

from the center of the pit. In this case, the zone of ejectamenta is at least six times the radius of the crater. At all well-observed terrestrial meteoritic craters, there is a substantial production of rock flour, which is very light in color and highly reflective.

On the moon, the diameter of the visible halo of rays is found to be roughly proportional to the central crater diameter. The ratio is about 12 to 1. Rough scaling of representative systems from the *Photographic Lunar Atlas* of Kuiper yielded Table 32. The only ray crater which departs significantly from the mean line is Tycho, whose system of rays is unusually widespread.

A key observation is that the rays remain approximately the same width, regardless of the distance along the ray from the crater. Rays from larger craters are often wider than those from the smaller pits, but the rough constancy-of-width rule is generally observed.

This is interpreted to mean that an individual ray is not the result of a single ejection of material from a crater, which might be expected to spread but that each ray is composed of a series of component parts, all roughly alike. The materials for each ray were ejected at the time of the impact. Discrete bombs and accompanying dust were projected at low angles, with slight spreads in velocity and/or slight variations in altitude.

When these rays are closely studied, they are found to be composed of long, narrow, elliptical sections, often with a small crater or elongated groove in the white region. Pickering (31) observed that the long pair of Tycho rays directed toward Bullialdus were made up of a series of end-to-end components never longer than 15–80 km. Billerbeck-Gentz (32) reached similar conclusions about another of the Tychonian rays.

In some of the Copernican rays, the ray proper is not aligned with the crater, but all parts or segments of the ray point back to the crater. While the individual segments are diffuse, it appears that they point to the area of the crater rather than to the geometric center (33).

The velocities with which the bombs and ray materials were ejected must have been in the range of up to nearly 1 mile per second. The circular velocity at the moon's surface is 1.04 miles per second. This maximum velocity is lower than the minimum speed with which meteoritic bodies could strike the moon. Such a velocity of ejection is thus compatible with a meteoritic-collision mechanism, but this is not definitive, because certain terrestrial volcanoes have been observed to eject matter, occasionally horizontally, at comparable velocities (34–36).

Fielder (37) states that if ash were ejected explosively from a summit crater, one would expect to observe the streams pointing toward that crater, whereas, during a collisional process, matter would be projected from no clear focus.

In Shoemaker's paper on the ballistics of the Copernican ray system, he made the following statement (38):

It may be noted from Equation 14 that the range, as defined, can be set independently of the total energy E and the size of the crater, if the linear dimensions of the shock scale as the cube root of the energy. Thus we would expect the trajectories from small craters formed by small bolides to be almost as long as those from large craters formed by large bolides, if the impact velocities are similar. However, there is a rough correlation between size of crater and length of observable rays on the moon. This can be interpreted to mean that the rays are visible only out to the point where the density of ejected material is so sparse that it can no longer be photographed. The smaller craters would have shorter observable rays because the quantity of ejected debris is less.

This interpretation may be questioned. Figure 44 shows that, to a fair approximation, the diameters of ray systems are directly proportional to the diameter of the central crater. Table 11 (p. 148) shows that, to a first approximation, the amount of material moved from a crater and not deposited in the rim is slightly greater at small craters than at large craters. Observations indicate that all dimensions of rays from small craters are proportionately less than

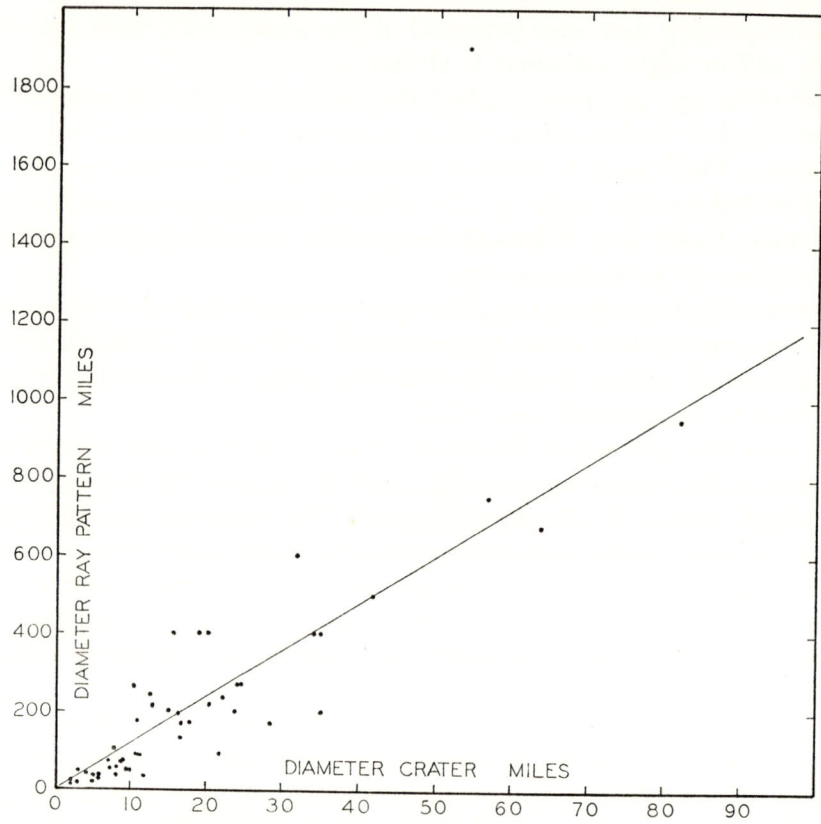

Fig. 44.—Relationship between diameter of ray pattern and diameter of central crater

at larger craters. In other words, the small ray systems are simply scaled-down versions of the large ray systems, and this scaling-down must have applied equally to the velocities with which the material was ejected.

If Shoemaker's model were correct, the velocities with which ray materials were ejected from small craters would be equal to the velocity of ejection from large craters if the meteoritic-impact velocities were the same, i.e., the ejection velocity is always proportional to the shock-wave velocity. This would not lead to ray systems scaled in proportion to the parent crater diameter.

It is concluded that the average ejection velocity of those masses which passed beyond the crater rims is a simple function of the total applied energy, not the shock-wave velocity. The scaling relationship is probably very similar to that for the craters proper, in which case the scaling exponent would be less than one-third.

In 1952, Urey (39) raised the question of possible circumlunar rays, with particular reference to a ray of Tycho running southward from the southeast wall of the crater. Urey attempted to determine the period of axial rotation of the moon when Tycho was formed from the lateral displacement of this ray. He estimated that the period of rotation was greater than 12 days at that time.

Examination of photographs shows that the material of this ray was traveling away from Tycho when it struck. If Urey's mechanism is correct, it implies that the ejected material traveled more than once around the moon. There are such severe dynamical objections to the formation of such a ray that acceptance of this interpretation does not seem warranted.

Giamboni (40) recently attempted to explain the doubling of certain rays in terms of an ejection at elevations both above and below that for maximum range. He concluded that, although no rays have been found which exhibited complete twin branches, partial twin branches have been found, and thus some of the rays could have been formed in high-angle and low-angle ejection.

This conclusion is highly questionable. If any material were ejected at high angles, it would fall at high angles to the horizontal. There would be little tendency for such bombs and powder to form long, elliptical, ray segments. Rather, they would be apt to produce circular or nearly circular craterlets, perhaps with a surrounding nimbus. In all suspected pairs of high- and low-angle rays, the eastern, or high-angle ray, is composed of segments effectively identical in appearance and structure with the low-angle ray. It must be concluded that the rays were always formed by low-angle ejections.

The shadow effects from ridges near Kepler and Proclus, for example, clearly indicate that the rays were ejected at very low angles. Similarly, the brightening of the limb region has been attributed to ray material which has been intercepted by ridges and crater walls.

Surrounding Tycho, there are plenty of examples of craterlets produced by material shot upward at very high angles. These craterlets are basically round. The appearance of this region resembles closely the Burton-on-Trent crater environs (Pl. XIX, p. 122).

Giamboni also tried to determine the period of rotation of the moon at the time when Tycho was formed, based on the observed offset of some of the rays to the east. He concluded that the period was between 0.5 and 6.8 days and that this period of rotation was achieved less than 10^8 years after the position of closest approach of the moon to the earth. This date is far earlier than can be assigned to the formation of the ray craters from any other point of view.

The conclusions of Urey and Giamboni that the rotation of the moon at the time of formation of a major ray system can be determined from the curvature and displacement of the rays are not accepted here. The conclusion here reached, and often stated previously, is that the rays are composed of light and dark material ejected at very low angles from the impact regions of meteorites on an airless moon. The bright material is composed of rock flour, whose color is determined primarily by the nature of the material pulverized. The Tycho rays are redder than those of craters formed on the maria (41).

Possibly some glasslike material is included with the rock flour. The Henbury main crater impact did produce small, tear-shaped drops of fused sandstone and some long, glassy threads, all with a high gloss surface (42).

Some of the light matter of the rays remains on the surface, some is partially hidden in the interiors of depressions, cracks, and other low regions.

The rays are subject to being covered by later-arriving material and probably to a darkening due to solar and cosmic radiations of all types (43). They must be considered to be transient features which appear only around relatively young impact craters.

Recently Cohen (44) has suggested that the discovery of megashatter cones (45)—these should be called mega "thrust" cones in Dietz's terminology—at the central uplift of the Kentland structure leads to a plausible mechanism for the production of certain lunar rays. Vand (46) had previously suggested that lunar rays were formed from dust ejected in narrow, hollow cones of roughly circular cross-section, inclined at small angles to the horizontal.

My own interpretation is that the lunar rays are from jets produced at the moon's surface at the instant of collision, before the crater was formed and before the meteorite had buried itself. Some solid material was shot out with the jets and produced "on-the-ray" craters. The offsets of certain rays would be due to rapid variations in azimuth and altitude of the stream of matter. Certain formations, such as the great elliptical rays of Copernicus, could have been formed from ejections coming shortly after the penetration of the lunar surface by the meteorite had begun. Some of the lunar rays may have been produced by dust generated from the moon's surface by the impact of solid materials ejected from the central crater, but it seems more likely that most of the dust was ejected as rock flour produced in the more violent main explosion. Where solid blocks were ejected with the ray dust, secondary impact craters would be formed. These secondary craters might also produce rays that would be concentrated on the side away from the primary crater. Numerous examples may be noted, particularly those associated with Copernicus and Tycho. Rosse is an excellent example of a secondary ray crater in the Tycho system.

References

1. FIELDER, G. "Some Aspects of Some Lunar Ray Systems and the Discrete Nature of the Rays," *J. Brit. Astr. Assoc.*, **66,** 223, 1956.
2. BALDWIN, R. B. *The Face of the Moon,* p. 159. Chicago: University of Chicago Press, 1949.
3. *Ibid.,* p. 158.
4. TOMKINS, H. G. "Note on the Bright Rays on the Moon," *J. Brit. Astr. Assoc.,* **18,** 126, 1908.
5. ———. *Ibid.,* p. 178.
6. ———. *Ibid.,* p. 215.
7. ———. *Ibid.,* p. 361.
8. ———. *Ibid.,* p. 386.
9. FAUTH, P. *The Moon in Modern Astronomy.* London: Owen, 1907.
10. NASMYTH, J., and CARPENTER, J. *The Moon.* New York: Scribner & Welford, 1885.
11. DARNEY, M. "Vers une explication rationelle des rayonnements lunaires," *Astronomie,* October, 1934, p. 154.
12. ALTER, D. "Nature of the Lunar Rays," *Pub. A.S.P.,* **67,** 237, 1955.
13. ———. Paper read at joint meeting A.S.P. and Section D, A.A.A.S., Berkeley, December, 1954.
14. Ref. 12. p. 245.
15. DIETZ, R. S. "Astroblemes," *Sci. American,* **205,** 51, 1961.
16. MACDONALD, T. L. "Those Lunar Rays," *J. Brit. Astr. Assoc.,* **55,** 161, 1945.
17. PICKERING, W. H. "The Bright Streaks upon the Moon," *J. Brit. Astr. Assoc.,* **17,** 25, 1907.
18. LENHAM, A. P. "The Copernicus Ray System," *J. Brit. Astr. Assoc.,* **65,** 241, 1955.
19. BUELL, E. N., and STEWART, J. Q. "A Laboratory Duplication of the Lunar Rays," *Pop. Astr.,* **40,** 264, 1932.
20. FIELDER, G. *Structure of the Moon's Surface,* p. 156. London: Pergamon Press, 1961.
21. SCHOENBERG, E. "Untersuchung zur Theorie der Beleuchtung des Mondes auf Grund photometrischer Messung," *Acta Soc. Sci. Fenn.,* **50,** 9, 1925.
22. ———. "Eine neue Beleuchtungstheorie des Mondes," *Hdb. d. Ap.,* **2,** Part 1, 76, 1929.
23. BARABASHOV, N. "Über die Reflexion des Lichtes an der Mondoberfläche und an porösen Flächen," *A.N.,* **221,** 289, 1924.
24. BENNETT, A. L. "A Photovisual Investigation of the Brightnesses of 59 Areas on the Moon," *Ap. J.,* **88,** 1, 1938.
25. O'KEEFE, J. A. "Lunar Rays," *Ap. J.,* **126,** 466, 1957.
26. DIGGELEN, J. VAN. *Rech. Obs. Utrecht,* **14,** 93, and dissertation, Utrecht, 1959.
27. TOMKINS, H. G. "The Bright Rays on the Moon," Part 1, *J. Brit. Astr. Assoc.,* **16,** 359, 1906.
28. Ref. 2, p. 163.

29. Schwinner, R. "Die hellen Strahlensysteme des Mondes, geologisch gedeutet," *A.N.*, **274,** 137, 1943.
30. Spurr, J. E. *Geology Applied to Selenology,* **3,** 90. Concord, N.H.: Rumford Press, 1948.
31. Pickering, W. H. "An Investigation of the Systems of Bright Streaks Visible upon the Full Moon," *A.N.*, **130,** 225, 1892.
32. Billerbeck-Gentz, F. "Unterbrechungserscheinungen am Mondstrahl Tycho-Polybius-Fracastor-Rosse," *A.N.*, **274,** 140, 1943.
33. Fielder, G. "On the Origin of Lunar Rays," *Ap. J.*, **134,** 425, 1961.
34. Wright, F. E. "The Surface Features of the Moon," *Ann. Rept. Smithsonian Inst.*, p. 169, 1935.
35. Marshall, R. K. "The Origin of Lunar Craters," *Pop. Astr.*, **51,** 415, 1943.
36. Hacker, S. G., and Stewart, J. Q. "Remarks on Lunar Ray Craters," *Ap. J.*, **81,** 37, 1935.
37. Fielder, G. *Structure of the Moon's Surface,* p. 152. London: Pergamon Press, 1961.
38. Shoemaker, E. M. "Ballistics of the Copernican Ray System," *Proc. Lunar and Planetary Exploration Colloquium,* **2,** 13, 1960.
39. Urey, H. C. *The Planets.* Oxford: Oxford University Press, 1952.
40. Giamboni, L. A. "Lunar Rays: Their Formation and Age," *Ap. J.*, **130,** 324, 1959.
41. Barabashov, N. P., and Chekirda, A. T. "Otsvete svetlykh luchey Kraterov Tikho, Kopernik i Kepler" ("The Color of the Light-colored Rays of Tycho, Copernicus, and Kepler"), *Tsirk. Khar'kovskoy Astr. Obs.*, No. 13, p. 3, 1955.
42. Alderman, A. R. "Meteorite Craters at Henbury, Central Australia," *Mineralog. Mag.*, **23,** 19, 1932.
43. Stair, R., and Johnston, R. "Ultraviolet Spectral Radiant Energy Reflected from the Moon," *J. Res. Nat. Bur. Stand.*, **51,** 81, 1953.
44. Cohen, A. J. "Central Uplifts of Terrestrial and Lunar Craters. II. Megashatter Cone Mechanism for Ray Formation," paper presented at first western national meeting, Planetary Sciences Committee, American Geophysical Union, Los Angeles, California, December, 1961.
45. Cohen, A. J., Reid, A. M., and Bunch, T. E. "Central Uplifts of Terrestrial and Lunar Craters. I. Kentland and Serpent Mound Structures," *ibid.*, 1961.
46. Vand. *J. Brit. Astr. Assoc.*, **55,** 47, 1945.

20

THE CENTRAL PEAKS OF LUNAR CRATERS

Invariably, when a volcanic caldera on the earth possesses a so-called central peak, this peak is in the form of a reasonably symmetrical cone. Erosion may modify the cone in time, but the form is standard. Usually the "central peak" is placed very eccentrically in the crater or caldera. Wizard Island in Crater Lake, Oregon, is a beautiful example. All such cones are crowned with a relatively tiny crater. Structures of this type may easily be distinguished from those like the Halemaumau pit in the floor of Kileaua.

The literature on the craters of the moon is replete with references to similar structures in the centers of many of the lunar craters. Only the careful observers have noted in print that the central peaks are not counterparts of the terrestrial volcanic forms. Wilkins and Moore, in their book *The Moon* (1), which describes their great map, give numerous accurate descriptions of the lunar craters and their peaks. They point out that very often the peaks are multiple. Even so, they feel that the peaks are volcanic.

The symmetrical form is nearly absent on the moon. Alpetragius stands almost alone in this respect; and its peak is a dome, not a cone, and even here there is a blister-like subsidiary dome to the north.

The lunar central peaks are not only multiple, they are jagged in the extreme. Plate C2d in the *Photographic Lunar Atlas* clearly shows Aristillus to have a cluster of seven major peaks covering about one-quarter of its floor. These peaks give the impression of a great jumble of angular blocks. The description is very comparable with that for the central uplift area of the Serpent Mound impact structure in Ohio.

In no case is a central peak, multiple or single, of such a height that it reaches to the original ground level. The highest peaks, such as that of Moretus, are about 7,500 feet high, much less than the true depth of the crater. All degrees of complexity and of prominence exist. Certain craters have no central

TABLE 33

CENTRAL PEAK DATA FOR CLASS 1 CRATERS

Diameter (Miles)	Group	Total Craters	No Peak	Per Cent with No Peak	Single Peak	Per Cent with Single Peak	Multiple Peaks	Per Cent with Multiple Peaks	Questionable Peak	Per Cent Questionable Peak	Total Peaks	Eccentric Single Peak	Eccentric Multiple Peaks
150+	1												
100.1–150	2												
80.1–100	3	2	0	0	0	0	1	50	1	50	1	0	1
70.1– 80	4	3	0	0	1	33	2	67	0	0	3	1	2
60.1– 70	5	1	0	0	0	0	1	100	0	0	1	0	0
55.1– 60	6	4	1	25	0	0	3	75	0	0	3	0	0
50.1– 55	7	6	0	0	1	17	5	83	0	0	6	1	2
45.1– 50	8	5	1	20	2	40	2	40	0	0	4	1	1
40.1– 45	9	14	2	14	7	50	5	36	0	0	12	4	4
35.1– 40	10	9	0	0	1	11	8	89	0	0	9	0	3
30.1– 35	11	8	0	0	3	38	5	62	0	0	8	2	1
25.1– 30	12	22	4	18	12	55	6	27	0	0	18	6	5
20.1– 25	13	26	2	8	10	38	13	50	1	4	23	3	4
15.1– 20	14	25	3	12	14	56	6	24	2	8	20	3	5
10.1– 15	15	17	3	18	9	53	5	29	0	0	14	2	2
5.1– 10	16	29	20	69	5	17	1	3	3	11	6	0	0
– 5	17	36	11	31	0	0	0	0	25	69	0	0	0

peaks. This does not mark a new or different type of crater but is only one extreme of a whole gamut of crater forms. The central peak massif of Theophilus, containing four major mountains, completely covers the floor area of the crater.

Table 7 (Appendix 2) contains coded descriptions of the central peaks of over 300 craters. This table is by no means complete, but it contains almost all the known geometrical data concerning lunar craters. There probably is a sufficient number of examples of craters with and without central peaks in all size ranges that general relationships may be derived. Tables 33–38 summarize the data from Table 7.

First and foremost, it is noted that many of the central peaks or multiple central peaks are not placed exactly in the center of the pit. It is suggested that this eccentricity is a measure of the angle of impact of the object which produced the crater. It is also suggested that these peaks are rebound phenomena from the transfer of momentum of the meteorite to the slightly compressible layers of the moon. The rebound became fixed as a dome, often intensely shattered. This dome is believed to overlie a shattered and brecciated uplifted lens such as has been regularly observed at large, old terrestrial impact structures. Cohen (2) has suggested that the entire central peak structure at lunar craters is a

TABLE 34

CENTRAL PEAK DATA FOR CLASS 2 CRATERS

Group	Total Craters	No Peak	Single Peak	Multiple Peaks	Questionable Peak	Total Peaks	Eccentric Single Peak	Eccentric Multiple Peaks
1								
2	2	0	0	2	0	2	0	1
3	2	1	0	1	0	1	0	1
4	1	0	0	1	0	1	0	0
5	1	1	0	0	0	0	0	0
6	3	1	1	1	0	2	0	0
7	4	0	1	3	0	4	1	1
8	3	1	2	0	0	2	0	0
9	5	1	1	2	1	3	0	1
10	2	1	1	0	0	1	1	0
11	4	2	1	1	0	2	1	0
12	4	1	0	3	0	3	0	1
13	7	2	2	2	1	4	1	1
14	2	1	0	1	0	1	0	0
15	1	0	0	1	0	1	0	1
16	1	0	0	1	0	1	0	1
17								

TABLE 35

CENTRAL PEAK DATA FOR CLASS 3 CRATERS

Group	Total Craters	No Peak	Single Peak	Multiple Peaks	Questionable Peak	Total Peaks	Eccentric Single Peak	Eccentric Multiple Peaks
1								
2	1	0	0	1	0	1	0	1
3	2	1	0	0	1	0	0	0
4	3	2	0	1	0	1	0	1
5	3	1	2	0	0	2	1	0
6	3	0	1	2	0	3	1	2
7	2	1	0	1	0	1	0	0
8	1	0	1	0	0	1	0	0
9	3	2	1	0	0	1	1	0
10	2	1	0	1	0	1	0	0
11	3	0	1	1	1	2	0	1
12	1	1	0	0	0	0	0	0
13								
14								
15								
16								
17								

megamegashatter cone (or megamega"thrust" cone). Personally, I prefer the concept of a rebound mechanism for the production of central peaks to any other mode.

Figure 45 shows in graphic form some of the relationships which are imbedded in Table 33 for the Class 1 craters. It is shown that there is essentially no difficulty in telling whether or not a crater has a central peak until the crater

TABLE 36

CENTRAL PEAK DATA FOR CLASS 4 CRATERS

Group	Total Craters	No Peak	Single Peak	Multiple Peaks	Questionable Peak	Total Peaks	Eccentric Single Peak	Eccentric Multiple Peaks
1	.	.	.	.	.	.	.	.
2	1	1	0	0	0	0	0	0
3	1	0	0	1	0	1	0	1
4	.	.	.	.	.	.	.	.
5	1	0	1	0	0	1	0	0
6	.	.	.	.	.	.	.	.
7	.	.	.	.	.	.	.	.
8	.	.	.	.	.	.	.	.
9	.	.	.	.	.	.	.	.
10	.	.	.	.	.	.	.	.
11	.	.	.	.	.	.	.	.
12	.	.	.	.	.	.	.	.
13	.	.	.	.	.	.	.	.
14	.	.	.	.	.	.	.	.
15	.	.	.	.	.	.	.	.
16	.	.	.	.	.	.	.	.
17	.	.	.	.	.	.	.	.

TABLE 37

CENTRAL PEAK DATA FOR CLASS 5 CRATERS

Group	Total Craters	No Peak	Single Peak	Multiple Peaks	Questionable Peak	Total Peaks	Eccentric Single Peak	Eccentric Multiple Peaks
1	1	1	0	0	0	0	0	0
2	3	2	0	1	0	1	0	1
3	5	2	1	2	0	3	1	2
4	6	1	2	3	0	5	2	3
5	4	1	2	1	0	3	0	1
6	2	1	1	0	0	1	0	0
7	3	1	1	1	0	2	1	1
8	4	2	1	1	0	2	0	0
9	1	1	0	0	0	0	0	0
10	6	4	2	0	0	2	1	0
11	4	1	1	2	0	3	0	1
12	6	4	1	1	0	2	0	1
13	9	5	2	2	0	4	1	0
14	4	1	1	2	0	3	0	1
15	1	0	1	0	0	1	0	0
16	.	.	.	.	.	.	.	.
17	.	.	.	.	.	.	.	.

diameter becomes less than 5 miles. A great many of the large craters possess central peaks. The percentage of those which do not increases rather steadily with decreasing crater size.

Among craters between 5.1 and 10 miles in diameter, a very great majority do not possess such peaks. No certain peaks have been detected in 36 examples of craters 5 miles and less in diameter. It may be that a few have escaped discovery, but basically these smaller lunar craters are peak-free. This observation ties in very nicely with the observed fact that terrestrial meteoritic craters in the same size ranges do not show central peaks. The only exception is a tiny irregularity in the center of the Kaali Järv.

TABLE 38

ECCENTRIC CENTRAL PEAK DATA FOR CLASSES 1–5 CRATERS

Group	Total Peaks	Total Eccentric Peaks	Per Cent Eccentric Peaks	Total Single Peaks	Total Eccentric Single Peaks	Per Cent Total Eccentric Single Peaks	Total Multiple Peaks	Total Eccentric Multiple Peaks	Per Cent Total Eccentric Multiple Peaks
1									
2	4	3	75	0	0	0	4	3	75
3	6	6	100	1	1	100	5	5	100
4	10	9	90	3	3	100	7	6	86
5	7	2	29	5	1	20	2	1	50
6	9	3	33	3	1	33	6	2	33
7	13	7	54	3	3	100	10	4	40
8	9	2	22	6	1	17	3	1	33
9	16	10	62	9	5	56	7	5	71
10	13	5	38	4	2	50	9	3	33
11	15	6	40	6	3	50	9	3	33
12	23	13	56	13	6	46	10	7	70
13	31	10	33	14	5	36	17	5	29
14	24	9	38	15	3	20	9	6	67
15	16	5	31	10	2	20	6	3	50
16	7	1	14	5	0	0	2	1	50
17									

When we examine the craters which possess single central peaks, we find a strange distribution. The larger craters rarely show single peaks. There are some exceptions, such as Moretus. The percentage of craters with single peaks increases steadily with decreasing crater size to a crater diameter of about 30 miles. Here about 40 per cent of the pits show a single central peak. At still smaller craters, the percentage drops and reaches zero at about a 5-mile crater.

The curve is somewhat different for those craters with multiple central peaks. At the larger craters, the percentage seems to average higher than 50 per cent, although relatively few craters are involved. This percentage of multiple central peaks increases to about a diameter of 60 miles, where perhaps 80 per cent of the craters show these peaks. Thereafter, the percentage declines steadily and reaches zero at a 5-mile crater.

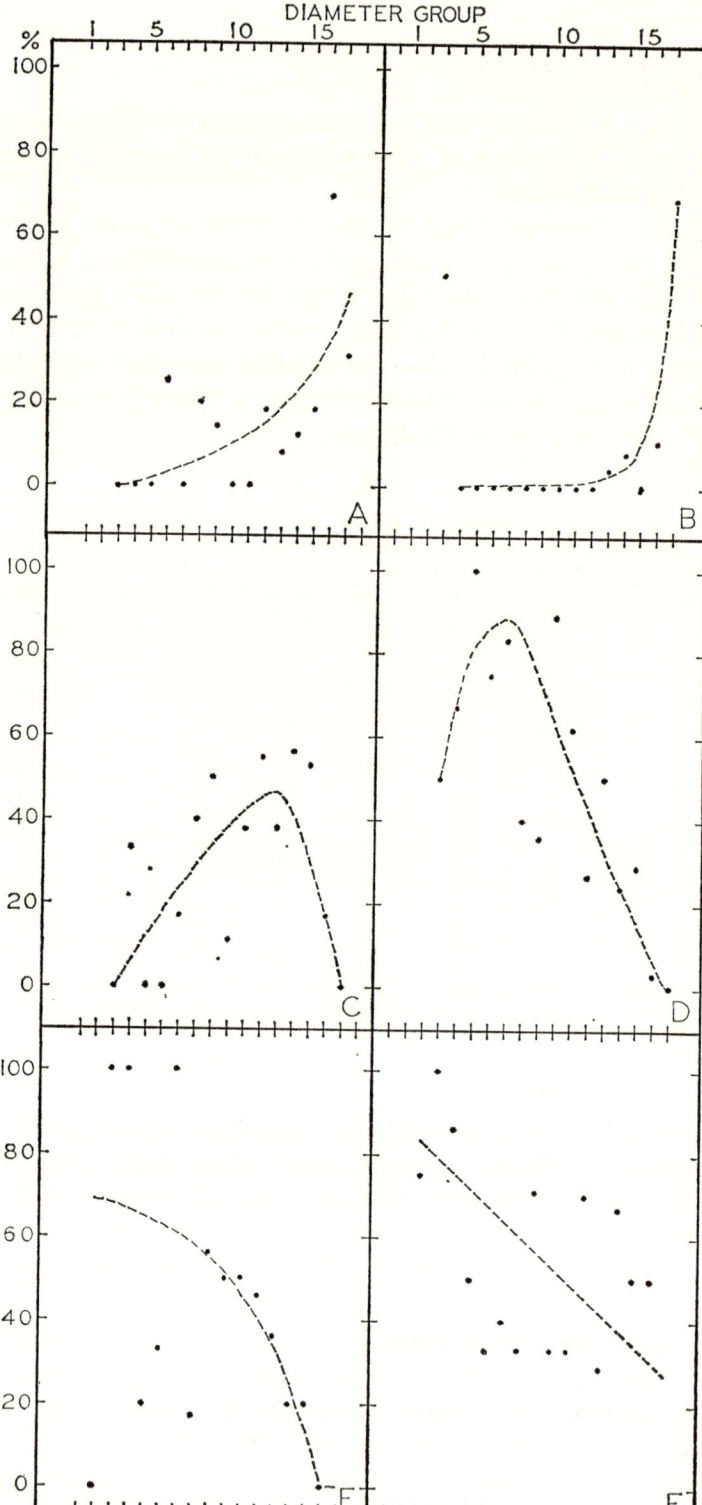

FIG. 45.—Correlations between crater diameter and central peak statistics. Abscissa for each panel is diameter group from Table 33. *A*: percentage of craters which do not possess a central peak; all questionable cases omitted. *B*: percentage of craters where the existence of a central peak is questionable. *C*: percentage of craters which have a single central peak. *D*: percentage of craters which have a multiple central peak. *E*: percentage of craters with single central peak where the peak is eccentrically placed. *F*: percentage of crater with multiple central peaks where the peaks are eccentrically placed.

It is noted at all large craters that the frequency of central peaks declines from a maximum as the crater size is increased. This is consistent with the observation that none of the circular maria shows a central peak. It is tentatively suggested that, at each of these great craters, the impact lasted so long, because of the size of the colliding body, that a rebound was constrained by the continuing impact of the rear portion of the meteorite.

A slight difference between the single- and multiple-peak craters is noticed in the percentages of peaks that are eccentrically placed in the crater. At the single-peak craters, a high percentage of the large pits have eccentrically placed peaks. The percentage drops steadily with decreasing crater size and reaches zero—i.e., the peaks are central—at small craters. As a general rule, for any size crater, a larger percentage of the multiple peaks are eccentrically placed than the single peaks.

The data for the conclusions on the eccentric nature of all central peaks were derived from Table 38, which contains combined data from all five classes of craters. From these data, the percentages of the craters in Classes 1 through 5 larger than 5 miles in diameter which do show central peaks are, respectively, 75, 67, 54, 33, and 54. The Class 1 craters are the newest, Class 4 the oldest, while the Class 5 craters are lava-filled. The general average for all classes combined is 68 per cent. These data are from personal examination of the craters in Table 7 on all photographs of the *Photographic Lunar Atlas*.

The idea that the central peak of lunar craters is a volcanic phenomenon is an attractive one, and many observers have accepted it. It received strong backing after it was found that, on occasion, there was a small craterlet near the top of the peak. At this point, it is pertinent to remember that shadows exaggerate the vertical scale of these mountains and that, while they may be high and jagged, the average slope is relatively low.

Campbell (3) seized upon the discovery of the central peak craterlets to state: "Must we not agree that the unquestioned existence of craterlets in the summits of central crater peaks is absolutely fatal to the impact theory of the origin of those peaks and at the same time in full and complete harmony with the hypothesis of the volcanic upbuilding of those peaks?" This is a conclusion which sounds logical, but, like so many of the quick conclusions regarding lunar phenomena, it does not stand up under analysis.

Moore, who is a volcanic enthusiast, has made a thorough study of the central peaks of lunar craters. He has endeavored to discover as many as possible of the central peak craterlets. He reports that over sixty are known (4). His claim that these little pits are symmetrically placed on the central peak is sometimes correct but often is not. The central peak craterlets in Timocharis and Regiomontanus, for examples, are not central. Fielder (5) even questions the existence of a summit crater in the small central block of Tycho. The great majority of these summit craters are very small and require rather large telescopes to see.

Young (6) has compiled a listing of 1,281 craters between 15 and 135 km in diameter (10–85 miles), lying within 70° of the moon's center and named craters outside this limit. Fielder (7) indicates that the total of such craters over the lunar disk should be multiplied by about 3. This gives a total of approximately 3,843 visible craters in this range of sizes.

Earlier in this chapter, it was shown that about 68 per cent of craters show central peaks. Because many of the unlisted craters are very old, Classes 3 and 4, or lava-filled, let us arbitrarily drop the percentage to 50 per cent for this calculation. On this basis, there are about 1,900 craters with central peaks.

The sizes of the summit craterlets are in the range of $\frac{1}{3}$–2 miles. Consequently, the area of the top of the central peak must be of the order of 2–3 square miles; it can scarcely be less. Multiplying 1,900 by 2 or 3 square miles, we find 3,800–5,700 square miles of central peak top lands over the visible half of the moon.

Now if we had a means of determining the frequency of small impact craters on the moon within the range of sizes indicated, we would have a measure of the frequency with which small craters were formed and thus a measure of the frequency with which such craters would be found on a central peak of a larger crater. The requirement would be a moderately flat area, unmarked by ancient craters. The floors of the great impact craters, Ptolemaeus and Alphonsus, seem ideal.

Intercomparison between the photographs of these two craters on Plate D5a of the *Photographic Lunar Atlas* and Dinsmore Alter's Plate 277, taken in infrared light on October 26, 1956, and the famous September 15, 1919, Mount Wilson plate gave evidence that at least 195 of these tiny craters may be identified on the floor of Ptolemaeus and 76 on the floor of Alphonsus. Since the areas of the lava-covered floors of these craters are 4,340 and 2,050 square miles, respectively, the observed frequency of these tiny pits is one per 22 square miles for Ptolemaeus and one per 27 square miles for Alphonsus.

The evidence is that many of these craters are meteoritic in nature. Some possess small ray systems. Other small craters may be of internal origin or are older than the lava; they show no rims. Let us assume one small meteoritic crater larger than about $\frac{1}{2}$ mile wide in each 50 square miles.

From these data, it is concluded that if the area of the central peak top is 2 square miles, the number of meteoritic craters of the requisite size which would be formed on a larger crater's central peak summit since the lava flows appeared would be $3,800/50 = 76$ craterlets, and if the summit area is 3 square miles, the number would be raised to $5,700/50 = 114$.

Presumably, the Class 1 craters and these tiny meteoritic craters have been formed continuously since the maria appeared. In this event, the above numbers might have to be reduced slightly but not more than 10 per cent.

The existence of small summit craters to a number of the order of more than 100 is a necessary requirement of the meteoritic-impact theory based on the observed numbers of small impact craters per unit area. Their presence cannot be used as an argument for a volcanic origin for the larger craters or their central peaks.

From this analysis, it would seem that Wilkins, Moore, and others have not found enough central peak craters by possibly a factor of 2. More work should be done in this field. An impact crater formed on a slope may be considerably distorted and hence difficult to recognize.

This argument has been considerably minimized here. To give a truer picture, we would have to consider that about half the central peaks of craters are multiple, and so the number of available summits must be multiplied by at least 2. Conversely, if the peaks with summit craters be considered as volcanic peaks, why are they so jagged and irregular, and why do only a small percentage of central peaks show such crowning craters? If they are volcanic, they should all have summit craters.

The arguments of this chapter should not be taken as indicating that no volcanic cones exist on the moon. Not one has ever been positively identified as a true volcano, although various such suggestions have been made, and it is entirely possible that some volcanic cones do exist.

There are other types of central peak craters or crater-like formations. There are the purely random larger craters, which, by accident, have been formed in the same general area as an earlier crater; and, on occasion, the newer and smaller crater lies where the central peak of the larger crater ought to be. Examples of this are Zagut and Cassini, but numerous others are known.

A second variety is typified by Plinius, Bohnenberger, Timocharis, and Lambert. A crater-like structure, perhaps up to 2 or $2\frac{1}{2}$ miles in diameter, has partially eliminated part of the central peak of each of these craters. Possibly some of these objects may be attributed to later impacts, but many must be of different origin. Perhaps they fall into the following category.

About half the central peaks are found to be multiple. Often the peaks are arranged roughly in a triangle or box or part of a circle around a center near the center of the crater. Examples are Atlas, Posidonius, Aristoteles, Manilius, Maurolycus, and Copernicus.

Intermediate between this class and the one above, are such craters as Eratosthenes and Theophilus, in which non-symmetrical valleys in the multiple central peaks resemble craters.

It is entirely possible that the violence of the impact and rebound has simply led to a cluster of blocks being raised in lieu of a central peak and distributed in the fashions observed; but another possibility exists.

Hendriks (8), in his very thorough study of the Crooked Creek structure in Missouri, showed that a rebound occurred, and the rocks became fixed as a structural dome, but, a section at the top and center of the dome, a mile in diameter, then collapsed to form a large (relatively) and shallow crater.

It would seem entirely possible that a similar action occurred at many of the larger lunar craters and that the rebound dome center collapsed and led to the variety of multiple central peaks that we find with low central regions and occasionally the appearance of an actual central crater, nearly obliterating the central peak.

Alternatively, it is noticed that the Timocharis and Plinius type of crater with large central peak craters all lie on the lava flows. If they were formed soon after the hardening of the lavas, escape of gases may have contributed to their modifications.

References

1. WILKINS, H. P., and MOORE, P. *The Moon*. London: Faber & Faber, Ltd., 1955.
2. COHEN, A. J. "Central Uplifts of Terrestrial and Lunar Craters. II. Megashatter Cone Mechanism for Ray Formation," paper presented at first western national meeting, Planetary Sciences Committee, American Geophysical Union, Los Angeles, 1961.
3. CAMPBELL, W. W. "Notes on the Problem of the Origin of the Lunar Craters," *Pub. A.S.P.*, **32**, 126, 1920.
4. MOORE, P. *Guide to the Moon*, p. 114. London: Collins, 1953.
5. FIELDER, G. *Structure of the Moon's Surface*, p. 233. London: Pergamon Press, 1961.
6. YOUNG, J. "Statistical Investigation of the Diameters and Distribution of Lunar Craters," *J. Brit. Astr. Assoc.*, **50**, 309, 1940.
7. Ref. 4, p. 219.
8. HENDRIKS, H. E. *The Geology of the Steelville Quadrangle, Missouri*. State of Missouri, Dept. of Bus. and Admin., Div. of Geol. Surv. and Water Resources, 1954.

21

RILLES, WRINKLES, AND FAULTS

The lunar rilles have a message to tell. They are great valleys whose parallels have not been found on earth. The system of rift valleys in Africa and the beautiful Finger Lake basins in New York State have certain points of resemblance to the rilles, but the terrestrial grabens are substantially different. Although their lengths are similar, their relationships to each other are quite distinct. There are no known terrestrial equivalents to the rilles.

A few estimates of the sizes of these lunar rilles have been made. Schmidt (1) said that they were usually 20–100 miles long and $\frac{1}{3}$–$2\frac{1}{2}$ miles wide and 300–1,300 feet deep. Later Schmidt (2) found the south part of Schröter's valley to be 1,600 feet deep. Fielder (3–6) has measured the Ariadaeus Rille to be between 1,900 and 3,800 feet deep at a point near the middle. The width at the same place is about 3 miles. Pickering's (7) earlier values for this rille are in good agreement.

Some observers have thought that the edges of the rilles are raised (8–10), but Fielder (5) reports that there is no concrete evidence that the walls of normal rilles, as distinguished from crater chains or crater rilles, rise above the general level of the surrounding ground. Fielder has also shown that the rilles are always shallow in depth relative to their widths. He has drawn the conclusion from this that the rilles are not cracks. A key observation is that in several places the large rilles are not continuous but are *en echelon*. The Ariadaeus and Hippalus Rilles are prime examples.

Fielder has been one of the most active proponents of the lunar grid system. He has lumped together the lunar rilles, the numerous faults, and the ridges into a non-homogeneous system. He feels that there is a fundamental connection between the rilles, faults, and mountain striae (5, 11–13). It is his hypothesis that the rilles follow pre-existing thrust and wrench faults.

A fault may be classified as a thrust, strike-slip, or normal fracture. If the lunar crust is subjected to stress, this stress may be resolved into three components, two horizontal and one vertical. A thrust fault occurs when one of the horizontal components is strongest. In this case, a portion of the surface may ride up over another. A strike-slip fault is of similar origin, but here one portion of the surface will slide past another adjacent area. A normal fault occurs when the strongest component is the vertical one. This leads to the raising or lowering of one part of the surface relative to a neighboring portion.

The fracture plane is always inclined less than 45° to the direction of the strongest stress component. Fielder notes that the angle is about 23° for granites and basalts. At normal faults, the plane of the throw is between 45° and 90° to the horizontal in a direction perpendicular to the least principal stress. Normal faults involve tension, thrust faults involve compression, while the strike-slip fault results from combination of forces.

Because the two most prominent rilles on the moon—Ariadaeus and Hyginus—and their sibling rilles are not associated with high mountains, Fielder sought their interpretation in terms of thrust and wrench or strike-slip faults resulting from horizontal forces. To follow Fielder's argument, it is necessary to quote from one of his papers (3) and to reproduce one of his drawings as Figure 46.

Look first at the sketch of the region, noting how Ariadaeus and Hyginus are linked by another rill, which is a key feature in the interpretation I propose. Then observe how the rills have directions related to those of nearby ridges, which seem to form three families. The members of each family run roughly parallel, though the families themselves have three different directions.

Often, the ridges form parts of the walls of craters, many of which have been deformed almost beyond recognition. The broken curves in the sketch map indicate their probable original outlines. These ruined craters clearly antedate the distortions and must therefore be very ancient.

It seems plausible that these ridges or striae represent overthrusts or folds caused by unequal horizontal pressures. Heretofore several writers, including myself, have suggested that the prominent striae of the best-developed family (running roughly northeast to southwest) originated as part of a collisional explosion centered in Mare Imbrium. That hypothesis cannot account for all the parallel linear formations simply by the grooving action of projectiles and by the splashing silicates, since it is evident that many of the ridges have a less superficial origin.

A second, less well-developed family runs perpendicular to the first, while the third and possibly weakest set goes from north-northeast to south-southwest. We shall assume that these three families originated from transverse horizontal pressures, P_a, P_b, and P_c, respectively, each acting independently of the other and at different periods of lunar history. Their directions, indicated on the sketch, have been determined from the trends observed for the ridges and striae in the Ariadaeus-Hyginus region, but independently of the directions of the rills.

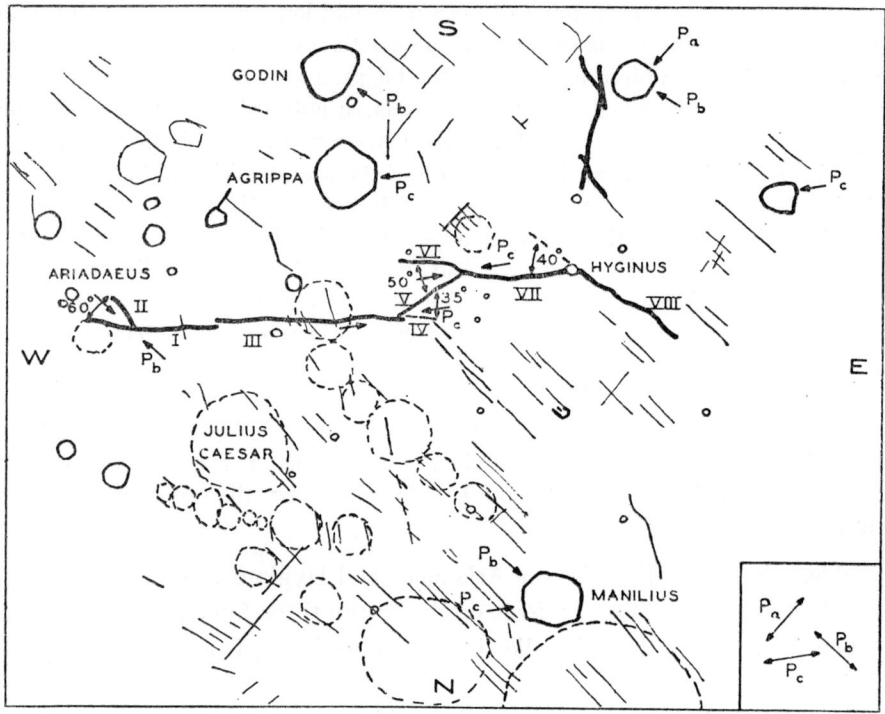

Fig. 46.—Sketch of Ariadaeus-Hyginus region. (Fielder, Ref. 3.)

In the sketch, parts of the two main rills have been labeled with Roman numerals, starting at the western end of Ariadaeus, where I and II join at an angle of about 60 degrees. Farther east IV and V meet at about 35 degrees, while V and VI (Hyginus rill) make an angle of about 50 degrees. The direction of the transverse pressure P_b approximately bisects the first angle (I–II), and P_c bisects each of the other two. If we consider a shifting of the lunar surface within these angles with respect to the surface outside of them, it is evident that the associated rills could follow wrench faults.

Such fractures are usually more nearly perpendicular to the surface than normal faults; thus, they may tap a deeper layer of the moon than other faults do. Presumably the moon contains, or used to contain, at least isolated chambers of magma. Can we consider rills to be dikes, where molten material has risen along these already existing fractures?

If so, the rills would appear to be of two kinds: those associated with normal fractures and the mountain masses, and the wide, prominent rills of the Ariadaeus-Hyginus system that originated deeper in the moon. In the two cases, the intruding magmas may have differed in chemical composition and physical conditions, yielding rills of different dimensions. On Earth, dikes are rarely intruded along wrench fractures, because the latter are held closed by compression, but such a fracture could be opened by subsequent tension of the surface layers.

After the fractures formed, relief of pressure would lower the melting point of any trapped magma. If its specific gravity was greater than that of the surrounding rocks, the melt would not, subject to equal horizontal pressures, reach the surface of the moon. The sides of the fractures would be displaced outward, yet no such movement of the *surface* has been detected. Whatever movement of this kind that may have occurred is probably not more than one-tenth of the widest rills's width.

Rock debris may be seen on the slopes and bottoms of the broader rills, and the longitudinal ridges within the rills often appear to be blocky and of irregular contour. It would seem that the rising magma had undermined the essentially fragmentary surface layer of the moon, and then subsided. Rocks were engulfed and detached, and there was further subsidence, so the rill became considerably wider than the original dike.

The ridges across the Ariadaeus rill slumped into it as a result of this subsidence. Meanwhile, the magma tended to a level consistent with hydrostatic equilibrium, thereby forcing up the central longitudinal ridges. The same mechanism probably arched the cross ridges to some extent. The diagram shows the proposed sequence of events.

Fielder's interpretation ties the Ariadaeus and Hyginus rille families into the same stresses as those that he feels produced the three families of ridges nearby. The strongest of these families is the one radiating from a center in Mare Imbrium.

I most definitely do not agree that these ridges or striae represent overthrusts or folds caused by unequal horizontal pressures. The entire system of northeast-southwest-trending ridges is a direct by-product of the explosion which produced Mare Imbrium. Most of this system was formed by flying debris, but there may have been some subsidence along the lines radial to Mare Imbrium. These lines are not parallel to any part of the great rille system, except possibly the eastern end of the Hyginus Rille. The other two systems are extremely weak and, as Fielder says, were produced at different periods of the lunar history.

If we take his hypothesis at face value, we would say that the west end of the Ariadaeus Rille was formed at one time and the portion at the eastern end and connections to the Hyginus Rille were produced at a different time. The strongest of the three ridge systems was good only for distorting certain craters such as Triesnecker.

Now, clearly, the system radial to Mare Imbrium was associated with the origin of Mare Imbrium. The origin of Mare Imbrium was prior to the lavas. Triesnecker, Manilius, Hyginus, and other similar craters were formed after the lavas. Therefore, it is quite difficult to picture a generic relationship between these rilles and the distortions of these craters and the great ridge and valley system radial to Mare Imbrium. Many of the radial valleys are partially lava-filled.

There is no more need to postulate that the distortions of lunar crater shapes are related to subsequent horizontal stresses than there is to postulate that the non-circular shapes of the Aouelloul and Arizona craters are due to later acting forces.

The numerous sharply detailed markings on the lunar surface yield no indication of horizontal movements as great as half a mile (13, 14). MacDonald (16) confirms this observation and suggests that the absence of horizontally displaced features on the surface of the moon means that there are no large-scale strike-slip faults or wrench faults on the moon. On the earth there are a number of such examples, in which relative horizontal displacements of tens or perhaps even hundreds of miles have taken place.

The Ariadaeus and Hyginus Rilles cannot be considered as isolated examples. Hundreds of similar structures are known in many places on the moon's surface. Figure 47 shows the major families of these rilles. The drawing is one made from examination of the various photographs of the *Photographic Lunar Atlas* edited by G. P. Kuiper. It is by no means complete but does include the major trenches. Rilles confined to a single crater are not included. The most obvious relationship among the objects is that the major rilles form one great family. They are closely associated with the edges of the maria.

The borders of Mare Serenitatis and Mare Humorum are particularly well defined. In each case, parallel rilles mark the borders. The Sirsalis Rille is distantly parallel to the edge of the latter sea. The rilles are among the latest objects on the moon's surface. They are somewhat later than the maria themselves.

The markings surrounding Mare Serenitatis, Mare Humorum, and indicated at Mare Imbrium and Mare Nectaris strongly suggest the ring faults surrounding many terrestrial meteoritic craters and the Vredefort structure. It is noted also that the Ariadaeus-Hyginus system is parallel to the southern edge of Mare Serenitatis and to the ridge interpreted as a shock wave from the Serenitatis impact.

Circumferential rilles probably mark the sites of great cracks formed at the times of the impacts causing the circular maria. They seem to mark the places where the material was stressed beyond its elastic limit. After the radial shock front had passed, a slight rebound occurred. Because the stresses were not uniform in all directions, the cracks resulted. The rilles subsequently developed in these weakened locations when isostatic adjustments caused crustal movements.

The pattern is not random. It is clearly associated with the lava sheets. In every major case, the rilles appear in places where the crust of the moon has bent in accomplishing its enforced isostatic adjustment. The rilles are tension cracks or tension faults.

How they reached their present appearance is a secondary matter. They are deep-seated phenomena which ignore surface details almost completely. Apparently they were formed over an extended period of time, for in the Triesnecker area are cracks of similar width which cross each other. Cracks or faults will not do this unless they originate at widely differing depths or unless, as seems more probable, the first set was old enough that its fill had become reconsolidated so as to transmit stress.

A simple experiment with modeling clay will duplicate many of the features of the Ariadaeus Rille system. Bend the clay slowly, and cracks will develop on the convex surface. If a very slight twist is incorporated into the bending, the cracks will appear in *en echelon* formation. The steps will be offset away from the greater bending stress. By analogy, the Ariadaeus Rille was formed by bending of the lunar surface downward in the northerly direction and was the result of tensions in the lunar surface. The greatest amount of bending was in the northwest. In this general direction lie the two great lava masses, Mare Tranquillitatis and Mare Serenitatis. They are each closer to the rille than is Mare Imbrium.

How the bottoms of the rilles developed is still not clear. Fielder is correct when he states that the rilles are relatively shallow, that they have central ridges on occasions, that there is no appreciable separation of the surface, and that the surface layers have slumped into the trench rather than having the rilles filled from below. Fielder (22) suggests:

A release of pressure occurred at some depth in the Moon. (That this process must have operated is evident in view of the great amount of faulting which has occurred.) This caused local melting and an appreciable increase in the volume of the material ensued [17] which was forced—partly by hydrostatic pressure—up fractures marking planes of faulting, and which began to crystallize during the ascent, with stoping and widening of the magma chamber [18].

In the case of normal rilles, the melt undermined the fragmentary surface layers, causing them to subside and produce rilles much wider than the original fractures. However, in the case of crater rilles, conditions in the melt were such that crystallization resulted in the confined accumulation of high pressures in the water vapour phase. These high pressures were released by explosions which produced craters within the rilles. Urey [19] has envisaged a similar type of mechanism for the crater rilles, and has correctly pointed out that craters produced by igneous explosions would probably be indistinguishable from meteoritic explosion craters. Pickering [7] noted that crater cones occurred on cracks in Kilauea "much the same as crater pits do on the Moon." Klein's [20] belief that the normal rilles and the crater rilles have evolved along very different lines is not supported. The author's theory has been expressed more fully elsewhere [3, 4], but the notion that the rilles are subsidence features—and not cracks —is considered to be an observational fact. This conclusion was reached before Kuiper [21] published a statement that the Ariadaeus Rille is a *graben* feature, and it is very pleasing to find agreement on this point.

Returning to the Ariadaeus Rille it is seen that, had the viscosity of the melt been such that it would not support the cross ridges, some slumping would have ensued, as observed. The convexities and central ridges within the normal rilles are thought [3, 4] to have arisen as a consequence of the pressure of the rising magma acting on the overlying materials.

The ascending magmas should have risen to greater altitudes in mountainous regions than elsewhere, but the narrowing and rough nature of rilles in such regions could be explained by a weakening of the underlying structure and the consequent settling down of solid blocks of rock.

Many, though not all, of the characteristics of rilles have been accounted for in the foregoing discussion. The association of rilles with faults has been explained satisfactorily, but the notion has emerged that the rilles themselves are not indicative of a surface-tensile phase of the Moon's history, as has been believed up to date.

All theories of terrestrial grabens require faults or cracks along which movement can occur. Presumably this is also true of the moon. The rilles give every indication of having been initiated by cracking or faulting of the surface.

Subsequent to the cracking, the surface collapsed. This can be accounted for by at least two mechanisms. The original cracks presumably were not perpendicular to the ground. The width of the crack may have been of the order of 100–500 feet. At a certain distance from the surface along the crack, the overhang became so great that it broke loose and slid downward, forming the rille. If the action were quick enough, a central ridge might have been produced in this ground slip by shock action.

It is more difficult to see how Fielder's stoping and widening theory can account for the great uniformity in width of a rille when the surface varies considerably in height along the length of the rille. It is also a question why the depths of the rilles seem to be proportional to their widths, regardless of their location on the moon.

In most cases there is no overlapping of the time when the rilles developed and the time when the lavas came, but there does appear to be one exception. A long rille parallels the front of the Apennines in Mare Imbrium. Where it crosses the blocky areas which have slipped from the Apennine scarp, the rille is deep and easily visible. When it crosses the lava of Palus Putredinis and nearby lava tongues, it practically disappears, although a very irregular rille still closer to the Apennines is clearly visible. Possibly this region marks one of the last lava flows on the moon.

Sometimes, but not always, related to the usual rilles are the so-called crater rilles. Numerous examples are known, and they are somewhat more diverse in apppearance than the rilles. The eastern branch of the Hyginus Rille seems to be formed, at least in part, from numerous crater-like forms. None of these is truly a crater. Hyginus itself does not come at the bend of the rille, as is often stated, but is a mile or two west of the bend. It is doubtful that Hyginus was instrumental in the formation of the trench.

The great and famous crater chain running from Stadius to Mare Imbrium is composed completely of small aligned craters. These craters were formed along an existing crack, and the eruptions which formed them probably were primarily gas venting or gas explosions. Little solid material was ejected in these eruptions, for no cinder cones were formed and there is no evidence of local lava flows from the vents.

Immediately south of Palisa and east of Ptolemaeus is a large and ancient ring. Crossing its center is a narrow rille with at least six small craters strung along its length. Apparently, Ptolemaeus G on its west end is not an associated crater. Five of the six craters are found on the lava-filled interior of the large ruined crater; the sixth is on the ruined rim. The best pictures of this object show dozens of small transverse markings, which are so numerous as to be almost continuous (Pl. XXV). On the east edge of Mare Humorum is a crater chain which blends into a rille.

The greatest of the crater-chain–rille combinations lies in the uplands and is quite informative (Pl. XXVII). It starts east of Herschel and crosses the north rim of Ptolamaeus. The feature can be traced for about 50 miles as it cuts through the Ptolemaeus rim. It is not prominent here but can be traced easily on Plate C5a of the *Photographic Lunar Atlas*. This observation dates the rille as younger than Ptolemaeus. West of Ptolemaeus the rille is replaced by five contiguous irregular craters in the same line. Perhaps the rille disappears as it passes north of Albategnius and on to the west, but if one sights along the same slightly curved line, there are indications of a depressed trench.

Immediately south of Abulfeda and going through its rim, the crater rille reappears in full strength. It continues in an almost straight line and crosses the Altai scarp some miles south of Tacitus. It is marked by numerous small craters. The larger crater superimposed on the rille near Almanon is probably a later formation. It is not quite central on the trench.

It is at the Altai scarp that the strangest part of this crater rille is found. The rille and associated small craters are normal east of the mountains, but, as the feature crosses the mountains, the rille and craters become uniformly smeared out until the whole feature is three times as wide as formerly. The widening is perpendicular to the direction of the rille. The rille also becomes offset to the north by a few miles just as it crosses the mountains. The interpretation of the Altai scarp is that of a great shock wave. The smearing may be interpreted as an effect of this shock wave crossing the fault.

The dating sequence would then be, first, Ptolemaeus and Abulfeda; then the formation of the rille; then the gas venting, which led to the small craters on the rille; then the Mare Nectaris impact, which caused the formation of the Altai Mountains and, in doing so, severely modified the great rille.

Observations of this strange feature at either end indicate that its mode of origin was different from the usual and more abundant rilles. It was formed before the lavas came and before the great circular maria were formed.

Although there are relatively few of these prelava crater rilles, they do exist and indicate an expansion of the moon and a degassing in the period before the lavas were extruded. Both mechanisms would be expected if the moon were becoming hotter in this period deep below the surface.

Warner (23) has recently discovered a series of 18 broad, shallow valleys which he calls "rilles." They are radial to Plato and Plato D. These objects are broad and long, but most of them are no more than a few hundred feet deep.

Inside many of the lava-filled craters are small rille systems, duplicating on a small scale the major systems on the edges of the maria.

The wrinkle ridges are formations almost the reverse of the rilles. True, both of them are phenomena of the lava flows and are approximately of the same age, but here the resemblance ceases. The rilles are trenches depressed below the surface; the wrinkle ridges are raised above the surface. The rilles usually are linear formations; the wrinkle ridges are often quite irregular. Each rille is reasonably constant in width; the wrinkle ridges are often quite variable in width. The rilles are usually individual structures, although they may appear as parts of a multiple system; the wrinkle ridges very often are complex rather than being single. The major rille systems are closely related to the birth processes of the maria and are generally found associated with the edges of the lava sheets. The wrinkle ridges also exhibit definite patterns but are more closely confined to the inner parts of the lava sheets.

Fielder (24, 25) and van Diggelen (26) have measured the heights of two of the wrinkle ridges. Fielder examined the ridge across the neck of lava joining Mare Imbrium and Mare Serenitatis. He found the top to be irregular and perhaps 800 feet high, and it is about a dozen miles wide. Van Diggelen made profile measures on a wrinkle ridge on Mare Imbrium northeast of Eratosthenes. His measures were photometric rather than geometric, but the height obtained was similar to Fielder's measure, 650 feet, and the width was nearly

20 miles. With such broad, low features, the slopes are very gentle, and it requires a low sun to make them appear prominently.

The drawing of Figure 48 was made from the *Photographic Lunar Atlas*. Undoubtedly, it is not a complete representation of all the large wrinkle ridges, but the major systems are clearly shown. One outstanding fact appears on this drawing. In three major and well-defined cases, the wrinkle ridges clearly outline places where high ridges of prelava ground are believed to exist under the lava.

Immediately north of Hellplain is an ancient and tremendous crater, well over 100 miles across. The western half is rather well defined and includes Thebit in its reaches.

It is well known that the ground slopes down toward the lava in this region, and hence one would expect the eastern half of the crater, if it still existed, to be buried below the lava. What we actually find is that the entire lost wall of this crater is outlined by a wrinkle ridge in the lava.

The second object—Sinus Iridum—is of comparable size. On our model, Sinus Iridum was formed on the edge of Mare Imbrium before the lavas came. Half the crater would be in the great pit, half on the high wall. The lost half of Sinus Iridum was drowned by the rising lavas and still remains lost to sight but is marked by a very definite series of wrinkle ridges, which outline the rim and, in doing so, go counter to the curvature of the Mare Imbrium ridges nearby.

The third and largest of these wrinkle-ridge systems lies in north-central Mare Imbrium. It has been suggested earlier that the original impact crater of Mare Imbrium was not the outline of the present lava flow but was a smaller ring, which was apparently marked by the circular ring of mountain peaks projecting above the lavas. This great ring of peaks is set off nicely by a nearly circular family of wrinkle ridges which lies slightly inside the peaks.

Inspection of Figure 48 shows that each of the circular maria—Mare Serenitatis, Mare Crisium, Mare Nectaris, and Mare Humorum—has a circular ring of wrinkle ridges essentially central in the lava flow. Mare Humboldtianum has a suggested ring of this type.

All the other maria are of the irregular type and do not have a regular central ring of wrinkles but show irregular ridges. The limb seas may be exceptions to this, but, because of their positions, they cannot be so studied. How can these observations be interpreted?

A wrinkle in a surface implies a compression effect. Previously we interpreted the rilles as tension by-products. These two observations would make the edges of the maria places of tension after the lava congealed and their interiors places of compression.

The contour map also shows that the lava flows are depressed areas, and, the farther from shore we get, the greater is the amount of depression. In no case is the depression relative to a sphere sufficient to make the lava flow a concave surface relative to a plane.

To make the method to be proposed to account for the wrinkle ridges palatable, we draw on two additional bits of information. In the region between Ukert and Eratosthenes, although by no means limited to this area, is a vast field where old markings are softened and muted. They look as though they had been partly melted and partly filled with a soft material (Pl. XXVIII).

It is suggested that the lavas were once sufficiently deep to cover this area and that they quickly withdrew back into the body of the moon before a significant crust could could be formed. In so withdrawing, the region shallowly covered would be drained like the sea beach at ebb tide. This withdrawal of lavas back into the earth is a well-known process. In discussing the recent eruption of Kilauea Iki, Eaton and Murata (27) say:

> At its highest stand, at the end of the eighth phase, the lava pond was 414 feet deep and contained 58 million cubic yards of lava. At the end of each phase the fountain died abruptly, and from the 2nd to the 16th phase, a mighty river of lava surged back down the vent as soon as the fountaining stopped. . . . Of the 133 million cubic yards of lava spewed out into Kilauea Iki crater during the eruption, only 48 million cubic yards remains in the 367-foot-deep pond. The other 85 million cubic yards poured back underground almost as soon as it collected in the Kilauea Iki lava pond, where its volume could be so conveniently measured.

There is similar evidence of such withdrawals at the great lava flows which formed the lava plateaus in Oregon.

Now if this process occurred on the moon, we have a mechanism to explain the observations in Sinus Aestuum and also one which may held to explain the sinking of the surface at the deeper lava flows. This could occur whether the lavas were made of one sheet of magma or of many.

Geometrically speaking and regardless of the local directions of gravity, the lava flows on the moon are great domes. If the crust bends under the weight of the lava, giving isostatic adjustments, we will have tensions at the edges; but whether the adjustment is isostatic or due to the withdrawal of still liquid lava beneath the crust, there must develop compressions in the central regions. The surface will tend toward a larger radius of curvature, and so the total area of the lava surface will become smaller, and wrinkle ridges must form. The wrinkle ridges were formed largely in those places where the lavas were thinner than nearby and where subsurface obstructions may have inhibited uniform adjustments to changing stresses.

These mechanisms operating concurrently—the bending of the crust, isostatic adjustments made necessary by the superposition of masses of dense lavas, and the partial withdrawal of some of the lavas into the body of the moon—seem both qualitatively and quantitatively able to account for the present variations in lava levels, the muted areas, the rilles, and the wrinkle ridges.

In general, the rilles avoid the areas of the wrinkle ridges and vice-versa, as we would expect from the difference in their modes of origin, but occasionally one seems to connect onto and turn into the other. Two examples are the Stadius Crater chain rille, which seems to change into a Mare Imbrium wrinkle ridge, and Fielder's wrinkle ridge at the passage between Mare Imbrium and Mare Serenitatis, which appears farther north as a rille.

Between Hercules and Eudoxus lies a small, lava-covered area which, strangely, has been given the name Lacus Mortis. In spite of the name, it is evident that it has been an extremely active part of the moon's surface. Its lava filling is less dark than most of the maria. A large crater, Bürg, was later formed near its center. Parts of the surface have sunk, and domelike features have been detected. The surface is covered with a fretwork of fine rilles. A relatively long one in the southern portion east of Bürg is of particular interest.

If we examine this feature, starting at the north end, it appears as a normal, narrow rille. After progressing about 20 miles in this fashion, the western side suddenly rises relative to the eastern and reaches a height of several hundred feet. It is unknown which side actually moved. The high offset continues for another 20 miles and then subsides into a normal rille just before leaving Lacus Mortis. There is a low peak at each end of the offset (28). A similar offset fault, associated with rilles, is to be seen in Boscovich (29).

The best-known example of faulting on the moon is the famous Straight Wall. This feature lies entirely within the tremendous unnamed ruined crater which includes Thebit within its bounds. The fault is high on the west. Ashbrook (30) has measured the wall height at seven points, finding that it rises abruptly out of the plain on the north and reaches a height of 270 meters (1,205 feet) at the approximate center and gradually becomes lower as the southern end is approached. The southern tip of the Straight Wall fines out in a raised section of the surface, apparently a part of a prelava crater wall. Schmidt's measures (31) at two points are confirmed by Ashbrook's work.

The line of the fault lies closely parallel to the general trend of the lava shore line. The lower side of the fault lies in the seaward direction. The sinking induced by the mass of the superimposed lava sheets has been interpreted as bringing into being the normal rilles. Apparently, the tensions induced here have split the crust, but, instead of resulting in a normal rille, the entire side of the fault sank, producing a tremendous wall. Somewhat farther east is a small rille, a bit irregular but lying roughly parallel to the Straight Wall.

Schmidt (32) measured the inclination of the Straight Wall to the horizontal as 40°. Patrick Moore (33) found it to be 41°.0. Ashbrook (30) finds that the

FIG. 47.—Major families of rilles

Fig. 48.—Major families of wrinkle ridges

Plate XXVII.—Ancient rille, Ptolemaeus-Altai region. (Mount Wilson.)

PLATE XXXVIII.—Muted area between Ukert and Eratosthenes. (Yerkes.)

slope lies between 36°.4 and 48°.2 and probably is about 41° ± 3°. As he points out, this is an extremely steep slope for any large-scale lunar feature.

The existence of such a fault scarp on the lava is still more evidence that there has been no continuing large-scale movement of dust into the maria since they were formed. The picture is consistent. The fault of the Straight Wall is a structure produced by the same tensional forces as those which resulted in the rilles.

On both sides of Cauchy are long normal rilles, each exhibiting the *en echelon* effect. They start in the mountains to the west of the crater and continue even farther beyond Cauchy. The southernmost of these rilles is the more interesting. It is important because it is the only prominent rille that shows a major difference in elevation of the two sides. The side toward Cauchy toward the shore, toward the north, is the higher. The lower side is directed toward the open mare, toward the greater weight of lava, and toward the area which has done the sinking. The Plates B4a and B4b of the *Photographic Lunar Atlas* show this difference in height clearly.

References

1. Schmidt, J. F. J. *Der Mond*, p. 82. Leipzig, 1856.
2. ———. *Über Rillen auf dem Monde.* Leipzig, 1866.
3. Fielder, G. "A Theory of the Origin of Lunar Rilles," *Sky and Telescope*, **19,** 334, 1960.
4. ———. *J.I.L.S.*, **1,** 166, 1960.
5. ———. *Structure of the Moon's Surface.* London: Pergamon Press, 1961.
6. ———. "Le Fond de la rainure 'Ariadaeus,'" *Proc. Soc. Astr. de France*, Juin, 1956.
7. Pickering, W. H. "Lunar and Hawaiian Physical Features Compared," *Mem. Amer. Acad. Arts Sci.*, **13,** 173, 1908.
8. Klein, H. J. "Verschiedene Typen der Mondrillen. Part I," *Sirius*, **32,** 7, 1899.
9. Smith, C. F. O. "Visibility of Lunar Clefts beyond the Apparent Terminator," *J. Brit. Astr. Assoc.*, **58,** 53, 1948.
10. Wilkins, H. P., and Moore, P. A. *The Moon*, pp. 70, 173. London, 1955.
11. Habakov, A. V. *Grundlegende Fragen der Entwicklung der Mondoberfläche.* Moscow.
12. Bülow, K. von, "Lunare und tellurische Fundamentaltektonik," *Wiss. Zs. U. Rostock, math.-naturwiss. Reihe*, Heft 1, Sec. 8, Jahrgang **19,** 1958.
13. Firsoff, V. A. *Strange World of the Moon.* London, 1959.
14. Baldwin, R. B. *The Face of the Moon.* Chicago: University of Chicago Press, 1949.
15. Urey, H. C. *The Planets.* New Haven, Conn.: Yale University Press, 1951.
16. MacDonald, G. J. F. "Stress History of the Moon," *Planetary Space Sci.*, **2,** 249, 1960. London: Pergamon Press.

17. YODER, H. S. "Change of Melting Point of Diopsode with Pressure," *J. Geol.*, **60**, 364, 1952.
18. DALY, R. *Igneous Rocks and the Depths of the Earth*. London, 1933.
19. UREY, H. C. Private discussion with G. FIELDER, March 7, 1957.
20. KLEIN, H. J. "Verscheidene Typen der Mondrillen. Part II," *Sirius*, **32**, 27, 1899.
21. KUIPER, G. P. *Vistas in Astronomy*, ed. M. ALPERIN and H. F. GREGORY, Vol. 2, Part V. London: Pergamon Press, 1959.
22. Ref. 5, p. 214.
23. WARNER, B. "Rilles near the Lunar Crater Plato," *J. Brit. Astr. Assoc.*, **70**, 299, 1961.
24. FIELDER, G. "Studies in Lunar Topography. IV. Measured Profiles of the Moon's Surface, and Estimates of the Magnitudes of the Errors in Relative Altitudes," *Tech. Sci. Note, No. 4, Contract No. AF 61 (052)-168, U.S.A.F.*, p. 49. Manchester, November, 1959.
25. Ref. 5, p. 24.
26. DIGGELEN, J. VAN. "A Photographic Investigation of the Slopes and the Heights of the Ranges of Hills in the Maria of the Moon," *Bull. Astr. Inst. Netherlands*, **11**, 283, 1951.
27. EATON, J. P., and MURATA, K. J. "How Volcanoes Grow," *Science*, **132**, 932, 1960.
28. HERRING, A. K. "Observing the Moon—Bürg and Lacus Mortis," *Sky and Telescope*, **19**, 357, 1960.
29. HERRING, A. K. "Observing the Moon—Boscovich," *Sky and Telescope*, **21**, 39, 1961.
30. ASHBROOK, J. "The Lunar Straight Wall," *Pub. A.S.P.*, **72**, 55, 1960.
31. SCHMIDT, J. F. J. *Charte der Gebirge des Mondes . . . Erläuterungsband*, pp. 40 and 58. Berlin: Dietrich Reimer, 1878.
32. Ref. 31, p. 190.
33. Quoted in Ref. 30, p. 58.

22

THE LUNAR GRID SYSTEM

Several authors have published charts of a "grid system" which is spread over the entire visible side of the moon. These charts were all produced independently, and all show certain points of agreement. They are by Habakov (1), von Bülow (2, 3), Firsoff (4), and Fielder (5, 6). Spurr (7) also shows a partial grid system.

Apparently, the only criterion used in the drawing of these charts[1] is that a nearly linear formation be found. Included in the same charts by the same symbol are elongated valleys, ridges, faults, rilles, and sections of crater walls. Habakov even included the pressure ridges on the maria.

Now it is entirely possible that there are generic relationships among these dissimilar formations, but, at the very least, they should be segregated and discussed independently and then correlated with conclusions reached elsewhere. There is no question that most of the features drawn do exist. All the charts show very prominently the great system of ridges and valleys which are radial to Mare Imbrium. Although the authors do not agree on the mode of origin, they do agree that the radial formations are related to the origin of Mare Imbrium.

Similar, but much less prominent, fans of markings are radial to Mare Humorum and Mare Crisium, and the vast system of giant valleys, including

[1] All are reproduced in *Structure of the Moon's Surface* by Fielder.

the Rheita Valley, is radial to Mare Nectaris. The Mare Nectaris valleys appear to be subsidence types as, perhaps, is the Alpine valley. Other markings around the circular maria are from flying debris. Both ridges and valleys are formed in this fashion. The ridges are often observed to be formed from existing crater walls and other formations which projected above ground.

Fielder (8) has objected to this interpretation as it applies in the Fra Mauro region, but I cannot accept his conclusions. The effect of moving matter ejected from the Imbrium explosion on the remnant walls of Fra Mauro and other nearby structures is exactly what would be expected. We must remember that the lava flows had not appeared at this time. The ridges are the portions of the walls remaining after the missiles had gouged out valleys between the ridges.

Included in the grid system are the sides of various craters. Often the terrestrial meteoritic craters are polygonal rather than truly circular. Many of the lunar craters are also polygonal. Four-, five-, and six-sided craters are common, or at least they tend toward such shapes.

Added to this is the effect of flying missiles from Imbrium. These will always act to create a spurious alignment of the crater rims on two sides parallel to the flight of the missiles, so that the two crater sides will appear to point toward Mare Imbrium. The walls of the crater on the sides toward and away from the explosive site will be furrowed. This has been discussed previously for Ptolemaeus, but it is equally true of many other craters, such as those in the Fra Mauro region.

Firsoff (9), Fielder (6, 10, 11), and others (12) have attempted to explain the polygonal characteristics of many lunar craters by distortions after they were formed, due to crustal movements.

This conclusion does not appear warranted, inasmuch as the previously discussed reasons are amply able to account for the polygonal shape of craters and also because there is no evidence of slippage along faults outside the craters. This would be a necessary accompaniment of such crustal movements. Shoemaker (13) has suggested that the sides of the Arizona Crater owe their alignment to pre-existing structural lines of weakness in the local rocks. If true and if applied to the lunar craters, it implies that certain polygonal shapes were conditioned by earlier structural lines of weakness, not later distortional forces. Fielder's assumption (18) that the numerous elongated, depressed features with raised rims in the general area of Mare Vaporum and Julius Caesar are ancient craters which have been distorted by the forces which produced the grid system is not sustained. Clearly, these markings were formed by masses ejected from the region of the explosion which produced Mare Imbrium. The markings have been called "splash craters" (19–20). They are elongated craters gouged out of the moon's surface by low-velocity, low-angle fragments.

They are post-Imbrium, prelava structures and clearly antedate the nearby rilles. They are similar to the great grooves which have been slashed through the Haemus Mountains by missiles from the same source. Their elongated forms are the same as when they were originally shaped. There is no suggestion of horizontal crustal movement along wrench faults in this area either before or after the lavas came.

The point is that there is a large number of markings radial to the circular maria and associated with the forces released at the origins of these maria and the ejectamenta from these explosions. If these radial markings are eliminated, there does not remain very much of the lunar grid system. The radial markings developed because of the tremendous impacts from space, not because of a settling or expansion of the entire moon.

Some of the remaining markings are real and associated. Starting at the north end of the Rheita valley is another prominent valley which makes an angle of about 30° with the larger formation. North of it and extending beyond Janssen is a series of nearly parallel subsidence valleys. They form an associated system which is only coincidentally in the same region as the Mare Nectaris system. The Russian photographs (14, 15) of the other side of the moon do not cover the region from which these marks may be thought to radiate.

Spurr (7) introduced the term "lattice pattern" to describe the intersecting ridge systems which may be found on the north and south sides of certain craters. Among these he listed Eratosthenes, Cyrillus, Rhaeticus, Sacrobosco, Pontanus, Gemma Frisius, Timocharis, Copernicus, Reinhold, Posidonius, and Aristoteles. Fielder concluded that these lattice patterns occurred even more frequently than one would be led to believe by reading Spurr's work, ". . . and, furthermore, that the patterns always form local portions of the general grid system" (16).

Close examination of the best photographs indicates that, in the regions surrounding Eratosthenes, Copernicus, Timocharis, Reinhold, and Aristoteles, the ridge pattern is radial to the main crater and was formed at the time the crater was formed. In the Timocharis and Reinhold areas are also to be found markings radial to Copernicus. The markings south of Cyrillus are radial to nearby Theophilus. Markings near Rhaeticus are in the Imbrium system. Sacrobosco, Pontanus, and Gemma Frisius are in the same general area, and faint markings radial to Mare Nectaris pass near them. Posidonius, on the north, shows markings radial to the nearby Class 1 crater, Daniell, while in the south are rilles of the Mare Serenitatis system.

Fielder (11, 17) shows a Pic-du-Midi photograph of the region south of Arzachel. There are clearly three sets of linear markings in this area. The best examples are from the Imbrium system, but the other two, at right angles to each other and at 45° from the Imbrium marks, are definite. They do not appear to be intimately associated with any nearby crater or mare. Some of them are valleys, and some are ridges.

A structure such as this, spread widely over the lunar surface, might legitimately be called a lunar grid system, but, to the present time, no evidence has been forthcoming to indicate the existence of such a universal grid. The moon is covered with a network of linear markings, but they are mostly associated with specific lunar craters and circular maria which were formed at widely differing times. In the case of the rilles, they were formed incident to the settling of the lava flows.

A moon-wide grid system may exist, but, if so, it is very faint and is subsidiary to the more prominent local systems. At present we are not warranted in drawing conclusions on lunar history from the relatively few and inconspicuous linear markings not identified as belonging to local systems.

A weakness in all the lunar grid charts published is that the markings are shown as straight lines, even out to the limbs. Consequently, the true delineation of the marks as drawn is that of curved lines in the limb regions, not parts of great circles, as appears to be the case nearer the center of the disk.

References

1. Habakov, A. V. *Grundlegende Fragen der Entwicklung der Mondoberfläche.* Moscow, 1949.
2. Bülow, K. von. "Tektonische Analyse der Mondrinde. Ein Versuch," *Geologie,* **6,** 565, 1957.
3. ———. "Lunare und tellurische Fundamentaltektonik," *Wiss. Zs. U. Rostock, math.-naturwiss. Reihe,* Heft 1, Sec. 8, Jahrgang **19,** 1958.
4. Firsoff, V. A. *Strange World of the Moon.* London, 1959.
5. Fielder, G. *Structure of the Moon's Surface.* London: Pergamon Press, 1961.
6. ———. *J. Brit. Astr. Assoc.,* **67,** 314, 1957.
7. Spurr, J. E. *Geology Applied to Selenology,* Vol. IV. Concord, N.H.: Rumford Press, 1949.
8. Ref. 5, p. 183.
9. Firsoff, V. A. "On the Structure and Origin of Lunar Surface Features," *J. Brit. Astr. Assoc.,* **66,** 314, 1956.
10. Fielder, G. "Dynamical Problems Connected with the Formation of the Lunar Ridges and Valleys," *J. Brit. Astr. Assoc.,* **67,** 60, 1957.
11. ———. "Lunar Lattice Patterns and Their Time of Origin," *Pub. A.S.P.,* **70,** 308, 1958.
12. Warner, B. "On the Lunar Grid System," *J. Brit. Astr. Assoc.,* **71,** 116, 1961.
13. Shoemaker, E. M. "Impact Mechanics at Meteor Crater, Arizona." Prepared on behalf of the U.S. Atomic Energy Commission, Open File Rept., July, 1959.

14. SYKES, J. B. (trans.). *The Other Side of the Moon*. London: Pergamon Press, 1960; issued by the U.S.S.R. Academy of Sciences.
15. KATZ, A. H. "Analysis of Lunik III Photographs," *Proc. Lunar and Planetary Exploration Colloquium*, **2**, 27, 1960.
16. Ref. 5, p. 191.
17. Ref. 5, pp. 192–93.
18. FIELDER, G. "The Contraction and Expansion of the Moon," *Planetary and Space Sci.*, **8**, 1, 1961.
19. BALDWIN, R. B. "The Meteoritic Origin of the Lunar Craters," *Pop. Astr.*, **50**, 365, 1942.
20. ———. "The Meteoritic Origin of Lunar Structures," *ibid.*, **51**, 117, 1943.

23

DOMES

Giant volcanoes, such as Aconcagua, Mayon, or Fujiyama, are completely missing on the moon. If present, they would stand out with great prominence. In view of the tremendous evidences of heat and igneous activities on the moon, it is surprising that volcanoes of this type are not found. Wilkins and Moore (1) claim that tiny crater cones do exist:

> Cratercones are quite distinct from the preceding cavities, as they consist of steep conical hills with minute central orifices, and strongly resemble terrestrial volcanoes. Although Fauth denied the very existence of these features, they have been too well established for controversy; they abound in the neighbourhood of Stadius, between Eratosthenes and Copernicus, and there are two well-known specimens near Bailly. In the majority of cases the central orifices are so small that they are difficult to detect except in a giant telescope; those that can be seen appear to be of inverted conical or cup shape. In all cases, the floors of these objects stand high above the surrounding country, and whether central peaks exist it is quite impossible to determine.

The area described near Stadius is a highly disturbed one, and it is perhaps to be expected that crater cones would appear there if anywhere on the moon, but the entire region is covered by ejectamenta from Copernicus and Eratosthenes. With the Dearborn Observatory $18\frac{1}{2}$-inch refractor, I have never been able to identify any crater cones for sure, but there are many blocky masses which could be so interpreted. At times of low sun, the tiny normal craters with raised rims are seen by irradiation to appear disproportionately high. Pickering (2) noted the distinction in 1908, when he said that *crater cones* occurred on cracks in Kilauea, "Much the same as *crater pits* [italics mine] do on the Moon."

A different type of volcano is well known on earth, the shield volcano. Mauna

Loa and Mauna Kea on Hawaii are perhaps the best known and largest. This type of structure is built up more gently than the strato-volcano. The lavas are highly fluid and spread out as thin sheets for relatively great distances. By the accumulation of successive flows in various directions, a broad dome with gentle slopes is formed. The slopes rarely exceed 6°–8°. The craters on these shield volcanoes are calderas of subsidence. Often subsidiary craters develop on the sides of the main shield volcano.

Spread widely over the moon, but only in the lava-covered areas, are many domes, often with exactly central craters. They do resemble shield volcanoes. Most of them are found in Mare Tranquillitatis and Mare Nubium and Oceanus Procellarum. None is found in the central impact areas of the circular maria. Five domes appear north of Hortensius, and two are near Milichius. A particularly well-developed one is near Kies, while others are found south of Cauchy. Several dozen are known in all. Where a central craterlet appears, it does not seem to have a raised rim and is relatively shallow. The symmetrical domes range up to perhaps 10 or 11 miles in diameter and are a few hundred to a few thousand feet high. My measures on the dome near Kies, for example, yield a diameter of 8.5 miles and a height of 2,250 feet, with an average wall slope of 3°.

Other structures may be of similar origin. Twenty-five miles from the Kies dome is an irregular object which may be two domes partially coalesced. Near Arago are two formations which are usually lumped together with the more symmetrical domes. Forty miles south of Stadius is what Alter (3) calls a dome, but it, too, is quite irregular.

Way out in the wide-open spaces of Oceanus Procellarum, not far from the eastern limb of the moon, lies the peculiar formation, Rümker (4). It is a very irregular plateau which rises abruptly out of the surrounding plain in a nearly circular area. The diameter is roughly 30 miles, and the height is about 2,500 feet. Rümker is visible for only a short time about $2\frac{1}{2}$ days before full moon. After the sun reaches an altitude of about 15°, it becomes nearly invisible. Herring makes the point that it is composed of numerous rounded, domelike hills, which give the formation a decidedly lumpy appearance. On the assumption that these formations are all of similar genesis, they range downward from Rümker to gently rounded domes at the limit of resolution on the best photographic plates.

While no such structure as the symmetrical domes has been identified in the bright uplands (5), these regions are so broken up and irregular that they would hide a multitude of irregular blisters. Alter (6) stated that, while domes might exist in the rough areas, he doubted that he could observe them, even with a 60-inch telescope. The nearest approach to an upland dome is one discovered by Barker in 1932 (7). It lies in the great lava-filled crater, Darwin.

With one exception, none of the domes, regular or irregular, is found on a fault, crack, or rille. The single exception lies on the shelf area in southern Mare Serenitatis, somewhat east and north of Menelaus. Paralleling the shore in this region is a series of peculiar rilles or cracks. One of these splits a dome into two equal halves. Barker described the Darwin dome as "a huge cinder-heap, a lunarian dust-heap which bristles with roughness—like a selenite slag-heap."

Under most angles of illumination, the domes appear to be of a color identical with that of the surrounding lavas, but Cooke (8) has noted that under a low sun all domes appear dark. He suggests that they are seamed with minute fissures which are shadow-filled under oblique lighting. Moore (9), who has studied them carefully, agrees with this explanation.

In 1908, Pickering (2, 10) advanced the idea that these domes were volcanoes of the shield type. Certainly, domelike structures could be produced in this fashion, but it is questionable that Rümker is so explicable. The same objection might be made here as against the volcanic hypothesis of the origin of the central peaks of craters. How do we explain the domes which do not show a summit craterlet?

Shaler (11), in 1903, and Spurr (12) revived the ancient suggestion of Hooke (13) in 1667, which postulated a large-scale degassing of the moon, with part of the gas trapped beneath a surface layer which then was up-arched into the observed dome. This mechanism could explain the symmetrical structures, with or without a central crater, but is completely impossible physically. Because of the finite strengths of the lunar rocks, the gases would break through and escape long before the domes of the sizes observed could be formed. Presumably, the surface layers would subside when the gas pressure was released.

A modification of the preceding hypothesis substitutes molten magma for the rising gas. Such a process produces a laccolith, many of which are known to exist on earth. Gilbert (14) described one in the Henry Mountains of Utah. Tomkins (15) in 1927, Marshall (16) in 1943, and Spurr (12) in 1945 endeavored to explain various lunar formations by a laccolithic mechanism. Marshall and Spurr specifically applied it to the domes. Herring (4) suggested that Rümker was a composite structure formed by multiple laccolithic intrusions. In this mechanism, the central craterlet could be formed either by a collapse mechanism or by explosion.

In 1954 and 1955, Hess (17, 18) proposed a mechanism to account for terrestrial domes and uplifts. In his second paper he presented evidence to show that the material below the Mohorovičić discontinuity is peridotite and that, for a considerable distance beneath the discontinuity, the temperature will be below 500° C. He then used the Bowen and Tuttle (19) data on the MgO-

SiO_2-H_2O system to show that the serpentinization of olivine would occur at temperatures of 500° C or less, provided that water could be introduced into the system. An approximate equation for the reaction is

$$\text{Olivine} + \text{water} \rightleftarrows \text{serpentine} + \text{heat}.$$

When olivine is changed to serpentine by this process, there is approximately a 25 per cent increase in volume.

Salisbury (20, 21) has recently applied this mechanism to the origin of the lunar domes. As he notes, in order to have this same uplift process active on the moon, we must assume that peridotitic materials are present in temperature regions below 500° C and that water vapor must later be introduced.

It is important to recognize that if Salisbury's mechanism did operate on the moon, it implies that the moon had become differentiated chemically and that the lava flows contained ultrabasic materials close to the surface. This, in turn, is consistent with the thesis that the moon was once at least partially melted.

The conversion of olivine to serpentine releases heat. If the temperature locally rises above 500° C, the reaction will stop and reverse. Salisbury suggests that this self-limiting phenomenon might account for the presence of central craterlets in many cases. He also suggested that the limitation of the domes to the lava-covered areas might be due to the greater possibilities of the release of water in these more active regions.

In summary, then, of the four proposed models, the gas dome is physically impossible. The shield volcano is a possible type, but two factors are opposed to it: there is no evidence of lava flows from the domes, and their surfaces appear to be rough and fissured, probably because of fracturing on expansion.

The other two models—the laccolithic and the mineral phase change—are possible. Both will permit the type of surface observed on the domes. In fact, the required expansion of the surface should demand a fissured top. Both will permit the domes to be regular or irregular. Both will permit the presence or absence of a central craterlet. Neither requires the presence of visible lava flows from a crater.

Based on the evidence, either or both models should be considered as possible mechanisms for the production of lunar domes. Both models call only upon internal forces.

REFERENCES

1. WILKINS, H. P., and MOORE, P. *The Moon*, p. 28. London: Faber & Faber Ltd., 1955.
2. PICKERING, W. H. "Lunar and Hawaiian Physical Features Compared," *Mem. Amer. Acad. Arts Sci.*, **13**, 173, 1908.
3. ALTER, D. "The Nature of the Domes and Small Craters of the Moon," *Pub. A.S.P.*, **69**, 245, 1957.

4. HERRING, A. K. "Observing the Moon—Rümker," *Sky and Telescope,* **20,** 219, 1960.
5. BALDWIN, R. B. *The Face of the Moon,* p. 61. Chicago: University of Chicago Press, 1949.
6. ALTER, D. In discussion of Ref. 21, p. 24.
7. MOORE, P. *Guide to the Moon,* p. 56. London: Collins, 1956.
8. COOKE, S. R. B. "Lunar Domes near Wagner and Birt," *J. Brit. Astr. Assoc.,* **66,** 216, 1956.
9. Ref. 7, p. 57.
10. PICKERING, W. H. "The Origin of the Lunar Formations," *Pub. A.S.P.,* **32,** 116, 1920.
11. SHALER, N. S. "Comparison of the Features of the Earth and Moon," *Smithsonian Contr. to Knowledge,* **34,** 1, 1903.
12. SPURR, J. E. *Geology Applied to Selenology,* Vol. 1. Lancaster, Pa.: Science Press, 1945.
13. HOOKE, R. *Micrographia.* 1667.
14. GILBERT, G. K. "Report on the Geology of the Henry Mountains," *U.S. Geog. and Geol. Survey of the Rocky Mountain Region.* Washington: Department of the Interior; Government Printing Office, 1877.
15. TOMKINS, H. G. "The Igneous Origin of Some of the Lunar Formations," *J. Brit. Astr. Assoc.,* **37,** 161, 1927.
16. MARSHALL, R. K. "The Origin of the Lunar Craters," *Pop. Astr.,* **51,** 415, 1943.
17. HESS, H. H. *Proc. R. Soc., London, A,* **222,** 341, 1954.
18. ———. *Geol. Soc. Amer. Spec. Paper No. 62,* p. 391, 1955.
19. BOWEN, N. L., and TUTTLE, O. F. *Bull. Geol. Soc. America,* **60,** 439, 1949.
20. SALISBURY, J. W. "The Origin of Lunar Domes," *Ap. J.,* **134,** 126, 1961.
21. ———. "Origin of Lunar Domes," *Proc. Lunar and Planetary Exploration Colloquium,* **2,** 22, 1960.

24

THE HEAT BALANCE OF THE MOON

Is the moon hot or cold on the inside? This is the key question in the understanding of its history. When this problem is definitely solved, we shall be on reasonably solid ground in depicting the processes which have brought the moon to its present state of dilapidated grandeur. In this chapter we shall discuss certain principles and certain possibilities which pertain to the actual heat balance of the moon.

It is now rather generally agreed that the moon was built up by the accretion of small cold bodies. Urey (1, 2) was not the first to suggest this, but he certainly should be given the credit for demonstrating its essential correctness.

The surface of the moon shows real variations in height, deviations from a sphere, of several miles. Dynamical considerations show that there is a bulge aligned with the earth. The new contour map is consistent in showing that the radius pointing toward the earth is slightly elongated.

Based on the accretion model of the moon's origin and the apparent departure from isostatic equilibrium, Gilbert (3) and Urey have concluded that the moon was formed cold and remained cold, although perhaps warming, throughout its history. Urey is willing to concede that the moon may be composed of hot silicates below the melting point. Many others, including the author (4–6), have postulated a moon which, at one time, was hot enough to produce lava flows. Obviously, both conclusions cannot be correct.

Urey (7) has presented arguments to show that the parent bodies of the meteorites had accumulated at temperatures below 600° K. He has also suggested that the moon is one of the type of primary objects from which the earth accumulated and that, as a body, it is older than the earth (8).

There are several possible sources of energy for heating the moon. The planetesimals were cold. That much is certain, but they were moving and thus possessed energies which could be released when the smaller body coalesced with the larger. In the early stages of formation, the combined masses of moon and planetesimal were very small, and the velocities with which they collided were low. The minimum possible impact velocities slowly increased as the moon grew, until at present the velocity of escape is 2.4 km/sec.

The gravitational energy released by the accumulation of these myriads of smaller bodies into a sphere of uniform density is

$$\text{Energy} = \tfrac{16}{15} G\pi^2 \rho^2 a^5, \qquad (24\text{-}1)$$

and the energy per gram is

$$\frac{\text{Energy}}{\text{Mass}} = \tfrac{4}{5} G\pi \rho a^2, \qquad (24\text{-}2)$$

where G is the gravitational constant, ρ the density, and a the moon's radius.

The average energy of accumulation of the moon is 1,685 joules per gram, if the objects always struck with exactly the escape velocity and no fragments were ejected from the moon. At the present time or at the end of the accretion process, the accumulation energy under these conditions would be 2,900 joules, or 690 calories per gram.

Let us take the heat capacities for silicate materials equal to that of pyroxenes (9) to 1,250° C and 0.3 calories per gram per degree above this temperature. The heat of fusion is 100 calories per gram. The available energy of impact at the escape velocity would be sufficient now to raise the temperature of such materials from 0° to about 2,100° C. Thus gravitational energy could melt lunar materials during the terminal phases of collection, particularly if the surface were hotter than 0° C.[1]

It is not known how much accumulation heat was actually absorbed by the body of the moon and how much was radiated out into space. It certainly depends on the rapidity with which new particles accumulated and buried the earlier particles. Much heat must have been lost to space, but if the process were rapid enough, the primitive moon could have retained substantial amounts of heat.

---

[1] Much of the preceding few paragraphs was summarized by H. C. Urey (8).

It is probable that the planetesimals themselves were of quite low density. In the early stage of accumulation, they would tend to remain in a low-density condition, perhaps with voids. As the moon grew and the pressure increased, the central volume would become reduced, thus raising the internal temperature.

Chemical heating has been proposed by Donn and Urey (10–12) to account for the very rapid increase in size of comets and for the igneous silicate minerals and the iron-nickel phases of meteorites. Urey (8) says:

> This source of energy may have melted the materials of the moon and in fact this source of heating appears to be the only source so far suggested which accounts for the conditions on the formation of the meteorites particularly within a reasonable time schedule. For our purposes here, this method of heating could produce a moon with a center at low temperatures and with higher temperatures up to around 500° C at larger radii.

Fish, Goles, and Anders (13) have questioned the elaborate series of generations of bodies required by Urey to produce meteoritic bodies. Their conclusions were that a single small body could be so constructed as to explain the observed meteoritic characteristics. In order to agree with the observations, the parent bodies must have passed through a hot, liquid phase. The parent bodies of the meteorites were presumably smaller than the moon, so similar conclusions may be drawn concerning the early history of the moon.

Beyond the methods just described by which the moon could be heated, there is still another—radioactive heating. There are many radioactive isotopes in the earth and meteorites, and so probably they are or were present in the early moon. The temperature in a radioactively heated object can be computed as a function of time for a given initial temperature and composition and distribution of radioactive materials. Certain simplifying assumptions must be made to permit the calculation (1, 14).

In the smaller bodies, if once melted, cooling would be controlled primarily by conduction, but in the larger objects, such as the moon, the molten cores would have a high Rayleigh number and would be convective. Convection would tend to equalize the temperature throughout the liquid volume.

Outside the molten zone would be a sintered shell of materials which were consolidated by the heat and probably an unconsolidated exterior. The larger the body, the greater the insulating effect.

Now it is true, I think, that if the moon were originally cold, it could now be cold, warm, or hot, depending on the speed with which it collected, the amount of effective chemical action, and the amount and type of distribution of the radioactive elements. If, on the contrary, the moon were formed hot, it is still a relatively hot body, with much of its volume close to the melting point.

Calculations by Kuiper (15) have shown that the age of a moon collected cold and heated only by radioactive decays of long-lived isotopes must be greater than 5.5 billion years, in order to produce major melting. This age does not at present appear possible, and so the moon cannot have been melted by this process alone. The higher the original temperature, however, the shorter the time required to melt the moon by long-lived radioactive elements.

If short-lived radionuclides were important in heating the moon, the latter must have been formed within a few tens of millions of years after nucleogenesis. These short-lived radioactive materials would now be extinct and could be detected only by some peculiar ratio of end products. Their activity on the moon is still hypothetical.

Of possible nuclides with half-lives between 10^5 and 10^8 years, Strominger, Hollander, and Seaborg (16) have identified 27 as suitable for such consideration. Analyses of their decay energies and estimated abundances (17–19) have permitted Fish, Goles, and Anders (13) to eliminate all but 8 as unimportant.

On further analysis, they have shown that only Fe^{60}, Cl^{36}, and Al^{26} appear to have sufficiently high heat outputs to cause central melting in planetesimals 100–150 km in radius. Similar conclusions may be drawn with respect to the moon, for heating by this mechanism would be rather rapid or else not effective. A central temperature of the order of $3{,}000°$ K could be reached under certain conditions.

It was also pointed out that, unless the formation of the solar nebula and the accretion of objects a few kilometers in radius occurred in less than a million years, the rapid decay of Fe^{60} and Cl^{36} would not have permitted them to be important. The most likely candidate, thus, is Al^{26} or possible undiscovered short-lived nuclides. Most modern work is consistent with a rapid accumulation of the moon and planets in a period shortly after the end of nucleogenesis, and an originally hot moon is becoming more probable. Several recent theories of the origin of the sun suggest that it was much brighter at about this stage in lunar history and may have aided in keeping the early moon hot.

Among the long-lived radioactive elements, only four may be considered as able to heat the moon effectively. They are K^{40}, Th^{232}, U^{235}, and U^{238}.

Birch in 1951 (20) and 1954 (21) and Aldrich and Wetherill in 1958 (22) have given detailed reviews of the data concerning the decay constants of heat-producing radioactive isotopes. Birch (23) noted that the rate of heat production of an earth composed of stony meteorites calculated from data obtained by neutron-activation analysis is equal within a factor of 2 to the rate at which heat is being lost by the earth.

It is only within the last six years that it has been found that, at high temperatures, large amounts of heat may be transported through solids by radiation. Clark (24–26) first noted the importance of radiative transfer in problems related to the earth's thermal history, and he has begun extensive laboratory investigations of such radioactive transfer through silicate materials. Lubimova, in 1958 (27), used an analogue computer in solving equations on the earth's thermal history, taking this factor into account. The observed flow of heat over both continents and oceans averages about 50 ergs/cm² sec, although large regional variations exist. It is known from seismologic observations that the crust and mantle of the earth are solid to a depth of 2,900 km and that the material immediately below the mantle is liquid.

MacDonald (28) made the point that heat is transported in the earth by radiation, conduction, and convection and that if currents of molten material exist with velocities as low as 10^{-10} cm/sec, convection will dominate the other two methods. He says (29):

> The major limitation of treating only conduction and radiation in dealing with the earth's thermal problems is emphasized by these small velocities. As will be shown, it is possible, assuming only conduction, to construct models that explain the major thermal features of the earth. However, these models are not simple, and the difficulties associated with them suggest that a detailed study of convective heat transport would be most profitable.

MacDonald has assumed that the earth (28) and the moon (30) are each of chondritic composition and has carried out calculations of their thermal histories for several models. The justifications for such an assumption were reviewed by MacDonald in 1959 (31).

On the average, chondrites now release about one-eighth as much heat as they did 4.5×10^9 years ago. The disintegration of K^{40} produces about three-fifths of the heat, with uranium and thorium each contributing about one-fifth of the total heat now being produced. Because of the shorter half-life of U^{235} and K^{40}, their contributions of heat were much greater, absolutely and relatively, 4.5×10^9 years ago than now.

Analyses have been made of the radioactive content of the earth's crustal rocks. Two factors stand out strongly. If the earth as a whole had as much heat-producing radioactivity throughout as it does in the crustal layers, it would be a boiling mass. The second conclusion is that the more acid an igneous rock is, the more radioactive material it contains. Tables 39 and 40 are taken from MacDonald (28). Dunite is a peridotitic or ultrabasic rock. Table 40 lists measures of the uranium content of dunite.

MacDonald stresses that, unlike meteorites, the principal heat producers in crustal rocks are uranium and thorium. They produce approximately equal amounts of heat, while potassium yields about one-third as much as either. On this basis, uranium and thorium have been concentrated far more in the crust than has potassium. The degree of differentiation of uranium and thorium is much greater than that which has affected potassium. It appears that the heat production of ultrabasic dunite is of the order of 1–5 per cent of that of the basic basalts. MacDonald, from the data of Tables 39 and 40, has calculated the heat production by igneous rocks as listed in Table 41.

Work by Birch and Clark in 1940 (32) showed that the thermal conductivities of a wide variety of rocks exhibited rather small variations with composition. A few of these rocks, such as the feldspar-rich materials, show increasing

TABLE 39

AVERAGE CONTENT OF URANIUM AND THORIUM IN IGNEOUS ROCKS

Type of Rock	No. of Samples	$U \times 10^8$ (Gm/Gm)	$Th \times 10^8$ (Gm/Gm)
Granites	9	380	1,030
Acidic	1,257	380–400	1,310–1,350
Intermediate	6	140	440
Intermediate	297	230–300	930–1,050
Basalts	8	83	500
Basic rocks	27	95	380
Basic lavas		60–110	
Hualalai basalts	1	46–50	

TABLE 40

URANIUM CONTENT OF DUNITES

Dunite	$U \times 10^8$ (Gm/Gm)	Analytical Method
Twin Sisters	0.10–0.12	Neutron activation
Twin Sisters	1.6	Mass spectrometer
Twin Sisters	2.4	Radium
Balsam Gap, N.C.	0.9–1.2	Radium
Dun Mountain	0.6	Radium
Addie, N.C.	2.1	Radium
Webster, N.C.	0.9	Radium

conductivities up to at least 400° C, while the conductivities of most materials decrease as the temperature increases. The phonon theory of Peierls (33) shows that the conductivity of rock should decrease approximately inversely with the temperature, but there are no experimental studies on the variation of conductivity with pressure. Lubimova has suggested, on the basis of a lattice-conduction theory, that conductivity should increase with pressure.

Clark (25) gives an equation which illustrates the relationship between thermal conductivity and radiation through solids:

$$K_R = \frac{16 n^2 s T^3}{3\epsilon}, \qquad (24\text{-}3)$$

where n is the refractive index of the material, s is the Stefan-Boltzmann constant, ε is the sum of the absorption and scattering coefficients averaged over all wave lengths, and T is the temperature in degrees Kelvin. The contribution of radiation to the conductivity is strongly temperature-dependent, and, unless the variations in the index of refraction and the opacity do not overwhelm the T^3 dependence, there must exist a temperature above which the contribution of radiation to the heat transfer will be greater than that due to conduction. For many rocks the radiative conductivity is equal to, or greater than, the ordinary conductivity at temperatures of the order of 1,500°–2,000° K and is nearly three times greater at 2,500° K.

TABLE 41

HEAT PRODUCTION BY IGNEOUS ROCKS

	Heat Produced by U (Ergs/Gm Year)	Heat Produced by Th (Ergs/Gm Year)	Assumed Content of K (10^{-4} Gm/Gm)	Heat Produced by K (Ergs/Gm Year)	Total (Ergs/Gm Year)
Granites	117	84	300	34	235
Acidic	126	109	340	38	273
Intermediate	43	36	263	29	108
Intermediate	81	81	263	29	191
Basalts	25	41	57	6.4	72
Basic lavas	26	28*	49	5.5	59
Hualalai basalt	15	16*	56	6.3	37
Twin Sisters dunite (neutron activation)	0.034	0.036*	0.1	0.01	0.08
Dunites	0.42	0.44*	0.1	0.01	0.87

* Calculated on basis of Th/U = 4 (34).

MacDonald has also emphasized that a knowledge of the electrical conductivity of the materials that make up the earth's mantle is important in discussions of the thermal state of the earth for two reasons: (1) the conductivity may give information as to the distribution of temperature, provided that the mechanism of electrical conduction is known; (2) the electrical conductivity enters into the expression for the variation of thermal conductivity with temperature.

From a study of geomagnetic transient variations, one can infer the electrical conductivity of the earth's mantle. The periodic solar daily variations and the aperiodic magnetic storms can be separated by a Gaussian analysis into induced and exciting components. The maximum depth penetrated by the induced currents may be computed from knowledge of the total field at the earth's surface. From the depth of penetration of the induced currents over a wide range of frequencies, the distribution of the electrical conductivity in the mantle can be calculated.

From the short-period variations in the magnetic field, Lahiri and Price (35) have estimated the conductivity to a depth of about 800 km. While they showed that a large number of interpretations are compatible with the magnetic variations, they also showed that there is a very rapid increase in electrical conductivity within the outer few hundred kilometers of the earth. Rikitake (36) has supported this conclusion.

MacDonald (28) has extended the analysis of Lahiri and Price to longer-period variations of the internal origin of the magnetic field and was thus able to make estimates of the conductivity at the core-mantle boundary. He found a continuous increase with depth in the lower mantle to a value of the order of 2 ohm^{-1} cm^{-1}. His values of the electrical conductivity near the surface corroborate those found by Lahiri and Price. The variation is of the order of 10^5.

Several of these effects work in opposite directions. For example, although semiconductors show a marked increase in electrical conductivity with temperature, if this increase is interpreted in terms of intrinsic conduction by free electrons, then the opacity will markedly increase with an increase in temperature. This tends to minimize the cubic dependence of the contribution of radiation upon the temperature. However, taking all these factors into consideration and since the thermal conductivity is directly related to the electrical conductivity, it is very probable that the conduction of heat from the interior of the moon may be substantially greater at high temperatures than elementary theory would indicate. This can significantly modify conclusions which have previously been drawn concerning the heat balance of the moon.

In 1959 and 1960 MacDonald (28, 30) made a series of calculations of the thermal state of the moon's interior, using only radiation and ordinary thermal conduction as the mechanisms of heat transfer. He assumed that the moon originally had a chondritic composition, was undifferentiated, and was throughout at 0° C. If the opacity is taken at the relatively low value of 100 cm^{-1}, radiation will make a noticeable contribution to the transfer of energy. His resulting temperature curve as a function of depth in the moon is shown in Figure 49, as Moon *I*. Next, he assumed an initial temperature of 600° C and an opacity of 1,000 cm^{-1}. Figure 49, Moon *II*, shows the results of this calculation. The two curves are quite similar. The central temperature of the originally cold moon is 1,800° C and that of the warm moon is 2,400° C. The difference in central temperatures is the difference in the original temperatures of the two models.

In the case of the earth, we know that the mantle is solid. If we knew the exact relationships between the pressure and the melting points of various substances in the mantle, we would have a determination of the maximum possible temperature at any point. Unfortunately, we do not possess much of this needed information. This, of course, also limits us in our interpretations of the thermal condition of the moon.

Heat Balance of the Moon

A new experimental approach to this subject is being undertaken by the General Electric Company and the Geophysical Laboratory of the Carnegie Institution of Washington. They have provided data which permit an estimate of the initial slope of the melting-point curves. Use of Simon's semiempirical equation permits extrapolation of these initial slopes to conditions deeper in the earth:

$$P = \frac{a}{B}\left[\left(\frac{T}{T_0}\right)^B - 1\right], \qquad (24\text{-}4)$$

where a and B are empirical constants and T_0 is the melting temperature at 1-bar pressure. The initial slope of the melting-point curve can be used to determine the constant a, and B is determined by the initial curvature.

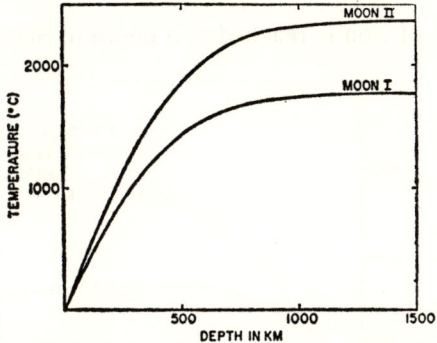

Fig. 49.—Temperature distribution in the moon after 4.5×10^9 years, for a cold moon (*I*) and a hot moon (*II*). (MacDonald, Ref. 30.)

The Simon equation is consistent (37–40, 55, 56) with the Debye theory of simple monatomic solids, but its applicability to silicates has not been justified experimentally. It can be only approximate, but it does suggest the form that the melting-point curves for various solids should take.

Yoder (41) showed that the melting point of diopside, a variety of pyroxene, increased with pressure at the rate of 13° C per 1,000 bars for pressures up to 5,000 bars. Boyd and England (42) carried Yoder's curve to 30,000 bars. At the high-pressure end of the curve, from 20,000 to 30,000 bars, the average slope is 10.3° C per 1,000 bars. LeComte (43) found an initial slope of 11° ± 2° C per 1,000 bars for the melting curve of albite, a triclinic feldspar often found in granite. Strong (44) recently determined that the melting point of iron is increased by 190° ± 20° C at 96,000 atmospheres. The pressure at the center of the moon is 46,000 atmospheres.

Figure 50 illustrates the results of MacDonald's calculations for a cold moon as a function of time. It may be seen that, after 3 billion years, iron would begin to melt at a depth of 1,700 km, or near the center. At this time, there would be a tendency toward differentiation if a metallic phase were present within the moon.

Figure 51 gives the similar relationship for an originally warm moon. Here the melting point of iron at a depth of 1,700 km would be reached after 1.6×10^9 years and at 500 km shortly thereafter. At present, the outer 500 km of a warm moon are cooling off, while the inner 1,200 km continue to warm up on this model. Table 42 summarizes MacDonald's calculations on several different models of a homogeneous moon.

If the moon were originally composed of chondritic material, these data show very clearly that the melting points of various materials are reached at shallow depths. Figure 52 shows the two curves for an originally cold and warm moon superimposed on the melting-point curves for iron and the silicate diopside.

The melting point of iron is reached at a depth of 500 km on the cold model

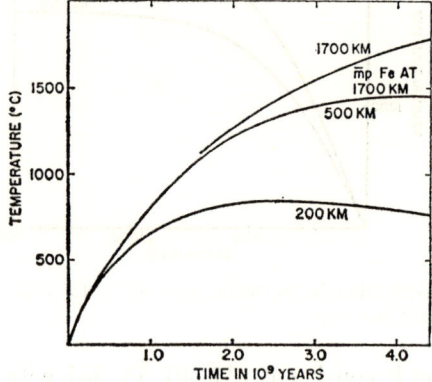

FIG. 50.—Development of temperature with time at various depths for a cold moon. (MacDonald, Ref. 30.)

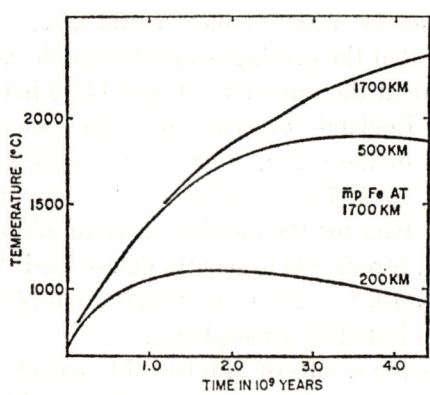

FIG. 51.—Development of temperature with time at various depths for a hot moon. (MacDonald, Ref. 30.)

Heat Balance of the Moon

and at about 400 km in the warm moon. Diopside would not be melted in the first case but would be melted below 400 km in the second. Consequently, it appears unavoidable that the moon became melted below 400 km, if it were originally as warm as 600° C after consolidation. If the amount of radioactivity is reduced by a factor of 2, the above conclusions are not invalidated if the original temperature were as high as 1,000° C.

Urey (45), in a prepublication copy of some of his work, has derived results quite comparable with MacDonald's, although he did not consider the effects of heat transfer by radiation.

It is essentially certain that if large portions of the moon became melted, convection currents would be set up. However, these currents would probably be extremely slow. An effect of the convection currents would be to equalize the temperature variations in the liquid zone, i.e., to lower the central temperature and to raise the temperature of the liquid of the outer portions of the melt.

TABLE 42

RESULTS OF INVESTIGATIONS OF VARIOUS MODELS OF HOMOGENEOUS MOONS

Opacity (Cm⁻¹)	Initial Temp. (° C)	Present Central Temp. (° C)	Surface Heat Flow (Erg/Cm² Sec)
100	0	1,780	10.3
1,000	600	2,380	12.4
1,000	1,200	2,980	14.6
10	1,200	2,700	16.4
10	1,200	1,160	4.2 (No radioactivity)

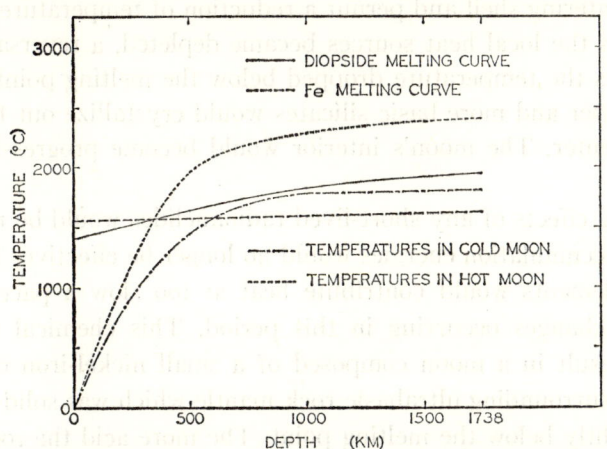

FIG. 52.—Melting point of diopside (42) and iron (44) as a function of depth within the moon. (Redrawn from MacDonald, Ref. 30.)

If the moon were ever substantially liquefied and heat were removed from the central region by convection, then three effects should occur. The liquid iron would probably drain toward the center, forming a core. In the case of the moon, Urey (8) has concluded that, because the moon's density is 3.34 and it is not highly concentrated toward the center, it can contain but little metallic nickel-iron phase.

The low-iron group of meteorites contains an average of 22.33 per cent of iron with 7.14, 3.66, and 11.63 per cent in the metal, iron sulfide, and silicate phases, respectively, and a mean density of 3.51 gm cm^{-3}. The high-iron group data are 28.58, 17.76, 3.62, and 7.20 per cent and a density of 3.66 gm cm^{-3}.

From these data, using either the high- or low-iron group of meteorites, Urey calculated that about 18 per cent of the moon is iron. This places it close to the low-iron group of chondritic meteorites in total percentage of iron.

Second, as the moon became hotter, various types of silicates would melt at different temperatures. Vogt (46) lists the following temperatures at which crystals of differing silicates will melt: granite, 1,000° C; syenite, 1,100°; diorite, 1,200°; gabbro, 1,250°; and dunite, 1,500°–1,600°. The more basic and ultrabasic a rock is, the higher is its melting point in general.

These melting points are altered, usually lowered, by the admixture of small quantities of other substances. Appreciable quantities of solute water are particularly effective, but there are reasons for believing that little water exists within the moon. The step-by-step melting of the various silicates would occur in a warming moon, and the lower-density liquids would be extruded upward as they melted.

Once the melting was completed at about 1,600° C in a large central volume, convection, conduction, and radiation would carry heat away from the center toward the sintering shell and permit a reduction of temperature to occur near the center. As the local heat sources became depleted, a reversal of processes could occur as the temperature dropped below the melting point for each silicate. The denser and more basic silicates would crystallize out first and settle toward the center. The moon's interior would become progressively differentiated.

The heating effects of any short-lived radionuclides would be rapidly declining, and the accumulation energies would no longer be effective. The long-lived radioactive elements would contribute heat at too slow a pace to upset the rather rapid changes occurring in this period. This chemical differentiation would then result in a moon composed of a small nickel-iron core, probably liquid, and a surrounding ultrabasic rock mantle which was solid at a temperature only slightly below the melting point. The more acid the rock, the higher the region in the moon it would attain.

The total amount of heat available in the early stages of the moon's history would determine whether the moon remained solid, whether it melted in the central regions only, whether a large portion or even all of it melted.

If the moon were entirely melted by a source or sources of heat which disappeared quickly, then the outer layers, which now show as the continental regions, might now be an acid silicate, perhaps comparable with a granite. If the moon did not become completely melted, then the continental regions probably are chondritic in composition. If differentiation did occur in the silicates, it is almost certain that the radioactive isotopes will be found to have moved upward and will be concentrated in the top layers of the mantle of the moon, just as they have done in the case of the earth.

Many gases, liquid water, and various hydrocarbons, if present, will have been extruded upward in a melting moon and may well have reached the surface. In this case, they will have been lost because of the moon's low gravity. The chondritic meteorites contain only minute traces of water. Ultrabasic rocks, such as olivine, do not normally contain significant quantities of water.

Because of the small size of the moon relative to the earth, any light liquids or gases would be markedly reduced in quantity per unit area, because the moon's surface in relation to its volume is so much larger than the earth's. Regardless of any differentiation, most of the moon would be ultrabasic in character, and there could be only a thin wash of the lighter and more acid rocks on top.

On earth, the mantle is considered to be ultrabasic, the crustal layers above that to be basic rocks, and the more acid continents are thought to have been formed gradually, primarily by magmatic action. Wilson summarizes the consensus when he says (47):

It has been suggested that before the beginning of the geological era, about three billion years ago, the earth had no continents but only a recently solidified cover of ultrabasic rocks, overlain perhaps by a few kilometers of basalt. The surface temperature had reached equilibrium with the sun's radiation, but the temperature gradient just below the surface was much greater than at present.

And (48):

In the mantle below the ocean floors, acid solutions, gases, and accompanying radioactive elements are trapped and can rise only along the arcuate fractures formed as continents grow. The rise of radioactive elements to the surface of shields, where their heat readily escapes, is assumed to explain why shields are inert (volcanically). This movement of heat-producing elements to the surface may have disturbed the rate at which the earth cools.

If we can reason by analogy, if the moon were melted early in its history, we might expect that the brighter upland regions could be basaltic in character rather than granitic, because the moon apparently never produced a true continent.

A fairly high percentage of the long-lived radioactive isotopes might well be trapped below the visible crust of the moon. In this case, after the rapid changes of the early history of the moon, there might be secondary heating cycles which originated in the upper mantle.

MacDonald (30) carried his studies one step further. While he did not take into account convection in a liquid zone or the effects of concentration of radioactivity in the upper layers, he does show that, as the moon warms up from a cold condition, the radius increases.

Figure 53 is representative of his originally cold moon and Figure 54 of his warm moon. The increase in radius is greater for the cold moon, but in each case the whole moon has expanded several kilometers. At present, both models predict a slight surface cooling and a small contraction. The contraction could

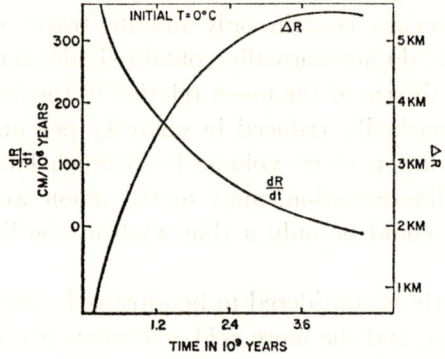

Fig. 53.—Change of radius of a moon initially at 0° C 4.5 × 10⁹ years ago. Opacity, 100 cm⁻¹; ordinary lattice conductivity, 0.025 joule per gram per second per degree centigrade. Final central temperature and surface heat flow, 1,780° C and 10.3 ergs per square centimeter per second, respectively. (MacDonald, Ref. 30.)

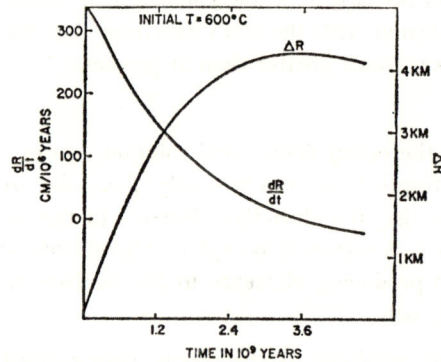

Fig. 54.—Change of radius of a moon initially at 600° C 4.5 × 10⁹ years ago. Opacity, 1,000 cm⁻¹. Final central temperature and surface heat flow, 2,380° C and 12.4 ergs per square centimeter per second, respectively. (MacDonald, Ref. 30.)

not have proceeded far enough to cause significant surface wrinkling. In each model, the moon has had essentially its present radius throughout much of geologic time.

If the moon has become differentiated and the radioactive elements concentrated in the upper mantle, we should expect a slight expansion of the moon and a second degassing in a possible second increasing-temperature phase, as well as possible production of liquid magmas in the radioactive zone. Such a stage could be equated with the formation of the maria.

Fielder (49–51) has recently proposed that the grid system is the result of two opposed factors working at the same time. He suggested that the early moon possessed a tidal bulge or a protuberance which gradually settled. This process would lead to a compression of the surface layers. At the same time, the moon was warming up most rapidly in the innermost parts. This resulted in an expansion of the core and tension in the outer layers. At some time in the past, the two forces resulted in a state of zero stress, and thereafter the tensional forces predominated. He thus differs in the last stage from MacDonald, who proposes that at present the moon's radius is nearly constant, with a possible slight contraction during the last billion years. Hédervári (52–54) has also considered the effects of an expansion of the moon.

In any case, if the moon were originally of chondritic composition, we would expect to find the moon at present hot at all depths well below the surface, with the temperature slightly below the melting point. The moon should be solid, with the possible exception of a small liquid core. Being solid, the moon should have a good deal of rigidity, even though still hot. Being hot, the moon should occasionally emit gases as a remnant of an earlier heat cycle or as a forerunner of a new cycle. Being hot and solid, with buried radioactivity, it is possible that the moon is preparing to undergo another, or third, heat cycle.

These seem to be the broad-view possibilities. It remains to be seen whether or not the observations of the moon's surface are consistent with any, or several, of these possible models.

REFERENCES

1. UREY, H. C. *The Planets.* New Haven, Conn.: Yale University Press, 1952.
2. ———. *Geochim. et Cosmochim. Acta*, **1**, 209, 1951.
3. GILBERT, G. K. "The Moon's Face," *Bull. Phil. Soc. Washington*, **12**, 241, 1893.
4. BALDWIN, R. B. *The Face of the Moon.* Chicago: University of Chicago Press, 1949.
5. ———. "The Meteoritic Origin of Lunar Craters," *Pop. Astr.*, **50**, 1, 1942.
6. ———. "The Meteoritic Origin of Lunar Structures," *ibid.*, **51**, 1, 1943.
7. UREY, H. C. *Ap. J. Suppl.*, **1**, 147, 1954.

8. Urey, H. C. Revised edition of *The Planets,* prepublication copy, 1961.
9. Landolt-Bornstein. *Tabellen,* **2,** 1959.
10. Donn, B., and Urey, H. C. *Ap. J.,* **123,** 339, 1956.
11. Urey, H. C., and Donn, B. *Ap. J.,* **124,** 307, 1956.
12. Donn, B., and Urey, H. C. "Chemical Heating Processes in Astronomical Objects," *Mém. Soc. R. Sci. Liège,* quatrième série, **18** (fascicule unique), 124, 1957.
13. Fish, R. A., Goles, G. G., and Anders, E. "The Record in the Meteorites. III. On the Development of Meteorites in Asteroidal Bodies," *Ap. J.,* **132,** 243, 1960.
14. Allan, D. W., and Jacobs, J. A. *Geochim. et Cosmochim. Acta,* **9,** 256, 1956.
15. Kuiper, G. P. "On the Origin of the Lunar Surface Features," *Proc. Nat. Acad. Sci.,* **40,** 1096, 1954.
16. Strominger, D., Hollander, J. M., and Seaborg, G. T. *Rev. Mod. Phys.,* **30,** 585, 1958.
17. Suess, H. E., and Urey, H. C. *Rev. Mod. Phys.,* **28,** 53, 1956.
18. Burbidge, E. M., Burbidge, G. R., Fowler, W. A., and Hoyle, F. *Rev. Mod. Phys.,* **29,** 547, 1957.
19. Hamaguchi, H., Reed, G. W., and Turkevich, A. *Geochim et Cosmochim. Acta,* **12,** 337, 1957.
20. Birch, F. "Recent Work on the Radioactivity of Potassium and Some Related Geophysical Problems," *J. Geophys. Res.,* **56,** 107, 1951.
21. ———. "Heat from Radioactivity," in *Nuclear Geology,* ed. H. Faul, p. 148. New York: John Wiley & Sons, 1954.
22. Aldrich, L. T., and Wetherill, G. "Geochronology by Radioactive Decay," *Ann. Rev. Nuclear Sci.,* **8,** 257, 1958.
23. Birch, F. "Differentiation of the Mantle," *Bull. Geol. Soc. America,* **69,** 483, 1958.
24. Clark, S. P. "Effect of Radiative Transfer on Temperatures in the Earth," *Bull. Geol. Soc. America,* **67,** 1123, 1956.
25. ———. "Radiative Transfer in the Earth's Mantle," *Trans. Amer. Geophys. Union,* **38,** 931, 1957.
26. ———. "Absorption Spectra of Some Silicates in the Visible and Near Infrared," *Amer. Mineralogist,* **42,** 732, 1957.
27. Lubimova, H. A. "Thermal History of the Earth with Consideration of the Variable Thermal Conductivity of the Mantle," *Geophys. J. R. Astr. Soc.,* **1,** 115, 1958.
28. MacDonald, G. J. F. "Calculations on the Thermal History of the Earth," *J. Geophys. Res.,* **64,** 1967, 1959.
29. *Ibid.,* p. 1968.
30. MacDonald, G. J. F. "Interior of the Moon," *Science,* **133,** 1045, 1961.
31. ———. "Chondrites and the Chemical Composition of the Earth," in *Researches in Geochemistry,* ed. P. H. Abelson, p. 476. New York: John Wiley & Sons, 1959.

32. BIRCH, F., and CLARK, H. "The Thermal Conductivity of Rocks and Its Dependence upon Temperature and Composition," *Amer. J. Sci.*, **238**, 529-58, 613-35, 1940.
33. PEIERLS, R. E. *Quantum Theory of Solids*. London: Oxford University Press, 1955.
34. MARSHALL, R. R. "Isotopic Composition of Common Leads and Continuous Differentiation of the Crust of the Earth from the Mantle," *Geochim. et Cosmochim. Acta*, **12**, 225, 1957.
35. LAHIRI, B. N., and PRICE, A. T. "Electromagnetic Induction in Non-uniform Conductors and the Determination of the Conductivity of the Earth from Terrestrial Magnetic Variations," *Phil. Trans. R. Soc. London, A*, **237**, 509, 1939.
36. RIKITAKE, T. "Electromagnetic Shielding within the Earth and Geomagnetic Secular Variation," *Bull. Earthquake Res. Inst., Tokyo U.*, **29**, 263, 1951.
37. GILVARRY, J. J. "The Lindemann and Grüneisen Laws," *Phys. Rev.*, **102**, 308, 1956.
38. ———. "Grüneisen's Law and the Fusion Curve at High Pressure," *ibid.*, p. 317.
39. ———. "Equation of the Fusion Curve," *ibid.*, p. 325.
40. ———. "Grüneisen Parameter for a Solid under Finite Strain," *ibid.*, p. 333.
41. YODER, H. S. "Change of Melting Point of Diopside with Pressure," *J. Geol.*, **60**, 364, 1952.
42. BOYD, F. R., and ENGLAND, J. L. "Melting of Diopside under High Pressure," *Carnegie Inst. Washington Year Book*, **57**, 173, 1958.
43. LECOMTE, P. Personal communication to G. J. F. MACDONALD.
44. STRONG, H. M. "The Experimental Fusion Curve of Iron to 96,000 Atmospheres," *J. Geophys. Res.*, **64**, 653, 1959.
45. UREY, H. C. "Origin and History of the Moon," prepublication copy, 1961.
46. VOGT, J. H. L. *Econ. Geol.*, **21**, 207, 1926.
47. WILSON, J. T. "The Development and Structure of the Crust," in *The Earth as a Planet*, ed. G. P. KUIPER, p. 205. Chicago: University of Chicago Press, 1954.
48. ———. *Ibid.*, p. 206.
49. FIELDER, G. "Stresses in the Moon," *Planet. Space. Sci.*, **5**, 286, 1961.
50. ———. "The Contraction and Expansion of the Moon," *ibid.*, **8**, 1, 1961.
51. ———. *Ibid.*; also *Comm. U. London Obs.*, No. 45, 1961.
52. HÉDERVÁRI, P. "A Holdfelszín Morfológiája és Képzódményeinek Eredete," *Künönnyomat a Földrajzi Közlemények*, Év12. Számából, 1959.
53. ———. *Die Sterne.*, **36**, 1960.
54. ———. *Hungarian Phys. Rev.*, **8**, 261, 1960.
55. GILVARRY, J. J. "Temperatures in the Earth's Interior," *Nature*, **178**, 1249-50, 1956.
56. ———. "Temperatures in the Earth's Interior," *J. Atm. Terr. Phys.*, **10**, 84-94, 1957.

25

MAGNETIC FIELD OF THE MOON

We do not yet know whether the moon has a magnetic field, but we are quite certain that the earth does have one. To the first approximation, the magnetic field of the earth is like that which would be produced either by a dipole at the center or by uniform magnetization of the material. At present, the axis of the dipole is tilted $11°.4$ to the earth's axis of rotation, and the geomagnetic pole is in latitude $78°.6$ north and $70°.1$ west. The magnetic field has a strength of 0.63 gauss at the geomagnetic pole and 0.31 on the equator (1).

In addition to the dipole field, there are local anomalies due to the presence of magnetic ores in the crust. These anomalies are usually small but occasionally are strong enough even to reverse the field locally. Superimposed on the dipole field and the locally perturbed field are departures covering large areas. These departures have no close relationship to the distribution of continents, oceans, or major geologic structures.

When such a field is analyzed in spherical harmonics, comparison of the coefficients derived from the horizontal and vertical components allows a determination to be made on which parts of the field are produced within the earth and which parts are of external origin. Within the accuracy with which the analysis can be made, the whole of the field is of internal origin, except for the diurnal variations, magnetic storms, and short-term phenomena that have their origin in the ionosphere.

The magnetic field is not constant but suffers slow secular changes. Vestine and his associates (2, 3) thoroughly studied the magnetic field of the earth and its variations between 1905 and 1945. Examination of the residual magnetization of both igneous and sedimentary rocks in past geologic ages (4–9) gives reasonably clear evidence that the magnetic field of the earth was oriented to the continents at quite different angles than at present. Tentative conclusions are that this is associated with wanderings of the pole or with continental drift.

The lack of correlation of the earth's magnetic field with geologic features, coupled with the fact that the field changes by a large fraction of itself in 100 years, and its internal origin practically exclude an origin in the crust. Since the mantle is solid, there does not seem to be anything in it which could change appreciably in a few years. Therefore, geologists believe that the core is the only place where relatively rapid changes can occur and where a magnetic field might originate. The core is fluid and is likely to be a conductor of electricity, and therefore its materials probably carry electric currents. The mechanism by which the core could produce a magnetic field is not thoroughly understood, but this view has been developed over the last fifteen years by Elsasser (10–13) and others (14–19). There are several other theories of the earth's magnetic field, but none has received general acceptance.

We cannot detect the presence of a magnetic field near the moon from observations made near the earth. The only direct observation is that made by the Russian rocket Lunik II. This object struck the moon on September 13, 1959, at 21:02:23, U.T. (20–22). It carried a magnetometer aboard. The threshold sensitivity of this instrument was such that only an upper limit of the moon's magnetic field on the sunlit side could be determined. This was 50 gamma (1 gamma $= 10^{-5}$ gauss). It is not certain that any magnetic field was detected.

If the moon does have a magnetic field, the nature of this field must be determined in the reasonably near future by instruments carried in circumlunar satellites or by instruments put on the moon by soft landings of rockets.

From the Russian data, it is clear that the moon does not possess a magnetic field comparable in strength with that of the earth. However, if the moon does possess a permanent magnetic field, the field could well serve as a clue to the manner in which the moon was formed and to the internal temperature and the presence or absence of a liquid core. The early history of the moon might thus be clarified.

Altshuler and his associates (23), in discussing the scientific observations of the ABLE-5 program, which is designed to place a space station in a circumlunar orbit, make the point that if the moon were once considerably closer to the earth than it now is and if the moon cooled under the influence of the earth's magnetic field, the lunar field would probably be a dipole aligned antiparallel to that of the earth, while its magnitude would depend on the composition of the moon. If, on the other hand, the moon's magnetic moment is not aligned with that of the earth, the possibility would suggest that large blobs of solar magnetism transported from sunspot emissions determine the magnetism of the moon.

A complicating factor is that the moon's magnetic field on the sunlit side may be confined to a region very close to the lunar surface because of the action of a ring current produced by the solar wind. The magnetic field on the night side of the moon might extend considerably farther from the surface. If this mechanism is valid, the field near certain portions of the lunar surface could be as high as 1,000 gammas without contradicting the Lunik II data.

Hopefully, this problem will be resolved in the next few years because of its importance in the delineation of the moon's early history.

References

1. BULLARD, SIR E. C. "The Interior of the Earth," in *The Earth as a Planet*, ed. G. P. KUIPER, p. 57. Chicago: University of Chicago Press, 1954.
2. VESTINE, E. H., LAPORTE, L., COOPER, C., LANGE, I., and HENDRIX, W. C. *Description of the Earth's Main Magnetic Field and Its Secular Change, 1905–1945.* ("Carnegie Institution of Washington Publications," No. 578.) Washington, D.C., 1947.
3. VESTINE, E. H., LAPORTE, L., LANGE, I., and SCOTT, W. E. *The Geomagnetic Field: Its Description and Analysis.* ("Carnegie Institution of Washington Publications," No. 580.) Washington, D.C., 1947.
4. JOHNSON, E. A., MURPHY, T., and TORRESON, O. W. *Terr. Mag.*, **53**, 349, 1948.
5. GRAHAM, J. W. *J. Geophys. Res.*, **54**, 131, 1949.
6. KAWAI, N. *J. Geophys. Res.*, **56**, 73, 1951.
7. HOSPERS, J. *Proc. Kon. Ned. Akad. Wetensch.*, B, **56**, 468, 1953.
8. ———. *Ibid.*, p. 477.
9. CLEGG, J. A., ALMOND, M., and STUBBS, P. H. S. *Phil. Mag.*, **7**, 583, 1954.
10. ELSASSER, W. M. *Phys. Rev.*, **70**, 106, 1946.
11. ———. *Ibid.*, p. 202.
12. ———. *Ibid.*, **72**, 821, 1947.
13. ———. *Rev. Mod. Phys.*, **22**, 1, 1950.
14. FRENKEL, J. *C. R. Acad. Sci. U.R.S.S.*, **49**, 98, 1945.
15. BULLARD, SIR E. C. *M.N., Geophys. Suppl.*, **5**, 248, 1948.
16. ———. *Proc. R. Soc. London*, A, **197**, 433, 1947.
17. ———. *Ibid.*, **199**, 413, 1949.
18. ———. *Phys. Today*, **2**, 6, 1949.
19. ———. *Phil. Trans. R. Soc. London*, A, 1954.
20. ANON. "Lunik II's Landing on the Moon," *Sky and Telescope*, **20**, 265, 1960.
21. DETRE, L. *Contr. Budapest-Szabadsaghegy Obs.*, No. 45, 1959.
22. MELIN, M. "Touching the Moon," *Sky and Telescope*, **19**, 11, 1959.
23. ALTSHULER, S., LINDNER, J., SCHWEIZER, F., WAGNER, R., and CONDON, R. "The Scientific Objectives of the ABLE-5 Program," *Phys. Today*, **14**, 20, 1961.

26

RECENT CHANGES ON THE MOON

With the myriads of markings, both meteoritic and igneous, visible on the moon, it has been a common expectation that sooner or later some change would be noticed. In the three and a half centuries of telescopic study, no new crater has been found. It is probably foolish to expect to see one formed, for the visible craters, spread uniformly over the life of the moon, would call for only one to have been formed since the pyramids were built. The Russian Lunik II, which was relatively large as compared with most meteorites, could only have produced a crater at least an order of magnitude too small to be seen.

Strangely, the most controversial of the claimed changes on the moon has been the disappearance of the crater Linné in Mare Serenitatis. In 1788 a professional artist, John Russell, drew Linné as a white patch, much as it appears today. This drawing came to light only in 1903 (1), after the controversy had started. In 1823, Lohrmann; in 1831, Mädler; and in 1841–43, Schmidt frequently described Linné as a crater 5 miles in diameter and very deep (2–4). In 1866, Schmidt (5) found that Linné appeared as a white patch with a tiny black spot near the center. Later and with better instruments, it was found that there is a tiny crater in the area and some ridges. The group lies on a dome about 300 feet high and perhaps 3,000 feet in diameter.

Fauth (6) studied the problem and the writings on it thoroughly and concluded that Lohrmann, Mädler, and Schmidt were in error and that Schmidt

415

compounded the mistake when he accepted their observations. Fauth wrote: "Without this error, lunar literature would be free from many of its fantasies; as it is, proofs of 'changes' on the moon spring up like weeds. We witness over and over again how 'scientific' methods and time and effort are squandered on unprofitable problems" (7).

Ashbrook (8) has analyzed the original drawings and references and concluded that there is no real evidence of change. The descriptions are ambiguous, and later astronomers have misinterpreted them.

A completely different, but indubitably real, change on the moon has recently been detected. For many years observations have been made on several large craters which seemed to vary in appearance. At least some of the smaller interior craters and markings have been clear and distinct at times, and at other times they have been reported as hazy. Plato, Eratosthenes, and Alphonsus are often mentioned in this connection.

Beginning in April, 1954, systematic lunar photographic observations were undertaken by Alter (9) at the Cassegrainian focus of the Mount Wilson 60-inch telescope in an attempt to establish the existence or non-existence of such obscurations. Some of the best photographs that have ever been taken are in the blue and infrared exposures of this series. Certain of the pictures taken in the blue showed less detail in Alphonsus than in Arzachel relative to that which appeared in adjacent infrared pictures.

Alter resisted the temptation to interpret these results as being due to a thin atmosphere, but he did comment that a slow leakage of gas from some of the small craters in the moon would explain the effects observed. He closed with a statement that the series of photographs demanded a more intensive search of this and of other areas when near the terminator.

Because of Alter's observations, the Russian astronomer Kosyrev (10) decided to investigate Alphonsus spectrographically and optically with the 50-inch Poulkova reflector. This spectrograph gave a linear dispersion of 23 A/mm at $H\gamma$. On many occasions he photographed the spectrum of the central peak of Alphonsus without finding anything unusual, but, while taking a spectrogram, he noticed at 01^h U.T. on November 3, 1958, that the central peak became "strongly washed out and of an unusual reddish hue."

At 3^h00^m U.T., Kosyrev commenced a second spectrogram. The exposure was $\frac{1}{2}$ hour. During this period the central peak looked unusually bright and white. Toward the end of the exposure the peak resumed its usual appearance. A third spectrogram was taken immediately afterward.

The three plates were then developed, and the first two plates were quite different from normal. The first spectrum was unusually faint in the violet, compared with adjacent crater areas. The second spectrum showed broad emission bands. On the next night Alphonsus appeared normal; that is, there were no permanent changes in the peak's structure or brightness.

Kosyrev has proposed the following interpretation: First there was an ejection of a reddish-colored volcanic ash or dust from a point near the central peak. This was then followed by an efflux of gas over a period of a few hours. The gas was responsible for the emission spectrum. Kosyrev identified a prominent emission structure as the Swan band of C_2 with maximum intensities at 4,737 and 4,715 A. Other C_2 bands were identified in the $H\delta$ region. A faint series of bands due to the C_3 molecule was also noticed, but the CN molecule was not detected.

Kuiper (11) has questioned these interpretations, claiming that, even if the observations were real, the C_2 band should appear in absorption rather than emission. Kosyrev (12) has answered these objections, stating that the microphotometer tracings of the original spectrum clearly show the Swan band in emission and that the maxima at 4,715 and 4,737 A are within 1 A of the proper places. At the same time, he said that undoubtedly other bands of unknown origin are superimposed on the Swan band. In this answer he demonstrated that Kuiper was in error in assuming that the Swan band should be in absorption rather than in emission.

Kosyrev estimated that the total amount of gas released from the central peak was of the order of a few hundred thousand cubic meters, or perhaps a cubic quarter of a mile.

Gaydon and Learner (13) showed that even cold gas could show an emission spectrum under the influence of sunlight, so that, in spite of newspaper articles, this cannot be considered a volcanic eruption but was only a mild emission of gas.

The origin of the C_2 molecules in lunar gases is a problem. Urey (14) has tried to develop a theory but has not succeeded definitely. He noted that volatilization of graphite would require very high temperatures.

Since the density of the gas was very low, the C_2 could hardly have been formed by collisions between molecules containing only one atom of carbon. Hence methane and other molecules containing only one atom of carbon could not be the parent molecules. Acetylene, C_2H_2, is very unstable and could scarcely have been stored for a long period of time, although if calcium carbide, CaC_2, did exist on the moon in the absence of water and if water then penetrated to the CaC_2, then acetylene might appear at the lunar surface.

Ferrosilicon and silicon carbide are found in meteorites; and the conditions for producing these substances from silicon dioxide and calcium dioxide from calcium oxide are similar. Thus Urey concludes that acetylene may be the parent of the C_2 molecule.

He also made the interesting point that small black areas exist in the Alphonsus crater and that they have small associated craters. Acetylene explodes under appropriate conditions into graphite and hydrogen. These small craters may be due to the explosion of accumulations of acetylene gas.

Urey also considered ethylene, C_2H_4, and ethane, C_2H_6, and higher hydrocarbons. Because of their lack of stability at fairly low temperatures, these compounds are not possible parents of the observed C_2. He suggested that tests be made for calcium carbide in the lunar-surface materials when instruments are landed on the moon.

Although the parent substances probably have not been positively identified and, as Kosyrev states, the C_2 molecules probably did not come to more than 1 per cent of the total gas emission, the fact remains that gas did escape from the moon at the central peak of a crater. This location is the most probable one, inasmuch as the central peak marks the top of a rebound dome and is situated in the center of a subcrater volume of brecciated material. It is above the crater's structural weak spot.

Occurrences of this type are not volcanic eruptions as usually understood but are a phase of degassing of the moon. Such degassing probably implies a rising temperature. The magnitude of the rise is unknown, but its duration must be measured in many tens of millions of years. Processes which affect the temperatures of large parts of the moon are extremely slow to act.

Since we have real proof of a degassing process acting in Alphonsus, we should expect similar occurrences to develop elsewhere, including Plato, but observations of this type, to avoid confusion, should be buttressed by photographic and spectrographic evidence. Other gases than C_2 and C_3 may well be detected in such emissions.

References

1. RAMBAUT, A. A. "Two Drawings of Mare Serenitatis by John Russell, R.A., Affording Some Hitherto Unpublished Evidence as to the Appearance of Linné in the Year 1788," *M.N.*, **64**, 156, 1903.
2. *1867 Report of the British Association.*
3. WEBB, T. W. *Celestial Objects for Common Telescopes*, p. 96. 4th ed. London: Longmans, Green & Co., 1881.
4. WILKINS, H. P., and MOORE, P. *The Moon*, p. 96. London: Faber & Faber, 1955.
5. SCHMIDT, J. F. J. *Der Mond.* Leipzig, 1866.
6. FAUTH, P. *Astr. Rundschau*, **3**, 172, 1901.
7. ———. *Unser Mond.* Breslau: Eschenhagen, 1936.
8. ASHBROOK, J. "Linné in Fact and Legend," *Sky and Telescope*, **20**, 87, 1960.

9. ALTER, D. "A Suspected Partial Obscuration of the Floor of Alphonsus," *Pub. A.S.P.*, **69,** 158, 1957.
10. KOSYREV, N. A. "Observations of a Volcanic Process on the Moon," *Sky and Telescope*, **18,** 184, 1959.
11. KUIPER, G. P. Letter in *Sky and Telescope*, **18,** 307, 1959.
12. KOSYREV, N. A. Letter in *Sky and Telescope*, **18,** 561, 1959.
13. GAYDON, A. G., and LEARNER, R. C. M. Letter in *Nature*, **183,** 37, 1959.
14. UREY, H. C. "On Possible Parent Substances for the C_2 Molecules Observed in the Alphonsus Crater," *Ap. J.*, **134,** 268, 1961.

27

SUMMARY AND CONCLUSIONS

The data of Chapters 1–8 were used to describe the effects that meteorites, employing the word in its broadest sense, have had on the earth and moon. Relationships have been empirically established which show within rather close limits what will happen if a given mass strikes a rocky solid at velocities between 2 and 50 miles per second. The equations are valid for the newborn crater.

Inasmuch as the modern theories of the origin and building of the moon all suggest that it was an accretion process, these data on meteoritic impact are of prime importance. They may also be used by geologists in studies of terrestrial meteoritic craters, new and old, and to aid in identification of such structures on the earth.

In Chapter 9, key observations were reported. In the earlier chapters, the listed craters of the moon had been classified on the basis of their appearance as belonging to Classes 1–5. The Class 1 craters are almost all postmare. They are essentially identical with newborn craters and have not been distorted by subsequent craters formed in the same area. Classes 2–4 were progressively more and more distorted by the later overlapping and overlying craters. It is logical to think of the sequence as an age sequence; that is, the Class 1 craters are the newest, and the Classes 2, 3, and 4 craters are progressively older. The statistics of this situation are such that no one crater can be absolutely placed in Classes 2–4, but, on the average, the Class 3 craters are older than those of Class 2, etc. The Class 5 craters are lava-filled.

Summary and Conclusions

The interesting thing is that the craters were placed in this sequence on the basis of their appearance, but that this is an age sequence is confirmed by an entirely different set of facts. The older the crater, the shallower it is; the older the crater, the lower is its rim.

The loss of height in the rim, as well as the geometry of the crater, is such that the materials lost from the rim, even if all were transported into the crater, would not be sufficient to raise the crater bottom by the amount observed in any class of crater. Actually, erosion of the rim would send a good deal of material away from the crater pit.

The conclusion, then, is that the forms of the ancient craters have been modified by a type of isostatic adjustment. The weight of the superimposed load of the rim has caused a downward adjustment in that region, while the relief of load of the crater has permitted the central zone to rise. Associated with this is the observation that the older a crater is, the less chance there is for it to have a central peak. The very old central peaks seem to have become minimized as the crater floors rose.

When an isostatic adjustment occurs, it progresses to the point where only a trace of the original structure is still visible and then either stops or the rate of progress is so slow as to be imperceptible. It is so on the moon. Certain structures, such as the large ring just west of Hipparchus and the ancient giant dry crater just east of the Altai Mountains, are even older than the Class 4 craters. They have reached the terminal condition of an isostatic adjustment.

Classes 4, 3, and 2 show decreasing amounts of adjustment, while the Class 1 craters, at least in the smaller sizes, are completely consistent in shape with extrapolations from terrestrial meteoritic craters. The larger Class 1 craters, those above 20 miles in diameter, are slightly more shallow than would be expected.

We now have clear evidence that the major period of crater formation on the moon covered an extended period of time, a period which is of unknown length but which was long enough to permit significant changes to occur in the strengths of the lunar rocks. The terminal phases of the moon's accretion mark the beginning of this era, and the lava flows mark its end, or at least they are reasonably close to its end.

In this early phase of the moon's history, the surface layers were hard and fairly thick. Craters of Class 1 could be formed any time in this period. After they were formed, later impacts occurred which allow us to date the age of the crater on a relative scale, based on the number of later craters superimposed. In this period, the older the crater, the more it was distorted by isostatic adjustments.

The conclusion is drawn that, at the terminal phases of the building of the moon, the moon had developed a hard crust but that the interior was hot enough to weaken the regions of the moon corresponding to the upper mantle on the earth. Isostatic adjustments then took place wherever changes in load occurred on the surface.

The amount of isostatic adjustment observed was greatest at the largest craters and least at the small craters. It was greatest at the oldest craters and grew progressively less as time passed, so that the latest premare craters, those of Class 2, show little modification of form and the postmare Class 1 craters are distorted very little, if at all. The crust was becoming thicker and more rigid during this period.

Inasmuch as the relief of load on the small scale of a crater could trigger an isostatic adjustment, the entire moon must have been able to adjust its form to exterior influences; and hence, during this period, the moon was close to the earth and was tidally distorted, for even to this day it still shows a bulge.

As time went on, after the terminal period of the building of the moon, the depth to which the moon was hard and rigid became greater and greater. This implies that the outer layers of the crust and upper mantle were becoming cooler. At some indefinite period after the maria developed, the local isostatic adjustments ceased, and the later craters, few in number, remained in Class 1.

In Chapters 10 and 11, the problems of the moon's very early history and shape were treated. The new contour map is definitive in showing that the lava-covered areas are, without exception, low. They average about 2 miles deeper than the uplands. In all cases, the places where the lavas are thickest are the lowest spots relative to a sphere. The lava flows are all of the same general age, as shown by crater counts.

Two possibilities exist. The surface of the moon could have had the observed contours and the lavas merely came from below and filled the low lands, or the lavas came from below and filled whatever low spots then existed and formed much of the great basin system by isostatic adjustments due to the superimposed load of dense lava. The isostatic adjustments forced the shore areas of the maria to downwarp.

Inasmuch as the last premare craters, those of Class 2, occurred when the moon could still adjust isostatically, the probabilities are very great that the lava flows did cause similar adjustments in height to occur.

Since isostatic adjustments could still happen, the moon was still close to the earth at the time the maria appeared. This is also apparent when we note that the maria are not distributed around the limb. We would expect them to be close to the limb if the moon did not possess an equipotential surface when they were formed. The maria appear in widely separated regions of the moon and give no evidence of thicker lava on the sides away from the center of the disk.

When the maria developed, the moon had become considerably more rigid in its outer layers than earlier but was still incapable of maintaining a bulge such

Summary and Conclusions

as we observe today. The earth was responsible for holding it up at that time. Therefore, the moon became more rigid after the development of the maria.

Because the moon's outer layers solidified a very long time ago and the bulge still exists, it implies that the viscosity of the lunar upper mantle is very high, which is consistent with the observed fact that the craters have continued to exist with only minor changes for billions of years. However, if it can be shown that the moon is still convective, then these conclusions regarding the bulge will be subject to modification.

If we determine the shape of the moon from the available height measures on the bright parts of the face and of the limb which are distant from the mountains and the downwarping of the maria, we find that it is a triaxial ellipsoid. It is in good agreement with the theoretical bulge which would be produced by the tidal effects of the earth and the centrifugal forces of the moon's motions, provided that the earth and moon were only about 69,000 miles apart when the outer layers solidified to the extent that the moon could no longer adjust in shape as it receded from the earth. Of course, it is possible that the isostatic adjustments in the regions of the lava flows caused certain compensatory adjustments in the uplands, but this does not seem to be a major effect.

Inasmuch as the moon's outer layers solidified a very long time ago and the bulge still exists, the implication is that the viscosity of the lunar upper mantle is higher than Haskell's theory would demand.

Chapters 12 and 13 are concerned with the nature of the surface materials of the moon. All observations of the moon's crust are consistent in showing that it has very highly insulating material on the outside, but closely below the surface the lunar rocks are to be found. The surface is probably dust, very finely pulverized, but certain amounts of reconsolidation may have occurred, possibly by a sputtering process. Little bare rock is exposed, and both maria and uplands are dust-covered. The dust must have been produced largely *in situ*, but each crater's formation broadcast some dust to land elsewhere on the moon or even to leave the moon. Because considerable degassing has occurred and the moon's surface is exposed to vacuum, it is probable that the surface rocks are quite porous.

Different colors are to be found, indicating a variety of lunar materials, but no positive identifications with terrestrial minerals have yet been made. The range of faint colors observed and the multitude of brightness changes are evidence of an extremely complex surface. The moon is not a simple, uncomplicated body as it is often thought to be. Certain areas even have been observed to fluoresce, and it is possible that the surface contains some chemically active free radicals.

The major slopes are generally small, while, on a smaller scale, the moon is very rough. Optically, the moon's disk very nearly obeys a Lambert's scattering law, and hence it is quite rough on a microscopic scale.

Radar reflections have shown that the moon is a poor reflector at long wave lengths but is a remarkably good reflector for wave lengths comparable to 10 cm. This implies that, at least on a scale of a few centimeters, the moon is quite smooth. This is to be expected if there is a slow, continuing production of dust on the surface.

In Chapter 15, the observations were compared with theories concerning the moon which had previously been advanced by other authors. None of the older theories has been found to be in full accord with the facts.

Gold's hypothesis that the maria are great fields of deep dust is rejected *in toto,* although there is indubitably a good deal of dust on the moon.

Urey's suggestion that the lavas were produced from the bodies of the great objects which produced the circular maria has been shown to be incorrect. The circular maria were produced dry and existed for a substantial period of time before the lavas came.

The crater-forming period could not have been as short as the thousands of years postulated by several, because of the need to account for changes in the rigidity of the moon's crust during this period.

Tremendous evidences of melting and erosion have been detected on the edges of the lava flows, illustrating the high temperature of magma.

Kuiper's model, in which the impacts of the objects which produced the circular maria released liquid magma from immediately below a very thin crust, must also be rejected on the same basis as Urey's model. A rather long period of time elapsed between the impacts and the release of lava.

The conclusion here reached is that the lavas developed deep below the crust and were extruded upon the surface after a period when the upper layers of the moon were cooling and becoming stronger. The lava surface, exposed to a near-vacuum as it cooled, probably became highly vesicular.

In Chapter 16, the circular maria were described and discussed. Their forms are logical extensions of the small craters formed from lesser impacts. Shock waves were sent out from each and traveled hundreds of miles before freezing into position as scarps. Fragments were sent from the impact areas to build extensive radial ridges and gouge great valleys. Apparently, some fracturing of the crust occurred on radii also, for structures which appear to be subsidence valleys are found.

In all cases, the circular maria are surrounded by mountainous borders and moderately high shock scarps. The Apennines, which are a raised area with many scattered high peaks, extend upward to about 20,000 feet. The argument has previously been used that such peaks and ranges could not remain standing if the moon were hot but would have disappeared quickly. Now there is no

Summary and Conclusions

question that, at the time these scarps and mountain ranges were formed, the moon must have been able to adjust itself to changing loads.

When the oldest of the circular maria are studied—Nectaris, Humorum, Serenitatis, and Crisium—it is found that they possess low scarps and low crater borders. Apparently, they have undergone some vertical adjustments of height. At Mare Imbrium the mountains are still high, but this is the youngest and largest of these objects and occurred nearly, but not quite, at the end of the period of isostatic adjustments. Pertinent questions are: How high were the Imbrium borders immediately after the impact? How deep are their roots? Possibly they were in or nearly in isostatic balance shortly after they were formed. If not, the thickness and strength of the lunar crust beneath them were sufficient to support them.

In Chapter 18 we noted that the lunar atmosphere is practically non-existent. It cannot affect the surface significantly, but firsthand study in the future may elicit certain information concerning the present state and past history of the moon.

The non-original conclusion is drawn in Chapter 19 that the rays are dust ejected from impact craters at very low angles at the times the craters were formed. Only relatively modern craters show rays, and older rays are either gradually covered up or fade away over an extended period of time.

All non-explosive theories and all volcanic theories of the origin of the rays are not in accord with the observations. It is concluded that there is no evidence of rays having been formed by high-angle ejection but that much matter was ejected at high angles from the original explosions and that it rained down and stippled the surfaces near the crater of origin. It is also concluded that we cannot specifically date the times when the ray craters were formed or determine the rotation of the moon from the offset of particular rays from true alignment with the central crater.

In Chapter 20 the data on central peaks of craters were summarized. It was shown that not all craters possessed central peaks but that it was a normal thing for certain craters to be peak-free. The very large and the very small craters are less likely to have peaks than those of intermediate sizes. This agrees with the observed terrestrial meteoritic craters, which are small and do not show central peaks, and also with the large circular maria, which do not show central peaks. The central peaks are considered to be rebound structures.

It was noticed that many of the craters have eccentrically placed interior mountains. The presence or absence of a central peak, the eccentricity of its position, and the complexity of the central massif may be functions of the angle of fall of the meteorite and the momentum.

Arguments were presented to show that these structures are not of volcanic nature, even though some of the peaks show summit craterlets. Statistically, even more central peak craterlets should exist than are now known. Certain complex central peaks may indicate a collapse of the rebound dome.

The lunar rilles are placed in such a definitive characteristic pattern around the edges of the irregular maria and concentrically with the circular maria that they can be interpreted only as surface expressions of great tensional forces. These forces are identified with the gradual isostatic adjustments which continued long after the lavas of the maria had hardened. The width of the rilles seems to be the result of a collapse mechanism. The rilles concentric with the circular maria may have formed from cracks which were the result of the impact. They would then correspond to ring faults which were opened by later isostatic adjustments.

Some of the rilles show crater-like structures along their length. They are interpreted as the result of powerful outpourings of gas along these deep lines of weakness.

Similarly, there are a few great faults in the upland areas. Along these are crater-like formations. One in particular extends across the highest part of the moon from Ptolemaeus to the Altai scarp. It can be dated as older than the lava flows. Such a structure implies an expansion of the crust and an outpouring of gases in the period before the lavas came. This presumably marks a stage when the deeper sections of the moon were heating up preparatory to extruding the lavas. This warming-up region was deeper than the thick crust, which was cooling in the same period.

The wrinkle ridges are found mainly on the interior part of lava flows and mark compressional regions formed as the isostatic adjustments proceeded. They were often formed in such a way as to outline crater rims and ridges which existed beneath the lava cover.

Only moderate seismic activity is evident on the moon's surface. The faults which are found are mostly normal faults with an occasional vertical throw, as at the Straight Wall. Essentially, no faults with horizontal displacements have been discovered.

There are thousands of linear markings of various sorts scattered over the entire face of the moon. Certain students have attempted to fit these into a lattice pattern or a grid system and to draw conclusions from this on the general history of the moon.

The most prominent of the markings are associated with the birth pangs of Mare Imbrium, a single isolated event. Many of the others can be attributed equally well to the origins of other craters and circular maria. These structures are from happenings thrust on the moon from the outside. They bear no relationship, or very little, to a moon-wide system of linear markings produced by some forces acting on the moon as a whole.

It is also true that certain of these linear markings cannot now be attributed

Summary and Conclusions

to specific craters or maria. They may form part of a lunar grid system; but, if it exists, it is very weak and very minor. The polygonalism of many lunar craters is a normal aspect and is not a part of the lunar grid system.

Distributed widely over the dark lava flows are numerous low domes. These are strictly of interior origin, not exterior. They probably are surface expressions of lunar laccoliths, although other explanations have been offered.

No normal strato-volcanoes exist on the moon, but the existence of small cinder cones is not thereby denied.

MacDonald has calculated from various assumptions what the heat balance of the moon might be. Urey has also made similar calculations.

It is unknown how much radioactivity there is in lunar rocks. It is also unknown what the initial temperature of the moon was and how it varied in temperature throughout.

In spite of differing assumptions, the results are quite similar and informative. Unless the radioactivity were markedly less than that determined from the earth and from chondritic meteorites, the center of the moon is now hot and above the melting point of iron. Probably, then, the moon has a liquid metallic core which is considerably smaller in proportion than the earth's core. In these calculations no attempt was made to estimate the original temperature distribution of the moon.

If the original temperature were uniform and 600° C, then the center is now very hot, and the moon might be melted out to within about 500 km of the surface, although it probably is now completely solid except for a liquid metallic core. It would be very difficult to devise a model of the moon which is now cool in the interior.

Various sources of energy could have been effective in raising the temperature in these early stages. As the moon accreted, some of the energy of the infalling bodies could remain with the moon. The rate of accretion is important here. Chemical energies could contribute some heat. Adiabatic compression undoubtedly added its share of heat. Short-lived radioactive nuclides, now extinct, may have been quite effective in raising the central temperature. Long-lived radioactive nuclides added their portion, but their major effects were slow in appearing.

When all these effects are added together, it does not seem possible that the moon began as a cold body. It very probably was warm or hot throughout, and very early in the moon's history it became completely melted, except perhaps for a thin sintered layer of less dense materials which floated on the surface of the liquid.

It is also probable that the liquid interior was convective. If so, heat was translated from inside a core of reasonably uniform temperature to the outer layers, where it became lost to space. This loss of heat could not have been replaced in a short period of time, as adiabatic heating, accretion energy, chemical action, and short-lived nuclides would all have become quickly exhausted and the long-lived nuclides could not add much new heat in the period of rapid loss of heat.

As heat became lost, the average temperature of the liquid declined, and selective crystallization occurred. Some molten iron descended to the center, but the most ultrabasic silicates would crystallize out first, and solidification would develop from the center outward. The great part of the moon, judging from the earth, would be of ultrabasic rock with a thin wash of basalt on top of the olivine and possibly some of the more acid rocks above that. These rocks would presumably be beneath a layer of mixed composition sintered by the heat, and there probably was a thin layer of unconsolidated rubble above that. This is the layer which appears in the continental area. If the moon were ever completely melted, the uplands should be acidic in nature, or possibly basaltic if there were not sufficient time to permit a more complete differentiation. If the outer layers did not become melted, they should be more like chondritic meteorites in composition. In either case, the light-colored surface materials are distinctly different in density and composition from the lower levels and from the lava flows. The moon has been chemically differentiated.

Under these conditions, the moon would become solid in a very few thousand years, with the possible exception of a liquid iron core. At each point, the temperature would be just slightly below the freezing point of the particular silicates peculiar to those depths.

In these years of differentiation, it may be expected that the large molecules containing the long-lived radioactive nuclides of uranium and thorium and, to a lesser extent, potassium were selectively transported into the upper half of the moon's mantle. On this model and for a long period of time we would have a moon which was solid from the core to the surface. It was hot and capable of adjusting isostatically. The outer layers were continuously and fairly rapidly cooling and becoming more rigid.

The craters, which were formed when the outer crust became strong enough to show them, were soon almost completely eliminated by the isostatic adjustments. As time went on, the depth of cooling grew greater, and the new craters became less and less distorted. Finally, the crust became deep enough and strong enough that when the tremendous impacts of the bodies which produced the circular maria came, the crust was not shattered, but great dry craters appeared. Presumably these bodies had also been growing larger by accretion in this period and then finally were destroyed when they struck the moon.

It has been noticed that the oldest craters are smaller than the circular maria

Summary and Conclusions

in general; this may be interpreted to mean that the average size of the planetesimals grew, up to the time when the Imbrium planetesimal fell, and thereafter only the smaller bodies were left to fall. These had been left over when the earth, moon, and larger planetesimals grew, and they were relatively few in number. At least 23 craters larger than 100 miles in diameter were formed before the lavas came, none since.

It is felt that most of the postmare craters resulted from impacts of the higher-velocity objects from the asteroid belt rather than from the fragments left over from the formation of the earth and moon.

During the late stages before the lavas came, numerous craters were produced all over the moon's surface, even inside the dry circular maria. In this period, small isostatic adjustments continued. The outer layers became cooler and more rigid, and the deep lower layers, where the long-lived radioactives were concentrated, were becoming warmer. It was only a little below the melting point anyway, and the entire region below about 200–300 miles probably was hotter than $1,400°$ K.

Under these conditions, the moon must have begun to expand, even though the crustal areas were still cooling. Such expansion would result in great cracks leading down to the hotter regions. Much degassing would be bound to occur.

Eventually, the melting point would be reached deep below the crust, and the resulting increase in volume and lowering of density would force hot basic magmas toward the surface from reservoirs perhaps several hundred miles below. This process was not simply the formation of isolated pools of lava in the body of the moon, although it may have started that way. It developed into almost a moon-wide spasm, in which boiling hot magmas, probably loaded with gases, were forced upward through great swarms of relatively narrow cracks.

Inevitably, these cracks would open into the weak and brecciated zones beneath craters and circular maria. In these cases, individual craters would be filled from below. Close study indicates that most of the lava-filled craters are filled to less than normal ground level and that they are distributed primarily in the low lands. These are the regions most accessible to the rising liquids.

Pressures would vary in different places. Wargentin was filled to the brim, while nearby craters showed much less lava. The lavas which filled Wargentin must have come from an isolated pipe.

Hundreds and thousands of flows were thrust upward and out over the surface until vast areas were covered to depths counted in thousands of feet. The lavas must have advanced and retreated many times.

Just as on the earth, the lavas at different places and different times were of different chemical compositions. In general, the lavas were of basic character. This is concluded for two reasons. The depth from which these lavas had to come demands that they were basic, unless the moon contains far more acidic materials than does the earth. Second, there are no evidences of terminal walls at the edges of what appear to be distinct flows.

A basic lava at the temperatures which would obtain at depth, particularly if it contained appreciable amounts of gas, would be very much more fluid than an acid lava, and hence the flows would fine out to low edges, which could not be detected from the earth.

The viscosity of a silicate melted at a given sealed temperature is determined entirely by the ratio of the metal ions, Mg, Fe, Ca, etc., to the silicon atoms. Melts in which the ratio of metal ions to silicon atoms is small are highly viscous for a wide range of temperatures above the melting temperature. A typical example is feldspar, in which the ratio of sodium to silicon plus aluminum, aluminum acting as silicon, is one-fourth, and it is almost impossible to form a homogeneous melt without repeated melting and grinding.

A melt of the composition of olivine is extremely fluid at a temperature just above the melting point and would fit the observations well, but it is extremely difficult to produce an ultrabasic magma. The lighter silicates would be melted first and extruded upward. If the lava were ultrabasic, the temperature would, of course, have had to be some $200°–300°$ C higher than in the case of the basalt. When any such high-temperature lavas reached the surface of the moon, the radiational losses would be greater, but a thin frozen layer would be formed, insulating the lava from the ground surface, and an outside crust would be formed, insulating the lava from space. The material in between would pour out the end, continuously forming new crusts. The lava would thus flow essentially in a two-dimensional tunnel relatively insulated from its surroundings. In such a case, with very fluid lavas, the front at the end of the flow might be so small that it would be below resolution from the earth.

At numerous places on the lava flows, particularly near Langrenus and Copernicus, there are tremendous evidences of degassing. Probably all over the moon, but particularly on the dark areas, there are thousands of relatively small craters formed by this process. Near Stadius they are often aligned along faults.

At this point of the moon's history we find that tremendous amounts of heat were transported by liquid and by gas to the surface. The heat supply of the deep layers was seriously depleted, and they resolidified. This led to a compaction of the subsurface layers. Adjustments occurred over broad areas where the superimposed weight of layers of lava forced a sinking of the crust. Probably other areas were forced upward by this process. Some lavas were undoubtedly withdrawn back below the ground.

During this period, there never was a thick, solid layer floating on a liquid; for, as fast as liquids were produced, they were driven upward, melting and

Summary and Conclusions

widening the cracks formed by the expansion and then solidifying, only to be split and penetrated by new lavas.

When the very hot lavas did finally reach the outside, they spread far and melted many parts of the earlier surface. Anywhere one looks at the edges of the maria, there may be seen craters, great and small, which have lost their seaward walls to the hot erosive lavas by melting or being overwhelmed.

This tremendous paroxysm apparently is the only one like it which has occurred in lunar history. In the case of the earth, the event on greater or lesser scale has occurred often.

The mild degassing recently observed in Alphonsus may be evidence that the moon is again regaining strength for another period of melting and expansion.

If the inequalities in the figure of the moon were once in isostatic equilibrium, as appears probable, then the variation in level between the maria and uplands does not place a stress difference on the interior. In this event the stress difference is only that due to the bulge, which is far higher than can be accounted for by the present tidal pull of the earth.

Jeffreys says: "For the moon, which appears to be a very homogeneous body, the elastic theory . . . indicates a stress difference at the centre of 2×10^7 dynes/cm.^{2}. With the modified stress distribution this can be reduced to about 1.3×10^7 dynes/cm.^{2}, say 13 atm. or the pressure of 130 m. of water, which would still need good masonry to hold it" (1).

The inequalities in the figure of the moon could be supported at the center by material with a strength at least an order of magnitude less than granite. If the bulge is supported by a layer 200 miles thick, it needs to have a strength capable of withstanding 130 bars, or about one-sixth that of granite. In all probability, the outer layers of the moon, above the region where the lavas of the moon developed, are sufficiently strong to support the observed bulge and the stress differences due to the maria, on the assumption that the maria never became fully compensated. The inner parts of the moon may still be weak, even though solid.

Conversely, if the lunar rocks cannot support the observed bulge, then some such mechanism as Runcorn's convection process must be operating to produce an elongation toward the earth. In this event, the interior of the moon must still be hot, and the moon is thus capable of adjusting its shape isostatically; but then we run head-on into the problem of why the Class 1 craters and, in particular, the older craters have not practically disappeared.

Based on the evidence only and without regard to theory, it looks very much as though the moon's interior is now hot, but not so hot as in earlier times, and the outer layers have a finite strength sufficient to maintain a small bulge and to preserve craters for billions of years when similar objects on the earth would quickly vanish.

* * *

As outlined in the first chapter, the scientific papers of the world on the moon and earth have been searched and studied over the last thirteen years. Work done by hundreds or perhaps thousands of men has been read and recorded. A great deal of new material has been derived by the author, and the preceding chapters contain a résumé of all the most important and most significant data which have come to my attention. A synthesis of pertinent facts has been made and the resulting conclusions summarized in this chapter. These conclusions differ in many ways from any broad theories which have been advanced previously.

At the present state of knowledge, this is the picture that emerges. It is certain that this synthesis will not please everyone. At best it will give a reasonable history of the moon from shortly after it was formed to the present day. Probably it will be changed in detail as new evidence is gathered. At worst, it may initiate sufficient new impetus to enable others to improve upon it.

References

1. JEFFREYS, H. *The Earth,* p. 192. 3d ed. Cambridge: Cambridge University Press, 1952.

APPENDIX 1

DERIVATION OF THE RELATIONSHIP BETWEEN THE DISTANCE OF THE MOON AND GEOLOGIC TIME

Let

P_E = Angular momentum of earth's rotation,
P_M = Angular momentum of moon's revolution,
m = Mass of moon,
v = Linear velocity of moon in its orbit,
r = Distance of moon in terms of its present distance,
c = Circumference of moon's orbit (circular),
S = Period of moon's revolution (seconds),
T = Time in billions of years,
A, B, C, D, E = Constants;

$$P_M = mrv.$$

Assume that

$$\frac{dP_E}{dt} = \frac{A}{r^6} = -\frac{dP_M}{dt};$$

now

$$v = \frac{c}{S} = \frac{2\pi r}{S};$$

but

$$S = Br^{3/2};$$

therefore,

$$v = \frac{2\pi r}{Br^{3/2}} = \frac{2\pi r^{-1/2}}{B};$$

433

therefore,
$$P_M = \frac{2m\pi r^{1/2}}{B} = Cr^{1/2},$$

$$\frac{dP_M}{dt} = \tfrac{1}{2}Cr^{-1/2}\frac{dr}{dt},$$

and
$$\frac{A}{r^6} = -\tfrac{1}{2}Cr^{-1/2}\frac{dr}{dt},$$

$$\frac{dr}{dt} = Dr^{-11/2},$$

$$r^{11/2}dr = D\,dt\,;$$

therefore,
$$2r^{13/2} = DT + E.$$

When T was equal to 0, the month was 4.8 hours long, and r was equal to 9,000 miles, or 0.0377. Therefore,
$$E = 0.000003,$$
which is negligibly small. When $r = 1$, $T = 4.5$; therefore
$$D = \frac{2}{58.5},$$

and
$$\frac{2r^{13/2}}{13} = \frac{2}{58.5}T,$$

or
$$T = 4.5r^{13/2}.$$

APPENDIX 2

TABLES

TABLE 4   Chemical and Nuclear Explosive Craters

TABLE 5   Heat of Explosion for Chemical Explosives

TABLE 6   Terrestrial Meteorite Craters

TABLE 7   Lunar Craters

TABLE 8   Scaled Crater Dimensions versus Scaled Depth of Burst

TABLE 23 Coordinates of Selected Lunar Features and Measured Heights Relative to a Sphere

TABLE 43 Original Measures of Positions of Selected Lunar Formations on Lick Observatory Photographs (See Table 23)

Table 4
Chemical and Nuclear Explosive Craters

No.	Explosive	Apparent Diameter (inches)	D	Apparent Depth (inches)	d	True Diameter (inches)	D_1	True Depth (inches)	d_1	T = Rim Height (inches)	R_H	Apparent Rim Width (inches)	R_W	E	Nature of Ground	12H (in.)	Notes
1	Dynamite	2 7/8	9.38		9/16	8.67	-	-	9/16	8.67	-	-	-	-	2.88	Sand	- 1	1
2	"	3 3/4	9.50	1 1/16	8.95	-	-	1 1/16	8.95	-	-	-	-	"	"	-1/2	1
3	"	4	9.52	15/16	8.89	-	-	13/16	8.93	1/8	8.02	-	-	"	"	-1/4	1
4	"	4 3/4	9.60	1 1/16	8.95	-	-	15/16	8.89	1/8	8.02	-	-	"	"	-1/8	1
5	"	2 1/2	9.32	5/8	8.72	-	-	5/8	8.72	-	-	-	-	"	Soil	0	1
6	"	4	9.52	3/4	8.80	-	-	3/4	8.80	-	-	-	-	"	"	0	1
7	"	4 3/4	9.60	1	8.92	-	-	1	8.92	-	-	-	-	"	"	0	1
8	"	5	9.62	7/8	8.86	4	9.52	3/4	8.80	1/8	8.02	-	-	"	"	0	1
9	"	5 1/2	9.66	1 3/8	9.06	4 1/2	9.57	1 1/4	9.02	1/8	8.02	-	-	"	"	0	1
10	"	5 3/4	9.68	7/8	8.86	4 3/4	9.60	3/4	8.80	1/8	8.02	-	-	"	"	0	1
11	"	5 3/8	9.65	1 1/8	8.97	-	-	1	8.92	1/8	8.02	-	-	"	Sand	0	1
12	"	5 1/8	9.63	7/8	8.86	-	-	11/16	8.76	3/16	8.19	2	9.22	"	"	0	1
13	"	5 3/8	9.65	1 1/2	9.10	-	-	1 3/8	9.06	1/8	8.02	-	-	"	"	0	1
14	"	5 3/8	9.65	1 3/8	9.06	-	-	1 3/16	9.00	3/16	8.19	1	8.92	"	"	0	1
15	"	5 3/4	9.68	1 1/8	8.97	-	-	1	8.92	1/8	8.02	-	-	"	"	0	1
16	"	5 5/8	9.67	1 1/4	9.02	-	-	1 1/8	8.97	1/8	8.02	1/2	8.62	"	"	0	1
17	"	5 1/2	9.66	1 1/8	8.97	-	-	1	8.92	1/8	8.02	3/4	8.80	"	"	0	1
18	"	5 1/2	9.66	1 1/8	8.97	-	-	1	8.92	1/8	8.02	-	-	"	"	0	1
19	Black Powder	4 1/2	9.57	1 3/4	9.16	-	-	1 9/16	9.12	3/16	8.19	7/8	8.86	-	"	-	1
20	Dynamite	4 1/2	9.57	1 5/8	9.13	-	-	1 3/8	9.06	1/4	8.32	2	9.22	"	"	0	1
21	"	6 1/8	9.71	1 9/16	9.12	-	-	1 5/16	9.04	1/4	8.32	1	8.92	2.88	"	1/8	1
22	"	5 1/2	9.66	1 1/4	9.02	-	-	1	8.92	1/4	8.32	-	-	"	Soil	1/8	1
23	"	5 3/4	9.68	1 1/4	9.02	-	-	1	8.92	1/4	8.32	-	-	"	"	1/4	1
24	"	6 1/2	9.74	1 1/4	9.02	6	9.70	1 1/16	8.95	3/16	8.19	-	-	"	"	1/4	1
25	"	6 3/4	9.75	1 5/8	9.13	5 3/4	9.68	1 1/2	9.10	1/8	8.02	-	-	"	"	1/4	1
26	"	7 3/4	9.81	1 1/2	9.10	7 1/4	9.78	1 5/16	9.04	3/16	8.19	-	-	"	"	1/4	1
27	"	5 7/8	9.69	1 1/2	9.10	-	-	1 1/4	9.02	1/4	8.32	1 1/8	8.97	"	Sand	1/4	1
28	"	8	9.82	2 1/8	9.25	-	-	1 3/4	9.16	3/8	8.49	1 1/4	9.02	"	"	3/8	1
29	"	6	9.70	1 1/4	9.02	-	-	1	8.92	1/4	8.32	5/8	8.71	"	Soil	3/8	1
30	"	6 1/4	9.72	1 5/8	9.13	5 1/2	9.66	1 1/2	9.10	1/8	8.02	-	-	"	"	1/2	1
31	"	6 3/4	9.75	1 1/4	9.02	-	-	7/8	8.86	3/8	8.49	1	8.92	"	"	1/2	1
32	"	7 1/2	9.79	2 1/4	9.27	6	9.70	2 1/4	9.24	3/16	8.19	-	-	"	"	1/2	1
33	"	7 1/2	9.79	2	9.22	5 3/4	9.68	1 3/4	9.16	1/4	8.32	-	-	"	"	1/2	1
34	"	8	9.82	2 1/8	9.25	-	-	1 3/4	9.16	3/8	8.49	1	8.92	"	Sand	1/2	1
35	"	7 7/8	9.82	1 7/8	9.19	-	-	1 5/8	9.13	1/4	8.32	1	8.92	"	"	5/8	1
36	"	4 1/2	9.57	1 1/8	8.97	-	-	7/8	8.86	1/4	8.32	3/4	8.80	"	Soil	5/8	1
37	"	6	9.70	1 1/8	8.97	-	-	1	8.92	1/8	8.02	3/4	8.80	"	"	5/8	1
38	"	7 1/2	9.79	1 5/8	9.13	-	-	1	8.92	5/8	8.72	1	8.92	"	"	3/4	1
39	"	9	9.88	2 1/4	9.27	7	9.77	2	9.22	1/4	8.32	-	-	"	"	3/4	1
40	"	9 1/4	9.89	2 1/4	9.27	7 1/2	9.79	2 1/16	9.24	3/16	8.19	-	-	"	"	3/4	1
41	"	9 3/4	9.91	2 1/4	9.27	7	9.77	2	9.22	1/4	8.32	-	-	"	"	3/4	1
42	"	9	9.88	2 1/2	9.32	-	-	2 1/8	9.25	3/8	8.49	1 1/8	8.97	"	Sand	3/4	1
43	"	9	9.88	2 1/4	9.27	-	-	1 7/8	9.19	3/8	8.49	1 1/4	9.02	"	"	7/8	1
44	"	6 1/4	9.72	1 3/8	9.06	-	-	1 1/8	8.97	1/4	8.32	5/8	8.71	"	Soil	7/8	1
45	"	7 1/4	9.78	1 3/4	9.16	-	-	1 1/4	9.02	1/2	8.62	1	8.92	"	"	1	1
46	"	10	9.92	2 1/8	9.25	8	9.82	1 7/8	9.19	1/4	8.32	1 1/2	9.10	"	"	1	1
47	"	10 1/4	9.94	2 1/4	9.27	9	9.88	2	9.22	1/4	8.32	-	-	"	"	1	1
48	"	11	9.96	2 1/4	9.27	9 3/4	9.91	2	9.22	1/4	8.32	1 1/2	9.10	"	"	1	1
49	"	9 3/4	9.91	2 3/4	9.36	-	-	2 3/8	9.30	3/8	8.49	1 1/2	9.10	"	Sand	1	1
50	"	8 1/4	9.84	3	9.40	-	-	2 3/4	9.36	1/4	8.32	2	9.22	"	"	1	1
51	"	10 3/8	9.94	2 5/8	9.34	-	-	2 1/8	9.25	1/2	8.62	2	9.22	"	"	1 1/8	1
52	"	9 3/4	9.91	2 5/16	9.28	-	-	1 13/16	9.18	3/8	8.49	1 1/2	9.10	"	"	1 1/4	1
53	"	10	9.92	2 1/8	9.24	8	9.82	1 3/4	9.16	3/8	8.49	1 1/2	9.10	"	Soil	1 1/4	1
54	"	8 7/8	9.87	3	9.40	-	-	2 5/8	9.34	3/8	8.49	1 1/2	9.10	"	Sand	1 3/8	1
55	"	9 3/8	9.89	2 7/8	9.37	-	-	2 1/2	9.32	3/8	8.49	2	9.22	"	"	1 3/8	1
56	"	10 1/2	9.94	2	9.22	9	9.88	1 1/2	9.10	1/2	8.62	1 1/2	9.10	"	Soil	1 1/2	1
57	"	7	9.77	2 1/2	9.32	-	-	2 1/4	9.27	1/4	8.32	1 1/2	9.10	"	Sand	2	1
58	"	7 3/4	9.81	2 1/2	9.32	-	-	2 1/4	9.27	1/4	8.32	1 1/2	9.10	"	"	3	1
59	"	8	9.82	2 1/2	9.32	-	-	2 1/4	9.27	1/4	8.32	-	-	"	"	3	1
60	"	8 1/4	9.84	2	9.22	-	-	1 5/8	9.13	3/8	8.49	2	9.22	"	"	3	1
61	"	7	9.77	2 3/4	9.36	-	-	2 1/4	9.27	1/2	8.62	2 1/2	9.32	"	"	3	1
62	"	6 5/8	9.74		7/8	8.86	-	-	3/4	8.80	1/8	8.02	-	-	3.31	"	-1 1/2	1
63	"	6 1/2	9.73		3/4	8.80	-	-	3/4	8.80	-	-	-	-	"	"	-1 1/8	1
64	"	6	9.70	1	8.92	-	-	13/16	8.83	3/16	8.19	1	8.92	"	"	-3/4	1
65	"	6	9.70	1 1/16	8.95	-	-	15/16	8.89	1/8	8.02	3/4	8.80	"	"	-5/16	1
66	"	5	9.62	1	8.92	-	-	1	8.92	-	-	-	-	"	Soil	0	1
67	"	5	9.62	3/4	8.80	-	-	3/4	8.80	-	-	-	-	"	"	0	1
68	"	5	9.62	7/8	8.86	-	-	7/8	8.86	-	-	-	-	"	"	0	1
69	"	7	9.77	1 1/2	9.10	-	-	1 3/8	9.06	1/8	8.02	-	-	"	Sand	0	1
70	"	7 5/8	9.80	1 1/2	9.10	-	-	1 5/16	9.04	3/16	8.19	1	8.92	"	"	0	1
71	"	7	9.77	1 7/16	9.08	-	-	1 3/16	9.00	1/4	8.32	1	8.92	"	"	0	1
72	"	6 3/4	9.75	1 13/16	9.18	-	-	1 9/16	9.12	1/4	8.32	1	8.92	"	"	0	1
73	"	6 3/4	9.75	1 7/16	9.08	-	-	1 1/16	8.95	3/8	8.49	1	8.92	"	"	0	1
74	"	6 1/4	9.72	1	8.92	-	-	13/16	8.83	3/16	8.19	1	8.92	"	"	0	1
75	"	7 5/8	9.80	1 7/8	9.19	-	-	1 1/2	9.10	3/8	8.49	1 1/4	9.02	"	"	1/8	1
76	"	5 1/2	9.66	1 1/8	8.97	-	-	7/8	8.86	1/4	8.32	5/8	8.71	"	Soil	1/8	1
77	"	7 3/4	9.81	1 7/8	9.19	-	-	1 5/8	9.13	1/4	8.32	1	8.92	"	Sand	1/4	1
78	"	6 1/4	9.72	1 1/4	9.02	-	-	1	8.92	1/4	8.32	3/4	8.80	"	Soil	5/16	1
79	"	8 3/8	9.84	1 15/16	9.21	-	-	1 9/16	9.12	3/8	8.49	1 1/2	9.10	"	Sand	5/16	1
80	"	9 1/2	9.90	2 1/4	9.27	-	-	1 3/4	9.16	1/2	8.62	3/4	9.10	"	"	1/2	1
81	"	6 3/8	9.73	1 1/2	9.10	-	-	1 1/4	9.02	1/4	8.32	1	8.92	"	Soil	1/2	1
82	"	7	9.77	1 3/8	9.06	-	-	1 1/4	9.02	1/8	8.02	1	8.92	"	"	11/16	1
83	"	10 3/4	9.95	3 1/8	9.42	-	-	2 5/8	9.34	1/2	8.62	1 1/4	9.02	"	Sand	11/16	1
84	"	8	9.82	2	9.22	-	-	1 3/4	9.16	1/4	8.32	1	8.92	"	Soil	13/16	1
85	"	11 1/2	9.98	2 3/4	9.36	-	-	2 3/8	9.30	3/8	8.49	2	9.22	"	Sand	13/16	1
86	"	12 1/4	0.01	3 1/8	9.42	-	-	2 5/8	9.34	1/2	8.62	2	9.22	"	"	1 1/16	1
87	"	8 1/4	9.85	2 1/4	9.27	-	-	2	9.22	1/4	8.32	1 1/2	9.10	"	Soil	1 1/16	1
88	"	11	9.96	3	9.40	-	-	2 1/2	9.32	1/2	8.62	3	9.40	"	"	1 1/4	1
89	"	13 1/4	0.04	3 1/8	9.42	-	-	2 5/8	9.34	1/2	8.62	1 1/2	9.10	"	Sand	1 1/4	1
90	"	12 7/8	0.03	3	9.40	-	-	2 3/4	9.30	5/8	8.72	2	9.22	"	"	1 7/16	1
91	"	12	0.00	3 5/8	9.48	-	-	3 1/8	9.42	1/2	8.62	1 1/2	9.10	"	"	1 5/8	1
92	"	12	0.00	3 7/8	9.51	-	-	3 3/8	9.45	1/2	8.62	2	9.22	"	"	1 11/16	1
93	"	12	0.00	4 3/8	9.56	-	-	3 5/8	9.48	3/4	8.80	2	9.22	"	"	2 3/16	1
94	"	8 1/4	9.84	3 7/8	9.46	-	-	-	-	-	-	-	-	"	"	3	1
95	"	8 1/2	9.85	3 1/4	9.43	-	-	2 7/8	9.38	3/8	8.49	2	9.22	"	"	4	1
96	"	9 1/2	9.90	3 1/2	9.46	-	-	3	9.40	1/2	8.62	2	9.22	"	"	5	1
97	"	10	9.92	2 1/2	9.32	-	-	2 1/8	9.25	3/8	8.49	2 1/2	9.32	"	"	5	1
98	Blast. Powder	8	9.82	1 7/16	9.09	-	-	1 7/16	9.09	-	-	-	-	3.78	Soil	0	1
99	"	9	9.87	2	9.22	-	-	2	9.22	-	-	-	-	"	"	0	1
100	"	13 1/4	0.03	2	9.22	13 1/4	0.03	2	9.22	0	-	-	-	"	Sand	0	1

Table 4
Chemical and Nuclear Explosive Craters

No.	Explosive	Apparent Diameter D (inches)	Apparent Depth d (inches)	True Diameter D_1 (inches)	True Depth d_1 (inches)	T = Rim Height (inches)	R_H	Apparent Rim Width R_W (inches)	E	Nature of Ground	12H (in.)	Notes
101	Blast. Powder	9 1/2	9.90	3 1/8	9.42	-	-	2 5/8	9.34	1/2	8.62	-	-	3.78	Clay Loam	2	1
102	"	17	0.15	4 3/4	9.60	-	-	4 1/8	9.54	5/8	8.72	5	9.62	"	Sand	2	1
103	"	11 1/2	9.98	5	9.62	-	-	4	9.57	1/2	8.62	-	-	"	"	4	1
104	"	13 1/4	0.04	4 1/2	9.57	-	-	4	9.52	1/2	8.62	-	-	"	"	4	1
105	"	15	0.10	5 1/4	9.64	-	-	4 1/2	9.57	3/4	8.80	4	9.52	"	"	4	1
106	"	14 1/2	0.08	5 1/2	9.66	-	-	4 1/2	9.57	1	8.92	6	9.70	"	"	6	1
107	"	12	0.00	3 1/4	9.43	-	-	-	-	-	-	-	-	"	"	7	1
108	"	10	9.92	2 1/4	9.27	-	-	2 1/4	9.27	-	-	-	-	4.05	Soil	0	1
109	"	18 1/4	0.18	6 5/8	9.74	-	-	4 7/8	9.61	1 3/4	9.16	6	9.70	"	Clay Loam	3 1/2	1
110	"	19	0.20	6	9.70	-	-	4 5/8	9.59	1 3/8	9.06	5	9.62	"	"	3 1/2	1
111	"	23	0.28	10	9.92	-	-	8 7/8	9.88	1 1/8	8.97	6	9.70	"	Sand	5	1
112	"	28	0.38	10 1/4	9.93	-	-	8 1/2	9.85	1 3/4	9.16	6	9.70	"	"	6	1
113	"	27	0.35	11	9.96	-	-	10	9.92	1	8.92	-	-	4.27	"	9	1
114	"	26	0.34	9 1/4	9.90	-	-	8	9.82	1 1/2	9.10	10	9.92	4.45	Clay Loam	6 1/2	1
115	"	27	0.35	11	9.96	-	-	9	9.88	2	9.22	6	9.70	"	"	7	1
116	"	40	0.52	16 3/4	0.14	-	-	15 1/4	0.10	1 1/2	9.10	10	9.92	"	Sand	8	1
117	"	44	0.56	12 1/4	0.02	-	-	9 3/4	9.91	2 1/2	9.32	14	0.07	"	"	10	1
118	"	40	0.52	12 1/2	0.01	-	-	11	9.96	1 1/2	9.10	10	9.92	"	"	12	1
119	"	36	0.48	14	0.07	-	-	12 1/2	0.01	1 1/2	9.10	11	9.96	4.63	"	10	1
120	"	33	0.44	12 1/2	0.02	-	-	9 1/2	9.90	3	9.40	6	9.70	"	Clay Loam	10	1
121	"	40 1/2	0.53	14	0.08	-	-	12 1/2	0.01	2	9.22	10	9.92	"	Sand	10	1
122	"	44	0.56	16	0.12	-	-	13 1/2	0.05	2 1/2	9.32	11	9.96	"	"	10	1
123	"	45	0.57	16	0.12	-	-	14	0.07	2	9.22	10	9.92	"	"	10	1
124	"	14 7/16	0.10	3 1/2	9.46	-	-	3 1/2	9.46	-	-	-	-	4.91	Soil	0	1
125	"	37	0.49	15	0.09	-	-	13	0.03	2	9.22	-	-	"	Clay Loam	9	1
126	"	45	0.57	16	0.12	-	-	13	0.03	3	9.40	12	0.00	"	Sand	12	1
127	"	54	0.65	17 3/4	0.17	-	-	15 3/4	0.12	2	9.22	15	0.10	"	"	12	1
128	"	48	0.60	13 1/2	0.05	-	-	9 1/2	9.90	4	9.52	12	0.00	"	"	16	1
		(feet)		(feet)		(feet)		(feet)		(feet)		(feet)				H (feet)	
129	TNT	3	0.48	1	0.00	-	-	1	0.00	-	-	-	-	7.32	Clay	0	2
130	"	20	1.30	6	0.78	-	-	-	-	-	-	-	-	"	"	-	41
131	"	21	1.32	6	0.78	-	-	-	-	-	-	-	-	"	Soil	-	41
132	"	21.2	1.33	9.2	0.96	-	-	7.5	0.88	1.7	0.23	-	-	"	"	-	41
133	"	21.8	1.34	7.2	0.86	-	-	5.2	0.72	2	0.30	-	-	"	"	-	41
134	"	22.5	1.35	6.6	0.82	-	-	4.6	0.66	2	0.30	-	-	"	"	-	41
135	"	23.4	1.37	6.3	0.80	-	-	4.6	0.66	1.7	0.23	-	-	7.58	"	0	2
136	"	7.4	0.87	2.5	0.40	-	-	2.5	0.40	-	-	-	-	"	"	0	2
137	Amatol	4.5	0.65	2	0.30	-	-	2	0.30	-	-	-	-	"	"	0	2
138	"	5	0.70	1.5	0.18	-	-	1.5	0.18	-	-	-	-	"	"	0	2
139	"	5	0.70	2	0.30	-	-	2	0.30	-	-	-	-	"	"	0	2
140	"	5	0.70	2	0.30	-	-	2	0.30	-	-	-	-	"	"	0	2
141	"	5	0.70	2.4	0.38	-	-	2.4	0.38	-	-	-	-	"	"	0	2
142	"	5.5	0.74	1.8	0.26	-	-	1.8	0.26	-	-	-	-	"	"	0	2
143	"	6	0.78	2	0.30	-	-	2	0.30	-	-	-	-	"	"	0	2
144	"	6	0.78	2	0.30	-	-	2	0.30	-	-	-	-	"	"	0	2
145	"	6.3	0.80	2	0.30	-	-	2	0.30	-	-	-	-	"	"	0	2
146	"	6.5	0.81	1.7	0.22	-	-	1.7	0.22	-	-	-	-	"	"	0	2
147	"	6.5	0.81	2	0.30	-	-	2	0.30	-	-	-	-	"	"	0	2
148	"	6.5	0.81	2	0.30	-	-	2	0.30	-	-	-	-	"	"	0	2
149	"	7	0.85	2.5	0.40	-	-	2.5	0.40	-	-	-	-	"	"	0	2
150	"	7	0.85	2.5	0.40	-	-	2.5	0.40	-	-	-	-	"	"	0	2
151	"	7	0.85	2	0.30	-	-	2	0.30	-	-	-	-	"	"	0	2
152	"	7	0.85	3	0.48	-	-	3	0.48	-	-	-	-	"	"	0	2
153	"	7.2	0.86	3	0.48	-	-	3	0.48	-	-	-	-	"	"	0	2
154	"	7.5	0.88	2	0.30	-	-	2	0.30	-	-	-	-	"	"	0	2
155	"	8	0.90	1.5	0.18	-	-	1.5	0.18	-	-	-	-	"	"	0	2
156	"	8	0.90	1.5	0.18	-	-	1.5	0.18	-	-	-	-	"	"	0	2
157	"	8	0.90	1.8	0.26	-	-	1.8	0.26	-	-	-	-	"	"	0	2
158	"	8	0.90	2	0.30	-	-	2	0.30	-	-	-	-	"	"	0	2
159	"	9	0.95	1.7	0.22	-	-	1.7	0.22	-	-	-	-	"	"	0	2
160	"	9	0.95	2.8	0.44	-	-	2.8	0.44	-	-	-	-	"	"	0	2
161	"	9	0.95	3	0.48	-	-	3	0.48	-	-	-	-	"	"	0	2
162	"	9	0.95	1.7	0.22	-	-	1.7	0.22	-	-	-	-	"	"	0	2
163	"	9.5	0.98	2.7	0.43	-	-	2.7	0.43	-	-	-	-	"	"	0	2
164	"	10	1.00	2	0.30	-	-	2	0.30	-	-	-	-	"	"	0	2
165	TNT	23.4	1.37	-	-	-	-	-	-	1.4	0.15	4.7	0.67	"	Clay	-	41
166	"	23.4	1.37	5.6	0.75	-	-	-	-	-	-	-	-	"	Soil	-	41
167	"	24.2	1.38	9.5	0.98	-	-	7.5	0.88	2	0.30	-	-	"	"	-	41
168	"	25	1.40	7	0.85	-	-	-	-	-	-	-	-	"	"	-	41
169	"	25	1.40	8.5	0.93	-	-	-	-	-	-	-	-	"	"	-	41
170	"	25	1.40	8.5	0.93	-	-	4.0	0.60	4.5	0.65	-	-	"	"	-	41
171	"	25	1.40	8.5	0.93	-	-	4.0	0.60	4.5	0.65	-	-	"	"	-	41
172	"	26	1.41	8.3	0.92	-	-	5.8	0.76	2.5	0.40	-	-	"	"	-	41
173	"	27.1	1.43	9.4	0.97	-	-	7.7	0.89	1.7	0.23	-	-	"	"	-	41
174	"	28	1.45	8	0.90	-	-	6.2	0.79	1.8	0.26	-	-	"	"	-	41
175	"	28.2	1.45	8	0.90	-	-	6.2	0.79	1.8	0.26	-	-	"	"	-	41
176	"	29	1.46	6.5	0.81	-	-	-	-	-	-	-	-	7.59	"	-	41
177	"	24	1.38	7	0.85	-	-	-	-	-	-	-	-	7.61	Clay	0	2
178	"	3	0.48	2	0.30	-	-	2	0.30	-	-	-	-	"	"	-	41
179	"	20	1.30	5	0.70	-	-	-	-	-	-	-	-	"	"	-	41
180	"	23	1.36	7	0.85	-	-	-	-	-	-	-	-	"	"	-	41
181	"	24	1.38	5	0.70	-	-	-	-	-	-	-	-	"	"	-	41
182	"	24	1.38	7	0.85	-	-	-	-	-	-	-	-	"	"	-	41
183	"	25	1.40	8	0.90	-	-	-	-	-	-	-	-	"	"	-	41
184	"	27	1.43	8	0.90	-	-	-	-	-	-	-	-	"	"	-	41
185	"	5	0.70	2	0.30	-	-	2	0.30	-	-	-	-	7.66	"	0	2
186	"	26	1.41	7	0.85	-	-	-	-	-	-	-	-	7.67	"	-	41
187	"	26	1.41	8	0.90	-	-	-	-	-	-	-	-	"	"	-	41
188	"	29	1.46	8	0.90	-	-	-	-	-	-	-	-	7.79	"	-	41
189	"	5	0.70	2	0.30	-	-	2	0.30	-	-	-	-	7.93	"	0	2
190	"	7	0.85	0.8	9.92	-	-	0.8	9.92	-	-	-	-	"	Soil	0	2
191	"	11	1.04	3	0.48	-	-	3	0.48	-	-	-	-	"	"	0	2
192	"	12	1.08	2.8	0.45	-	-	2.8	0.45	-	-	-	-	"	"	0	2
193	"	14	1.15	3.3	0.51	-	-	3.3	0.51	-	-	-	-	"	"	0	2
194	"	15	1.18	3	0.48	-	-	3	0.48	-	-	-	-	"	"	0	2
195	"	29	1.46	8	0.90	-	-	-	-	-	-	-	-	"	"	-	41
196	"	32	1.51	10	1.00	-	-	-	-	-	-	-	-	"	"	-	41

Table 4
Chemical and Nuclear Explosive Craters

No.	Explosive	Apparent Diameter (feet)	D	Apparent Depth (feet)	d	True Diameter (feet)	D_1	True Depth (feet)	d_1	T = Rim Height (feet)	R_H	Apparent Rim Width (feet)	R_W	E	Nature of Ground	H (ft.)	Notes
197	TNT	32	1.51	7	0.85	-	-	-	-	-	-	-	-	7.93	Soil	-	41
198	"	33	1.52	10	1.00	-	-	-	-	-	-	-	-	"	"	-	41
199	"	35	1.54	9	0.95	-	-	-	-	-	-	-	-	"	"	-	41
200	"	33	1.52	7	0.85	-	-	5	0.70	2	0.30	5	0.70	"	"	-	41
201	"	33.4	1.52	13.5	1.13	-	-	10.5	1.02	3	0.48	-	-	"	"	-	41
202	"	34	1.53	11.5	1.06	-	-	7.5	0.88	4	0.60	-	-	"	"	-	41
203	"	7	0.85	3	0.48	-	-	3	0.48	-	-	-	-	"	"	-	41
204	"	29	1.46	8	0.90	-	-	-	-	-	-	-	-	7.99	Clay	0	2
205	"	30	1.48	9	0.95	-	-	-	-	-	-	-	-	"	"	-	41
206	"	25.1	1.41	5.5	0.74	-	-	5.5	0.74	-	-	-	-	8.01	Soil	0	2
207	"	31.7	1.50	7.7	0.88	-	-	5.8	0.76	2.0	0.29	8.5	0.93	"	"	-	41
208	"	8	0.90	0.5	9.70	-	-	0.5	9.70	-	-	-	-	8.16	"	-3.5	3
209	"	16.1	1.21	3.3	0.52	-	-	2.5	0.40	0.8	9.90	-	-	"	"	0	3
210	"	22.9	1.36	7.5	0.88	-	-	6.0	0.78	1.5	0.18	-	-	"	"	1.3	3
211	"	25.4	1.40	8.1	0.91	-	-	6.5	0.81	1.6	0.20	-	-	"	"	3.5	3
212	"	32.9	1.52	10.5	1.02	-	-	8.5	0.93	2.0	0.30	-	-	"	"	7	3
213	"	35.5	1.55	6.2	0.79	-	-	4.5	0.65	1.7	0.24	-	-	"	"	14	3
214	"	28.7	1.46	5	0.70	-	-	3.5	0.54	1.5	0.18	-	-	"	"	21	3
215	"	7	0.85	3	0.48	-	-	3	0.48	-	-	-	-	8.25	"	0	2
216	"	10	1.00	2	0.30	10	1.00	2	0.30	0	-	-	-	"	"	0	2
217	"	11	1.04	2.8	0.44	-	-	2.8	0.44	-	-	-	-	"	"	0	2
218	"	12	1.08	3.5	0.54	-	-	3.5	0.54	-	-	-	-	"	"	0	2
219	"	12	1.08	1.7	0.22	-	-	1.7	0.22	-	-	-	-	"	"	0	2
220	"	12	1.08	3.2	0.51	12	1.08	3.2	0.51	0	-	-	-	"	"	0	2
221	"	15	1.18	4	0.60	-	-	3.5	0.54	0.5	9.70	6.5	0.81	"	"	0	2
222	"	32	1.51	8	0.90	-	-	-	-	-	-	-	-	"	Clay	-	41
223	"	34	1.53	10	1.00	-	-	-	-	-	-	-	-	"	"	-	41
224	"	35	1.54	10	1.00	-	-	-	-	-	-	-	-	"	"	-	41
225	"	36	1.56	10	1.00	-	-	-	-	-	-	-	-	"	"	-	41
226	"	37	1.57	10	1.00	-	-	-	-	-	-	-	-	"	"	-	41
227	"	38	1.58	10	1.00	-	-	-	-	-	-	-	-	"	"	-	41
228	"	9	0.95	4	0.60	-	-	4	0.60	-	-	-	-	8.31	"	0	2
229	"	36	1.56	10	1.00	-	-	-	-	-	-	-	-	"	"	-	41
230	"	37	1.57	10	1.00	-	-	-	-	-	-	-	-	"	Soil	-	41
231	"	39	1.59	9.8	0.99	-	-	8.3	0.92	1.5	0.18	-	-	"	"	-	41
232	"	40	1.60	12	1.08	-	-	8.5	0.93	3.5	0.54	-	-	"	"	-	41
233	"	41.3	1.62	12.5	1.10	-	-	10.2	1.01	2.3	0.37	-	-	"	Clay	-	41
234	"	13.6	1.13	2.7	0.43	-	-	2.7	0.43	-	-	-	-	8.33	Soil	0	2
235	"	18	1.26	4	0.60	-	-	4	0.60	-	-	-	-	"	"	0	2
236	"	20.5	1.31	4.8	0.68	-	-	4.8	0.68	-	-	-	-	"	"	0	2
237	"	28	1.45	5.7	0.75	-	-	5.7	0.75	-	-	-	-	"	"	0	2
238	"	43.3	1.64	11.7	1.07	-	-	9.1	0.96	2.6	0.41	8	0.91	"	"	-	41
239	Dynamite	12	1.08	4.5	0.65	-	-	4.5	0.65	-	-	-	-	8.47	"	-	4
240	TNT	9	0.95	4	0.60	-	-	4	0.60	-	-	-	-	8.57	Clay	-2	4
241	"	42	1.62	13	1.11	-	-	-	-	-	-	-	-	"	"	-	41
242	"	44	1.64	14	1.15	-	-	-	-	-	-	-	-	"	"	-	41
243	"	46	1.66	13	1.11	-	-	-	-	-	-	-	-	"	"	-	41
244	"	46.2	1.66	17.8	1.25	-	-	13.8	1.14	4	0.60	-	-	"	Soil	-	41
245	"	47	1.67	12	1.08	-	-	-	-	-	-	-	-	"	Clay	-	41
246	"	48	1.68	12	1.08	-	-	10	1.00	2	0.30	-	-	"	Soil	-	41
247	Gelatine	17	1.23	3	0.48	-	-	3	0.48	-	-	-	-	8.60	"	-8	5
248	TNT	18	1.26	3	0.48	-	-	3	0.48	-	-	-	-	8.61	"	0	2
249	"	18.5	1.27	3.5	0.54	-	-	3.5	0.54	-	-	-	-	"	"	0	2
250	"	19	1.28	5	0.74	-	-	5.5	0.74	-	-	-	-	"	"	0	2
251	"	20	1.30	4	0.60	-	-	4	0.60	-	-	-	-	"	"	0	2
252	"	21	1.32	5.5	0.74	-	-	5.5	0.74	-	-	-	-	"	"	0	2
253	"	21	1.32	4.5	0.65	-	-	4.5	0.65	-	-	-	-	"	"	0	2
254	"	24.2	1.38	9.5	0.98	-	-	9.5	0.98	-	-	-	-	"	"	0	2
255	"	24.7	1.39	6	0.78	-	-	6	0.78	-	-	-	-	"	"	0	2
256	"	33.4	1.52	13.5	1.13	-	-	13.5	1.13	-	-	-	-	"	"	0	2
257	"	11	1.04	4	0.60	-	-	4	0.60	-	-	-	-	"	Clay	0	2
258	Black Powder	17	1.23	4	-	-	-	-	-	-	-	-	-	"	Soil	-2	6
259	TNT	52.7	1.72	16.4	1.21	-	-	12.0	1.08	4.4	0.64	9.5	0.98	"	"	-	41
260	"	45	1.65	12	1.08	-	-	-	-	-	-	-	-	"	Clay	-	41
261	"	48	1.68	14	1.15	-	-	-	-	-	-	-	-	"	"	-	41
262	"	48.3	1.68	15.5	1.19	-	-	13.5	1.13	2	0.30	-	-	"	Soil	-	41
263	"	50	1.70	15	1.18	-	-	-	-	-	-	-	-	"	Clay	-	41
264	"	51.2	1.71	15	1.18	-	-	10.5	1.02	4.5	0.65	-	-	"	Soil	-	41
265	"	51.8	1.71	19.5	1.29	-	-	16.2	1.21	3.3	0.52	-	-	"	"	-	41
266	"	54	1.73	11	1.04	-	-	7.5	0.88	3.5	0.54	-	-	"	"	-	41
267	"	54	1.73	17	1.23	-	-	-	-	-	-	-	-	"	"	-	41
268	"	56	1.75	17	1.23	-	-	-	-	-	-	-	-	"	Clay	-	41
269	"	57	1.76	17	1.23	-	-	-	-	-	-	-	-	"	"	-	41
270	"	60	1.78	15	1.18	-	-	-	-	-	-	-	-	"	"	-	41
271	Nitroglycerine	20	1.30	6	0.78	-	-	6	0.78	-	-	-	-	8.67	Soil	-2	7
272	Gelatine	16	1.20	2	0.30	-	-	2	0.30	-	-	-	-	8.83	"	-2	8
273	Nitroglycerine	"	"	4	0.60	-	-	4	0.60	-	-	-	-	8.86	"	-2	9
274	"	30	1.48	4.5	0.65	-	-	4.5	0.65	-	-	-	-	"	"	-2	10
275	TNT	21	1.32	5.8	0.76	-	-	5.8	0.76	-	-	-	-	8.87	"	0	2
276	"	21.5	1.33	3.8	0.57	-	-	3.8	0.57	-	-	-	-	"	"	0	2
277	"	21.8	1.34	7.2	0.86	-	-	7.2	0.86	-	-	-	-	"	"	0	2
278	"	22	1.34	5.5	0.74	-	-	5.5	0.74	-	-	-	-	"	"	0	2
279	"	24	1.38	6	0.78	-	-	6	0.78	-	-	-	-	"	"	0	2
280	TNT	57	1.76	16	1.20	-	-	-	-	-	-	-	-	"	"	-	41
281	"	62	1.79	20	1.30	-	-	15.3	1.18	4.7	0.67	17.3	1.24	"	"	-	41
282	Black Powder	30	1.48	15	1.18	-	-	15	1.18	-	-	-	-	8.89	"	-8	11
283	TNT	40.4	1.61	10	1.00	-	-	6.8	0.83	3.2	0.50	-	-	9.06	"	2.6	3
284	"	42.4	1.63	14.5	1.16	-	-	12.0	1.08	2.5	0.40	-	-	"	Clay	2.6	3
285	"	52.5	1.72	10.3	1.01	-	-	7.5	0.88	2.8	0.44	-	-	"	Soil	7	3
286	"	63.5	1.80	16	1.20	-	-	12	1.08	4	0.60	-	-	"	"	7	3
287	"	45.6	1.66	15.3	1.18	-	-	12.5	1.10	2.8	0.45	-	-	"	Clay	7	3
288	"	55.1	1.74	17.8	1.25	-	-	15.0	1.18	2.8	0.45	-	-	"	"	7	3
289	"	48.7	1.69	18	1.26	-	-	15.5	1.19	2.5	0.40	-	-	"	"	7	3
290	"	48.7	1.69	15.9	1.20	-	-	13.5	1.13	2.4	0.38	-	-	"	"	7	3
291	Dynamite	-	-	4	0.60	-	-	4	0.60	-	-	-	-	9.09	Soil	-1	12
292	TNT	17	1.23	2.8	0.44	-	-	2.8	0.44	-	-	-	-	9.17	"	0	2
293	"	18	1.26	3	0.48	-	-	3	0.48	-	-	-	-	"	"	0	2
294	"	26	1.41	4	0.60	-	-	4	0.60	-	-	-	-	"	"	0	2
295	"	31	1.49	6	0.78	-	-	6	0.78	-	-	-	-	"	"	0	2
296	"	32	1.51	6	0.78	-	-	6	0.78	-	-	-	-	"	"	0	2

Table 4

Chemical and Nuclear Explosive Craters

No.	Explosive	Apparent Diameter (feet)	D	Apparent Depth (feet)	d	True Diameter (feet)	D_1	True Depth (feet)	d_1	T = Rim Height (feet)	R_H	Apparent Rim Width (feet)	R_W	E	Nature of Ground	H (ft.)	Notes
297	TNT	38	1.58	7.5	0.88	-	-	7.5	0.88	-	-	-	-	9.17	Soil	0	2
298	"	76	1.88	24.1	1.38	-	-	18.4	1.26	5.7	0.76	19.9	1.30	"	"	-	41
299	Nitroglycerine	18	1.26	3	0.48	-	-	3	0.48	-	-	-	-	9.18	"	-2	13
300	Black Powder	25	1.40	3.2	0.51	-	-	3.2	0.51	-	-	-	-	9.27	"	-2	14
301	"	40	1.60	9	0.95	-	-	9	0.95	-	-	-	-	9.40	"	-3	15
302	Dyn. & Nitro.	14	1.15	2.5	0.40	-	-	2.5	0.40	-	-	-	-	9.53	"	-2	16
303	Gunpowder	171	2.23	36	1.56	-	-	28	1.45	8	0.90	-	-	9.60	"	-	-
304	Black Powder	-	-	6	0.78	-	-	6	0.78	-	-	-	-	9.84	"	-3	17
305	Nitroglycerine	40	1.60	5	0.70	-	-	5	0.70	-	-	-	-	9.85	"	-2	18
306	Smokeless Pwdr	50	1.70	18	1.26	-	-	18	1.26	-	-	-	-	9.93	"	0	-
307	TNT	50	1.70	15	1.18	-	-	15	1.18	-	-	-	-	10.00	"	0	-
308	Black Powder	-	-	45	1.65	-	-	-	-	-	-	-	-	10.02	"	15	42
309	Dynamite	60	1.78	10	1.00	-	-	10	1.00	-	-	-	-	10.04	"	-4	19
310	Ammonal plus	139	2.14	21	1.32	105	2.02	14	1.15	7	0.85	38	1.58	10.11	"	55	49
311	"	159	2.20	30	1.48	141	2.15	25	1.40	5	0.70	21	1.32	10.12	"	55	50
312	Dynamite	75	1.88	30	1.48	-	-	30	1.48	-	-	-	-	10.15	"	-8	20
313	"	40	1.60	15	1.18	-	-	15	1.18	-	-	-	-	"	"	-4	21
314	Black Powder	50	1.70	15	1.18	-	-	15	1.18	-	-	-	-	10.16	"	0	22
315	Dynamite	90	1.95	15	1.18	-	-	15	1.18	-	-	-	-	"	"	-8	23
316	Ammonal	235	2.37	25.5	1.41	195	2.29	22	1.34	3.5	0.54	44	1.64	"	"	60	51
317	Dynamite	120	2.08	20	1.30	-	-	20	1.30	-	-	-	-	10.24	"	-8	24
318	"	108	2.03	20	1.30	-	-	20	1.30	-	-	-	-	10.25	"	0	-
319	"	40	1.60	5.5	0.74	-	-	5.5	0.74	-	-	-	-	10.26	"	-4	25
320	TNT	160	2.20	33	1.52	-	-	23	1.36	10	1.00	-	-	"	"	17.5	3
321	"	138	2.14	50.8	1.71	-	-	42	1.62	8.8	0.94	-	-	"	Clay	17.5	3
322	Dynamite & BP	50	1.70	15	1.18	-	-	15	1.18	-	-	-	-	10.30	Soil	-4	-
323	Ammonal	242	2.38	37	1.57	202	2.31	30	1.48	7	0.85	45	1.66	10.34	"	50	52
324	"	263	2.42	50	1.70	217	2.34	40	1.60	10	1.00	52	1.72	"	"	62	53
325	Ammonal plus	237	2.37	54	1.73	175	2.24	49	1.69	5	0.70	69	1.84	"	"	70	54
326	"	279	2.45	50	1.70	217	2.34	46	1.66	4	0.60	69	1.84	"	"	57	55
327	"	233	2.37	38	1.58	183	2.26	29	1.46	9	0.95	55	1.74	10.40	"	60	56
328	Ammonal	198	2.30	12	1.08	182	2.26	10	1.00	2	0.30	17	1.23	10.42	"	75	57
329	Black Powder	75	1.88	25	1.40	-	-	25	1.40	-	-	-	-	"	"	0	26
330	TNT	270	2.43	70	1.85	220	2.34	55	1.74	15	1.18	90	1.95	10.44	Chalk	52	43
331	Ammonal	268	2.43	32	1.51	228	2.36	28	1.45	4	0.60	49	1.69	10.46	Soil	75	58
332	"	177	2.25	58	1.76	130	2.11	40	1.60	18	1.26	79	1.90	"	"	65	59
333	Ammonal plus	285	2.45	43	1.63	235	2.37	34	1.53	9	0.95	55	1.74	10.55	"	57	60
334	Ammonal	252	2.40	20	1.30	210	2.32	16	1.20	4	0.60	45	1.65	10.56	"	76	61
335	Ammonal plus	253	2.40	53.5	1.73	204	2.31	41	1.61	12.5	1.10	26	1.41	10.59	"	90	62
336	Ammonal	206	2.31	4	0.60	200	2.30	0	-	4	0.60	7	0.85	10.64	"	103	63
337	"	340	2.53	85.5	1.93	273	2.44	63	1.80	22.5	1.35	39	1.59	10.71	Sd,Cl	100	44
338	50/50 Amatol	102	2.01	13.1	1.12	80	1.90	7.7	0.89	5.4	0.73	28	1.43	10.75	"	-2	27
339	"	115	2.06	15.1	1.18	88	1.94	10.9	1.04	4.2	0.62	10	1.00	10.75	"	-2	27
340	Ammonal plus	268	2.43	55	1.74	240	2.38	46	1.66	9	0.95	31	1.49	10.79	Soil	70	64
341	Ammonal	306	2.49	53	1.72	250	2.40	40	1.60	13	1.11	62	1.79	10.82	"	88	65
342	Ammonal plus	261	2.42	35	1.54	205	2.31	23	1.36	12	1.08	62	1.79	10.84	"	100	66
343	Ammonal	224	2.35	25	1.40	176	2.25	17	1.23	8	0.90	53	1.72	"	"	125	67
344	Smokeless Pwdr	150	2.18	60	1.78	-	-	60	1.78	-	-	-	-	10.87	"	-4	28
345	Dynamite	50	1.70	22	1.34	-	-	22	1.34	-	-	-	-	11.03	"	-4	29
346	TNT	127	2.10	18.6	1.27	112	2.05	14.9	1.17	3.7	0.57	22	1.34	11.06	So,Cl	-2	30
347	"	159	2.20	15.8	1.20	122	2.09	9.6	0.98	6.2	0.79	31	1.49	"	"	-2	31
348	"	180	2.26	13.4	1.13	137	2.14	9.0	0.95	4.4	0.64	30	1.48	"	"	-8	32
349	"	257	2.41	77	1.89	240	2.38	60	1.78	17	1.24	-	-	11.16	Clay	35	3
350	"	207	2.32	59	1.77	194	2.29	47	1.67	12	1.08	-	-	11.30	Sdstn	35	3
351	Dynamite	120	2.08	12	1.08	-	-	12	1.08	-	-	-	-	"	Soil	-4	33
352	Torpex	165	2.22	22.8	1.36	142	2.15	17.8	1.25	5	0.70	29	1.46	11.31	So,Cl	-8	34
353	"	183	2.26	16.2	1.21	146	2.16	12	1.08	4.2	0.62	30	1.48	"	"	-8	34
354	TNT	184	2.26	21.4	1.33	138	2.14	16	1.20	5.4	0.73	29	1.46	11.36	"	-2	35
355	"	201	2.30	18.6	1.27	165	2.22	13.5	1.13	5.1	0.71	22	1.34	"	"	-8	36
356	"	210	2.32	14.8	1.17	177	2.25	10	1.00	4.8	0.68	25	1.40	"	"	-8	36
357	Amm. Nitrate	150	2.18	30	1.48	-	-	30	1.48	-	-	-	-	11.45	Soil	-5	37
358	Black Powder	158	2.20	45	1.65	-	-	30	1.48	15	1.18	-	-	11.57	"	0	38
359	TNT	338	2.53	98	1.99	315	2.49	77	1.89	21	1.33	-	-	11.66	"	125	46
360	Note 39	428	2.63	71	1.85	398	2.60	61	1.79	10	1.00	-	-	12.18	Soil	-20	39
361	TNT	328	2.52	148	2.17	-	-	-	-	-	-	-	-	12.38	"	?	40
362	Amatol plus	800	2.90	108	2.03	741	2.87	89	1.95	19	1.28	200	2.30	12.39	Gyps.	95	45
363	Nuclear	339	2.53	104	2.02	289	2.46	91	1.96	13	1.11	-	-	12.08	Alluv.	67	47
364	"	275	2.44	60	1.78	248	2.39	44	1.64	12	1.08	-	-	"	"	18	48
365	"	255	2.41	-	-	212	2.33	35	1.54	-	-	-	-	-	Tuff	110	68

NOTES FOR TABLE 4

D = logarithm to the base 10 of apparent crater diameter in feet

d = logarithm of apparent depth (feet)

D_1 = logarithm of the true diameter (feet)

d_1 = logarithm of true depth (feet)

R_H = logarithm of rim height (feet)

R_W = logarithm of apparent rim width (feet)

E = logarithm of effective energy of explosion in calories

H = the distance from the center of explosion to the ground level in feet. For Craters Nos. 1–128, the tabular value is expressed in inches. Minus value indicates an above-ground burst

1. Experiment by author, 1956–57.
2. Bomb—instantaneous fuze—corrected for effects of casing.
3. Nearly spherical charge.
4. Jeanette, Pa., 1910. 500 lb. dynamite (47).
5. Perranporth, Cornwall, 1902. 533 lb. gelatin.
6. Faversham, England, 1880. 720 lb. black powder.
7. Uplee's Marshes, Faversham, 1903. 650 lb. nitroglycerin.
8. Upton Towans, Cornwall, 1899. 900 lb. gelatin.
9. Lower Hope Point, England, 1902. 1,000 lb. nitroglycerin.
10. Uplee's Marshes, Faversham, 1903. 1,000 lb. nitroglycerin.

439

11. Winsted, Conn., 1892. 2,500 lb. black powder.
12. Hazelton, Pa., 1905. 2,100 lb. dynamite.
13. Umbogintwini, Natal, 1909. 2,100 lb. nitroglycerin.
14. Cabot, Pa., 1910. 6,000 lb. black powder. Crater $20 \times 30 \times 3\frac{1}{4}$ feet.
15. Mt. Carmel, Pa., 1907. 8,000 lb. black powder.
16. Marquette, Mich., 1905. 6,080 lb. dynamite and nitroglycerin.
17. Bridgeport, Conn., 1906. 22,300 lb. black powder.
18. Barksdale, Wis., 1906. 9,670 lb. nitroglycerin.
19. Newburgh Heights, Ohio, 1912. 18,000 lb. dynamite.
20. Tellico, Tenn., 1906. 22,500 lb. dynamite.
21. Yreka, Calif., 1906. 24,300 lb. dynamite.
22. McAlester, Okla., 1908. 44,950 lb. black powder.
23. Council Bluffs, Iowa, 1881. 25,000 lb. dynamite.
24. Reddick, Ill., 1907. 30,000 lb. dynamite.
25. Kimberley, South Africa, 1884. 31,500 lb. dynamite.
26. Erith, England, 1864. 83,500 lb. black powder.
27. Arco, Idaho, 1945. 125,000 lb. 50/50 amatol in revetment.
28. Batuco, Chile, 1908. 176,000 lb. smokeless powder.
29. Highland Station, Calif., 1892. 207,100 lb. dynamite.
30. Arco, Idaho, 1945. 250,000 lb. TNT in stack.
31. Arco, Idaho, 1946. 250,008 lb. TNT in revetment.
32. Arco, Idaho, 1946. 250,170 lb. TNT in igloo.
33. Manila, P.I., 1924. 346,000 lb. dynamite.
34. Arco, Idaho, 1945. 250,000 lb. torpex in igloo.
35. Arco, Idaho, 1946. 300,005 lb. TNT in revetment.
36. Arco, Idaho, 1946. 500,340 lb. TNT in igloo.
37. Perth Amboy, N.J., 1918.
38. Morgan, N.J., 1918. 1,000,000 lb. ammonium **nitrate**.
39. Oppau, Germany, 1921, 4,500 tons ammonium **nitrate** plus other chemicals. Primarily $2NH_4NO_3$-$(NH_4)SO_4$. Crater slightly elliptical. Average values taken from contour map.
40. Heligoland Explosion—destruction of fortifications. Was not a point source. Not used in analysis.
41. Bomb—delay fuze—corrected for effect of casing. Depth of burst unknown.
42. Fort Lyons, Washington, D.C., 1863. 32,000 lb. black powder in underground magazine.
43. La Boiselle—military mine—World War I. Rim unusually wide. 2 charges 50 feet apart.
44. Hill 60B, Caterpillar—military mine—World War I.
45. Burton-on-Trent, England (or Fauld), 1944. Amatol 54 per cent. TNT or TNT/aluminum 39 per cent. More powerful fillings 7 per cent. Total effective energy $= (9.1–11.8) \times 10^{19}$ ergs. Bombs were stored in an old gypsum mine with about 90 feet of head cover, consisting of 4 feet subsoil, 60 feet marl, 10 feet gravel, and up to 20 feet of gypsum beds. Crater was oval, 720×900 feet. Data furnished by D. E. Jarrett, Ministry of Supply, Armament Research and Development Establishment, Woolwich, S.E. 18, England.
46. Data in letter from Gerald W. Johnson, associate director, Lawrence Radiation Laboratory. Test conducted as part of "Plowshare Program."
47. Nuclear Test Teapot Ess. Dimensions scaled from Shoemaker's drawing (33), 1.2 ± 0.05 metric kilotons.
48. Nuclear Test Jangle U. Dimensions scaled from Shoemaker's drawing (33), 1.2 ± 0.05 metric kilotons.
49. Hollandscheschuur #2, military mine, World War I. 12,500 lb. ammonal, 2,400 lb. blastine.
50. Hollandscheschuur #3, military mine, World War I. 15,000 lb. ammonal, 2,500 lb. blastine.
51. Trench 122 #5 left, military mine, World War I. 20,000 lb. ammonal.
52. Kruisstraat #3, military mine, World War I. 30,000 lb. ammonal.
53. Kruisstraat #2, military mine, World War I. 30,000 lb. ammonal.
54. Petit Bois #1 right, military mine, World War I. 21,000 lb. ammonal, 9,000 lb. blastine.
55. Petit Bois #2 left, military mine, World War I. 21,000 lb. ammonal, 9,000 lb. blastine.
56. Hollandscheschuur #1, military mine, World War I. 30,000 lb. ammonal, 4,200 lb. blastine.
57. Trench 127 #7 left, military mine, World War I. 36,000 lb. ammonal.
58. Trench 122 #6 right, military mine, World War I. 40,000 lb. ammonal.
59. Hawthorn Ridge Redoubt, military mine, World War I. 40,000 lb. ammonal.
60. Kruisstraat #1–4, military mine, World War I. 30,000 lb. ammonal, 18,500 lb. ammonal, 1,000 lb. gun cotton.
61. Trench 127 #8 right, military mine, World War I. 50,000 lb. ammonal.
62. Hill 60A left, military mine, World War I. 45,700 lb. ammonal, 7,800 lb. gun cotton.
63. Ontario Farm, military mine, World War I. 60,000 lb. ammonal.
64. Peckham, military mine, World War I. 65,000 lb. ammonal, 15,000 lb. blastine, 7,000 lb. gun cotton.
65. Spanbroekmolen, military mine, World War I. 91,000 lb. ammonal.
66. Maedelstede Fm, military mine, World War I. 90,000 lb. ammonal, 4,000 lb. gun cotton.
67. St. Eloi, military mine, World War I (45). 95,600 lb. ammonal.
68. Neptune nuclear explosion, October 14, 1958, Nevada Test Site (46).

Reference numbers in these notes refer to authors cited in Chap. 6.

Table 5

Heat of Explosion for Chemical Explosives

Explosive	Heat of Explosion (Cal/gm)	Value Used	Notes
Ammonium Nitrate	660	660	1
Black Powder	680-720	700	2,3
Blasting Powder	700ca	700	2
Gun Powder	738	738	4
Explosive "D"	800ca	800	2,3
Nitrocellulose (13% N_2)	931	931	4
Amatol (20-50% TNT)	920-1004	960	2,3
TNT	925-1080	1000	2,3
Dynamite (40% straight)	1255	1265	3
Dynamite (75/25 NG/Kieselguhr)	1290	1265	2,3
RDX	1300-1370	1335	2,3
Ammonal (73/23.5/4.5 NH_4NO_3/Al/C)	1578-1600	1600	2,4
Nitroglycerine	1600-1652	1620	2,3,4
Blasting Gelatine (NG/NC 93/7)	1640	1640	2,3
Torpex	1800ca	1800	2

NOTES FOR TABLE 5

1. Averaged from several sources.
2. Letter from Robert Frye, Assistant, Samuel Feltman Ammunition Laboratories, Picatinny Arsenal, Dover, N.J.
3. Letter from George E. King, Captain, U.S. Navy, Commanding Officer, U.S. Naval Powder Factory, Indian Head, Md.
4. *Smithsonian Physical Tables,* Ninth Revised Edition, Forsythe, Washington, D.C., 1956. Table 177, p. 183.

Table 6

Terrestrial Meteorite Craters

Name	Apparent Diameter (feet)	D	True Diameter (feet)	D_1	Apparent Depth (feet)	d	True Depth (feet)	d_1	Rim Height (feet)	R_H	Apparent Rim Width (feet)	R_W	Nature of Ground	Notes
New Quebec	11290	4.05	9115	3.96	1355	3.13	855	2.93	500	2.70	2000	3.30	Granite	
Arizona	4000	3.60	3250	3.51	680	2.83	500	2.70	180	2.26	800	2.90	Sandstone	
Wolf Creek	2700	3.43			(200)	—			100	2.00	470	2.67	Quartzite	
Clot de Cabrerolles	722	2.86			164	2.21							Schist	
Odessa #1	550	2.74			(106.7)	(2.03)	81.5	1.91	(25.2)	(1.40)			Sedimentary Rock	1
Wabar #1	319	2.50			53.4	1.73							Sand over rock	
Kaali Jarv	319	2.50	263	2.42	50	1.70	36.4	1.56	17	1.23	75	1.88	Dolomite	2
Chagigan Toushtou #1	260	2.41			75	1.88					63	1.80	Limestone	
Faugères	197	2.29			20	1.30							Schist	
Sikhote Alin #1	90	1.95			17	1.23							Clay and rock	
Odessa #2	70	1.85			10	1.00							Soil	
Henbury #13	30	1.48											Sandstone	

Where
  D = Logarithm of apparent crater diameter in feet
  d = Logarithm of apparent depth in feet
  D_1 = Logarithm of true diameter in feet
  d_1 = Logarithm of true depth in feet
  R_H = Logarithm of rim height in feet
  R_W = Logarithm of apparent rim width in feet

NOTES
1. Rim height estimated from eq. (7-6). From true depth of 81.5 feet the apparent depth of uneroded crater thus found.
2. Assuming 4 feet of muck on bottom.

Table 7

Lunar Craters

Name	Class	Apparent Diameter (miles)	D (feet)	Apparent Depth (feet)	d (feet)	Rim Height (feet)	R_H (feet)	Apparent Rim Width (miles)	R_W (feet)	Central Peak	Notes
Bailly	5	183.5	5.986	13000	4.11					0	
Clavius	2	144	5.881	16100	4.21	5400	3.73	35.3	5.27	2,4,5	
Unnamed	4	139	5.866	(5000)	(3.70)					0	Near Walter
Schickard	5	134	5.850	8500	3.93	(6400)	(3.81)	41.7	5.34	0	
W. Humboldt	2	130	5.836	14000	4.15					4,6	
Grimaldi	5	127	5.826	8700	3.94					0	
Maginus	3	116	5.787	14800	4.17			19.5	5.01	2,4,5	
Schiller	3	112	5.772	13000	4.11	(9000)	(3.95)			0	Multiple Crater?
Petavius	5	110	5.764	8000	3.90					2,4,7	
Riccioli	3	99	5.718	(11200)	(4.05)					?	Hill probably not CP
Hipparchus	5	95.2	5.701	(6900)	(3.84)					2,4,5	
Vendelinus	5	93.5	5.693	8400	3.92	(4900)	(3.69)			0	
Ptolemaeus	5	93.3	5.693	4000	3.60	(8200)	(3.91)			0	
Longomontanus	2	92.4	5.688	14800	4.17	(6100)	(3.79)	17.6	4.97	2,4,5	
Newton	1	85	5.652	20000	4.30					?	Multiple Crater?
Stoflerus	2	84.4	5.649	12300	4.09	(4800)	(3.68)			0	
Albategnius	5	83.0	5.642	9400	3.97					1,2,6	
Walter	4	82.2	5.638	9800	3.99	(6700)	(3.83)			2,4,7	
Langrenus	1	82.2	5.638	13300	4.12	3600	3.56			2,4,7	
Cleomedes	5	82.0	5.636	9700	3.99	(5200)	(3.72)			2,4,5	
Furnerius	3	81.1	5.632	10000	4.00					0	
Pythagoras	1	80	5.626	16100	4.21					2,4,7	
Purbach	5	77.1	5.610	7400	3.87					2,4,6	CP modified by crater
Endymion	5	77.0	5.609	8400	3.92					0	
Neper	5	75	5.598	6000	3.78					2,4,7,8	
Hommel	3	74.8	5.597	(11000)	(4.04)					2,4,6	
Fracastorius	5	74.8	5.597	7700	3.89					1,2,5	
Phocylides	3	74.6	5.595	7700	3.89	(6600)	(3.82)			0	
Orontius	3	73.9	5.591	(10200)	(4.01)	(5900)	(3.77)			0	
Alphonsus	5	72.8	5.585	6400	3.81					1,2,5	CP modified by ridge
Moretus	1	72.7	5.584	14600	4.16			13.3	4.85	1,2,7	
Letronne	5	72.5	5.583	3300	3.52					2,4,5	
Maurolycus	2	72.3	5.582	14300	4.16					4,7	
Blancanus	1	72.0	5.580	12000	4.08					2,4,5	
Scheiner	2	71.4	5.576	14600	4.16					9	
Gassendi	5	68.6	5.559	6600	3.82					2,4,7	
Hevel	5	68.5	5.558	6000	3.78					1,5,8	
Theophilus	1	64.9	5.535	14400	4.16	3800	3.58	16.0	4.93	4,7	
Wilhelm I	3	63.9	5.525	10000	4.00					0	
Posidonius	5	62.9	5.521	6700	3.83	3300	3.52			1,5	
Plato	5	62.6	5.519	7900	3.90					0	
Catherina	4	62.5	5.519	9000	3.95					1,5	
Boguslawsky	2	60.9	5.507	11200	4.05					0	
Arzachel	3	60.8	5.507	10800	4.03	(5900)	(3.77)	10.0	4.72	1,2,7,8	
Rosenberger	3	60.8	5.507	(7100)	(3.85)					1,5	
Pontecoulant	5	60	5.501	6000	3.78					1,5	
Manzinus	1	60.0	5.501	12500	4.10					0	
Sacrobosco	3	60.0	5.501	12000	4.08					2,4,5	
Casatus	2	59.4	5.496	18000	4.26	7400	3.87			0	
Grueberger	2	58.1	5.487	14400	4.16					4,5	
Cyrillus	3	58.0	5.486	11600	4.06	4800	3.68			2,4,7	
Inghirami	2	57.2	5.480	12000	4.08					1,5	
Copernicus	1	56.7	5.476	11000	4.04	3300	3.52	14.0	4.87	4,6	
Piccolomini	1	55.7	5.468	12100	4.08	(3900)	(3.59)			4,7	
Vlacq	1	55.5	5.467	10700	4.03					4,6	
Anaximander	5	55.4	5.466	5900	3.77					0	
Gemma Frisius	3	55.2	5.462	15300	4.18					1,2,5	
Aristoteles	1	54.7	5.461	10000	4.00	(3600)	(3.56)			2,4,5	
Tycho	1	54.0	5.455	14000	4.15	7900	3.90	12.6	4.82	4,7	
Atlas	5	53.9	5.454	7700	3.89	4300	3.63			2,4,6	
Geminus	1	53.8	5.453	16000	4.20	3900	3.59			2,4,6,8	
Barocius	2	53.8	5.453	11800	4.07	10500	4.02			4,5	
Wurzelbauer	5	53.6	5.452	(5600)	(3.75)					1,2,5	
Metius	2	53.5	5.451	10200	4.01	(5700)	(3.76)			2,4,5,8	
Vieta	2	53.1	5.448	11300	4.05					4,5	
Wargentin	5	52.9	5.446	200	2.30					0	
Schomberger	1	52.1	5.439	14900	4.17					4,6	
Mersenius	3	51.1	5.431	7900	3.90	4300	3.63			4,5	
Byrgius	3	50.9	5.429	7000	3.85					0	
Pitiscus	2	50.7	5.428	10200	4.01					1,2,7	
Snellius	1	50.4	5.425	9700	3.99					4,5	
Aliacensis	1	50.1	5.422	13100	4.12					1,2,5	
Archimedes	5	50.0	5.422	6100	3.79	4800	3.68			0	
Anaximines	5	48.6	5.409	8000	3.90					0	
Condorcet	5	48.6	5.409	8500	3.93					1,5	
Cuvier	1	48.4	5.407	10200	4.01	(3400)	(3.53)			1,2,5	
Fabricius	2	48.1	5.405	11500	4.06					1,7	
Kircher	1	47.9	5.403	(14100)	(4.15)					0	
Mutus	2	47.4	5.398	11800	4.07					0	
Licetus	1	47.1	5.396	11600	4.06	(4900)	(3.69)			2,4,5	
Clairaut	2	46.9	5.394	8900	3.95					9	
Colombo	5	46.0	5.385	7200	3.86					4,6	
Stevinus	1	46.0	5.385	10200	4.01	(5600)	(3.75)			1,7	
Legendre	3	46.0	5.385	(7700)	(3.89)					1,5	
Hainzel	2	45.5	5.381	11800	4.07	(5200)	(3.72)			yes	Double Crater
Berosus	1	45.3	5.379	11800	4.07					4,5	Two small hills
Faraday	2	44.7	5.373	12600	4.10	(5700)	(3.76)			1,5	

Table 7

Lunar Craters

Name	Class	Apparent Diameter (miles)	D (feet)	Apparent Depth (feet)	d (feet)	Rim Height (feet)	R_H (feet)	Apparent Rim Width (miles)	R_W (feet)	Central Peak	Notes
Werner	1	44.2	5.368	14400	4.16	5700	3.76			2,4,5	
Gutenberg	5	44.2	5.368	5700	3.76					9	
Hase	2	44.1	5.367	7500	3.88					?	
Rheita	1	43.7	5.363	10200	4.01					1,2,7	
Philolaus	1	43.7	5.363	12000	4.08					4,5	
Baco	1	43.6	5.362	10200	4.01					1,2,5	Very small hill
Bettinus	1	43.4	5.360	12500	4.10					1,6	
Zach	2	43.2	5.358	11300	4.05					2,4,6	
Nearchus	1	43.0	5.356	10800	4.03					0	
Reichenbach	1	42.9	5.355	12000	4.08					1,5	
Lacaille	3	42.8	5.354	9000	3.95					0	
Olbers	1	41.8	5.344	10000	4.00					1,5	
Hercules	1	41.7	5.343	11000	4.04					2,4,5	Two small hills
Eudoxus	1	41.6	5.342	10500	4.02	8200	3.91			2,4,5	
Simpelius	1	41.5	5.341	14900	4.17					1,2,5	
Mercurius	2	41.5	5.341	7500	3.88					4,5	
Apianus	2	41.4	5.340	9000	3.95					0	
Abulfeda	1	41.3	5.339	9500	3.98					0	
Segner	3	41.3	5.339	8400	3.92	(6400)	(3.81)			1,2,5	
Jacobi	2	41.2	5.338	10300	4.01	3600	3.56			9	
Heinsius	3	40.9	5.334	9700	3.99					9	Probable CP
Watt	1	40.7	5.332	11000	4.04					2,4,6	
Stadius	5	40.6	5.331	130	2.11					0	
Santbech	1	40.5	5.330	11200	4.05	(3600)	(3.56)			1,2,5	
Fernelius	3	40.2	5.327	(5700)	(3.76)					0	
Capuanus	5	39.7	5.321	6000	3.78					0	
Macrobius	1	39.6	5.320	13000	4.11	(5100)	(3.71)			4,7	
Zuchius	1	39.0	5.314	10000	4.00					4,7	
Pictet	3	38.8	5.311	9700	3.99					4,6	
Lilius	1	38.4	5.307	10300	4.01					4,7	
Miller	1	38.3	5.306	10800	4.03					4,7	
LeMonnier	5	38.2	5.305	8000	3.90					0	
Cavalerius	1	38.0	5.302	10200	4.01					2,4,5	
Bullialdus	1	37.3	5.294	10500	4.02			9.5	4.70	2,4,6	
Guericke	5	36.8	5.288			2100	3.32			0	
Eratosthenes	1	36.3	5.283	10300	4.01	3300	3.52			4,7,8	
Short	1	36	5.279	14800	4.17					1,9	
Firmicus	5	35.8	5.277	5000	3.70					0	See Wilkins, Moore
Kaiser	3	35.6	5.274	(5700)	(3.76)					0	
Agatharchides	5	35.5	5.273	3800	3.58	2500	3.40			1,5	
Taruntius	5	35.3	5.270	3800	3.58	2100	3.32			1,2,6,8	
Thebit	2	35.3	5.270	9200	3.96					1,2,5	
Saussure	2	35.2	5.269	6200	3.79					0	
Cassini	5	35.2	5.269			3300	3.52			9	Probably did have CP
Aristillus	1	35.1	5.268	10300	4.01	4400	3.64	7.9	4.62	2,4,7	
Wrottesley	1	35.0	5.267	9000	3.95					4,6	
Pentland	1	34.9	5.265	12100	4.08					4,7	
Burckhardt	1	34.8	5.264	12700	4.10					1,2,6	
Scoresby	1	34.3	5.258	10500	4.02	3600	3.56			2,4,6	
Busching	3	34.0	5.254	4000	3.60					9	
Cavendish	3	34.0	5.254	6000	3.78					1,5	
Franklin	2	33.7	5.250	7900	3.90					1,2,5	
Fourier	2	32.9	5.240	7700	3.89					9	
Frauenhofer	3	32.9	5.240	5000	3.70					2,4,5	
Clavius B	1	32.8	5.238	13100	4.12	8000	3.90			4,7	
Nasireddin	1	32.5	5.235	9000	3.95					4,5	
Apollonius	5	32.4	5.233	5000	3.70					0	
Lindenau	1	32.4	5.233	8500	3.93					4,5	
Beaumont	5	32.4	5.233	5400	3.73	3000	3.48			4,6	
Neander	2	32.4	5.233	8200	3.91					4,7	
Anaxagoras	1	31.9	5.226	10000	4.00					1,5	
Almanon	2	31.2	5.217	6600	3.82					0	
Vendelinus B	3	31	5.214	(8400)	(3.92)					?	
Rost	2	31.0	5.214	7200	3.86					0	
Epigenes	5	30.5	5.207	5200	3.72					2,4,6	
Cysatus	1	30.3	5.204	11800	4.07					1,2,5	
Campanus	5	30.3	5.204	5900	3.77					1,5	
Azophi	1	29.9	5.198	10200	4.01	(3800)	(3.58)			0	
Cardanus	2	29.8	5.197	4000	3.60					2,4,5	
Bayer	1	28.9	5.184	8000	3.90					0	
Hansteen	5	28.9	5.184	3000	3.48	2600	3.41			2,4,5	
Delambre	2	28.8	5.182	9800	3.99			7.0	4.57	4,5,8	
Playfair	2	28.6	5.179	8200	3.91					0	
Seleucus	1	28.5	5.177	10000	4.00					1,2,5	
Billy	5	28.5	5.177	3600	3.56	2500	3.40			0	
Agrippa	1	28.5	5.177	7500	3.88	(3800)	(3.58)			1,7	
Geber	1	28.4	5.176	8400	3.92					0	See Wilkins, Moore
Mercator	5	28.2	5.173	4800	3.68					0	
Bernouilli	1	28.2	5.173	9800	3.99					2,4,5	
Parry	5	28.2	5.173	(4800)	(3.68)	(3400)	(3.53)			0	
Stiborius	1	27.9	5.168	9400	3.97					1,2,6	
Reinhold	1	27.8	5.167	9000	3.95	2300	3.36			1,2,5	
Vitello	5	27.8	5.167			4900	3.69			1,6,8	
Kies	5	27.8	5.167			2000	3.30			0	
Rothmann	1	27.0	5.154	8500	3.93					2,4,6	
Newton C	1	27.0	5.154	14100	4.15					0	
Baco B	1	26.9	5.152	9200	3.96					1,5	

Table 7

Lunar Craters

Name	Class	Apparent Diameter (miles)	D (feet)	Apparent Depth (feet)	d	Rim Height (feet)	R_H (feet)	Apparent Rim Width (miles)	R_W (feet)	Central Peak	Notes
Eimmart	1	26.8	5.151	10000	4.00					1,5	
Taylor	1	26.8	5.151	7400	3.87					1,2,7	
Plinius	1	26.7	5.149	(7400)	(3.87)	2000	3.30			1,2,6	See Wilkins, Moore
Abenezra	1	26.3	5.143	10300	4.01	(3800)	(3.58)			1,2,5	
Sirsalis	1	26.2	5.141	(10200)	(4.01)					1,6	
Newton A	1	26.0	5.138	9400	3.97					1,6	
Newcomb	1	25.6	5.131	12000	4.08					2,4,5	
Alpetragius	1	25.6	5.131	9000	3.95	3300	3.52			1,3,7,8	
Tisserand	2	25.6	5.131	9000	3.95					4,5	
Isidorus	3	25.4	5.127	(5200)	(3.72)					0	
Herschel	1	25.4	5.127	9400	3.97					4,6,8	
Mairan	1	25.2	5.124	8700	3.94	4300	3.63			2,4,5	See Wilkins, Moore
Tacitus	1	25.1	5.122	11000	4.04					2,4,5	
Vendelinus A	1	25	5.120	(6700)	(3.83)					?	
Harpalus	1	25.0	5.120	(7900)	(3.90)	(3100)	(3.49)			2,4,5	
Barocius B	1	24.9	5.119	7400	3.87					1,2,5	
Cichus	1	24.9	5.119	8000	3.90	6700	3.83			4,5	
Christian Mayer	2	24.8	5.117	(3900)	(3.59)					1,2,7	
Marius	5	24.7	5.115	4500	3.65					1,5	
Tralles	2	24.7	5.115	(9200)	(3.96)					2,4,7	
Lilius A	1	24.7	5.115	10000	4.00					0	
Aristarchus	1	24.5	5.112	6900	3.84	2600	3.41			1,6,8	
Cepheus	1	24.5	5.112	9200	3.96					1,2,5	
Ball	2	24.5	5.112	5000	3.70					1,7	
Römer	1	24.4	5.110	11600	4.06					1,2,7,8	
Baco A	1	24.4	5.110	8400	3.92	(4800)	(3.68)			1,5	
Autolycus	1	24.4	5.110	9500	3.98	4800	3.68	4.7	4.40	4,5	
Manilius	1	24.3	5.108	7700	3.89	3000	3.48			4,7	
Lansberg	1	24.3	5.108	7700	3.89	3000	3.48	6.3	4.52	4,7	
Bianchini	1	24.1	5.105	8400	3.92					2,4,5	
Colombo A	1	24.1	5.105	8000	3.90					1,5	
Sharp	1	24.1	5.105	9200	3.96					2,4,6	
Fermat	2	24.0	5.103	6000	3.78					0	
Reichenbach H	2	24	5.103	(7400)	(3.87)					?	Identification?
Fontenelle	5	24.0	5.103	6100	3.79					9	
Bürg	1	23.9	5.101	6200	3.79					4,7,8	
Magelhaens	5	23.9	5.101	4600	3.66					0	
Wurzelbauer D	1	23.7	5.097	8000	3.90					0	
Egede	5	23.5	5.094	400	2.60					0	
Democritus	1	23.3	5.090	7000	3.85					2,4,6	
Mason	5	23.3	5.090	(6100)	(3.79)	(3400)	(3.53)			0	
Schroter	5	23.1	5.086			(5100)	(3.71)			0	
Condamine	5	23.1	5.086	(3400)	(3.53)					0	Some low hills
Liebig	1	23.1	5.086	7700	3.89	5900	3.77			1,5	Very small CP
Herodotus	5	22.5	5.075	4400	3.64					1,2,5,8	
Godin	1	22.3	5.071	7700	3.89					1,7	
Gassendi A	1	22.0	5.065	10500	4.02					4,6	
Timocharis	1	21.9	5.063	7100	3.85	3400	3.53			1,5,8or9	
Halley	2	21.8	5.061	7500	3.88	(3600)	(3.56)			0	
Miller A	1	21.4	5.053	10000	4.00					4,5	
Davy	5	21.3	5.051	(4400)	(3.64)	3400	3.53			4,5	
Kant	1	20.5	5.034	7500	3.88					1,7	
Timeaus	2	20.3	5.030	(4800)	(3.68)					4,6	
Kepler	1	20.2	5.028	7500	3.88			5.6	4.47	4,5	
Archytas	1	20.1	5.026	6200	3.79					4,6	
Sabine	5	20.1	5.026	2800	3.45	(1300)	(3.11)			4,5	
Ritter	5	20.0	5.024	2500	3.40					4,5	
Calippus	1	19.2	5.006	9700	3.99	7900	3.90			2,4,5	
Archytas A	1	19	5.001	6200	3.79					?	Identification?
Horrocks	1	18.9	4.999	8000	3.90					2,4,5	
Reiner	1	18.6	4.992	6900	3.84					1,2,6	
Lambert	1	18.4	4.987	6600	3.82	2100	3.32			1,5,8or9	
Mädler	1	17.9	4.975	7500	3.88	3600	3.56			1,7	
Vitruvius	5	17.9	4.975	4400	3.64					1,5	
Clavius D	1	17.8	4.973	6000	3.78	3100	3.49			1,5	
Encke	5	17.4	4.963	2000	3.30					2,4,6	
Gemma Frisius D	1	17.3	4.961	8900	3.95					1,5	
Hind	1	17.2	4.958	7000	3.85					2,4,5	
Proclus	1	17.1	4.956	8900	3.95					4,5	
Triesnecker	2	17.1	4.956	(5400)	(3.73)	2300	3.36			4,6	
Hypatia	2	17.1	4.956	7000	3.85					0	
Grove	1	17.0	4.953	7000	3.85					1,5	
Euler	1	16.6	4.943	6700	3.83	2300	3.36			1,6	
Ross	1	16.6	4.943	(4800)	(3.68)					1,5	
Sömmering	5	16.6	4.943	(4800)	(3.68)	(3400)	(3.53)			0	
Mösting	1	16.5	4.940	6600	3.82	(1600)	(3.20)			1,5	
Arago	1	16.4	4.937	(5900)	(3.77)					2,4,6	
Ramsden	1	16.1	4.929			2000	3.30			0	
Menelaus	1	16.0	4.927	6600	3.82					2,4,5	
Newton B	1	16	4.927	7100	3.85					1,5	
Langrenus M	1	16	4.927	9500	3.98					?	
Macrobius A	1	15.7	4.919	9500	3.98					0	
Theatatus	1	15.6	4.916	8400	3.92	3400	3.53			1,5	
Delisle	1	15.6	4.916	6200	3.79	1800	3.26			1,2,5	
Helicon	1	15.5	4.913	5700	3.76					0	
Lalande	1	15.2	4.904	6000	3.78					1,2,5	
Schiaparelli	1	15.1	4.902	(6700)	(3.83)	1800	3.26			1,5	

Table 7

Lunar Craters

Name	Class	Apparent Diameter (miles)	D (feet)	Apparent Depth (feet)	d (feet)	Rim Height (feet)	R_H (feet)	Apparent Rim Width (miles)	R_W (feet)	Central Peak	Notes
Heinsius D	1	15.0	4.899	8700	3.94					2,4,5,8	
Picard	1	14.9	4.896	6400	3.81	2500	3.40			1,5	
Grimaldi B	1	13.9	4.866	10200	4.01					1,5	
Krieger	5	13.9	4.866			2300	3.36			1,5	
König	1	13.7	4.859	(5100)	(3.71)	(1300)	(3.11)			2,4,5	
Flamsteed	1	13.6	4.856	(6200)	(3.79)	1500	3.18			4,5	
Bullialdus B	1	13.6	4.856	(5200)	(3.72)	2800	3.45			1,2,5	
Conon	1	13.3	4.846	6400	3.81					1,5	
Le Verrier	1	13.2	4.843	6200	3.79	1500	3.18			1,5	
Dawes	1	12.0	4.802	(4800)	(3.68)	2600	3.41			1,5	
Peirce	1	11.8	4.795	6600	3.82	2500	3.40			1,5	
Pytheas	1	11.8	4.795	(4800)	(3.68)	2600	3.41			4,6	
Bode	1	11.6	4.787	5000	3.70					1,2,6	
Diophantus	1	11.3	4.776	7900	3.90	2600	3.41			4,6	
Macrobius B	1	11.3	4.776	(7700)	(3.89)					0	
Kunowsky	2	10.8	4.756			2000	3.30	2.1	4.04	2,4,5	
Cassini A	1	10.5	4.744	7400	3.87	(2800)	(3.45)			9	
Manners	1	10.5	4.744							1,5	
Birt	1	10.4	4.740	5900	3.77	2500	3.40			0	
Mairan A	1	10.2	4.731	4000	3.60					0	
Bessel	1	10.0	4.723	4300	3.63	1600	3.20			0	
Gay-Lussac A	2	10.0	4.723	4800	3.68					2,4,6	
Nicollet	1	9.6	4.705	4000	3.60					0	
Picard A	1	9.2	4.686	7500	3.88	2000	3.30			0	
Galileo	1	9.1	4.682			2100	3.32			0	
Piazzi Smyth	1	9.0	4.677	3500	3.54	2100	3.32			0	
Marius A	1	8.9	4.672	(7400)	(3.87)	2100	3.32			0	
Caroline Herschel	1	8.7	4.662	3000	3.48	2300	3.36			1,5	
Hortensius	1	8.1	4.631			1600	3.20	1.8	3.97	1,5	
Milichius	1	8.1	4.631	(3100)	(3.49)	(2000)	(3.30)	1.4	3.87	0	
Gwilt	1	8	4.626			1500	3.18			?	
Picard E	1	8	4.626	4400	3.64					?	Identification?
Kepler C	1	7.4	4.592			1800	3.26			0	
Galileo A	1	7.3	4.586			1600	3.20			0	
Euclides	1	7.2	4.580	2000	3.30					0	
Carlini	1	7.1	4.574	[2000]	[3.30]					1,5	True depth
Posidonius A	1	7	4.568	(3800)	(3.58)	(2000)	(3.30)			0	
Peirce A	1	7	4.568			1600	3.20			1,5	
Pico B	1	7.0	4.568			1800	3.26			1,5	
W.H.Pickering	1	6.5	4.536	6700	3.83	2100	3.32			4,5	
Brayley B	1	6.5	4.536	3000	3.48	1000	3.00			0	
Beer	1	6.4	4.529	(3000)	(3.48)	1500	3.18			0	
Kepler A	1	6.3	4.522					1.3	3.82	0	
Milichius A	1	6.3	4.522			3100	3.49	1.0	3.74	0	
Feuillé	1	6.2	4.515	2000	3.30					0	
Luther	1	5.9	4.493			1300	3.11			0	
Laplace A	1	5.8	4.486			1600	3.20			1,6	See Wilkins,Moore
Lansberg A	1	5.6	4.471					1.1	3.78	0	
Kies B	1	5.4	4.455	3600	3.56					0	
Murchison A	1	5	4.422	3000	3.48					0	
Bullialdus F	1	5	4.422	3800	3.58					0	
SE of Purbach	1	5	4.422	1650	3.21					0	
Hortensius A	1	4.9	4.413					1.3	3.82	0	
D east of Heinsius	1	4.7	4.395	3200	3.50					0	
Diophantus A	1	4.7	4.395	4800	3.68					0	
N of Lambert	1	4.6	4.385	2800	3.45					0	
Copernicus D	1	4.3	4.356	4200	3.62					0	
Herodotus B	1	4	4.325			1000	3.00			?	
Bullialdus E	1	4	4.325	3900	3.59					0	
Lubiniesky I	1	4	4.325	3800	3.58					?	Identification?
Birt D	1	4	4.325	1950	3.29					0	
Hortensius B	1	3.8	4.302	3200	3.50			1.0	3.71	0	
S of Plato F	1	3	4.200	1400	3.15					?	
E of Ptolemaeus	1	3	4.200	1800	3.26					?	
Ptolemaeus Y	1	3	4.200	1700	3.23					?	
Piton A	1	2.6	4.138	2000	3.30	560	2.75			?	
Archimedes D	1	2.6	4.138	1600	3.20					?	
Ptolemaeus D	1	2.3	4.084	1050	3.03					?	
Piton B	1	2.1	4.045	1460	3.16	1080	3.03			?	
W of Archimedes K	1	2	4.024	1450	3.16					?	
Ptolemaeus C	1	2.0	4.024	1000	3.00					?	
Ptolemaeus S	1	1.7	3.953	940	2.97					?	
Piazzi Smyth B	1	1.6	3.927	1240	3.09	520	2.72			?	
Betw. above & Piton	1	1.3	3.837	1030	3.01	300	2.48			?	
E of Piazzi Smyth	1	1.3	3.837	1240	3.09	300	2.48			?	
S of Kirch	1	1.2	3.802	980	2.99					?	
N of Piton	1	1	3.723	810	2.91					?	
N of Kirch	1	1	3.723	1000	3.00					?	
S of Birt	1	1	3.723	1150	3.05					?	
In Alphonsus	1	1	3.723	950	2.98					?	
In Purbach	1	1	3.723	940	2.97					?	

D = Logarithm of the diameter in feet of the apparent crater R_H = Logarithm of the rim height in feet
d = Logarithm of the depth in feet of the apparent crater R_W = Logarithm of the apparent rim width in feet

Class 1 craters are the newest-appearing and least deformed craters. Classes 2, 3, and 4 are progressively older in appearance. Class 5 craters have been partially filled with some dark substance.

The numbers in the column headed "Central Peak" signify the following:

0 = No visible central peak
1 = Single peak
2 = Peak or peaks not centrally placed in crater
3 = Peak or peaks centrally placed in crater
4 = Multiple peaks
5 = Small peak relative to crater
6 = Medium peak relative to crater
7 = Large peak relative to crater
8 = Has central peak craterlet
9 = Location of central peak hidden by later crater

Values in parentheses are considered less reliable than other measures.

Table 8

Scaled Crater Dimensions vs Scaled Depth of Burst

Observed								Corrected values which would obtain at scaled depths of burst indicated									
No.	$H/W^{1/3}$	$R/W^{1/3}$	$S/W^{1/3}$	R/S	$T/W^{1/3}$	E	Nature of Ground	$H/W^{1/3}$ = 0.00		0.10		0.25		0.50		0.00	0.10
								D	d	D	d	D	d	D	d	R_H	R_H
1	-0.70	1.01	0.40	2.5	-	2.88	Sand	9.53	8.95	9.56	9.00	9.61	9.08	9.69	9.21	-	-
2	-0.35	1.32	0.75	1.7	-	"	"	9.59	9.13	9.62	9.18	9.67	9.26	9.75	9.39	-	-
3	-0.18	1.41	0.66	2.1	0.09	"	"	9.57	8.99	9.61	9.04	9.65	9.12	9.73	9.25	8.24	8.35
4	-0.09	1.67	0.75	2.2	0.09	"	"	9.62	9.00	9.65	9.04	9.70	9.13	9.78	9.26	8.13	8.24
5	0.00	0.88	0.44	2.0	-	"	Soil	9.32	8.72	9.35	8.76	9.40	8.85	9.48	8.98	-	-
6	0.00	1.41	0.53	2.6	-	"	"	9.52	8.79	9.56	8.84	9.60	8.92	9.68	9.05	-	-
7	0.00	1.67	0.70	2.3	-	"	"	9.60	8.92	9.63	8.96	9.68	9.05	9.76	9.18	-	-
8	0.00	1.76	0.61	2.9	0.09	"	"	9.62	8.86	9.65	8.92	9.70	9.00	9.78	9.13	8.02	8.13
9	0.00	1.93	0.97	2.0	0.09	"	"	9.66	9.06	9.69	9.12	9.74	9.20	9.82	9.33	8.02	8.13
10	0.00	2.02	0.61	3.3	0.09	"	"	9.68	8.86	9.71	8.92	9.76	9.00	9.84	9.13	8.02	8.13
11	0.00	1.89	0.79	2.3	0.09	"	Sand	9.65	8.97	9.68	9.02	9.73	9.11	9.81	9.24	8.02	8.13
12	0.00	1.80	0.61	2.9	0.13	"	"	9.63	8.86	9.66	8.91	9.71	9.00	9.79	9.12	8.19	8.30
13	0.00	1.89	1.05	1.8	0.09	"	"	9.65	9.10	9.68	9.14	9.73	9.23	9.81	9.36	8.02	8.13
14	0.00	1.89	0.97	1.9	0.13	"	"	9.65	9.06	9.68	9.11	9.73	9.19	9.81	9.32	8.19	8.30
15	0.00	2.02	0.79	2.5	0.09	"	"	9.68	8.97	9.71	9.02	9.76	9.11	9.84	9.24	8.02	8.13
16	0.00	1.98	0.88	2.2	0.09	"	"	9.67	9.02	9.71	9.06	9.75	9.15	9.83	9.28	8.02	8.13
17	0.00	1.93	0.79	2.4	0.09	"	"	9.66	8.97	9.70	9.02	9.74	9.11	9.82	9.24	8.02	8.13
18	0.00	1.93	0.79	2.4	0.09	"	"	9.66	8.97	9.70	9.02	9.74	9.11	9.82	9.24	8.02	8.13
21	0.09	2.15	1.10	2.0	0.18	"	"	9.68	9.07	9.71	9.12	9.76	9.20	9.84	9.33	8.24	8.35
22	0.09	1.93	0.88	2.1	0.18	"	Soil	9.63	8.98	9.66	9.02	9.71	9.11	9.79	9.24	8.24	8.35
23	0.18	2.02	0.88	2.2	0.18	"	"	9.62	8.92	9.66	8.96	9.70	9.05	9.78	9.18	8.14	8.25
24	0.18	2.28	0.88	2.6	0.13	"	"	9.68	8.92	9.71	8.98	9.76	9.06	9.84	9.19	8.01	8.12
25	0.18	2.37	1.14	2.1	0.09	"	"	9.70	9.02	9.73	9.08	9.78	9.16	9.86	9.29	7.84	7.95
26	0.18	2.72	1.05	2.6	0.13	"	"	9.76	8.99	9.79	9.05	9.84	9.13	9.92	9.29	8.01	8.12
27	0.18	2.06	1.05	2.0	0.18	"	Sand	9.63	9.00	9.66	9.05	9.71	9.13	9.79	9.26	8.14	8.25
28	0.26	2.81	1.49	1.9	0.26	"	"	9.74	9.11	9.77	9.16	9.82	9.24	9.90	9.37	8.25	8.36
29	0.26	2.11	0.88	2.3	0.18	"	Soil	9.62	8.88	9.65	8.93	9.70	9.01	9.78	9.14	8.08	8.19
30	0.35	2.20	1.14	1.9	0.09	"	"	9.61	8.93	9.64	8.99	9.69	9.07	9.77	9.20	7.73	7.84
31	0.35	2.37	0.88	2.6	0.26	"	"	9.63	8.83	9.67	8.88	9.71	8.96	9.79	9.09	8.20	8.31
32	0.35	2.63	1.49	1.8	0.13	"	"	9.68	9.07	9.71	9.13	9.76	9.21	9.84	9.34	7.90	8.01
33	0.35	2.63	1.40	1.9	0.18	"	"	9.68	9.02	9.71	9.08	9.76	9.16	9.84	9.29	8.03	8.14
34	0.35	2.81	1.49	1.9	0.26	"	Sand	9.71	9.06	9.74	9.11	9.79	9.19	9.87	9.32	8.20	8.31
35	0.44	2.77	1.32	2.1	0.18	"	"	9.67	8.97	9.71	9.01	9.75	9.10	9.83	9.23	7.98	8.09
36	0.44	1.58	0.79	2.0	0.18	"	Soil	9.43	8.75	9.46	8.79	9.51	8.88	9.59	9.01	7.98	8.09
37	0.44	2.11	0.79	2.6	0.09	"	"	9.56	8.75	9.59	8.79	9.64	8.88	9.72	9.01	7.68	7.79
38	0.53	2.63	1.14	2.3	0.44	"	"	9.63	8.86	9.66	8.91	9.71	9.00	9.79	9.12	8.34	8.45
39	0.53	3.16	1.58	2.0	0.18	"	"	9.72	8.99	9.75	9.05	9.80	9.13	9.88	9.26	7.94	8.05
40	0.53	3.25	1.58	2.1	0.13	"	"	9.72	8.99	9.75	9.05	9.80	9.13	9.88	9.26	7.81	7.92
41	0.53	3.42	1.58	2.2	0.18	"	"	9.74	8.99	9.77	9.05	9.82	9.13	9.90	9.26	7.94	8.05
42	0.53	3.16	1.76	1.8	0.26	"	Sand	9.71	9.05	9.74	9.10	9.79	9.18	9.87	9.31	8.11	8.22
43	0.61	3.16	1.58	2.0	0.26	"	"	9.68	8.97	9.72	9.02	9.76	9.10	9.84	9.23	8.09	8.21
44	0.61	2.20	0.97	2.2	0.18	"	Soil	9.52	8.76	9.56	8.81	9.61	8.89	9.68	9.02	7.92	8.04
45	0.70	2.55	1.23	2.1	0.35	"	"	9.57	8.84	9.60	8.88	9.65	8.97	9.73	9.10	8.20	8.32
46	0.70	3.51	1.49	2.4	0.18	"	"	9.71	8.91	9.74	8.97	9.79	9.05	9.87	9.18	7.90	8.02
47	0.70	3.60	1.58	2.3	0.18	"	"	9.73	8.93	9.76	8.99	9.81	9.07	9.89	9.20	7.90	8.02
48	0.70	3.86	1.58	2.4	0.18	"	"	9.75	8.93	9.78	8.99	9.83	9.07	9.91	9.20	7.90	8.02
49	0.70	3.43	1.93	1.8	0.26	"	Sand	9.70	9.03	9.73	9.08	9.78	9.16	9.86	9.29	8.08	8.19
50	0.70	2.90	2.11	1.4	0.18	"	"	9.63	9.07	9.66	9.12	9.71	9.20	9.79	9.33	7.90	8.02
51	0.79	3.64	1.84	2.0	0.35	"	"	9.71	8.98	9.75	9.03	9.79	9.12	9.87	9.25	8.20	8.31
52	0.88	3.42	1.62	2.1	0.26	"	"	9.67	8.90	9.70	8.95	9.75	9.04	9.83	9.17	8.06	8.17
53	0.88	3.51	1.49	2.4	0.26	"	Soil	9.69	8.85	9.72	8.91	9.77	8.99	9.85	9.12	8.06	8.17
54	0.97	3.12	2.11	1.5	0.26	"	Sand	9.62	9.00	9.65	9.05	9.70	9.13	9.78	9.26	8.06	8.17
55	0.97	3.29	2.02	1.6	0.26	"	"	9.64	8.98	9.68	9.03	9.73	9.11	9.80	9.24	8.06	8.17
56	1.06	3.69	1.40	2.6	0.35	"	Soil	9.69	8.79	9.72	8.85	9.77	8.93	9.85	9.06	8.20	8.31
57	1.41	2.46	1.76	1.4	0.18	"	Sand	9.49	8.86	9.52	8.91	9.57	9.00	9.65	9.12	7.91	8.03
58	2.11	2.72	1.76	1.5	0.18	"	"	9.59	8.87	9.62	8.91	9.67	9.00	9.75	9.13	8.03	8.14
59	2.11	2.81	1.76	1.6	0.18	"	"	9.60	8.87	9.63	8.91	9.68	9.00	9.76	9.13	8.03	8.14
60	2.11	2.90	1.41	2.0	0.27	"	"	9.61	8.77	9.65	8.82	9.69	8.90	9.77	9.04	8.20	8.31
61	2.81	2.46	1.93	1.3	0.35	"	"	9.63	8.97	9.66	9.02	9.71	9.11	9.79	9.24	8.60	8.71
62	-0.76	1.67	0.44	3.7	0.06	3.31	"	9.87	9.09	9.91	9.15	9.97	9.25	0.07	9.41	8.42	8.53
63	-0.57	1.64	0.38	4.3	-	"	"	9.85	9.00	9.89	9.06	9.95	9.17	0.05	9.32	-	-
64	-0.38	1.51	0.50	3.0	0.09	"	"	9.79	9.10	9.83	9.16	9.89	9.26	9.99	9.42	8.49	8.60
65	-0.16	1.51	0.54	2.7	0.06	"	"	9.75	9.02	9.79	9.08	9.85	9.19	9.94	9.34	8.21	8.32
66	0.00	1.26	0.50	2.5	-	"	Soil	9.62	8.92	9.66	8.98	9.72	9.08	9.81	9.24	-	-
67	0.00	1.26	0.38	3.3	-	"	"	9.62	8.79	9.66	8.86	9.72	8.96	9.81	9.11	-	-
68	0.00	1.26	0.44	2.8	-	"	"	9.62	8.86	9.66	8.92	9.72	9.03	9.81	9.18	-	-
69	0.00	1.77	0.76	2.3	0.06	"	Sand	9.77	9.10	9.80	9.16	9.86	9.26	9.96	9.42	8.02	8.13
70	0.00	1.92	0.76	2.5	0.10	"	"	9.80	9.10	9.84	9.16	9.90	9.26	0.00	9.42	8.19	8.30
71	0.00	1.77	0.73	2.4	0.13	"	"	9.77	9.08	9.80	9.14	9.86	9.24	9.96	9.40	8.32	8.43
72	0.00	1.70	0.91	1.8	0.13	"	"	9.75	9.18	9.79	9.24	9.85	9.34	9.94	9.50	8.32	8.43
73	0.00	1.70	0.72	2.3	0.19	"	"	9.75	9.08	9.79	9.14	9.85	9.24	9.94	9.40	8.49	8.60
74	0.00	1.58	0.50	3.1	0.09	"	"	9.72	8.92	9.75	8.98	9.82	9.08	9.91	9.24	8.19	8.30
75	0.06	1.92	0.95	2.0	0.19	"	"	9.78	9.15	9.81	9.21	9.88	9.31	9.97	9.47	8.43	8.54
76	0.06	1.39	0.57	2.4	0.13	"	Soil	9.64	8.93	9.67	8.99	9.74	9.09	9.83	9.25	8.26	8.37
77	0.13	1.95	0.95	2.0	0.13	"	Sand	9.76	9.11	9.79	9.17	9.86	9.27	9.95	9.43	8.19	8.30
78	0.16	1.58	0.63	2.5	0.13	"	Soil	9.65	8.91	9.69	8.97	9.75	9.08	9.85	9.23	8.14	8.25
79	0.16	2.11	0.98	2.1	0.19	"	Sand	9.78	9.10	9.82	9.16	9.88	9.26	9.97	9.42	8.33	8.44
80	0.25	2.40	1.13	2.1	0.25	"	"	9.80	9.11	9.84	9.17	9.90	9.27	9.99	9.43	8.39	8.50
81	0.25	1.61	0.76	2.1	0.13	"	Soil	9.63	8.93	9.66	9.00	9.73	9.10	9.82	9.25	8.09	8.20
82	0.35	1.77	0.69	2.5	0.06	"	"	9.63	8.82	9.66	8.89	9.73	8.99	9.82	9.15	7.73	7.84
83	0.35	2.71	1.58	1.7	0.25	"	Sand	9.81	9.18	9.85	9.25	9.91	9.35	0.01	9.50	8.33	8.44
84	0.41	2.02	1.01	2.0	0.13	"	Soil	9.66	8.95	9.70	9.01	9.76	9.11	9.86	9.27	8.00	8.11
85	0.41	2.90	1.39	2.1	0.19	"	Sand	9.82	9.09	9.85	9.15	9.92	9.26	0.01	9.41	8.17	8.28
86	0.54	3.09	1.58	2.0	0.25	"	"	9.80	9.08	9.84	9.14	9.90	9.25	0.00	9.40	8.23	8.34
87	0.54	2.14	1.13	1.9	0.13	"	Soil	9.65	8.94	9.68	9.00	9.74	9.10	9.84	9.26	7.93	8.04
88	0.63	2.77	1.51	1.8	0.25	"	"	9.73	9.02	9.77	9.08	9.83	9.18	9.93	9.34	8.21	8.33
89	0.63	3.34	1.58	2.1	0.25	"	Sand	9.81	9.04	9.85	9.10	9.91	9.20	0.01	9.36	8.21	8.33
90	0.73	3.25	1.51	2.2	0.31	"	"	9.78	8.98	9.82	9.04	9.88	9.15	9.98	9.30	8.30	8.42
91	0.82	3.03	1.83	1.6	0.25	"	"	9.74	9.03	9.77	9.09	9.84	9.20	9.93	9.35	8.19	8.30
92	0.85	3.03	1.95	1.6	0.25	"	"	9.73	9.05	9.77	9.11	9.83	9.21	9.93	9.37	8.19	8.30
93	1.10	3.03	2.21	1.4	0.38	"	"	9.72	9.04	9.75	9.10	9.82	9.20	9.91	9.36	8.38	8.49
94	1.51	2.08	1.77	1.2	-	"	"	9.55	8.90	9.58	8.96	9.65	9.06	9.74	9.21	-	-
95	2.02	2.14	1.64	1.3	0.19	"	"	9.60	8.93	9.63	9.00	9.69	9.10	9.79	9.25	8.19	8.30
96	2.52	2.40	1.77	1.4	0.25	"	"	9.72	9.11	9.75	9.17	9.81	9.27	9.91	9.43	8.46	8.57
97	2.52	2.52	1.26	2.0	0.19	"	"	9.74	8.96	9.77	9.03	9.84	9.13	9.93	9.28	8.33	8.44
98	0.00	1.42	0.51	2.7	-	3.78	Soil	9.82	9.08	9.85	9.14	9.88	9.21	9.94	9.33	-	-
99	0.00	1.58	0.72	2.2	-	"	"	9.88	9.22	9.90	9.28	9.93	9.36	9.99	9.48	-	-
100	0.00	2.32	0.72	3.2	0.00	"	"	0.04	9.22	0.06	9.28	0.10	9.36	0.16	9.48	-	-
101	0.70	1.67	1.10	1.5	0.18	"	Clay loam	9.74	9.08	9.76	9.14	9.80	9.21	9.86	9.33	8.20	8.32
102	0.70	2.99	1.67	1.8	0.22	"	"	9.99	9.26	0.02	9.32	0.05	9.40	0.11	9.52	8.30	8.42

Table 8
Scaled Crater Dimensions vs Scaled Depth of Burst

	Observed						Corrected values which would obtain at scaled depths of burst indicated									
						Nature	$H/W^{1/3}$ = 0.00		0.10		0.25		0.50		0.00	0.10
No.	$H/W^{1/3}$	$R/W^{1/3}$	$S/W^{1/3}$	R/S	$T/W^{1/3}$	E	of Ground									
							D	d	D	d	D	d	D	d	R_H	R_H
103	1.41	2.03	1.76	1.2	0.18	3.78	Sand	9.70	9.08	9.73	9.14	9.76	9.21	9.82	9.33	8.21	8.33
104	1.41	2.33	1.58	1.5	0.18	"	"	9.76	9.03	9.79	9.09	9.82	9.16	9.88	9.29	8.21	8.33
105	1.41	2.64	1.85	1.4	0.26	"	"	9.82	9.10	9.84	9.16	9.88	9.23	9.94	9.35	8.39	8.51
106	2.11	2.55	1.94	1.3	0.35	"	"	9.73	9.08	9.75	9.15	9.79	9.22	9.85	9.34	8.63	8.74
107	2.47	2.11	1.14	1.8	-	"	"	9.68	8.93	9.70	8.99	9.74	9.06	9.80	9.18	-	-
108	0.00	1.43	0.65	2.2	-	4.05	Soil	9.92	9.27	9.94	9.33	9.97	9.41	0.04	9.53	-	-
109	1.00	2.61	1.90	1.4	0.50	"	Clay loam	9.97	9.31	9.99	9.37	0.02	9.44	0.08	9.56	8.73	8.84
110	1.00	2.72	1.72	1.6	0.39	"	"	9.98	9.27	0.01	9.33	0.04	9.40	0.10	9.52	8.63	8.74
111	1.43	3.29	2.86	1.2	0.32	"	Sand	0.00	9.37	0.02	9.43	0.06	9.51	0.12	9.63	8.56	8.68
112	1.72	4.08	2.93	1.4	0.50	"	"	0.06	9.36	0.08	9.42	0.11	9.49	0.17	9.61	8.80	8.91
113	2.17	3.26	2.66	1.2	0.24	4.27	"	0.00	9.39	0.03	9.45	0.06	9.53	0.12	9.61	8.66	8.78
114	1.37	2.73	2.00	1.4	0.32	4.45	Clay loam	0.06	9.37	0.09	9.43	0.12	9.50	0.18	9.62	8.68	8.80
115	1.47	2.84	2.31	1.2	0.42	"	"	0.07	9.40	0.09	9.47	0.12	9.54	0.18	9.66	8.81	8.93
116	1.68	4.21	3.52	1.2	0.32	"	Sand	0.21	9.55	0.23	9.61	0.26	9.68	0.32	9.80	8.71	8.82
117	2.10	4.63	2.58	1.8	0.53	"	"	0.21	9.43	0.24	9.49	0.27	9.56	0.33	9.69	9.03	9.14
118	2.52	4.21	2.63	1.6	0.32	"	"	0.21	9.53	0.23	9.59	0.27	9.66	0.33	9.78	8.94	9.05
119	1.83	3.30	2.56	1.3	0.27	4.63	"	0.14	9.47	0.17	9.53	0.20	9.60	0.26	9.72	8.75	8.86
120	1.83	3.02	2.29	1.3	0.55	"	Clay loam	0.11	9.42	0.13	9.48	0.16	9.55	0.22	9.67	9.05	9.16
121	1.83	3.71	2.65	1.4	0.37	"	Sand	0.20	9.48	0.22	9.54	0.25	9.62	0.31	9.74	8.87	8.98
122	1.83	4.03	2.93	1.4	0.46	"	"	0.23	9.52	0.25	9.58	0.29	9.66	0.35	9.78	8.97	9.08
123	1.83	4.12	2.93	1.4	0.37	"	"	0.24	9.52	0.26	9.58	0.30	9.66	0.36	9.78	8.87	8.98
124	0.00	1.06	0.51	2.0	-	4.91	Soil	0.08	9.47	0.10	9.53	0.14	9.60	0.20	9.72	-	-
125	1.33	2.74	2.22	1.2	0.30	"	Clay loam	0.22	9.57	0.24	9.63	0.28	9.70	0.34	9.83	8.80	8.92
126	1.77	3.33	2.37	1.4	0.44	"	Sand	0.25	9.53	0.27	9.59	0.31	9.66	0.37	9.78	9.03	9.14
127	1.77	3.99	2.63	1.5	0.30	"	"	0.33	9.57	0.35	9.63	0.38	9.71	0.44	9.83	8.85	8.96
128	2.37	3.55	2.00	1.8	0.59	"	"	0.27	9.52	0.29	9.58	0.33	9.66	0.39	9.78	9.33	9.44
136	0.00	0.84	0.57	1.5	-	7.58	Soil	0.87	0.40	0.93	0.46	1.00	0.59	1.13	0.76	-	-
137	0.00	0.51	0.46	1.1	-	"	"	0.65	0.30	0.71	0.36	0.79	0.49	0.91	0.66	-	-
138	0.00	0.57	0.34	1.7	-	"	"	0.70	0.18	0.76	0.24	0.83	0.37	0.96	0.54	-	-
139	0.00	0.57	0.46	1.2	-	"	"	0.70	0.30	0.76	0.36	0.83	0.49	0.96	0.66	-	-
140	0.00	0.57	0.46	1.2	-	"	"	0.70	0.30	0.76	0.36	0.83	0.49	0.96	0.66	-	-
141	0.00	0.57	0.55	1.0	-	"	"	0.70	0.38	0.76	0.44	0.83	0.57	0.96	0.74	-	-
142	0.00	0.63	0.41	1.5	-	"	"	0.74	0.26	0.80	0.32	0.88	0.45	1.00	0.62	-	-
143	0.00	0.69	0.46	1.5	-	"	"	0.78	0.30	0.83	0.36	0.91	0.49	1.04	0.66	-	-
144	0.00	0.69	0.46	1.5	-	"	"	0.78	0.30	0.83	0.36	0.91	0.49	1.04	0.66	-	-
145	0.00	0.72	0.46	1.6	-	"	"	0.80	0.30	0.86	0.36	0.93	0.49	1.06	0.66	-	-
146	0.00	0.74	0.39	1.9	-	"	"	0.81	0.23	0.87	0.29	0.95	0.42	1.07	0.59	-	-
147	0.00	0.74	0.46	1.6	-	"	"	0.81	0.30	0.87	0.36	0.95	0.49	1.07	0.66	-	-
148	0.00	0.74	0.46	1.6	-	"	"	0.81	0.30	0.87	0.36	0.95	0.49	1.07	0.66	-	-
149	0.00	0.80	0.57	1.4	-	"	"	0.85	0.40	0.90	0.46	0.98	0.59	1.10	0.76	-	-
150	0.00	0.80	0.57	1.4	-	"	"	0.85	0.40	0.90	0.46	0.98	0.59	1.10	0.76	-	-
151	0.00	0.80	0.46	1.7	-	"	"	0.85	0.30	0.90	0.36	0.98	0.49	1.10	0.66	-	-
152	0.00	0.80	0.69	1.2	-	"	"	0.85	0.48	0.90	0.54	0.98	0.67	1.10	0.84	-	-
153	0.00	0.82	0.69	1.2	-	"	"	0.86	0.48	0.91	0.54	0.99	0.67	1.11	0.84	-	-
154	0.00	0.86	0.46	1.9	-	"	"	0.88	0.30	0.93	0.36	1.01	0.48	1.13	0.66	-	-
155	0.00	0.91	0.34	2.7	-	"	"	0.90	0.18	0.96	0.24	1.04	0.37	1.16	0.54	-	-
156	0.00	0.91	0.34	2.7	-	"	"	0.90	0.18	0.96	0.24	1.04	0.37	1.16	0.54	-	-
157	0.00	0.91	0.41	2.2	-	"	"	0.90	0.26	0.96	0.32	1.04	0.45	1.16	0.62	-	-
158	0.00	0.91	0.46	2.0	-	"	"	0.90	0.30	0.96	0.36	1.04	0.49	1.16	0.66	-	-
159	0.00	1.03	0.39	2.6	-	"	"	0.95	0.23	1.01	0.29	1.09	0.42	1.21	0.59	-	-
160	0.00	1.03	0.64	1.6	-	"	"	0.95	0.45	1.01	0.51	1.09	0.64	1.21	0.81	-	-
161	0.00	1.03	0.69	1.4	-	"	"	0.95	0.48	1.01	0.54	1.09	0.67	1.21	0.84	-	-
162	0.00	1.03	0.39	2.6	-	"	"	0.95	0.23	1.01	0.29	1.09	0.42	1.21	0.59	-	-
163	0.00	1.09	0.62	1.7	-	"	"	0.98	0.43	1.03	0.49	1.11	0.62	1.23	0.79	-	-
164	0.00	1.14	0.46	2.4	-	"	"	1.00	0.30	1.06	0.36	1.14	0.49	1.26	0.66	-	-
190	0.00	0.61	0.14	4.4	-	7.93	"	0.85	9.90	0.90	9.96	0.98	0.09	1.10	0.26	-	-
191	0.00	0.96	0.52	1.8	-	"	"	1.04	0.48	1.10	0.54	1.18	0.67	1.30	0.84	-	-
192	0.00	1.05	0.49	2.1	-	"	"	1.08	0.45	1.14	0.51	1.21	0.64	1.34	0.81	-	-
193	0.00	1.22	0.58	2.1	-	"	"	1.15	0.52	1.20	0.58	1.28	0.71	1.40	0.88	-	-
194	0.00	1.31	0.52	2.5	-	"	"	1.18	0.48	1.23	0.54	1.31	0.67	1.43	0.84	-	-
206	0.00	2.06	0.90	2.2	-	8.01	"	1.40	0.74	1.46	0.80	1.53	0.93	1.66	1.10	-	-
208	-0.51	0.59	0.07	8.4	-	8.16	"	1.15	0.05	1.21	0.11	1.29	0.24	1.41	0.41	-	-
209	0.00	1.18	0.48	2.4	0.12	"	"	1.21	0.52	1.26	0.58	1.34	0.71	1.46	0.88	9.90	0.01
210	0.19	1.68	1.10	1.5	0.22	"	"	1.25	0.73	1.31	0.79	1.39	0.93	1.51	1.10	9.99	0.10
211	0.51	1.86	1.19	1.6	0.24	"	"	1.14	0.54	1.20	0.60	1.28	0.73	1.40	0.90	9.84	9.95
212	1.03	2.41	1.54	1.6	0.29	"	"	1.00	0.53	1.06	0.61	1.14	0.74	1.26	0.86	9.82	9.93
213	2.05	2.60	0.91	2.8	0.25	"	"	1.01	0.44	1.07	0.52	1.15	0.65	1.27	0.80	9.85	9.96
214	3.08	2.10	0.73	2.8	0.22	"	"	1.24	0.84	1.30	0.92	1.38	1.05	1.50	1.20	0.14	0.25
215	0.00	0.48	0.41	1.2	-	8.25	"	0.85	0.48	0.90	0.54	0.98	0.67	1.10	0.84	-	-
216	0.00	0.68	0.27	2.5	0.00	"	"	1.00	0.30	1.06	0.36	1.14	0.49	1.26	0.66	-	-
217	0.00	0.75	0.38	2.0	-	"	"	1.04	0.45	1.10	0.51	1.18	0.64	1.30	0.81	-	-
218	0.00	0.82	0.48	1.7	-	"	"	1.08	0.54	1.14	0.60	1.21	0.74	1.34	0.91	-	-
219	0.00	0.82	0.23	3.6	-	"	"	1.08	0.23	1.14	0.29	1.21	0.42	1.34	0.59	-	-
220	0.00	0.82	0.44	1.9	0.00	"	"	1.08	0.51	1.14	0.57	1.21	0.70	1.34	0.87	-	-
221	0.00	1.03	0.54	1.9	0.07	"	"	1.18	0.60	1.23	0.66	1.31	0.79	1.43	0.96	9.70	9.81
234	0.00	0.87	0.35	2.5	-	8.33	"	1.13	0.43	1.19	0.49	1.27	0.62	1.39	0.79	-	-
235	0.00	1.16	0.51	2.2	-	"	"	1.26	0.60	1.31	0.66	1.39	0.79	1.51	0.96	-	-
236	0.00	1.32	0.62	2.1	-	"	"	1.31	0.68	1.37	0.74	1.45	0.87	1.57	1.04	-	-
237	0.00	1.80	0.73	2.4	-	"	"	1.45	0.76	1.50	0.82	1.58	0.95	1.70	1.12	-	-
239	-0.23	0.69	0.52	1.3	-	8.47	"	1.21	0.81	1.27	0.87	1.35	1.00	1.47	1.17	-	-
247	-0.84	0.89	0.31	2.9	-	8.60	"	1.52	0.92	1.58	0.98	1.66	1.11	1.78	1.28	-	-
248	0.00	0.93	0.31	3.0	-	8.61	"	1.26	0.48	1.31	0.54	1.39	0.67	1.51	0.84	-	-
249	0.00	0.96	0.36	2.7	-	"	"	1.27	0.54	1.32	0.60	1.40	0.74	1.52	0.91	-	-
250	0.00	0.99	0.57	1.7	-	"	"	1.28	0.74	1.33	0.80	1.41	0.93	1.54	1.10	-	-
251	0.00	1.04	0.42	2.4	-	"	"	1.30	0.60	1.36	0.66	1.44	0.79	1.56	0.96	-	-
252	0.00	1.09	0.57	1.9	-	"	"	1.32	0.74	1.38	0.80	1.46	0.93	1.58	1.10	-	-
253	0.00	1.09	0.47	2.3	-	"	"	1.32	0.65	1.38	0.71	1.46	0.84	1.58	1.01	-	-
254	0.00	1.25	0.99	1.2	-	"	"	1.38	0.98	1.44	1.04	1.52	1.17	1.64	1.34	-	-
255	0.00	1.28	0.62	2.0	-	"	"	1.39	0.78	1.45	0.84	1.53	0.97	1.65	1.14	-	-
256	0.00	1.73	1.40	1.2	-	"	"	1.52	1.13	1.58	1.19	1.66	1.32	1.78	1.49	-	-
258	-0.21	0.88	-	-	-	"	"	1.35	-	1.40	-	1.48	-	1.60	-	-	-
271	0.00	0.99	0.59	1.7	-	8.67	"	1.41	0.91	1.46	0.97	1.54	1.10	1.66	1.27	-	-
272	-0.17	-	0.17	-	-	8.83	"	-	0.42	-	0.48	-	0.61	-	0.78	-	-
273	-0.17	0.68	0.34	2.0	-	8.86	"	1.30	0.72	1.35	0.78	1.43	0.91	1.55	1.08	-	-
274	-0.17	1.28	0.38	3.3	-	"	"	1.57	0.77	1.63	0.83	1.70	0.96	1.83	1.13	-	-
275	0.00	0.89	0.49	1.8	-	8.87	"	1.32	0.77	1.38	0.82	1.46	0.95	1.58	1.12	-	-
276	0.00	0.91	0.32	2.8	-	"	"	1.33	0.58	1.39	0.64	1.47	0.77	1.59	0.94	-	-
277	0.00	0.93	0.61	1.5	-	"	"	1.34	0.86	1.39	0.92	1.47	1.05	1.60	1.22	-	-
278	0.00	0.93	0.47	2.0	-	"	"	1.34	0.74	1.40	0.80	1.48	0.93	1.60	1.10	-	-
279	0.00	1.02	0.51	2.0	-	"	"	1.38	0.78	1.44	0.84	1.52	0.97	1.64	1.14	-	-

Table 8
Scaled Crater Dimensions vs Scaled Depth of Burst

	Observed					Nature of Ground	Corrected values which would obtain at scaled depths of burst indicated										
							$H/W^{1/3}$ = 0.00		0.10		0.25		0.50		0.00	0.10
No.	$H/W^{1/3}$	$R/W^{1/3}$	$S/W^{1/3}$	R/S	$T/W^{1/3}$	E										
							D	d	D	d	D	d	D	d	R_H	R_H
282	-0.67	1.25	1.25	1.0	-	8.89	Soil	1.75	1.57	1.81	1.63	1.89	1.76	2.01	1.93	-	-
283	0.19	1.48	0.73	2.0	0.23	9.06	"	1.50	0.86	1.56	0.92	1.63	1.05	1.76	1.22	0.31	0.42
285	0.51	1.93	0.76	2.5	0.21	"	"	1.46	0.65	1.52	0.71	1.59	0.84	1.72	1.01	0.08	0.19
286	0.51	2.33	1.17	2.0	0.29	"	"	1.54	0.84	1.60	0.90	1.68	1.03	1.80	1.20	0.24	0.35
291	-0.07	-	0.29	-	-	9.09	"	-	0.66	-	0.72	-	0.85	-	1.02	-	-
292	0.00	0.57	0.19	3.0	-	9.17	"	1.23	0.45	1.29	0.51	1.37	0.64	1.49	0.81	-	-
293	0.00	0.61	0.20	3.0	-	"	"	1.26	0.48	1.31	0.54	1.39	0.67	1.51	0.84	-	-
294	0.00	0.88	0.27	3.2	-	"	"	1.41	0.60	1.47	0.66	1.55	0.79	1.67	0.96	-	-
295	0.00	1.05	0.40	2.6	-	"	"	1.49	0.78	1.55	0.84	1.63	0.97	1.75	1.14	-	-
296	0.00	1.08	0.40	2.7	-	"	"	1.51	0.78	1.56	0.84	1.64	0.97	1.76	1.14	-	-
297	0.00	1.28	0.51	2.5	-	"	"	1.58	0.88	1.64	0.94	1.71	1.07	1.84	1.24	-	-
299	-0.13	0.60	0.20	3.0	-	9.18	"	1.33	0.57	1.39	0.63	1.46	0.76	1.59	0.93	-	-
300	-0.12	0.78	0.20	3.9	-	9.27	"	1.47	0.59	1.52	0.65	1.60	0.78	1.72	0.95	-	-
301	-0.17	1.13	0.51	2.2	-	9.40	"	1.69	1.07	1.75	1.13	1.83	1.26	1.95	1.43	-	-
302	-0.10	0.36	0.13	2.8	-	9.53	"	1.21	0.47	1.26	0.53	1.34	0.66	1.46	0.83	-	-
304	-0.12	-	0.24	-	-	9.84	"	-	0.86	-	0.92	-	1.05	-	1.22	-	-
305	-0.08	0.80	0.20	4.0	-	9.85	"	1.70	0.81	1.76	0.88	1.84	1.01	1.96	1.18	-	-
306	0.00	0.94	0.68	1.4	-	9.93	"	1.70	1.26	1.76	1.32	1.83	1.45	1.96	1.62	-	-
307	0.00	0.89	0.53	1.7	-	10.00	"	1.70	1.18	1.76	1.24	1.83	1.37	1.96	1.54	-	-
308	0.53	-	1.60	-	-	"	"	-	1.28	-	1.34	-	1.47	-	1.64	-	-
309	-0.14	1.05	0.35	3.0	-	10.02	"	1.86	1.10	1.91	1.16	1.99	1.30	2.11	1.47	-	-
310	1.91	2.41	0.73	3.3	0.24	10.04	"	1.71	0.94	1.77	1.02	1.85	1.15	1.97	1.30	0.53	0.64
311	1.81	2.62	0.99	2.6	0.16	10.11	"	1.77	1.07	1.83	1.15	1.91	1.28	2.03	1.43	0.34	0.45
312	-0.26	1.22	0.98	1.2	-	10.12	"	2.02	1.64	2.08	1.70	2.16	1.84	2.28	2.01	-	-
313	-0.13	0.64	0.48	1.3	-	10.15	"	1.68	1.27	1.73	1.33	1.81	1.46	1.93	1.63	-	-
314	0.00	0.79	0.48	1.6	-	"	"	1.70	1.18	1.76	1.24	1.83	1.37	1.96	1.54	-	-
315	-0.25	1.42	0.47	3.0	-	10.16	"	2.10	1.33	2.15	1.39	2.23	1.52	2.35	1.69	-	-
316	1.89	3.69	0.80	4.6	0.11	"	"	1.94	1.02	2.00	1.10	2.08	1.23	2.20	1.38	0.21	0.32
317	-0.24	1.78	0.59	3.0	-	10.24	"	2.21	1.44	2.27	1.50	2.35	1.63	2.47	1.80	-	-
318	0.00	1.59	0.59	2.6	-	10.25	"	2.03	1.30	2.09	1.36	2.17	1.49	2.29	1.66	-	-
319	-0.12	0.58	0.16	3.6	-	10.26	"	1.67	0.82	1.73	0.88	1.81	1.01	1.93	1.18	-	-
320	0.51	2.34	0.96	2.4	0.29	"	"	1.94	1.15	2.00	1.21	2.08	1.34	2.20	1.51	0.64	0.75
322	-0.11	0.71	0.42	1.7	-	10.30	"	1.76	1.26	1.82	1.32	1.90	1.45	2.02	1.62	-	-
323	1.38	3.33	1.02	3.3	0.19	10.34	"	1.95	1.09	2.01	1.17	2.09	1.30	2.21	1.45	0.44	0.55
324	1.71	3.62	1.38	2.6	0.28	"	"	1.98	1.28	2.04	1.36	2.12	1.49	2.24	1.64	0.61	0.72
325	1.92	3.25	1.48	2.2	0.14	"	"	1.94	1.35	2.00	1.43	2.08	1.56	2.20	1.71	0.38	0.49
326	1.56	3.82	1.37	2.8	0.11	"	"	2.01	1.24	2.07	1.32	2.15	1.45	2.27	1.60	0.19	0.31
327	1.58	3.07	1.00	3.1	0.24	10.40	"	1.93	1.14	1.99	1.22	2.07	1.35	2.19	1.50	0.54	0.66
328	1.94	2.56	0.31	8.2	0.05	10.42	"	1.87	0.70	1.93	0.78	2.01	0.91	2.13	1.06	9.98	0.09
329	0.00	0.97	0.65	1.5	-	"	"	1.88	1.40	1.93	1.46	2.01	1.59	2.13	1.76	-	-
330	1.32	3.44	1.78	1.9	0.38	10.44	Chalk	2.01	1.36	2.07	1.44	2.15	1.57	2.27	1.72	0.76	0.88
331	1.88	3.35	0.80	4.2	0.10	10.46	Soil	2.00	1.12	2.06	1.20	2.14	1.33	2.26	1.48	0.28	0.39
332	1.62	2.21	1.45	1.5	0.45	"	"	1.81	1.32	1.87	1.40	1.95	1.53	2.07	1.68	0.86	0.98
333	1.33	3.32	1.00	3.3	0.21	10.55	"	2.03	1.14	2.09	1.22	2.17	1.35	2.29	1.50	0.54	0.65
334	1.76	2.92	0.46	6.3	0.09	10.56	"	1.96	0.89	2.02	0.97	2.10	1.10	2.22	1.25	0.22	0.33
335	2.04	2.86	1.21	2.4	0.28	10.59	"	1.98	1.37	2.04	1.45	2.12	1.58	2.24	1.73	0.80	0.91
336	2.25	2.25	0.09	25.0	0.09	10.64	"	1.93	0.30	1.99	0.38	2.07	0.51	2.19	0.66	0.36	0.47
337	2.07	3.53	1.77	2.0	0.47	10.71	Sand & Clay	2.12	1.58	2.18	1.66	2.26	1.79	2.38	1.94	1.06	1.17
338	-0.04	1.02	0.26	3.9	0.11	10.75	Soil,Gravel,Clay	2.04	1.16	2.09	1.22	2.17	1.35	2.29	1.52	0.78	0.89
339	-0.04	1.15	0.30	3.8	0.08	"	"	2.09	1.22	2.14	1.28	2.22	1.41	2.35	1.58	0.67	0.78
340	1.36	2.62	1.07	2.4	0.18	10.79	Soil	2.00	1.26	2.06	1.34	2.14	1.47	2.26	1.62	0.54	0.65
341	1.67	2.91	1.67	1.7	0.25	10.82	"	2.05	1.28	2.11	1.36	2.19	1.49	2.31	1.64	0.72	0.84
342	1.88	2.45	0.66	3.7	0.23	10.84	"	1.99	1.15	2.05	1.23	2.13	1.36	2.25	1.51	0.76	0.87
343	2.34	2.09	0.47	4.4	0.15	"	"	1.98	1.13	2.04	1.21	2.12	1.34	2.24	1.49	0.70	0.81
344	-0.07	1.37	1.10	1.2	-	10.87	"	2.22	1.84	2.28	1.90	2.36	2.03	2.48	2.20	-	-
345	-0.06	0.40	0.36	1.1	-	11.03	"	1.74	1.39	1.80	1.45	1.88	1.58	2.00	1.75	-	-
346	-0.03	1.00	0.29	3.4	0.06	11.06	Soil,Gravel,Clay	2.13	1.29	2.18	1.35	2.26	1.48	2.38	1.65	0.61	0.72
347	-0.03	1.25	0.25	5.0	0.10	"	"	2.22	1.21	2.28	1.28	2.36	1.41	2.48	1.58	0.83	0.94
348	-0.13	1.42	0.21	6.7	0.07	"	"	2.35	1.22	2.41	1.28	2.49	1.41	2.61	1.58	0.80	0.91
351	-0.05	0.79	0.16	4.9	-	11.30	Soil	2.15	1.13	2.20	1.19	2.28	1.32	2.40	1.49	-	-
352	-0.10	1.08	0.30	3.6	0.07	11.31	Soil,Gravel,Clay	2.28	1.43	2.33	1.49	2.41	1.62	2.54	1.79	0.82	0.93
353	-0.10	1.19	0.21	5.6	0.05	"	"	2.32	1.28	2.38	1.34	2.46	1.47	2.58	1.64	0.74	0.85
354	-0.03	1.15	0.27	4.2	0.07	11.36	"	2.29	1.35	2.34	1.41	2.42	1.54	2.55	1.71	0.77	0.88
355	-0.10	1.26	0.23	5.4	0.06	"	"	2.36	1.34	2.42	1.40	2.50	1.53	2.62	1.70	0.84	0.95
356	-0.10	1.32	0.19	6.9	0.06	"	"	2.38	1.24	2.44	1.30	2.52	1.43	2.64	1.60	0.80	0.91
357	-0.06	0.88	0.35	2.5	-	11.45	Soil	2.22	1.53	2.28	1.59	2.36	1.72	2.48	1.89	-	-
358	0.00	0.84	0.48	1.8	0.16	11.57	"	2.20	1.65	2.26	1.71	2.33	1.84	2.46	2.01	1.18	1.29
359	1.25	1.69	0.98	1.7	0.21	11.66	Desert Alluvium	2.11	1.50	2.17	1.58	2.25	1.71	2.37	1.86	0.92	1.03
360	-0.13	1.47	0.48	3.0	0.07	12.18	Soil	2.71	1.94	2.76	2.00	2.84	2.13	2.96	2.30	1.16	1.27
362	0.54	2.28	0.62	3.6	0.54	12.39	Gypsum & Soil	2.63	1.51	2.69	1.57	2.77	1.70	2.89	1.87	0.89	1.00
363	0.50	1.27	0.78	1.6	0.10	12.08	Desert Alluvium	2.27	1.66	2.33	1.74	2.41	1.87	2.53	2.02	0.76	0.87
364	0.13	1.02	0.45	2.3	0.09	"	"	2.36	1.68	2.42	1.76	2.50	1.89	2.62	2.04	0.95	1.06
365	1.80	2.09	-	-	-	11.02	Tuff	-	-	-	-	-	-	-	-	-	-

NOTES FOR TABLE 8

$H/W^{1/3}$ = Scaled depth of burst
$R/W^{1/3}$ = Scaled radius of apparent crater
$S/W^{1/3}$ = Scaled depth of apparent crater
R/S = Ratio of radius to depth for apparent crater
$T/W^{1/3}$ = Scaled rim height
E = Logarithm of explosive energy (calories)
D = Logarithm of apparent crater diameter (feet)
d = Logarithm of apparent crater depth (feet)
R_H = Logarithm of rim height (feet)

Table 23
Coordinates of Selected Lunar Features and Measured Heights Relative to a Sphere

No.	Name	ξ_1	ξ_2	ξ_3	ξ_4	ξ_5	$\bar{\xi}$	Schrutka ξ	Franz ξ	Saunder ξ	η_1	η_2	η_3	η_4	η_5
1	Blancanus F	-.19261	-.19320				-.19290				-.90591	-.90600			
2	Kircher B	-.28691	-.28854				-.28772				-.90603	-.90564			
3	Bettinus A	-.31944	-.32040				-.31992				-.90452	-.90406			
4	Bailly F	-.35693	-.35865				-.35779				-.92221	-.92159			
5	Bailly G	-.35348	-.35541				-.35444				-.90972	-.90936			
6	Weigel A	-.31778	-.31837				-.31808				-.85235	-.85234			
7	Weigel E	-.36600	-.36630				-.36615				-.83618	-.83635			
8	Segner A	-.39386	-.39479				-.39432				-.83924	-.83928			
9	N of Weigel E	-.38265	-.38313				-.38289				-.82393	-.82368			
10	Bayer M	-.32716	-.32781				-.32748				-.77247	-.77280			
11	Bayer K	-.35708	-.35762				-.35735				-.76814	-.76822			
12	Hainzel P	-.34732	-.34763				-.34748				-.71770	-.71776			
13	Bayer L	-.37373	-.37378				-.37376				-.73649	-.73647			
14	Schiller A	-.41387	-.41416				-.41402				-.73219	-.73231			
15	E of Schiller A	-.43823	-.43869				-.43846				-.74437	-.74442			
16	Nögerath F	-.48705	-.48775				-.48740				-.74272	-.74239			
17	Lagalla N	-.31053	-.31063				-.31058				-.70570	-.70525			
18	S of Hainzel A	-.34722	-.34723				-.34722				-.69931	-.69898			
19	Hainzel E	-.36700	-.36735				-.36718				-.70071	-.70050			
20	Hainzel A	-.36506	-.36499				-.36502				-.68378	-.68357			
21	Hainzel F	-.43464	-.43447				-.43456				-.68506	-.68489			
22	Hainzel I	-.46228	-.46273				-.46250				-.64479	-.64495			
23	Hainzel B	-.43414	-.43471				-.43442				-.61365	-.61396			
24	Hainzel O	-.48645	-.48730				-.48688				-.62368	-.62351			
25	Schickard H	-.63998	-.64051				-.64024				-.68767	-.68850			
26	Schickard N	-.64302	-.64292				-.64297				-.65492	-.65583			
27	Lehmann D	-.64819	-.64809				-.64814				-.63504	-.63616			
28	Lehmann H	-.65921	-.65999				-.65960				-.62313	-.62415			
29	Ramsden G	-.42731	-.42715				-.42723				-.57706	-.57728			
30	(A peak)	-.43323	-.43305				-.43314				-.52251	-.52202			
31	(A peak)	-.43689	-.43635				-.43662				-.51518	-.51461			
32	(A peak)	-.43323	-.43244				-.43284				-.51355	-.51340			
33	Mercator A	-.40061	-.40104				-.40082				-.50830	-.50850			
34	Vitello B	-.49594	-.49602				-.49598				-.51674	-.51689			
35	Vitello P	-.53086	-.53044				-.53065				-.51740	-.51495			
36	Lee C	-.57000	-.57063				-.57032				-.53530	-.53569			
37	Vitello E	-.50958	-.50919				-.50938				-.48581	-.48670			
38	Campanus A	-.43031	-.42985				-.43008				-.43698	-.43669			
39	Palmieri E	-.65350	-.65394				-.65372				-.48729	-.48830			
40	Fourier C	-.69108	-.69266				-.69187				-.47685	-.47723			
41	Mare Humorum D	-.53093	-.52984				-.53038				-.43281	-.43342			
42	W of Mare Humorum D	-.56514	-.56437				-.56476				-.43154	-.43245			
43	Mare Humorum J	-.59764	-.59695				-.59730				-.41323	-.41409			
44	Mare Humorum K	-.59504	-.59378				-.59441				-.40566	-.40579			
45	Mare Humorum L	-.59487	-.59443				-.59465				-.39952	-.40038			
46	Mare Humorum F	-.57506	-.57404				-.57455				-.39575	-.39642			
47	Mare Humorum R	-.56764	-.56703				-.56734				-.37222	-.37246			
48	Gassendi J	-.55930	-.55884				-.55907	-.55919			-.36700	-.36733			
49	Mare Humorum Y	-.58034	-.57926				-.57980				-.35526	-.35553			
50	Mare Humorum O	-.53238	-.53198				-.53218				-.37253	-.37290			
51	Mersenius E	-.66427	-.66405				-.66416				-.38133	-.38147			
52	Gassendi L	-.62276	-.62203				-.62240				-.34675	-.34709			
53	Mersenius S	-.68999	-.68997				-.68998	-.68997			-.32782	-.32822			
54	Gassendi G	-.67165	-.67140				-.67152	-.67184			-.28668	-.28713			
55	Gassendi ζ	-.65029	-.65117				-.65073	-.65181	-.65246		-.28244	-.28233			
56	Lubiniezky C	-.42531	-.42455				-.42493				-.24298	-.24311			
57	Herigonius C	-.48693	-.48658				-.48676				-.22890	-.22912			
58	Herigonius E	-.56508	-.56490				-.56499				-.23766	-.23793			
59	Gassendi F	-.68180	-.68270				-.68225				-.25804	-.25883			
60	Crüger C	-.84350	-.84524				-.84437				-.28937	-.29025			
61	Herigonius B	-.49401	-.49364				-.49382				-.20328	-.20275			
62	(A spot)	-.66062	-.66049				-.66056				-.18442	-.18457			
63	Euclides	-.48831	-.48826				-.48828	-.48771		-.48870	-.12813	-.12770			
64	(A peak)	-.56839	-.56795				-.56817				-.11476	-.11488			
65	Wichmann	-.61081	-.61098				-.61090				-.12994	-.13019			
66	W of Wichmann	-.62610	-.62525				-.62568				-.12298	-.12305			
67	Letronne A	-.67399	-.67368				-.67384				-.13586	-.13645			
68	(A peak)	-.60666	-.60623				-.60644				-.09051	-.09051			
69	Flamsteed F	-.65430	-.65430				-.65430				-.08147	-.08173			
70	Flamsteed B	-.68726	-.68726				-.68726				-.10212	-.10226			
71	Flamsteed C	-.71840	-.71773				-.71806				-.09514	-.09523			
72	(A spot)	-.75193	-.75169				-.75181				-.12246	-.12277			
73	Damoiseau D	-.88593	-.88456				-.88524				-.11050	-.10894			
74	Damoiseau C	-.87657	-.87679				-.87668				-.04500	-.04510			
75	Hermann	-.84097	-.84065				-.84081				-.01409	-.01443			
76	Flamsteed D	-.70367	-.70276				-.70322				-.05458	-.05520			
77	C NE of Flamsteed	-.65147	-.65075				-.65111				-.05944	-.05983			
78	NE of Flamsteed	-.64983	-.64834				-.64908				-.03970	-.03980			
79	Lansberg D	-.50830	-.50783				-.50806				-.05197	-.05162			
80	Lansberg B	-.47028	-.47016				-.47022				-.04295	-.04269			
81	Lansberg	-.44761	-.44727				-.44744				-.00523	-.00529			
82	Lansberg G	-.49148	-.49056				-.49102				-.01037	-.01067			
83	Lansberg A	-.51658	-.51719				-.51688	-.51637			.00365	.00374			
84	Encke C	-.59335	-.59262				-.59298				.01206	.01205			
85	Encke E	-.64423	-.64385				-.64404				.00673	.00679			
86	Hortensius B	-.48962	-.49023				-.48992				.09222	.09214			
87	Hortensius	-.46558	-.46566				-.46562				.11317	.11333			
88	Kepler A	-.58419	-.58347				-.58383	-.58442			.12495	.12451			
89	Möstlin	-.64887	-.64882				-.64884				.08599	.08607			
90	S of Suess D	-.71349	-.71352				-.71350				.06973	.06939			
91	Suess D	-.72253	-.72217				-.72235				.08213	.08185			
92	Suess	-.73608	-.73600				-.73604				.07681	.07681			
93	Reiner E	-.75975	-.75947				-.75961				.03390	.03377			
94	Reiner C	-.77942	-.77989				-.77966				.06192	.06178			
95	Kepler F	-.62302	-.62250				-.62276				.14506	.14519			
96	Milichius	-.49499	-.49471				-.49485	-.49490			.17418	.17414			
97	Kepler C	-.65617	-.65598				-.65608				.17434	.17432			
98	SW of Marius D	-.69969	-.69941				-.69955				.18329	.18336			
99	Reiner H	-.80472	-.80450				-.80461				.15882	.15914			
100	Marius A	-.70110	-.70079				-.70094	-.70116			.21856	.21864			

No.	$\bar{\eta}$	Schrutka η	Franz η	Saunder η	h 1,2	h 2,3	h 3,4	h 3,5	$\bar{h}$	Diameter in miles	Rim Height	For Lunar Surface h	Schrutka h	Franz h	Saunder h	Notes
1	-.90596				81				81	5.1	14	67				1
2	-.90584				211				211	6.3	17	194				1
3	-.90429				121				121	16.0	38	83				1
4	-.92190				83				83	9.8	25	58				1
5	-.90954				154				154	11.3	28	126		-71		1
6	-.85234				97				97	9.3	24	73				1
7	-.83626				36				36	6.8	18	18				1
8	-.83926				131				131	6.0	16	115				1
9	-.82380				96				96	4.2	12	84				1
10	-.77264				112				112	5.7	16	96				1
11	-.76818				108				108	10.0	25	83				1
12	-.71773				69				69	9.1	23	46				1
13	-.73648				15				15	8.3	22	-7				1
14	-.73225				51				51	6.2	17	34				1
15	-.74440				87				87	5.2	14	73				1
16	-.74256				148				148	5.3	15	133				1
17	-.70548				76				76	7.0	19	57				1
18	-.69914				38				38	8.1	21	17				1
19	-.70060				109				109	8.7	23	86				1
20	-.68368				5				5	9.1	23	-18				1
21	-.68498				-23				-23	7.8	21	-44				1
22	-.64487				95				95	6.4	17	78				1
23	-.61380				120				120	9.1	23	97				1
24	-.62360				226				226	7.6	20	206				1
25	-.68808				26				26	9.0	23	3				1
26	-.65538				-66				-66	8.2	21	-87				1
27	-.63560				-83				-83	8.2	21	-104				1
28	-.62364				62				62	7.8	21	41				1
29	-.57717				-72				-72	6.9	19	-91				2
30	-.52226				6				6	-	(10)	-4				2,4
31	-.51490				-92				-92	-	(21)	-113				2,4
32	-.51348				-220				-220	-	(21)	-241				2,4
33	-.50842				103				103	5.0	14	89				
34	-.51682				6				6	7.0	19	-13				
35	-.51618				161				161	5.3	15	146				1
36	-.53550				119				119	8.9	23	96				1
37	-.43626				-210				-210	4.6	13	-223				
38	-.43684				-111				-111	6.8	18	-129				2
39	-.48780				11				11	8.3	22	-11				1
40	-.47704				301				301	8.7	23	278		-44		1
41	-.43312				-397				-397	4.9	14	-411				
42	-.43200				-323				-323	2.1	7	-330				
43	-.41366				-280				-280	2.9	9	-289		-10		
44	-.40572				-368				-368	3.9	11	-379				
45	-.39995				-219				-219	2.0	6	-225				
46	-.39608				-376				-376	2.4	8	-384				
47	-.37234				-211				-211	2.4	8	-219				
48	-.36716	-.36736			-181				-181	6.1	17	-198	34			
49	-.35540				-355				-355	2.8	8	-363				
50	-.37272				-170				-170	7.2	19	-189				
51	-.38140				-70				-70	6.5	18	-88				
52	-.34692				-245				-245	2.6	8	-253				
53	-.32802	-.32819			-44				-44	9.2	24	-68	-69			1
54	-.28690	-.28730			-116				-116	4.9	14	-130	52			1
55	-.28238	-.28235	-.28330		258				258	-	(0)	258	-6	-134		1,3
56	-.24304				-293				-293	10.0	25	-318		24		2
57	-.22901				-152				-152	6.5	18	-170				2
58	-.23780				-92				-92	4.6	13	-105				2
59	-.25844				161				161	4.1	12	149				1
60	-.28981				234				234	6.7	18	216				1
61	-.20302				-54				-54	6.7	18	-72				2
62	-.18450				-58				-58	-	(0)	-58				2,3
63	-.12792	-.12813		-.12810	44				44	7.2	19	25	-29			2
64	-.11482				-163				-163	-	(17)	-180				2,4
65	-.13006				21				21	5.6	15	6				2
66	-.12302				-275				-275	2.3	7	-282				2
67	-.13616				-160				-160	7.2	19	-179				2
68	-.09051				-137				-137	-	(13)	-150				2,4
69	-.08160				-32				-32	2.6	8	-40				2
70	-.10219				-15				-15	5.4	15	-30				2
71	-.09518				-194				-194	5.2	14	-208				2
72	-.12262				-93				-93	-	(0)	-93				2,3
73	-.10972				-133				-133	7.5	20	-153		-201		1
74	-.04505				32				32	7.2	19	13				2
75	-.01426				-92				-92	9.8	25	-117				2
76	-.05489				-327				-327	3.8	11	-338		-390		2
77	-.05966				-274				-274	2.1	7	-281				2
78	-.03975				-474				-474	2.4	8	-482				2
79	-.05180				-117				-117	6.4	17	-134		-227		2
80	-.04282				-6				-6	5.6	15	-21				2
81	-.00526				-134				-134	24.3	54	-188		-131		2
82	-.01052				-375				-375	5.4	15	-390				2
83	.00370	.00365			229				229	5.6	15	214	121	-225		2
84	.01206				-239				-239	4.9	14	-253				2
85	.00676				-111				-111	5.7	16	-127		-235		2
86	.09218				207				207	3.8	11	196				2
87	.11325				52				52	8.1	21	31				2
88	.12473	.12466			-292				-292	6.3	17	-309	17			2
89	.08603				-4				-4	3.8	11	-15				2
90	.06956				-28				-28	2.3	7	-35				2
91	.08199				-128				-128	3.3	10	-138				2
92	.07681				-21				-21	4.9	14	-35				2
93	.03384				-83				-83	2.0	6	-89				2
94	.06185				101				101	3.4	10	91				2
95	.14512				-145				-145	3.6	10	-155				2
96	.17416	.17405			-103				-103	8.1	21	-124	29			2
97	.17458				0				0	7.4	20	-20				2
98	.18332				-70				-70	2.1	7	-77				2
99	.15898				-21				-21	3.4	10	-31				2
100	.21860	.21822			-78				-78	8.9	23	-101	-126			2

Table 23
Coordinates of Selected Lunar Features and Measured Heights Relative to a Sphere

No.	Name	ξ_1	ξ_2	ξ_3	ξ_4	ξ_5	$\bar{\xi}$	Schrutka ξ	Franz ξ	Saunder ξ	η_1	η_2	η_3	η_4	η_5
101	E of Marius C	-.71017	-.70854				-.70936				.23917	.23964			
102	Marius E	-.77638	-.77631				-.77634				.21058	.21007			
103	Galileo	-.87259	-.87200				-.87230	-.87314			.18312	.18246			
104	Galileo A	-.87069	-.87072				-.87070				.20347	.20333			
105	Galileo E	-.85427	-.85362				-.85394				.24112	.24098			
106	SW of Marius P	-.79262	-.79255				-.79258				.27388	.27460			
107	N of Marius P	-.79560	-.79468				-.79514				.28402	.28503			
108	Bessarion B	-.63576	-.63514				-.63545				.28974	.29007			
109	Bessarion A	-.61132	-.61042				-.61087				.29370	.29375			
110	(A peak)	-.48316	-.48265				-.48290				.33550	.33590			
111	Brayley F	-.52230	-.52067				-.52148				.35978	.36126			
112	Brayley B	-.52617	-.52630				-.52624				.35377	.35402			
113	Brayley C	-.59085	-.59110				-.59098				.36463	.36547			
114	Bessarion D	-.62533	-.62488				-.62510				.33842	.33852			
115	Diophantus E	-.50753	-.50637				-.50695				.41775	.41791			
116	Diophantus A	-.52754	-.52713				-.52734				.46300	.46330			
117	Diophantus B	-.46874	-.46863				-.46868				.48558	.48622			
118	Angstrom	-.57495	-.57354				-.57424				.49796	.49893			
119	Wollaston	-.62775	-.62692				-.62734				.50869	.50824			
120	Aristarchus	-.67380	-.67568				-.67469	-.67467	-.67552		.40274	.40166			
121	Seleucus A	-.80452	-.80445				-.80448				.37513	.37602			
122	Schiaparelli A	-.81149	-.81072				-.81110				.38991	.39057			
123	SE of Lichtenburg A	-.76353	-.76246				-.76300				.46593	.46651			
124	Lichtenburg A	-.75643	-.75601				-.75622				.48399	.48450			
125	(A peak)	-.70581	-.70578				-.70580				.46163	.46188			
126	Wollaston C	-.66567	-.66475				-.66521				.52660	.52787			
127	Gruithuisen	-.53590	-.53582				-.53586				.54239	.54283			
128	Heis D	-.43865	-.43818				-.43842				.52406	.52460			
129	Caroline Herschel	-.42699	-.42626				-.42662				.56521	.56590			
130	Car. Herschel C	-.42813	-.42808				-.42810				.60382	.60471			
131	Gruithuisen B	-.50724	-.50735				-.50730				.58205	.58296			
132	Laplace A	-.32722	-.32543				-.32632				.68954	.69091			
133	Sharp A	-.45556	-.45509				-.45532	-.45608	-.45663		.73762	.73814			
134	(A peak)	-.55622	-.55531				-.55576				.66473	.66585			
135	E of Naumann B	-.66771	-.66647				-.66709				.60442	.60496			
136	Naumann B	-.69004	-.68819				-.68912				.60742	.60858			
137	Repsold E	-.54836	-.54678				-.54757				.77209	.77378			
138	Bouguer A	-.33809	-.33659				-.33734				.79150	.79339			
139	Harpalus C	-.40005	-.39945				-.39975				.82328	.82390			
140	Harpalus B	-.38341	-.38259				-.38300				.82942	.83045			
141	South B	-.37887	-.37827				-.37857				.84186	.84341			
142	Horrebow B	-.35260	-.35248				-.35254				.85325	.85449			
143	Pythagoras A	-.42198	-.42300				-.42249				.85698	.85795			
144	Pythagoras C	-.43072	-.43155				-.43114				.85683	.85728			
145	Condamine F	-.27724	-.27695				-.27710				.83991	.84118			
146	J. Herschel C	-.29734	-.29730				-.29732				.88399	.88495			
147	Gruemberger F		-.10822	-.10753			-.10788					.91144	.91093		
148	Gruemberger F		-.03530	-.03508			-.03519					.88848	.88811		
149	Cysatus A		-.00560	-.00518			-.00539					.89996	.89946		
150	E of Clavius		-.03731	-.03651			-.03691					.86811	.86779		
151	Pentland F		.09253	.09277	.09230		.09259					.88281	.88232	-.88290	
152	W of Jacobi		.08349	.08440	.08410		.08410					.83971	.83952	-.84025	
153	Lilius		.06222	.06187			.06204		.06205			.81259	.81273		
154	Lilius O		.03563	.03576			.03570					.82279	.82261		
155	Lilius M		.01717	.01726			.01722					.83144	.83122		
156	Clavius BB		-.08666	-.08570			-.08618					.81274	.81250		
157	Clavius CA		-.13015	-.12961			-.12988					.78854	.78894		
158	Clavius R		-.15891	-.15816			-.15854					-.79991	-.80005		
159	Street J		-.15570	-.15518			-.15544					-.75076	-.75049		
160	Street M		-.15330	-.15291			-.15310					-.73718	-.73665		
161	S of Wilhelm I E		-.21977	-.21951			-.21964					-.70358	-.70359		
162	Wilhelm I E		-.22147	-.22067			-.22107					-.69597	-.69600		
163	Lilius S		.06205	.06254	.06197		.06228		.06200			-.79529	-.79511	-.79571	
164	Licetus D		.05600	.05645			.05622		.05600			-.77186	-.77170		
165	E of Lilius W		.09556	.09549	.09527		.09545					-.80347	-.80338	-.80393	
166	Cuvier M		.11285	.11295	.11294		.11292					-.80158	-.80141	-.80171	
167	Jacobi T		.14681	.14704	.14708		.14699					-.82864	-.82832	-.82853	
168	Jacobi J		.14050	.14052	.14051		.14051					-.81280	-.81286	-.81298	
169	Cuvier R		.14302	.14321	.14349		.14323					-.77684	-.77685	-.77704	
170	Baco D		.17462	.17464	.17454		.17461					-.78336	-.78331	-.78341	
171	Licetus Q		.11415	.11472	.11448		.11452					-.73302	-.73299	-.73340	
172	SE of Licetus G		.13826	.13853	.13844		.13844					-.71451	-.71456	-.71496	
173	Licetus H		.03834	.03846			.03840		.03815			-.71769	-.71755		
174	Licetus M		.02222	.02270			.02246		.02270			-.72815	-.72792		
175	Huggins A		-.02949	-.02970			-.02960		-.02935			-.64924	-.64906		
176	Orontius D		-.08277	-.08251			-.08264		-.08300			-.63405	-.63386		
177	Sasserides B		-.14947	-.14878			-.14912					-.63560	-.63523		
178	Heinsius A		-.23134	-.23116			-.23125	-.23130				-.63799	-.63796		
179	Barocius EB		.25336	.25373	.25345		.25357					-.72645	-.72644	-.72666	
180	Barocius N		.24698	.24742	.24722		.24726					-.68327	-.68313	-.68321	
181	Buch D		.21828	.21873	.21881		.21864					-.63752	-.63729	-.63785	
182	Maurolycus G		.14595	.14631	.14604		.14615					-.62428	-.62425	-.62466	
183	Maurolycus H		.14152	.14177	.14186		.14173					-.61854	-.61828	-.61881	
184	Maurolycus K		.15085	.15128	.15072		.15103					-.60799	-.60780	-.60810	
185	Gemma Frisius F		.14516	.14529	.14547		.14530					-.58472	-.58449	-.58507	
186	Stöfler K		.05637	.05652			.05641		.05670			-.63478	-.63480	-.63534	
187	Nonius A		.07914	.07936	.07909		.07924					-.57791	-.57798	-.57832	
188	Nonius G		.08193	.08218	.08236		.08216					-.57084	-.57030	-.57160	
189	Aliacensis G		.06877	.06893	.06884		.06887					-.54919	-.54918	-.54958	
190	Aliacensis D		.09987	.10016	.09971		.09998					-.54636	-.54620	-.54680	
191	Miller		.01010	.00960			.00985					-.63250	-.63316		
192	Walter S		.00913	.00923			.00918		.00970			-.59272	-.59306		
193	Walter N		-.00259	-.00261			-.00260		-.00915			-.55377	-.55378		
194	Lexell		-.05828	-.05825			-.05826		-.00250			-.58061	-.58104		
195	Lexell E		-.08807	-.08809			-.08808		-.05840			-.56681	-.56700		
196	Hell E		-.09538	-.09522			-.09530					-.47247	-.47252		
197	Gauricus D		-.16177	-.16141			-.16159					-.57481	-.57488		
198	Pitatus H		-.18581	-.18535			-.18558					-.53526	-.53569		
199	Wurzelbauer M		-.26213	-.26183			-.26198					-.56665	-.56692		
200	Cichus D		-.29896	-.29908			-.29902					-.50904	-.50951		

No.	$\bar{\eta}$	Schrutka η	Franz η	Saunder η	h 1,2	h 2,3	h 3,4	h 3,5	$\bar{h}$	Diameter in Miles	Rim Height	For Lunar Surface h	Schrutka h	Franz h	Saunder h	Notes
101	.23940				-382				-382	3.3	10	-392				2
102	.21032				172				172	2.6	8	164				2
103	.18279	.18224			-136				-136	9.1	23	-159	-121			2
104	.20340				-4				-4	7.3	20	-24				2
105	.24105				-119				-119	3.6	10	-129				2
106	.27424				41				41	3.3	10	31				2
107	.28452				-111				-111	2.8	8	-119				2
108	.28990				-141				-141	6.8	18	-159				2
109	.29372				-264				-264	7.9	21	-285				2
110	.33570				-116				-116	-	(19)	-135				4
111	.36052				-326				-326	3.5	10	-336				
112	.35390				74				74	6.5	18	56				
113	.36505				168				168	5.6	15	153				
114	.33847				-115				-115	5.7	16	-131				2
115	.41783				-339				-339	3.9	11	-350				
116	.46315				206				206	4.7	13	193				
117	.48590				42				42	4.2	12	30				
118	.49844				-269				-269	5.9	16	-285				
119	.50846				-235				-235	6.1	17	-252				
120	.40243	.40243	.40201		362				362	24.5	55	307	-322			
121	.37558				43				43	3.6	10	33				2
122	.39024				-86				-86	5.6	15	-101				2
123	.46622				-144				-144	4.1	12	-156				2
124	.48424				-39				-39	5.6	15	-54				2
125	.46176				13				13	-	(23)	-10				4
126	.52724				-92				-92	6.4	17	-109				2
127	.54261				24				24	9.7	25	-1				
128	.52433				-78				-78	4.7	13	-91				
129	.56556				-131				-131	8.7	23	-154				
130	.60426				83				83	4.7	13	70				
131	.58250				118				118	5.9	16	102				
132	.69022				-332				-332	5.8	16	-348		-273		
133	.73788	.73779	.73763		-54				-54	10.2	26	-80	-172	-200		
134	.66529				-95				-95	-	(20)	-115				2,4
135	.60469				-172				-172	3.3	10	-182				2
136	.60800				-209				-209	6.5	18	-227				2
137	.77294				-115				-115	7.8	21	-136				2
138	.79244				-158				-158	5.7	16	-174				
139	.82359				-59				-59	6.5	18	-77				2
140	.82994				-69				-69	4.9	14	-83				2
141	.84264				-3				-3	9.1	23	-26				
142	.85387				52				52	7.2	19	33				1
143	.85746				148				148	19.6	45	103		-203		1
144	.85706				104				104	7.5	20	84				1
145	.84054				39				39	4.0	12	27				
146	.88447				46				46	7.5	20	26				1
147	-.91118				156				156	7.3	20	136				1
148	-.88830				70				70	3.8	11	59				1
149	-.89971				117				117	8.8	23	94				1
150	-.86795				201				201	7.6	20	181				1
151	-.88259				85	110			98	7.6	20	78				1
152	-.83975				234	115			174	3.9	11	163				1
153	-.81266			-.81360	-102				-102	-	(-30)	-72				5
154	-.82270				48				48	4.0	12	36				1
155	-.83133				38				38	7.7	20	18				1
156	-.81262				265				265	7.0	19	246				1
157	-.78874				109				109	3.4	10	99				1
158	-.79998				180				180	4.9	14	166				1
159	-.75062				168				168	3.4	10	158				1
160	-.73692				160				160	5.0	14	146				1
161	-.70358				77				77	3.3	10	67				1
162	-.69598				232				232	8.2	21	211				1
163	-.79530		-.79635		147	163			155	8.4	22	133				1
164	-.77178		-.77265		142				142	-	(-31)	173				5
165	-.80354				-9	113			52	3.1	9	43				1
166	-.80153				41	49			45	3.6	10	35				1
167	-.82845				83	26			54	3.9	11	43				1
168	-.81288				46	22			34	8.1	21	13				1
169	-.77690				51	-1			25	4.3	12	13				1
170	-.78335				56	27			42	5.0	14	28				1
171	-.73310				171	108			140	5.1	14	126				1
172	-.71465				76	88			82	4.2	12	70				1
173	-.71762		-.71830		51				51	6.6	18	33		-198		1
174	-.72804		-.72870		166				166	5.2	14	152				1
175	-.64915		-.64965		-48				-48	4.9	17	-65		-145		6
176	-.63396		-.63455		106				106	9.8	25	81				1
177	-.63542				263				263	4.0	12	251				1
178	-.63798	-.63787			59				59	12.4	30	29	-144	38		1
179	-.72650				103	73			88	6.7	18	70				1
180	-.68318				147	41			94	6.4	17	77				1
181	-.63749				170	104			137	4.3	12	125				1
182	-.62436				95	128			112	5.1	14	98				1
183	-.61848				117	101			109	4.6	13	96				1
184	-.60792				141	150			146	6.7	18	128				1
185	-.58469				71	101			86	5.8	16	70				1
186	-.63493			-.63540	47	132			90	11.6	29	61			-92	1
187	-.57805				70	122			96	6.7	18	78				1
188	-.57076				150	269			210	3.9	11	199				1
189	-.54928				59	107			83	4.8	14	69				1
190	-.54639				124	214			169	5.8	16	153				1
191	-.63283		-.63390		-244				-244	-	(-60)	-184				5
192	-.59289		-.59370		-5				-5	7.0	19	-24				1,7
193	-.55378		-.55465		-11				-11	3.5	10-(54)	38				5
194	-.58082		-.58165		-40				-40	-	(-17)	-23				8
195	-.56690				-30				-30	6.0	16	-46				
196	-.47240				27				27	4.2	12	15				
197	-.57484				114				114	8.0	21	93				1
198	-.53548				113				113	4.2	12	101				
199	-.56678				70				70	5.6	15	55				
200	-.50928				-92				-92	4.6	13	-105				2

Table 23
Coordinates of Selected Lunar Features and Measured Heights Relative to a Sphere

No.	Name	ξ_1	ξ_2	ξ_3	ξ_4	ξ_5	$\bar{\xi}$	Schrutka ξ	Franz ξ	Saunder ξ	η_1	η_2	η_3	η_4	η_5
201	Kies B	-.32671	-.32640				-.32656					-.47992	-.48026		
202	Pitatus B	-.26718	-.26704				-.26711					-.45511	-.45523		
203	Goodacre O	.24084	.24125	.24070			.24101			-.26720		-.54461	-.54455	-.54523	
204	Pontanus J	.19665	.19683	.19678			.19677					-.50091	-.50084	-.50124	
205	Apianus L	.16485	.16499	.16474			.16489					-.48587	-.48579	-.48619	
206	Aliacensis P	.13043	.13048	.13065			.13051					-.52775	-.52771	-.52817	
207	Apianus E	.12508	.12519	.12522			.12517					-.48140	-.48171	-.48193	
208	Apianus F	.09767	.09817	.09778			.09795					-.47051	-.47077	-.47089	
209	Aliacensis α	.07587	.07518				.07552			.07590		-.50586	-.50613		
210	Aliacensis E	.03537	.03549				.03543					-.50544	-.50561		
211	Werner α	.05423	.05390				.05406			.05415		-.46824	-.46869		
212	Werner	.04569	.04528				.04548			.04550		-.46492	-.46542		
213	Purbach A	-.02965	-.02994				-.02980			-.02955		-.43961	-.44010		
214	Lippershey	-.16125	-.16124				-.16124					-.43623	-.43659		
215	Nicollet	-.19997	-.20011				-.20004	-.19970				-.37261	-.37286		
216	Zagut F	.26029	.26066	.26043			.26051					-.50429	-.50417	-.50469	
217	Pontanus N	.25452	.25439	.25442			.25443					-.45486	-.45455	-.45504	
218	Sacrobosco L	.23592	.23618	.23561			.23597					-.43150	-.43163	-.43174	
219	Sacrobosco O	.21876	.21956	.21926			.21928					-.43912	-.43893	-.43923	
220	Sacrobosco K	.19785	.19814	.19793			.19802					-.43403	-.43411	-.43409	
221	Sacrobosco N	.21746	.21812	.21759			.21782					-.41697	-.41681	-.41713	
222	W of Pontanus	.19718	.19750	.19744			.19740					-.46833	-.46786	-.46889	
223	Donati α	.08379	.08355				.08367			.08375		-.35212	-.35208		
224	Faye	.06209	.06168				.06188			.06185		-.36455	-.36508		
225	Faye G	.03263	.03255				.03259					-.34971	-.34994		
226	Arzachel C	.02033	.01997				.02015			.02015		-.31759	-.31751		
227	Arzachel L	.01558	.01554				.01556					-.30933	-.30936		
228	Purbach D	-.02444	-.02484				-.02464					-.38801	-.38819		
229	Thebit E	-.07382	-.07404				-.07393					-.39177	-.39188		
230	Thebit L	-.08700	-.08620				-.08660					-.36522	-.36517		
231	Thebit A	-.07886	-.07860				-.07873	-.07943		-.07885		-.36702	-.36723		
232	Birt	-.13727	-.13717				-.13722	-.13732				-.37957	-.37979		
233	Opelt E	-.29306	-.29338				-.29322					-.29220	-.29217		
234	Darney	-.38590	-.38580				-.38585	-.38573				-.25131	-.25144		
235	Almanon L	.27021	.27065	.27040			.27048					-.32488	-.32481	-.32511	
236	Geber E	.21023	.21041	.21033			.21034					-.35013	-.35057	-.35039	
237	Abenezra D	.15593	.15638	.15652			.15630					-.36996	-.37006	-.37046	
238	Abenezra E	.14376	.14402	.14390			.14392					-.37031	-.37052	-.37106	
239	Airy L	.12322	.12341	.12351			.12339					-.34885	-.34903	-.34898	
240	Abulfeda A	.17958	.17948	.17953			.17952	.17942				-.28224	-.28234	-.28247	
241	Abulfeda E	.16923	.16914	.16884			.16909	.16882				-.28784	-.28778	-.28793	
242	Airy P	.13995	.14046	.14013			.14025					-.27315	-.27319	-.27343	
243	Airy F	.12049	.12022	.12049			.12036					-.31172	-.31204	-.31203	
244	Airy	.09475	.09451				.09463					-.31000	-.31039		
245	Argelander D	.07459	.07463	.07412			.07449	.07453		.07455		-.30236	-.30243	-.30284	
246	Argelander	.09735	.09698				.09716			.09745		-.28441	-.28451		
247	Argelander A	.11267	.11344	.11316			.11318					-.28378	-.28385	-.28395	
248	Albategnius S	.05998	.05991	.06008			.05997					-.27390	-.27378	-.27385	
249	Alphonsus H	-.00847	-.00853				-.00850					-.26850	-.26857		
250	Alphonsus α	-.04580	-.04666				-.04623	-.04578		-.04660		-.23109	-.23073		
251	Alphonsus A	-.04161	-.04204				-.04182					-.25945	-.26038		
252	Alphonsus B	-.11469	-.11502				-.11486			-.11505		-.26003	-.26006		
253	Davy B	-.15226	-.15190				-.15208					-.18834	-.18829		
254	Guericke C	-.19547	-.19580				-.19564	-.19580				-.19979	-.19983		
255	Guericke D	-.24592	-.24644				-.24618					-.20683	-.20721		
256	Parry A	-.27104	-.27093				-.27098	-.27079				-.16454	-.16456		
257	Tacitus B	.33952	.33999	.33943			.33973					-.24238	-.24200	-.24238	
258	Tacitus F	.28874	.28942	.28892			.28912					-.29355	-.29323	-.29343	
259	Abulfeda R	.21943	.21942	.21933			.21940					-.22081	-.22072	-.22082	
260	Ritchey C	.15692	.15716	.15720			.15711					-.18969	-.18926	-.18933	
261	Ritchey D	.15788	.15768	.15795			.15795					-.17768	-.17745	-.17759	
262	Hipparchus Q	.04978	.05041				.05010					-.14754	-.14704		
263	Müller A	.03713	.03732				.03722					-.14148	-.14103		
264	Ptolemaeus A	-.01382	-.01401				-.01392	-.01380		-.01400		-.14771	-.14756		
265	Palisa D	-.11829	-.11860				-.11844					-.15041	-.15054		
266	Palisa C	-.11129	-.11084				-.11106					-.13411	-.13395		
267	S of Ptolemaeus H	-.09298	-.09329				-.09314					-.12888	-.12856		
268	Ptolemaeus H	-.09391	-.09374				-.09382					-.12358	-.12346		
269	N of Ptolemaeus H	-.09017	-.09017				-.09017					-.11872	-.11877		
270	Lalande D	-.12853	-.12907				-.12880					-.10731	-.10756		
271	Lalande B	-.16840	-.16851				-.16846	-.16857				-.11494	-.11515		
272	Fra Mauro A	-.35546	-.35554				-.35550					-.09456	-.09433		
273	Dollond	.24516	.24591	.24531			.24557			.24580		-.18155	-.18145	-.18147	
274	Andél F	.19022	.19050	.19067			.19047					-.14458	-.14469	-.14442	
275	Hipparchus L	.15592	.15569	.15580			.15578			.15635		-.11898	-.11893	-.11884	
276	Hind G	.12839	.12882	.12842			.12861					-.15113	-.15103	-.15104	
277	Hipparchus G	.12881	.12885	.12858			.12877	.12878		.12910		-.08728	-.08718	-.08723	
278	Hipparchus N	.08722	.08697	.08687			.08701			.08720		-.08387	-.08392	-.08401	
279	Hipparchus F	.04342	.04354				.04348					-.07261	-.07282		
280	Hipparchus H	.03962	.04000	.03973			.03984					-.09508	-.09485	-.09510	
281	Herschel C	-.05495	-.05498				-.05496	-.05496				-.08669	-.08680		
282	Mösting A	-.08973	-.08998				-.08986	-.08992	-.09000	-.09055		-.05565	-.05570		
283	Mösting B	-.12779	-.12816				-.12798					-.04713	-.04718		
284	Lalande B	-.15635	-.15602				-.15618					-.05429	-.05431		
285	Mösting C	-.14051	-.14037				-.14044	-.14010				-.03104	-.03103		
286	Turner	-.22853	-.22847				-.22850	-.22824				-.02372	-.02383		
287	Turner F	-.24342	-.24381				-.24362					-.02799	-.02779		
288	Gambart A	-.32083	-.32102				-.32092	-.32082				.01713	.01727		
289	Taylor D	.26979	.27004	.26987			.26994					-.09234	-.09266	-.09270	
290	Hind H	.19511	.19510	.19530			.19515					-.11545	-.11564	-.11547	
291	E. Pickering B	.12913	.12898	.12876			.12896					-.03629	-.03625	-.03621	
292	E. Pickering	.12216	.12206	.12175			.12201	.12187				-.04956	-.04975	-.04971	
293	Seeliger	.05255	.05251	.05249			.05252			.05270		-.03820	-.03826	-.03820	
294	Rhaeticus A	.09055	.09114	.09059			.09086	.09057				-.03039	-.03051	-.03048	
295	Blagg	.02569	.02580				.02574					.02149	.02161		
296	Bruce	.00653	.00699				.00676	.00696		.00665		.02071	.02065		
297	Gambart G	-.20826	-.20857				-.20842					.03426	.03463		
298	Gambart B	-.20049	-.20026				-.20038					.03794	.03809		
299	Gambart C	-.20398	-.20402				-.20400					.05830	.05841		
300	Theon Sr. A	.26571	.26572	.26578			.26573					-.00290	-.00291	-.00307	

454

No.	$\bar{\eta}$	Schrutka η	Franz η	Saunder η	h 1,2	h 2,3	h 3,4	h 3,5	$\bar{h}$	Diameter in Miles	Rim Height	For Lunar Surface h	Schrutka h	Franz h	Saunder h	Notes
201	-.48009				67				67	5.4	15	52				
202	-.45517			-.45530	37				37	6.2	17	20				2
203	-.54474				146	240			193	7.0	19	174				1
204	-.50096				75	104			90	5.3	15	75				1
205	-.48591				60	140			100	2.8	8	92				1
206	-.52784				26	80			53	4.2	12	41				1
207	-.48169				-3	48			22	5.3	15	7				1
208	-.47074				152	96			124	3.2	10	114				1
209	-.50600			-.50630	-289				-289	-	(-79)	-210				5
210	-.50552				24				24	5.4	15	9				1
211	-.46846			-.46890	-182				-182	-	(-82)	-100				9
212	-.46517			-.46590	-220				-220	-	(-82)	-138				9
213	-.43986			-.44070	-177				-177	4.6	13-(130)	-60			-346	10
214	-.43641				-40				-40	4.0	12	-52				2
215	-.37274	-.37272			-83				-83	9.6	24	-107	-63			2
216	-.50433				147	159			153	5.1	14	139				1
217	-.45475				-6	113			54	3.5	10	44				1
218	-.43162				80	125			102	5.3	15	87				1
219	-.43905				317	127			222	6.5	18	204				1
220	-.43408				96	31			64	6.2	17	47				1
221	-.41693				265	173			219	6.5	18	201				1
222	-.46824				180	262			221	4.5	13	208				1
223	-.35210			-.35245	-90				-90	-	(-60)	-30				5
224	-.36482			-.36495	-234				-234	-	(-30)	-204				5
225	-.34982				-62				-62	6.4	17	-79				
226	-.31755			-.31775	-131				-131	-	(-28)	-103				5
227	-.30934				-18				-18	4.5	13	-31				
228	-.38810				-182				-182	7.0	19	-201				
229	-.39182				-96				-96	4.8	14	-110				
230	-.36520				315				315	6.0	16	299				11
231	-.36712	-.36715		-.36745	74				74	12.0	30	44	6	-68/-90	-46	
232	-.37966	-.37963			0				0	10.4	26	-26	-34			2
233	-.29218				-118				-118	4.8	14	-132				2
234	-.25138	-.25132			18				18	8.8	23	-5	80			2
235	-.32490				179	120			150	4.2	12	138				1
236	-.35042				7	-32			-12	3.8	11	-23				1
237	-.37014				160	80			120	5.0	14	106				1
238	-.37060				70	161			116	3.6	10	106				1
239	-.34897				48	-31			8	4.9	14	-6				
240	-.28235	-.28223			-52	26			-13	8.6	22	-35	57			1
241	-.28783	-.28788			-24	96			36	3.7	11	25	-80			1
242	-.27324				195	129			162	4.8	14	148				1
243	-.31196				-151	-54			-102	3.0	9	-111				
244	-.31020			-.31020	-152				-152	-	(-60)	-92		-48		5
245	-.30252	-.30232		-.30255	4	206			105	7.5	20	85	155			1
246	-.28446			-.28455	-161				-161	-	(-50)	-111				5
247	-.28386				294	79			186	5.9	16	170				1
248	-.27383				-11	-11			-11	5.7	16	-27				1
249	-.26854				-33				-33	4.9	14	-47				1
250	-.23091	-.23035		-.23135	-294				-294	-			172			5
251	-.25992				-305				-305	4.6	13	-318		-30		12
252	-.26004			-.26045	-139				-139	6.7	18	-157			-150	2
253	-.18832				150				150	4.6	13	137				
254	-.19981	-.19982			-134				-134	6.8	18	-152	-52			2
255	-.20702				-254				-254	4.8	14	-268				2
256	-.16455	-.16479			17				17	8.1	21	-4	34	33		2
257	-.24219				230	203			216	7.4	20	196				1
258	-.29336				301	144			222	5.8	16	206				1
259	-.22077				7	44			26	4.5	13	13				1
260	-.18938				160	9			84	3.1	9	75				1
261	-.17754				77	46			62	4.0	12	50				1
262	-.14729				332				332	4.2	12	320				1
263	-.14126				142				142	6.2	17	125				
264	-.14764	-.14760		-.14790	-56				-56	6.2	17-(40)	-33	-34	-148/-150	-183	13
265	-.15048				-147				-147	4.6	13	-160				
266	-.13403				206				206	5.0	14	192				
267	-.12872				-82				-82	3.8	11	-93				
268	-.12352				85				85	4.4	13	72				
269	-.11874				-7				-7	4.6	13	-20				
270	-.10744				-255				-255	4.2	12	-267		-98/-34	-162	
271	-.11504	-.11508			-73				-73	8.2	21	-94	132	-74		2
272	-.09444				0				0	6.2	17	-17				2
273	-.18148			-.18170	309	121			215	6.9	19	196				1
274	-.14460				95	-110			-8	6.1	17	-25				1
275	-.11891			-.11895	-82	-42			-62	8.1	21	-83				1
276	-.15106				188	82			135	4.6	13	122				1
277	-.08722	-.08704		-.08705	30	66			48	8.9	23	25	195	11/74	-53	1
278	-.08393			-.08420	-110	48			-31	3.8	11-(40)	-2			-70	14
279	-.07272				19				19	5.9	16	3				
280	-.09497				188	125			156	2.8	8	148				1
281	-.08674	-.08687			26				26	6.6	18	8	190	-32/53	19	
282	-.05568	-.05551	-.05565	-.05560	-108				-108	8.2	21	-129	80	-93/-100	-91	
283	-.04716				-159				-159	4.4	13	-172				2
284	-.05430				131				131	4.9	14	117				
285	-.03104	-.03108			55				55	2.0	6	49	75	-210		2
286	-.02378	-.02390			10				10	7.2	19	-9	-69			2
287	-.02789				-126				-126	4.6	13	-139				
288	-.01720	.01720			-116				-116	7.2	19	-135	57	-123		2
289	-.09259				49	45			47	4.2	12	35				1
290	-.11555				-33	-86			-60	3.9	11	-71				1
291	-.03625				-56	36			-10	3.7	11	-21				
292	-.04969	-.04969			-64	50			-7	9.9	25	-32	195		-143	
293	-.03823			-.03830	-26	-14			-20	5.6	15	-35		-59/17	-11	
294	-.03047	.03064			256	118			187	6.8	18	169	109			2
295	-.02155				64				64	3.6	10	54				
296	-.02068	.02074		.02055	176				176	4.4	13	163	115	25/-12	63	2
297	-.03444				-71				-71	3.7	11	-82				2
298	-.03802				113				113	6.8	18	95		-212		2
299	-.05836				-4				-4	7.5	20	-24				2
300	-.00295				0	34			17	3.4	10	7				

455

Table 23

Coordinates of Selected Lunar Features and Measured Heights Relative to a Sphere

No.	Name	ξ_1	ξ_2	ξ_3	ξ_4	ξ_5	$\bar{\xi}$	Schrutka ξ	Franz ξ	Saunder ξ	η_1	η_2	η_3	η_4	η_5
301	De Morgan	.25675	.25690	.25690			.25686					.05809	.05815	.05819	
302	Cayley	.26003	.26017	.26010			.26012	.26005				.06917	.06922	.06918	
303	Cayley B	.25913	.25916	.25912			.25914					.08563	.08604	.08579	
304	Godin A	.16792	.16797	.16804			.16798					.04706	.04694	.04684	
305	Godin D	.14347	.14391	.14367			.14374					.01718	.01709	.01706	
306	Dembowsky A	.11285	.11310	.11278			.11296					.05273	.05308	.05289	
307	Dembowsky D	.10393	.10362	.10361			.10370					.06117	.06134	.06103	
308	Hyginus A	.09825	.09858	.09845			.09846					.11061	.11057	.11047	
309	Hyginus	.10887	.10839	.10877			.10860	.10860				.13537	.13558	.13517	
310	Hyginus B	.08795	.08777	.08784			.08783					.13268	.13287	.13274	
311	Triesnecker	.06278	.06372	.06285			.06327	.06306				.07328	.07335	.07304	
312	Chladni	.02000	.01971	.02000			.01986	.01992		.01980		.06975	.06990	.07015	
313	Pallas	-.02825	-.02910				-.02868			-.02835		.09521	.09513		
314	Pallas A	-.04048	-.04017				-.04032					.10441	.10449		
315	Bode	-.04172	-.04189				-.04180	-.04203		-.04210		.11706	.11767		
316	Bode A	-.01981	-.02003				-.01992	-.01973		-.02005		.15684	.15687		
317	Bode B	-.05288	-.05298				-.05293	-.05286		-.05300		.15218	.15231		
318	Bode J	.10505	.10506				.10506					.14809	.14844		
319	Ritter D	.32097	.32093	.32090			.32093					.06423	.06401	.06402	
320	Schmidt	.32212	.32227	.32199			.32216					.01709	.01697	.01683	
321	Manners B	.35468	.35498	.35481			.35486					.06037	.06014	.06021	
322	Maclear	.33748	.33800	.33798			.33786	.33786				.18267	.18262	.18297	
323	Maclear C	.31777	.31808	.31783			.31794					.20164	.20134	.20177	
324	Silberschlag A	.22725	.22720	.22715			.22720	.22707				.12101	.12092	.12108	
325	Boscovich A	.21615	.21597	.21583			.21598					.16472	.16494	.16479	
326	Manilius C	.17549	.17572	.17589			.17570					.20947	.20952	.20977	
327	Hyginus C	.14374	.14327	.14362			.14348					.13398	.13411	.13411	
328	Manilius D	.11841	.11848	.11882			.11855					.22907	.22918	.22943	
329	Hyginus D	.07456	.07396	.07441			.07422					.19800	.19795	.19807	
330	Conon B	-.03124	-.03088				-.03106					.29563	.29575		
331	Conon A	-.03265	-.03285				-.03275					.25721	.25700		
332	Conon D	-.06323	-.06344				-.06334					.25784	.25768		
333	Conon C	-.08439	-.08418				-.08428					.24263	.24264		
334	Bode C	-.08111	-.08076				-.08094					.21218	.21178		
335	Eratosthenes A	-.13699	-.13737				-.13718					.31460	.31475		
336	Eratosthenes C	-.20426	-.20505				-.20466					.29043	.29030		
337	Eratosthenes B	-.14276	-.14302				-.14289					.32077	.32055		
338	Draper C	-.34973	-.35045				-.35009					.29380	.29320		
339	Draper	-.35249	-.35302				-.35276					.30206	.30159		
340	Taquet A	.33527	.33522	.33539			.33528	.33541				.24767	.24760	.24811	
341	Maclear B	.33805	.33837	.33810			.33822					.19752	.19730	.19770	
342	A W of Taquet A	.30522	.30540	.30528			.30532					.23823	.23879	.23853	
343	Taquet	.31509	.31521	.31493			.31511	.31477				.28625	.28588	.28654	
344	Menelaus	.26387	.26385	.26374			.26383					.28005	.27972	.28006	
345	Menelaus A	.22146	.22110	.22159			.22131					.29364	.29354	.29399	
346	Menelaus B	.21379	.21343	.21380			.21361					.30877	.30886	.30917	
347	Bessel E	.25096	.25041	.25083			.25065					.33575	.33591	.33609	
348	Sulpicius Gallus	.19070	.19051	.19095			.19067					.33553	.33563	.33589	
349	Manilius A	.15071	.15090	.15123			.15094					.30271	.30297	.30305	
350	Manilius B	.12179	.12128	.12194			.12157					.28586	.28576	.28619	
351	Manilius	.15314	.15281	.15310			.15296			.15375		.24941	.24968	.24933	
352	Manilius F	.07824	.07834	.07880			.07843					.29196	.29156	.29192	
353	A S of Aratus A	.07217	.07240	.07213			.07228					.33677	.33680	.33661	
354	Aratus A	.08085	.08070	.08025			.08062					.37413	.37425	.37434	
355	Aratus	.07258	.07236	.07240			.07242	.07249		.07270		.40043	.40054	.40037	
356	Conon	.03202	.03208	.03217			.03209			.03245		.36935	.36930	.36946	
357	Timocharis A	-.23977	-.24050				-.24014					.41970	.41982		
358	Lambert γ	-.28330	-.28270				-.28300	-.28296				.44538	.44589		
359	Deseilliguy	.32792	.32791	.32806			.32795					.36049	.36030	.36051	
360	Bessel A	.32527	.32530	.32558			.32536	.32537				.41866	.41835	.41797	
361	Linné E	.25292	.25270	.25301			.25283					.44725	.44699	.44723	
362	Bessel F	.22258	.22274	.22294			.22275					.36235	.36233	.36255	
363	Aratus C	.13670	.13614	.13667			.13641					.41221	.41224	.41260	
364	Archimedes C	-.02216	-.02256				-.02236			-.02225		.52391	.52386		
365	Archimedes A	-.09786	-.09841				-.09814	-.09825		-.09830		.47009	.47009		
366	Beer	-.13990	-.14042				-.14016					.45501	.45454		
367	Feuillé	-.14507	-.14569				-.14538					.45965	.45907		
368	Spitzbergen A	-.10324	-.10381				-.10352					.54004	.54010		
369	Spitzbergen C	-.12790	-.12850				-.12820					.54243	.54260		
370	Timocharis F	-.21719	-.21817				-.21768					.51933	.51906		
371	Carlini D	-.23076	-.23123				-.23100					.54397	.54435		
372	Carlini	-.33842	-.33989				-.33916	-.33870				.55539	.55522		
373	Bessel H	.30813	.30777	.30810			.30794					.43352	.43318	.43339	
374	Bessel D	.30161	.30151	.30182			.30161					.45896	.45887	.45896	
375	Bessel N	.31158	.31140	.31178			.31154					.49528	.49527	.49518	
376	Posidonius E	.28967	.29039	.29020			.29016					.50810	.50749	.50815	
377	N of Posidonius E	.29202	.29275	.29278			.29258					.52422	.52375	.52452	
378	Bessel M		.26839	.26852			.26846						.45731	.45761	
379	Linné D	.25804	.25791	.25795			.25795					.48027	.48030	.48043	
380	Linné A	.21729	.21719	.21744			.21728					.48419	.48404	.48425	
381	Linné B	.21105	.21074	.21109			.21090					.50779	.50766	.50782	
382	Linné F	.20356	.20312	.20385			.20341					.53465	.53464	.53466	
383	Linné C	.16096	.16108	.16134			.16112					.50585	.50606	.50634	
384	Aristillus A	.06642	.06604	.06634			.06621					.55380	.55363	.55377	
385	Archimedes B	-.02708	-.02745				-.02726			-.02720		.55995	.57022		
386	Linné G	.18638	.18645	.18671			.18650					.58625	.58569	.58595	
387	Cassini F	.09598	.09593	.09628			.09603					.65945	.65945	.65451	
388	Cassini B	.05197	.05183	.05231			.05198			.05200		.64256	.64220	.64254	
389	Cassini M	.04937	.04917	.04971			.04936			.04915		.66030	.66014	.66083	
390	Piton B	-.00175	-.00199				-.00187					.63375	.63362		
391	Piton A	-.01246	-.01274				-.01260					.64001	.63979		
392	Piton	-.01219	-.01127				-.01173			-.01060		.65179	.65187		
393	Piazzi Smyth	-.04159	-.04184				-.04172			-.04190		.66736	.66717		
394	Kirch	-.07580	-.07573				-.07576			-.07545		.63244	.63245		
395	Pico E	-.12989	-.13069				-.13029					.68139	.68151		
396	Pico D	-.14144	-.14217				-.14180					.68713	.68729		
397	Kirch D	-.16364	-.16415				-.16390					.63866	.63905		
398	Leverrier E	-.21495	-.21585				-.21540					.67395	.67405		
399	Pico B	-.18179	-.18247				-.18213					.72484	.72478		
400	Eudoxus A	.23986	.23981	.23980			.23982	.23952				.71689	.71697	.71682	

No.	$\bar{\eta}$	Schrutka η	Franz η	Saunder η	h 1,2	h 2,3	h 3,4	h 3,5	$\bar{h}$	Diameter in Miles	Rim Height	For Lunar Surface h	Schrutka h	Franz h	Saunder h	Notes
301	.05814					70	-9		30	6.7	18	12				
302	.06920	.06904				66	25		46	9.1	23	23	6			
303	.08588					73	77		75	5.2	14	61				
304	.04694					0	14		7	6.3	17	-10				
305	.01710					162	57		110	3.0	9	101				
306	.05294					151	118		134	3.3	10	124				
307	.06122					-96	89		-4	3.9	11	-15				
308	.11056					131	55		93	4.9	14	79				2
309	.13542	.13540				-159	38		-60	6.3	17	-77	0	-26		2
310	.13279					-44	24		-10	3.4	10	-20				2
311	.07326	.07319				386	266		326	17.1	40	286	195	17/42	-7	15
312	.06992	.06995		.06985		-93	-130		-112	8.7	23	-135	86	-24/-45	18	2
313	.09517			.09510		-356			-356	-	(10)	-346				5
314	.10445					135			135	6.9	19	116				
315	.11736	.11734		.11750		22			22	11.6	29	-7	109	47/-69	125	
316	.15686	.15668		.15695		-83			-83	8.2	21	-104	161	-17/-33	-10	
317	.15224	.15236		.15225		-22			-22	6.5	18	-40	115	-17/-57	24	
318	.14826					50			50	3.4	10	40				2
319	.06407					-47	2		-22	4.3	12	-34				
320	.01696					41	90		66	7.1	19	47				
321	.06022					82	13		48	4.5	13	35				
322	.18272	.18265				158	-90		34	12.1	30	4	-80			
323	.20152					70	-67		2	2.9	9	-7				
324	.12098	.12103				-31	-37		-34	4.8	14	-48	-40	-92		
325	.16485					-38	70		16	3.9	11	5		102		
326	.20957					95	-102		-4	4.6	13	-17				2
327	.13408					-165	-73		-119	3.2	10	-129				2
328	.22922					44	-136		-46	3.3	10	-56		-104		2
329	.19799					-244	-122		-183	3.3	10	-193				2
330	.29569					158			158	4.3	12	146				
331	.25710					-109			-109	4.2	12	-121				
332	.25776					-105			-105	4.0	12	-117				
333	.24264					85			85	4.2	12	73				
334	.21198					79			79	4.8	14	65				2
335	.31468					-124			-124	3.6	10	-134				
336	.29036					-312			-312	3.0	9	-321				
337	.32066					-129			-129	3.6	10	-139				
338	.29350					-328			-328	4.0	12	-340				
339	.30182					-244			-244	4.6	13	-257				
340	.24774	.24749				-42	-162		-102	7.8	21	-123	-144	-263		
341	.19746					87	-54		16	3.3	10	6				
342	.23858					149	92		120	5.0	14	106				
343	.28614	.28615				-12	-119		-66	4.2	12	-78	-46			
344	.27989					-56	-69		-62	16.0	38	-100		-121		
345	.29368					-148	-213		-180	4.1	12	-192				
346	.30892					-119	-150		-134	4.2	12	-146				
347	.33592					-173	-126		-150	4.2	12	-162		-165		
348	.33567					-52	-149		-100	7.6	20	-120				
349	.30292					109	-85		12	6.2	17	-5				
350	.28589					-208	-246		-227	3.2	10	-237				2
351	.24952			.24985		-84	40		-22	24.3	54	-76				5
352	.29175					-16	-189		-102	5.2	14	-116				
353	.33674					92	100		96	4.4	13	83				
354	.37424					-37	63		13	5.7	16	-3				
355	.40047	.40058		.40065		-65	37		-14	6.7	18	-32	126			
356	.36935			.36980		15	-60		-22	13.3	32	-54				
357	.41976					-236			-236	4.2	12	-248				
358	.44564	.44582				258			258	-	(0)	258	98			16
359	.36040					-30	-80		-55	3.8	11	-66				
360	.41833	.41838				-30	44		7	4.7	13	-6	-149			
361	.44712					-112	-114		-113	3.4	10	-123		-292		
362	.36239					56	-95		-20	-	(0)	-20				3
363	.41232					-194	-193		-194	2.7	8	-202				
364	.52388			.52410		-138			-138	5.2	14	-152			-133	
365	.47009	.47011		.47015		-189			-189	8.2	21	-210	17	-61/-16	-105	
366	.45478					-239			-239	6.4	17	-256				
367	.45936					-291			-291	6.2	17	-308				
368	.54007					-178			-178	3.8	11	-189				
369	.54252					-170			-170	4.0	12	-182				
370	.51920					-348			-348	3.6	10	-358				
371	.54416					-102			-102	5.7	16	-118				
372	.55530	.55484				-450			-450	7.1	19	-469	-121			
373	.43332					-163	-107		-135	2.2	7	-142				
374	.45892					-47	-75		-61	3.5	10	-71				
375	.49525					-57	-42		-50	4.0	12	-62				
376	.50781					148	-119		14	2.0	6	8				
377	.52406					-163	-165		-164	2.2	7	-171				
378	.45746						-95		-95	-	(0)	-95				3
379	.48032					-39	-38		-38	2.6	8	-46				
380	.48413					-53	-95		-74	2.7	8	-82				
381	.50773					-122	-96		-109	3.2	10	-119				
382	.53465					-140	-127		-134	3.2	10	-144				
383	.50608					67	-111		-22	-	(0)	-22				3
384	.55371					-145	-82		-114	2.5	8	-122				
385	.57008			.57030		-83			-83	4.9	14	-97			-118	
386	.58590					-47	-102		-74	2.5	8	-82				
387	.65444					-19	-76		-48	4.2	12	-60				
388	.64238			.64275		-88	-149		-118	6.0	16	-134				
389	.66035			.66075		-73	-230		-152	5.6	15	-167			-156	
390	.63368					-84			-84	2.1	7	-91				
391	.63990					-107			-107	2.6	8	-115				
392	.65183			.65340		273			273	-	122	151				9
393	.66726			.66750		-90			-90	9.0	23	-113		-79		
394	.63244			.63260		22			22	7.1	19	3		-77		
395	.68145					-205			-205	5.8	16	-221				
396	.68721					-180			-180	3.6	10	-190				
397	.63886					-105			-105	5.5	15	-120				
398	.67400					-232			-232	3.4	10	-242				
399	.72481					-176			-176	7.0	19	-195				
400	.71691	.71666				-4	29		12	8.7	23	-11	52			1

457

Table 23
Coordinates of Selected Lunar Features and Measured Heights Relative to a Sphere

No.	Name	ξ_1	ξ_2	ξ_3	ξ_4	ξ_5	$\bar{\xi}$	Schrutka ξ	Franz ξ	Saunder ξ	η_1	η_2	η_3	η_4	η_5
401	Eudoxus D		.16646	.16686	.16700		.16680					.68638	.68642	.68606	
402	Egede B		.09874	.09828	.09890		.09855					.77162	.77164	.77167	
403	Trouvelot		.06620	.06569	.06639		.06599					.75794	.75772	.75805	
404	Plato K		-.03862	-.03867			-.03864					.72848	.72842		
405	Plato J		-.05171	-.05227			-.05199					.75444	.75451		
406	Plato D		-.16176	-.16249			-.16212					.76151	.76165		
407	Plato P		-.16243	-.16307			-.16275					.78186	.78222		
408	Plato M		-.15975	-.16083			-.16029					.79885	.79904		
409	Plato Y		-.16765	-.16818			-.16792					.79989	.79983		
410	Plato B		-.17770	-.17816			-.17793					.79941	.79906		
411	Plato C		-.19841	-.19854			-.19848					.80050	.80080		
412	Egede C		.14409	.14426	.14459		.14430					.76752	.76746	.76758	
413	Egede A		.11391	.11324	.11377		.11354	.11338				.78250	.78218	.78251	
414	Plato H		-.01960	-.02012			-.01986					.81999	.81988		
415	Plato Q		-.04844	-.04901			-.04872					.81463	.81441		
416	Galle C		.22116	.22094	.22143		.22112					.84534	.84519	.84538	
417	Sheepshanks C		.16903	.16910	.16942		.16916					.83831	.83821	.83815	
418	Sheepshanks A		.16228	.16275	.16303		.16270					.86576	.86557	.86533	
419	Fontenelle A		-.10533	-.10569			-.10551					.92398	.92403		
420	W.C. Bond C		.05942	.05908	.05967		.05931					.91126	.91103	.91091	
421	Anaxagoras A		-.03706	-.03599			-.03652					.95259	.95214		
422	W.C. Bond B		.05564	.05548	.05581		.05560	.05529				.90641	.90602	.90611	
423	Manzinus F			.16031	.15931		.15981						-.93700	-.93778	
424	Helmholtz B			.32579	.32567	.32556	.32570						-.90066	-.90031	-.90080
425	Helmholtz D			.31451	.31479	.31489	.31468						-.89415	-.89376	-.89384
426	Helmholtz F			.28008	.28034		.28021						-.89007	-.88926	
427	Helmholtz G			.28783	.28798		.28790						-.89269	-.89188	
428	Mutus L			.26758	.26742		.26750						-.87758	-.87748	
429	S of Hommel C			.26823	.26789		.26806						-.84274	-.84229	
430	Hommel M			.23168	.23208		.23188						-.86312	-.86284	
431	Baco N			.22951	.22986		.22968						-.82632	-.82580	
432	Baco L			.20379	.20367		.20373						-.84181	-.84169	
433	Baco K			.19558	.19555		.19556						-.84521	-.84521	
434	N of Manzinus H			.18142	.18156		.18149						-.88777	-.88826	
435	Manzinus M			.17312	.17293		.17302						-.89352	-.89367	
436	S of Biela B			.42022	.42131	.42114	.42072						-.85623	-.85520	-.85512
437	Biela K			.42897	.42939	.42943	.42919						-.81277	-.81214	-.81200
438	Hagecius D			.39651	.39808	.39631	.39685						-.83980	-.83930	-.83918
439	SE of Hagecius D			.39352	.39422	.39396	.39380						-.86597	-.86501	-.86524
440	Vlacq K			.37412	.37406	.37418	.37412						-.77872	-.77784	-.77839
441	Pitiscus L			.34714	.34670		.34692						-.77856	-.77837	
442	Pitiscus R			.32880	.32834		.32857						-.79317	-.79327	
443	Ideler L			.24683	.24682		.24682						-.77884	-.77875	
444	N of Pontecoulant C			.50160	.50252	.50325	.50224						-.81398	-.81333	-.81232
445	SW of Reimarus K			.55152	.55231	.55240	.55194						-.76711	-.76542	-.76521
446	SW of Reimarus J			.53929	.54056	.54039	.53988						-.77524	-.77361	-.77370
447	S of Watt D			.51642	.51625	.51623	.51633						-.78158	-.78006	-.77959
448	SW of Watt A			.46453	.46329	.46342	.46394						-.77648	-.77576	-.77565
449	Janssen K			.46579	.46610	.46596	.46591	.46572	.46643				-.71990	-.71917	-.71888
450	S of Janssen F			.42536	.42542	.42498	.42528						-.76967	-.76961	-.76626
451	Vlacq H			.38374	.38354		.38364						-.74118	-.74099	
452	SW of Nicolai L			.31409	.31392		.31400						-.70387	-.70327	
453	Pitiscus G			.28740	.28723		.28732						-.73836	-.73833	
454	Barocius EE			.26986	.26971		.26978						-.73659	-.73675	
455	Barocius EC			.25546	.25537		.25542						-.74410	-.74420	
456	S of Barocius F			.24011	.24023		.24017						-.72175	-.72231	
457	Barocius D			.22846	.22862		.22854						-.71841	-.71885	
458	Brisbane A			.61644		.61662	.61653						-.75105		-.74845
459	Metius D			.55118	.55112	.55112	.55115						-.67641	-.67596	-.67565
460	Metius C			.54169	.54184	.54209	.54183						-.69607	-.69548	-.69524
461	Metius G			.53526	.53535	.53557	.53536						-.71424	-.71412	-.71376
462	Metius H			.51066	.51067	.51051	.51062						-.71544	-.71569	-.71527
463	Fabricius F			.45784	.45717	.45709	.45748						-.65124	-.65170	-.65160
464	N of Nicolai R			.38930	.38921		.38926						-.64232	-.64280	
465	Nicolai R			.38294	.38258		.38276						-.66043	-.66056	
466	Lockyer A			.37043	.37015		.37029						-.69505	-.69500	
467	Nicolai G			.32422	.32399		.32410						-.62219	-.62273	
468	Nicolai A			.29586	.29567		.29576	.29560	.29599				-.67411	-.67411	
469	Young G			.61652	.61694	.61715	.61678						-.65034	-.64911	-.64874
470	Rheita A			.60454	.60437	.60455	.60450						-.61621	-.61664	-.61566
471	Rheita P			.55330	.55269	.55278	.55302						-.61377	-.61433	-.61379
472	Rheita O			.51216	.51128	.51151	.51178						-.58186	-.58210	-.58167
473	NW of Stiborius G			.47170	.47190		.47180						-.61054	-.61097	
474	Stiborius B			.43937	.43926		.43932						-.60564	-.60574	
475	Wöhler C			.40852	.40824		.40838						-.59777	-.59818	
476	Riccius P			.38277	.38248		.38262						-.58349	-.58386	
477	E of Rabbi Levi H			.28642	.28647		.28644						-.59446	-.59469	
478	Rabbi Levi H			.27814	.27777		.27796						-.59341	-.59393	
479	Busching C			.26674	.26663		.26668						-.60492	-.60527	
480	Reichenbach M			.60863	.60862	.60894	.60870						-.54434	-.54361	-.54309
481	NE of Reichenbach M			.60999	.60950	.60982	.60982						-.52960	-.52930	-.52876
482	Neander H			.56678	.56613	.56632	.56650						-.54470	-.54453	-.54409
483	Neander R			.52171	.52161	.52158	.52165						-.54886	-.54923	-.54914
484	Piccolomini P			.50548	.50570		.50559						-.50710	-.50710	
485	Rothmann E			.41029	.40999		.41014						-.54343	-.54375	
486	Lindenau G			.38421	.38388		.38404						-.54769	-.54790	
487	N of Rothmann			.39086	.39071		.39078						-.49634	-.49639	
488	W of Rothmann C			.36221	.36167		.36194						-.48193	-.48203	
489	Zagut G			.32128	.32074		.32101						-.47462	-.47522	
490	Zagut H			.30807	.30796		.30802						-.50024	-.50064	
491	Neander K			.59142	.59091	.59115	.59122						-.48222	-.48251	-.48218
492	Weinek B			.55197	.55199	.55192	.55196						-.45250	-.45258	-.45214
493	N of Weinek			.53513	.53526	.53520	.53518						-.43770	-.43773	-.43720
494	Piccolomini L			.49931	.49895		.49913	.49895					-.43919	-.43970	
495	N of Piccolomini L			.49687	.49654		.49670						-.42971	-.42992	
496	NW of Piccolomini L			.50191	.50143		.50167						-.41966	-.41993	
497	Piccolomini K			.44685	.44674		.44680	.44684					-.43311	-.43311	
498	Piccolomini D			.41795	.41780		.41788						-.45151	-.45159	
499	Piccolomini N			.39387	.39359		.39373						-.45905	-.45887	
500	Polybius E			.40231	.40212		.40222						-.41263	-.41303	

No.	$\bar{\eta}$	Schrutka η	Franz η	Saunder η	h 1,2	h 2,3	h 3,4	h 3,5	$\bar{h}$	Diameter in Miles	Rim Height	For Lunar Surface h	Schrutka h	Franz h	Saunder h	Notes
401	.68632					110	54		82	5.5	15	67				1
402	.77164					-103	-84		-94	5.1	14	-108				2
403	.75786					-144	-153		-148	5.6	15	-163				
404	.72845					-12			-12	4.4	13	-25				
405	.75448					-130			-130	5.2	14	-144				
406	.76158					-158	-158		-158	5.9	16	-174				
407	.78204					-111			-111	5.0	14	-125				
408	.79894					-214			-214	5.0	14	-228				
409	.79986					-116			-116	5.6	15	-131				
410	.79924					-125			-125	8.0	21	-146				
411	.80065					-4			-4	5.6	15	-19				
412	.76750					33	-62		-14	3.1	9	-23				
413	.78234	.78268				-180	-123		-152	8.2	21	-173	-46			2
414	.81994					-117			-117	6.6	18	-135		-89		
415	.81452					-138			-138	3.8	11	-149				
416	.84528					-48	-72		-60	6.6	18	-78				2
417	.83822					5	-24		-10	6.3	17	-27				2
418	.86556					62	5		34	3.8	11	23				2
419	.92400					-41			-41	12.8	31	-72				1
420	.91106					-60	-34		-47	4.0	12-(20)	-39				1,17
421	.95236					80			80	9.3	24	56			-11	1
422	.90614	.90590				-47	-38		-42	9.8	25-(20)	-47	-103			1,17
423	-.93689						186		186	5.6	15	171				1
424	-.90061					-20	20		0	7.3	20	-20				1
425	-.89398					-48	-45		-46	5.0	14	-60				1
426	-.88966					-93			-93	4.7	13	-106				1
427	-.89228					-83			-83	3.3	10	-93				1
428	-.87753					1			1	6.6	18	-17				1
429	-.84252					-27			-27	3.8	11	-38				1
430	-.86298					-67			-67	4.4	13	-80				1
431	-.82606					-106			-106	6.0	16	-122				1
432	-.84175					-5			-5	4.3	12	-17				1
433	-.84521					3			3	4.0	12	-9				1
434	-.88802					44			44	4.6	13	31				1
435	-.89360					30			30	3.0	9	21				1
436	-.85570					-141	-132		-136	6.2	17	-153				
437	-.81242					-98	-115		-106	11.7	29	-135				
438	-.83952					-157	-47		-102	10.2	26	-128				
439	-.86555					-117	-81		-99	6.1	17	-116				
440	-.77856					-34	-51		-42	7.3	20	-62				
441	-.78846					17			17	5.2	14	3				
442	-.79322					59			59	6.7	18	41				1
443	-.77880					-13			-13	3.9	11	-24				1
444	-.81340					-93	-214		-154	9.3	24	-178				
445	-.76621					-202	-222		-212	6.5	18	-230				
446	-.77445					-226	-203		-214	7.5	20	-234				
447	-.78070					-135	-176		-156	5.4	15	-171				
448	-.77609					17	-7		5	7.4	20	-15				
449	-.71946	-.71989	-.72018			-136	-161		-148	10.0	25	-173	-184	-63		
450	-.76880					-14	-412		-213	5.0	14	-227				
451	-.74108					-8			-8	7.1	19	-27				
452	-.70357					-86			-86	6.0	16	-102				
453	-.73834					14			14	10.0	25	-11				1
454	-.73667					46			46	5.3	15	31				1
455	-.74415					27			27	5.0	14	13				1
456	-.72203					85			85	5.4	15	70				1
457	-.71863					59			59	5.3	15	44				1
458	-.74975						-136		-136	11.4	28	-164				
459	-.67611					-55	-97		-76	7.1	19	-95				
460	-.69572					-91	-144		-118	7.0	19	-137				
461	-.71409					-23	-86		-54	11.6	29	-83				
462	-.71546					32	-9		12	11.8	29	-17				
463	-.65144					157	148		152	8.7	23	129				1
464	-.64256					100			100	3.6	10	90				1
465	-.66050					69			69	3.0	9	60				1
466	-.69502					25			25	6.0	16	9				1
467	-.62246					139			139	8.0	21	118				1
468	-.67426	-.67436	-.67494			82			82	8.8	23	59	-98	4		1
469	-.64963					-189	-252		-220	8.4	22	-242				
470	-.61613					50	-79		-14	9.0	23	-37				
471	-.61392					157	61		109	6.9	19	90				1
472	-.58187					152	48		100	6.7	18	82				1
473	-.61076					52			52	3.7	11	41				1
474	-.60569					33			33	5.8	16	17				1
475	-.59798					117			117	5.8	16	101				1
476	-.58368					115			115	6.4	17	98				1
477	-.59458					40			40	3.3	10	30				1
478	-.59367					166			166	4.2	12	154				1
479	-.60510					90			90	5.0	14	76			-32	1
480	-.54384					-117	-238		-178	6.9	19	-197				
481	-.52932					8	-119		-56	8.7	23	-79				
482	-.54450					51	-50		0	8.5	22	-22				
483	-.54902					94	67		80	7.3	20	60				
484	-.50710					-30			-30	6.6	18	-48				
485	-.54359					109			109	6.7	18	91				1
486	-.54780					91			91	6.6	18	73				1
487	-.49636					34			34	5.0	14	20				1
488	-.48198					108			108	3.3	10	98				1
489	-.47492					226			226	3.4	10	216				1
490	-.50044					109			109	3.8	11	98				1
491	-.48228					118	28		73	6.7	18	55				1
492	-.45243					13	-64		-26	6.6	18	-44				1
493	-.43760					0	-112		-56	5.9	16	-72				1
494	-.43974	-.43982				36			36	7.0	19	17	0			1
495	-.42982					95			95	4.9	14	81				1
496	-.41980					130			130	4.3	12	118				1
497	-.43311	-.43317				18			18	5.5	15	3				1
498	-.45155					10			10	5.5	15	-5				1
499	-.45896					3			3	5.3	15	-12				1
500	-.41283					123			123	5.3	15	108				1

Table 23
Coordinates of Selected Lunar Features and Measured Heights Relative to a Sphere

No.	Name	ξ_1	ξ_2	ξ_3	ξ_4	ξ_5	$\bar{\xi}$	Schrutka ξ	Franz ξ	Saunder ξ	η_1	η_2	η_3	η_4	η_5
501	SW of Polybius B			.38447	.38428		.38438						-.43388	-.43388	
502	Polybius M			.36387	.36337		.36362						-.45652	-.45656	
503	Biot			.71800	.71782	.71799	.71795						-.38502	-.38496	-.38400
504	Biot A			.69789	.69745	.69772	.69774						-.37699	-.37729	-.37690
505	E of Biot A			.69367	.69405	.69379	.69380						-.37100	-.37083	-.37098
506	Santbech B			.60345	.60310	.60299	.60325						-.41762	-.41786	-.41738
507	Polybius L			.43807	.43832		.43820						-.37369	-.37398	
508	Polybius J			.36777	.36777		.36777						-.38657	-.38712	
509	Polybius H			.36082	.36037		.36060						-.36122	-.36169	
510	Fermat E			.32070	.32031		.32050						-.34061	-.34109	
511	Cook G			.70997	.71055	.71077	.71032						-.32406	-.32363	-.32312
512	Bohnenberger D			.64274	.64260	.64261	.64267						-.31469	-.31445	-.31401
513	N of Bohnenberger C			.62365	.62390	.62413	.62383						-.31201	-.31190	-.31131
514	Bohnenberger G			.61485	.61525	.61516	.61503						-.29512	-.29461	-.29445
515	Rosse			.54519	.54509	.54514	.54515	.54495					-.30652	-.30673	-.30640
516	Rosse C			.53480	.53547		.53514						-.31685	-.31672	
517	N of Fracastorius			.50450	.50525		.50488						-.31668	-.31605	
518	N of Fracastorius			.49996	.50021		.50008						-.32519	-.32459	
519	Beaumont D			.42185	.42155		.42170	.42194					-.29303	-.29348	
520	Theophilus G			.43210	.43198		.43204						-.26932	-.26940	
521	Theophilus E			.41194	.41171		.41182						-.27352	-.27345	
522	Tacitus A			.33491	.33418		.33454						-.29970	-.29998	
523	Tacitus E			.33508	.33451		.33480						-.24052	-.24082	
524	Abulfeda BA			.28107	.28035		.28071						-.25266	-.25371	
525	W of Cook F			.79949	.80088	.80034	.80005						-.29737	-.29694	-.29631
526	Cook B			.74868	.74873	.74872	.74870						-.29723	-.29681	-.29609
527	Bohnenberger F			.61765	.61727	.61735	.61748						-.25309	-.25292	-.25270
528	In NE Mare Nectaris			.54746	.54816		.54781						-.24258	-.24191	
529	E of Daguerre			.53393	.53403		.53398						-.20350	-.20290	
530	Daguerre B			.54471	.54464		.54468						-.18699	-.18647	
531	Langrenus L			.86065	.86039	.86068	.86059						-.21877	-.21818	-.21795
532	E of Langrenus			.77090	.77123	.77113	.77104						-.15292	-.15261	-.15225
533	W of Magelhaens			.69897	.69915	.69899	.69902						-.20404	-.20343	-.20382
534	Gutenberg A			.63371	.63356	.63379	.63369	.63398					-.15624	-.15609	-.15619
535	Gutenberg J			.61916	.61900	.61911	.61911						-.19288	-.19278	-.19275
536	Capella D			.60469	.60470	.60449	.60464						-.11726	-.11725	-.11744
537	S of Capella F			.56297	.56332		.56314						-.17427	-.17367	
538	Isidorus K			.54335	.54294		.54314						-.15448	-.15422	
539	Zöllner K			.35367	.35341		.35354						-.11323	-.11305	
540	Zöllner J			.35174	.35147		.35160						-.10765	-.10760	
541	Langrenus C			.86229	.86202	.86246	.86226						-.09757	-.09718	-.09719
542	Messier G			.79279	.79324	.79331	.79303	.79320					-.09388	-.09322	-.09321
543	Messier A			.76432	.76453	.76457	.76444						-.12046	-.11987	-.11999
544	Messier D			.72116	.72143	.72138	.72128						-.06205	-.06180	-.06180
545	E of Messier D			.71067	.71079	.71071	.71071						-.05774	-.05759	-.05750
546	NE of Messier D			.69632		.69630	.69631						-.08260		-.08234
547	Lubbock D			.62904	.62939	.62921	.62917						-.07908	-.07880	-.07876
548	Lubbock G			.63112	.63116	.63160	.63138						-.06387	-.06417	-.06386
549	S of Lubbock F			.60671	.60704	.60712	.60690						-.07050	-.07040	-.07048
550	Lubbock F			.60656	.60671	.60671	.60664						-.05513	-.05486	-.05486
551	Censorinus D			.58564	.58550		.58556						-.03332	-.03320	
552	N of Isidorus D			.56408	.56414	.58544	.56411						-.06450	-.06420	-.03315
553	Censorinus B			.52135	.52119		.52127						-.03450	-.03439	
554	Torricelli			.48579	.48675		.48627						-.04532	-.04518	
555	Torricelli F			.48895	.48895		.48895						-.07311	-.07272	
556	Torricelli A			.49502	.49479		.49490						-.07859	-.07836	
557	S of Torricelli A			.49165	.49176		.49170						-.09012	-.08963	
558	Torricelli η			.43112	.43115		.43114						-.08597	-.08565	
559	S of Torricelli C			.42490	.42479		.42484						-.07001	-.06979	
560	SW of Torricelli C			.42659	.42665		.42662						-.05760	-.05762	
561	Torricelli C			.43780	.43777		.43778						-.04692	-.04668	
562	Alfraganus G			.36165	.36172		.36168						-.04641	-.04622	
563	Alfraganus F			.35564	.35530		.35547						-.06111	-.06116	
564	Alfraganus D			.34372	.34357		.34364						-.07011	-.07004	
565	W of Webb			.87291	.87250	.87251	.87271						-.02327	-.02301	-.02251
566	Webb M			.85036	.85030	.85049	.85038						-.01447	-.01354	-.01380
567	E of Webb			.82529	.82546	.82527	.82533						-.02170	-.02093	-.02150
568	Messier P			.78248	.78297	.78286	.78270						.00159	.00193	.00169
569	Messier K			.78216	.78269	.78240	.78235						.01186	.01202	.01202
570	Messier H			.76348	.76419	.76393	.76377						.00612	.00616	.00631
571	Messier B			.74282	.74322	.74302	.74297						-.01514	-.01501	-.01494
572	Messier			.73788	.73825	.73810	.73803	.73803					-.03244	-.03260	-.03243
573	W.H. Pickering			.72915	.73003	.72932	.72941	.72977					-.03479	-.03469	-.03459
574	NE of Messier			.71241	.71266	.71273	.71255						-.00238	-.00200	-.00236
575	ENE of Messier			.68947	.69008	.68999	.68975						-.01315	-.01266	-.01273
576	N of #575			.68435	.68458	.68443	.68445						.00162	.00167	.00177
577	NW of Censorinus N			.60919	.60893	.60874	.60901						-.01694	-.01754	-.01746
578	Maskelyne B			.48380	.48378		.48379						.03490	.03505	
579	Sabine G			.44858	.44883		.44870						.04072	.04064	
580	Moltke			.40942	.40945		.40944	.40941					-.00984	-.00986	
581	Delambre E			.34874	.34868		.34871						-.00546	-.00568	
582	Apollonius C			.83686	.83686	.83682	.83685						.05778	.05838	.05833
583	Taruntius O			.81040	.81085	.81076	.81060						.03914	.03960	.03974
584	Taruntius N			.80365	.80367	.80369	.80366						.04207	.04247	.04227
585	Taruntius E			.75896	.75937	.75936	.75916	.75909					.03261	.03322	.03284
586	S of Tarentius			.72520	.72552	.72526	.72530						.05813	.05810	.05799
587	Secchi A			.66101	.66116	.66080	.66100						.05740	.05749	.05733
588	Secchi B			.66137	.66141	.66104	.66130						.06453	.06445	.06403
589	Tarentius F			.64814	.64820	.64786	.64808						.06858	.06894	.06866
590	Tarentius E			.64238	.64275	.64246	.64249						.09643	.09646	.09624
591	Maskelyne H			.53211	.53210		.53210						.08604	.08578	
592	Maskelyne K			.49441	.49439		.49440						.05669	.05700	
593	NE of Maskelyne K			.46911	.46931		.46921						.09606	.09667	
594	E of #593			.46184	.46196		.46190						.09568	.09607	
595	Firmicus D			.89647	.89650	.89634	.89644						.10297	.10357	.10331
596	W of Apollonius A			.82705	.82674	.82686	.82692						.09245	.09262	.09270
597	Apollonius K			.81237	.81228	.81237	.81235						.09767	.09737	.09722
598	Tarentius A			.75817	.75827	.75810	.75818	.75817					.12655	.12681	.12671
599	Cauchy D			.63683	.63697	.63655	.63680						.17447	.17445	.17444
600	Cauchy			.61594	.61595	.61553	.61584						.16636	.16647	.16643

No.	$\bar{\eta}$	Schrutka η	Franz η	Saunder η	h 1,2	h 2,3	h 3,4	h 3,5	$\bar{h}$	Diameter in Miles	Rim Height	For Lunar Surface h	Schrutka h	Franz h	Saunder h	Notes
501	-.43388						30		30	4.4	13	17				1
502	-.45654						91		91	6.7	18	73				1
503	-.38475						11	-165	-77	8.2	21	-98				2
504	-.37704						105	4	54	9.1	23	31				
505	-.37095						-77	-18	-48	-	(29)	-77				2,4
506	-.41762						94	16	55	10.1	26	29				1
507	-.37384						26		26	4.3	12	14				1
508	-.38684						130		130	5.6	15	115				1
509	-.36146						193		193	5.3	15	178				1
510	-.34085						188		188	4.4	13	175				1
511	-.32372						-149	-295	-222	6.4	17	-239				2
512	-.31446						-27	-116	-72	7.9	21	-93				
513	-.31181						-58	-211	-134	5.4	15	-149				
514	-.29482						-164	-183	-174	7.4	20	-194				
515	-.30654	-.30644					63	-19	22	7.7	20	2	121			
516	-.31678						-134		-134	1.8	6	-140				
517	-.31636						-265		-265	1.7	6	-271				
518	-.32489						-178		-178	2.2	7	-185				
519	-.29326	-.29298					161		161	6.8	18	143	167			1
520	-.26936						41		41	4.9	14	27				1
521	-.27348						22		22	6.3	17	5				1
522	-.29984						202		202	6.0	16	186				1
523	-.24067						182		182	5.4	15	167				1
524	-.25318						406		406	7.2	19	387				1
525	-.29700						-209	-245	-227	3.3	10	-237				2
526	-.29684						-76	-196	-136	5.6	15	-151				2
527	-.25295						20	-36	-8	6.0	16	-24				
528	-.24224						-267		-267	2.3	7	-274				
529	-.20320						-157		-157	-	(0)	-157				3
530	-.18673						-109		-109	-	(0)	-109				3
531	-.21842						-51	-108	-80	7.6	20	-100				2
532	-.15268						-95	-144	-120	3.5	10	-130				2
533	-.20383						-143	-45	-94	-	(23)	-117				2,4
534	-.15619	-.15633					-8	-24	-16	9.7	25	-41	40			
535	-.19282						3	-21	-9	6.5	18	-27				
536	-.11730						-2	72	35	5.0	14	21				
537	-.17397						-194		-194	-	(0)	-194				3
538	-.15435						13		13	4.2	12	1				
539	-.11314						0		0	4.8	14	-14				
540	-.10762						38		38	6.4	17	21				
541	-.09738						-27	-70	-48	8.4	22	-70				2
542	-.09355	-.09376					-164	-176	-170	8.3	22	-192	-75			2
543	-.12020						-133	-115	-124	7.0	19	-143				2
544	-.06192						-88	-80	-84	4.9	14	-98				2
545	-.05764						-48	-53	-50	2.6	8	-58				2
546	-.08247							-50	-50	1.7	6	-56				
547	-.07893						-116	-97	-106	7.0	19	-125				
548	-.06394						-17	-77	-47	6.4	17	-64				
549	-.07047						-73	-70	-72	1.8	6	-78				
550	-.05500						-84	-84	-84	7.9	21	-105				
551	-.03325						-5	-6	-6	6.6	18	-24				
552	-.06435						-79		-79	1.3	4	-83				
553	-.03444						0		0	4.2	12	-12				
554	-.04525						-26		-26	4.4	13	-39				2
555	-.07292						-96		-96	4.8	14	-110				2
556	-.07848						-16		-16	6.2	17	-33				2
557	-.08988						-141		-141	1.3	4	-145				2
558	-.08581						-87		-87	-	(12)	-99				2,4
559	-.06990						-36		-36	3.2	10	-46				2
560	-.05761						-6		-6	4.9	14	-20				2
561	-.04680						-55		-55	6.9	19	-74				2
562	-.04632						-61		-61	3.8	11	-72				
563	-.06114						78		78	5.4	15	63				
564	-.07006						15		15	5.4	15	0				
565	-.02302						6	-63	-28	2.2	7	-35				2
566	-.01407						-130	-111	-120	2.8	8	-128				2
567	-.02146						-141	-30	-86	1.7	6	-92				2
568	-.00170						-122	-65	-94	4.6	13	-107				2
569	.01194						-95	-58	-76	3.4	10	-86				2
570	.00618						-101	-94	-98	6.0	16	-114				2
571	-.01506						-80	-65	-72	4.3	12	-84				2
572	-.03248	-.03262					-20	-33	-26	6.0	16	-42	-98			2
573	-.03472	-.03471					-142	-62	-102	6.5	18	-120	-103			2
574	-.00228						-111	-50	-80	4.4	13	-93				2
575	-.01292						-190	-163	-176	3.0	9	-185				2
576	.00167						-45	-44	-44	-	(0)	-44				2,3
577	-.01722						176	188	182	6.0	16	166				1
578	.03498						-34		-34	5.7	16	-50				2
579	.04068						-23		-23	3.7	11	-34				2
580	-.00985	-.00982					-2		-2	4.6	13	-15	-184			2
581	.00557						68		68	3.6	10	58				
582	.05807						-90	-78	-84	5.6	15	-99				2
583	.03940						-129	-141	-135	3.8	11	-146				2
584	.04222						-69	-37	-53	3.7	11	-64				2
585	.03282	.03263					-166	-94	-130	6.6	18	-148	-167			2
586	.05809						-38	19	-10	3.8	11	-21				2
587	.05733						-42	110	34	2.6	8	26				2
588	.06438						11	155	83	2.9	9	74				2
589	.06869						-85	25	-30	7.0	19	-49				2
590	.09639						-64	28	-18	6.9	19	-37				2
591	.08591						64		64	4.6	13	51				2
592	.05684						-72		-72	3.8	11	-83				2
593	.09636						-189		-189	-	(0)	-189				2,3
594	.09588						-118		-118	-	(0)	-118				2,3
595	.10320						-72	-28	-50	6.4	17	-67				1
596	.09256						8	-17	-4	6.1	17	-21				
597	.09748						59	72	66	6.5	18	48				
598	.12666	.12648					-61	-20	-40	7.9	21	-61	-138	-90		2
599	.17446						-18	49	16	6.1	17	-1				2
600	.16640						-24	49	12	7.5	20	-8				2

Table 23
Coordinates of Selected Lunar Features and Measured Heights Relative to a Sphere

No.	Name	ξ_1	ξ_2	ξ_3	ξ_4	ξ_5	$\bar{\xi}$	Schrutka ξ	Franz ξ	Saunder ξ	η_1	η_2	η_3	η_4	η_5
601	E of Cauchy			.54431	.54445		.54438						.17404	.17436	
602	N of #601			.54557	.54588		.54572						.17924	.17938	
603	Sina A			.53352	.53392		.53372						.13571	.13590	
604	Sina			.51817	.51788		.51802						.15424	.15438	
605	Sina E			.50800	.50788		.50794						.16787	.16787	
606	Jansen G			.43269	.43271		.43270						.16173	.16197	
607	Ross G			.41244	.41309		.41276						.18563	.18573	
608	Ross F			.40262	.40310		.40286						.18934	.18974	
609	Ross E			.38983	.38978		.38980						.19166	.19197	
610	Arago E			.38138	.38163		.38150						.14817	.14847	
611	Sosigenes A			.31416	.31406		.31411	.31384					.13547	.13536	
612	Mare Crisium Y			.86449	.86457	.86459	.86454						.22357	.22389	.22375
613	Mare Crisium N			.85580	.85578	.85623	.85590						.22722	.22756	.22716
614	E of Mare Crisium N			.84395	.84390	.84420	.84400						.22801	.22862	.22830
615	S of Mare Crisium N			.85789	.85868	.85862	.85827						.20991	.21018	.21024
616	In S Mare Crisium			.84078	.84118	.84109	.84096						.18717	.18741	.18752
617	In S Mare Crisium			.82965	.83023	.83018	.82993						.18747	.18837	.18793
618	In S Mare Crisium			.84000	.84032	.84031	.84016						.20222	.20276	.20207
619	In S Mare Crisium			.81438	.81435		.81436						.21876	.21972	
620	In S Mare Crisium			.80661	.80694	.80701	.80679						.20363	.20438	.20389
621	Lick D			.77362	.77407	.77419	.77388						.22760	.22842	.22835
622	W of Cauchy			.67436	.67437	.67398	.67427						.18764	.18753	.18757
623	Proclus C			.67170	.67222	.67182	.67186						.22464	.22431	.22396
624	Proclus A			.65418	.65431	.65377	.65411	.65394					.23056	.23095	.23059
625	Proclus E			.74289	.74258	.74276	.74278						.27374	.27448	.27428
626	Proclus			.70219	.70180	.70175	.70198	.70183	.70223				.27822	.27783	.27733
627	Franz D			.62570	.62610		.62580						.30050	.30000	.29938
628	Maraldi B			.58064	.57992	.57987	.58027						.24831	.24820	.24793
629	E of Maraldi B			.55114	.55109		.55112						.24047	.24049	
630	Jansen F			.50407	.50391		.50399						.21821	.21836	
631	Jansen L			.48488	.48497		.48492						.25391	.25403	
632	Jansen			.46595	.46613		.46604						.23456	.23488	
633	Jansen E			.45179	.45206		.45192						.25008	.25024	
634	Jansen D			.45826	.45860		.45843						.27096	.27119	
635	Jansen C			.46780	.46809		.46794						.27975	.27992	
636	Ross D			.38656	.38645		.38650						.21761	.21813	
637	Taquet C			.34970	.34980		.34975						.23306	.23318	
638	In Mare Crisium			.85780	.85794	.85804	.85790						.29440	.29453	.29461
639	Peirce B			.75614	.75681	.75672	.75645						.33195	.33156	.33127
640	Macrobius B			.61002	.61032	.60999	.61009	.61021	.61035				.33561	.33475	.33427
641	Maraldi A			.55603	.55605	.55585	.55599						.34311	.34253	.34204
642	Littrow B			.46128	.46160		.46144						.36962	.36944	
643	Littrow C			.41099	.41122		.41110						.38055	.38063	
644	Dawes			.42293	.42396		.42344	.42363					.29732	.29616	
645	On Eimmart C			.81071	.81063	.81048	.81063						.37533	.37531	.37546
646	SW of Cleomedes F			.77930	.77959	.77949	.77946						.38119	.38156	.38153
647	Cleomedes F			.77303	.77343	.77328	.77319						.38362	.38363	.38371
648	Roemer K			.53712	.53673	.53684	.53695						.38454	.38422	.38379
649	Roemer L			.52334	.52279	.52292	.52310						.39506	.39516	.39477
650	W of Roemer L			.53120	.53107	.53116	.53116						.39993	.39989	.39933
651	E of Roemer			.51373	.51343		.51358						.42778	.42746	
652	On Le Monnier			.44537	.44584		.44560						.46564	.46546	
653	Le Monnier B			.38520	.38582		.38551						.43206	.43224	
654	Cleomedes T			.81533	.81610	.81607	.81571						.43081	.43105	.43098
655	Tralles B			.68756	.68744	.68738	.68748						.45839	.45808	.45780
656	Tralles C			.67108	.67154	.67119	.67127						.46742	.46637	.46643
657	Burkhardt G			.64429	.64387	.64377	.64406						.51224	.51213	.51173
658	Bond A			.51189	.51132	.51131	.51160						.52419	.52398	.52384
659	Posidonius M			.41295	.41274		.41284						.56405	.56400	
660	Posidonius F			.38304	.38332		.38318						.54147	.54161	
661	Luther			.34257	.34243		.34250						.54724	.54710	
662	Posidonius H			.31222	.31319		.31270						.58829	.58821	
663	Maury B			.54816	.54802	.54801	.54810						.57495	.57504	.57482
664	Bond J			.48847	.48897	.48904	.48874						.58021	.58021	.58007
665	Maury D			.48048	.48107	.48106	.48077						.61957	.61900	.61897
666	Bond K			.45732	.45783		.45758						.58159	.58114	
667	E of Groves			.38166	.38159		.38162						.63556	.63550	
668	Luther D			.34856	.34800		.34828						.60176	.60218	
669	Struve			.66342	.66386	.66383	.66363						.68867	.68698	.68705
670	Shuckburgh C			.57601	.57683	.57669	.57638						.68794	.68827	.68845
671	Shuckburgh E			.60266	.60252	.60266	.60262						.69570	.69496	.69481
672	Cepheus A			.54663	.54717	.54712	.54689	.54718					.65716	.65620	.65602
673	Hercules H			.45313	.45365	.45353	.45336						.70469	.70413	.70386
674	Hercules J			.42567	.42634	.42641	.42602						.69594	.69571	.69541
675	Hercules C			.42414	.42506	.42515	.42462						.67850	.67823	.67819
676	Plana D			.32959	.32915		.32937						.66605	.66584	
677	Plana F			.31185	.31231		.31208						.63994	.64031	
678	Plana E			.30339	.30413		.30376						.65005	.64988	
679	Bürg B			.29239	.29265		.29252						.67865	.67818	
680	Atlas B			.55467	.55533	.55543	.55502						.70920	.70851	.70851
681	Hercules B			.40118	.40085	.40098	.40105						.74106	.74085	.74081
682	Hercules A			.37292	.37306		.37299						.72919	.72918	
683	Baily B			.36201	.36263		.36232						.77662	.77637	
684	Baily A			.34400	.34376		.34388						.75072	.75073	
685	Aristoteles N			.27202	.27237		.27220						.79705	.79694	
686	Aristoteles M			.27237	.27265		.27251						.80308	.80297	
687	Gärtner F			.27041	.27087		.27064						.84316	.84268	
688	Gärtner D			.29233	.29196		.29214						.85186	.85187	
689	Democritus B			.23903	.23938		.23920						.86624	.86600	
690	E of Democritus B			.23433	.23410		.23422						.86791	.86694	
691	Democritus A			.25556	.25557		.25556						.87947	.87895	
692	Democritus D			.23598	.23644		.23621						.88991	.88991	
693	Gärtner G			.34039	.33831	.33830	.33935						.88073	.88041	.88040
694	Schwabe G			.27797	.27807	.27807	.27802						.90942	.90936	.90939
695	Moigno C			.19829	.19836		.19832						.91336	.91290	
696	Arnold G			.20113	.20105		.20109						.92310	.92283	

Subscripts refer to Lick plates on which the measures were made.

ξ and η are the accepted co-ordinates, averaged from the individual plate measures.

h is a height measure referred to the mean lunar sphere; the unit is 10^{-5} lunar radii.

$\bar{h}$ is the average height determination from the preceding pairs of plates.

Diameter in miles refers to the tiny crater measured; the height measure is that of the crest of the **crater rim**.

Rim height is that which a Class 1 crater would have according to eq. (7-6) in units of 10^{-5} lunar **radii**.

The column headed "For Lunar Surface, h," gives the value of h minus rim height. This column gives the accepted deviation of the lunar surface from the mean sphere.

Values of ξ, η, and h are given from Schrutka, Franz, and Saunder for comparison purposes.

Two reductions of certain height measures were given by Franz and Saunder from Franz's measures.

No.	$\bar{\eta}$	Schrutka η	Franz η	Saunder η	h 1,2	h 2,3	h 3,4	h 3,5	$\bar{h}$	Diameter in Miles	Rim Height	For Lunar Surface h	Schrutka h	Franz h	Saunder h	Notes
601	.17420						-97		-97	2.0	6	-103				2
602	.17931						-78		-78	2.2	7	-85				2
603	.13580						-114		-114	2.8	8	-122				2
604	.15431						22		22	7.4	20	2				2
605	.16787						21		21	6.0	16	5				2
606	.16185						-64		-64	3.4	10	-74				2
607	.18568						-146		-146	2.5	8	-154				2
608	.18954						-189		-189	2.5	8	-197				2
609	.19182						-72		-72	2.3	7	-79				2
610	.14832						-124		-124	3.3	10	-134				2
611	.13542	.13529					48		48	7.6	20	28	46			2
612	.22370						-50	-31	-40	-	(0)	-40				3
613	.22729						-40	-33	-36	4.2	12	-48				
614	.22824						-76	-62	-69	2.5	8	-77				
615	.21006						-108	-110	-109	-	(0)	-109				3
616	.18732						-75	-80	-78	1.7	6	-84				
617	.18782						-193	-128	-160	1.7	6	-166				
618	.20232						-107	-11	-59	-	(0)	-59				3
619	.21924						-139		-139	-	(0)	-139				3
620	.20388						-152	-85	-118	2.3	7	-125				
621	.22799						-188	-192	-190	9.0	23	-213				
622	.18760						21	69	45	-	(24)	21				2,4
623	.22439						-8	116	54	5.7	16	38				1
624	.23066	.23074					-98	54	-22	8.5	22	-44	-115			1
625	.27406						-87	-76	-82	5.8	16	-98				
626	.27790	.27740	.27698				122	221	172	17.1	40	132	23			1
627	.30010						44	227	136	7.6	20	116				
628	.24819						138	203	170	4.6	13	157				2
629	.24048						3		3	3.3	10	-7				2
630	.21828						-8		-8	6.0	16	-24				2
631	.25397						-43		-43	5.0	14	-57				2
632	.23472						-110		-110	14.3	34	-144		16		2
633	.25016						-88		-88	4.4	13	-101				2
634	.27108						-116		-116	4.5	13	-129				2
635	.27984						-89		-89	5.8	16	-105				2
636	.21787						-112		-112	6.0	16	-128				2
637	.23312						-50		-50	3.6	10	-60				2
638	.29448						-26	-44	-35	2.8	8	-43				
639	.33168						-16	41	12	7.2	19	-7				
640	.33506	.33459	.33455				130	274	202	12.6	31	171	6	-182		1
641	.34270						122	255	188	5.2	14	174				
642	.36953						-11		-11	4.7	13	-24				
643	.38059						-57		-57	3.0	9	-66				
644	.29674	.29588					105		105	12.0	30	75	-75			
645	.37536						10	5	8	-	(50)	-42				18
646	.38137						-79	-65	-72	4.5	13	-85				
647	.38364						-42	-37	-40	7.5	20	-60				
648	.38427						150	202	176	7.4	20	156				1
649	.39501						63	127	95	6.5	18	77				1
650	.39977						28	132	80	4.6	13	67				1
651	.42762						112		112	5.6	15	97				1
652	.46555						-35		-35	2.8	8	-43				
653	.43215						-146		-146	3.2	10	-156				
654	.43091						-81	-72	-76	9.4	24	-100				
655	.45816						61	112	86	6.8	18	68				
656	.46691						113	144	128	4.0	12	116				1
657	.51208						65	138	102	8.6	22	80		-124		1
658	.52405						118	148	133	5.5	15	118				1
659	.56402						41		41	6.0	16	25				2
660	.54154						-72		-72	3.9	11	-83				2
661	.54717						52		52	5.9	16	36				
662	.58825						-131		-131	3.7	11	-142				
663	.57494						2	40	21	5.7	16	5				1
664	.58018						-67	-50	-58	5.0	14	-72				2
665	.61928						24	31	28	4.8	14	14				2
666	.58136						16		16	4.5	13	3				2
667	.63553						22		22	1.3	4	18				2
668	.60197						-2		-2	4.4	13	-15				
669	.68784						89	86	88	6.8	18	70				1
670	.68815						-109	-118	-114	7.4	20	-134				1
671	.69529						84	89	86	6.0	16	70				1
672	.65664	.65619					80	110	95	7.8	21	74	-17			1
673	.70434						27	80	54	4.8	14	40				1
674	.69575						-40	-2	-21	5.4	15	-36				2
675	.67836						-66	-70	-68	6.1	17	-85				2
676	.66594						100		100	4.7	13	87				1
677	.64012						-140		-140	3.3	10	-150				2
678	.64996						-72		-72	3.7	11	-83				1
679	.67842						54		54	3.3	10	44				1
680	.70886						23	15	19	8.0	21	-2				1
681	.74094						66	58	62	5.3	15	47				1
682	.72918						-15		-15	7.6	20	-35				2
683	.77650						-28		-28	4.9	14	-42				2
684	.75072						26		26	10.8	27	-1		-73		2
685	.79700						-21		-21	3.0	9	-30				2
686	.80302						-14		-14	5.3	15	-29				2
687	.84292						18		18	9.4	22	-4				
688	.85186						30		30	5.4	15	15				1
689	.86612						-2		-2	7.4	20	-22				
690	.86742						134		134	4.0	12	122				1
691	.87921						54		54	6.3	17	37				1
692	.88991						-35		-35	4.6	13	-48				1
693	.88057						151	153	152	6.4	17	135				1
694	.90940						-1	-3	-2	9.0	23	-25				1
695	.91313						38		38	5.2	14	24				1
696	.92296						27		27	6.0	16	11				1

1. Height measure used in determining shape of bulge of moon from upland areas
2. Height measure used in determining shape of bulge of moon from lava-covered areas
3. White spot
4. Peak
5. Central peak—omit
6. In Huggins—omit
7. In Walter—omit
8. In crater—omit
9. Peak—omit
10. In Purbach—omit
11. On rim—omit
12. In Alphonsus—omit
13. In Ptolemaeus—omit
14. In Hipparchus—omit
15. Omit
16. Diffuse White spot—omit
17. In W. C. Bond
18. Height correction estimated

Table 43

Original Measures of Positions of Selected Lunar Formations on Lick Observatory Photographs (See Table 23)

No.	Measured x (mm)					Measured y (mm)				
	1	2	3	4	5	1	2	3	4	5
1	90.0595	77.6710				19.4250	16.9595			
2	98.3305	84.4505				20.5190	14.8700			
3	101.2040	86.6800				21.0625	14.2985			
4	105.2395	87.7080				20.6930	12.2270			
5	104.4550	88.5485				21.2775	13.1650			
6	99.9220	89.1770				24.6365	18.2605			
7	103.8950	93.2395				26.4005	18.4290			
8	106.4175	95.0355				26.5495	17.5845			
9	105.1210	94.9670				27.4950	19.0300			
10	99.3645	93.1635				30.6550	24.1815			
11	101.9110	95.4795				31.3195	23.8700			
12	100.2930	96.6470				35.0610	28.0140			
13	102.8770	97.8440				33.9130	25.9740			
14	106.3350	100.8640				34.7095	25.3985			
15	108.6760	102.1070				34.0870	23.9180			
16	112.9980	105.5270				34.8530	23.0015			
17	96.9255	94.3910				35.5975	29.8225			
18	100.0370	97.2875				36.4815	29.4865			
19	101.8365	98.6925				36.5980	28.9165			
20	101.3530	99.1175				37.8790	30.2895			
21	107.4585	104.0430				38.5685	28.6345			
22	109.3210	107.4100				42.0195	31.0430			
23	106.4325	106.5070				44.1295	34.1680			
24	111.1615	109.9090				43.9570	32.2500			
25	125.9965	117.4440				41.3095	23.9040			
26	125.6955	119.0120				43.7610	26.3645			
27	125.8445	120.1290				45.3320	27.7740			
28	126.6690	121.3360				46.4095	28.4530			
29	105.3635	107.1635				46.9430	37.2130			
30	105.2065	109.3260				51.3510	41.4200			
31	105.4375	109.7935				51.9770	41.9285			
32	105.0990	109.5465				52.0675	42.1110			
33	102.1990	107.4115				52.1400	43.2020			
34	110.6235	114.0000				52.5165	40.4110			
35	113.7035	116.4895				52.8770	39.7970			
36	117.3915	118.6145				51.9430	37.2805			
37	111.4490	115.7945				55.1485	42.5130			
38	103.9605	111.6465				58.2010	48.2120			
39	124.2395	125.7700				56.8560	39.1700			
40	127.4985	128.6585				58.2220	39.1990			
41	112.7080	118.9300				59.6580	46.2230			
42	115.7065	121.3900				60.1660	45.5300			
43	118.3715	124.2010				62.0420	46.2525			
44	118.0550	124.2275				62.6200	46.9775			
45	117.9710	124.4305				63.1130	47.3895			
46	116.1765	123.1260				63.1760	48.1540			
47	115.2625	123.3260				64.9935	50.2005			
48	114.4720	122.8960				65.3190	50.7870			
49	116.1980	124.6670				66.5240	51.2640			
50	112.1660	120.8355				64.5560	50.9475			
51	123.9365	129.7475				65.4745	47.3430			
52	119.8580	127.8805				67.7355	50.9790			
53	125.6420	133.0360				70.1490	50.9710			
54	123.5630	132.9415				73.2540	54.6240			
55	121.6185	131.6935				73.3255	55.4485			
56	101.5000	116.5800				74.0130	63.6410			
57	106.7180	121.4745				75.8235	63.3455			
58	113.6405	126.8650				75.9710	60.8840			
59	124.1725	134.4825				75.7295	56.6105			
60	139.2145	143.9250				75.6210	50.6290			
61	107.0885	122.6670				78.0170	65.2755			
62	121.5570	134.9350				81.5155	62.9790			
63	105.8915	124.1620				84.1960	71.3545			
64	112.7705	130.2080				86.1925	70.5740			
65	116.6410	132.8710				85.4230	68.3930			
66	117.9260	134.0470				86.1845	68.6425			
67	122.2870	137.0630				85.7065	66.5015			
68	115.9190	133.5110				88.6580	71.6515			
69	120.0475	137.0765				89.9870	71.2740			
70	123.1235	138.8665				88.6755	68.9030			
71	125.8750	141.0740				89.6685	68.7970			
72	129.1330	142.6495				87.8550	65.8665			
73	141.3035	151.2390				90.9625	64.1090			
74	139.8270	152.3800				96.2315	69.3300			
75	136.2640	150.9470				98.2175	72.5325			
76	124.2000	141.0315				92.8605	72.3055			

Table 43

Original Measures of Positions of Selected Lunar Formations on Lick Observatory Photographs (See Table 23)

No.	Measured x (mm)					Measured y (mm)				
	1	2	3	4	5	1	2	3	4	5
77	119.6035	137.3560				91.7930	73.0890			
78	119.2880	137.6650				93.4255	74.7385			
79	106.9605	127.4170				90.7720	76.9585			
80	103.5720	124.8885				91.1240	78.5215			
81	101.2840	124.0825				94.0555	82.0185			
82	105.1400	127.1205				94.0820	80.6070			
83	107.2135	129.3820				95.5320	81.1505			
84	113.8705	134.9635				97.1005	80.1110			
85	118.4015	138.4345				97.2545	78.5440			
86	104.1535	129.4055				102.7190	88.8020			
87	101.9005	128.0740				104.2415	91.0510			
88	112.1785	136.7930				106.5400	89.2750			
89	118.1790	140.5460				104.0020	84.7460			
90	124.0525	144.6060				103.4435	81.9770			
91	124.7650	145.4590				104.6120	82.7780			
92	126.0200	146.2735				104.3435	82.0705			
93	128.4845	146.8770				101.0435	78.1275			
94	130.0395	148.8285				103.6865	79.9110			
95	115.4495	139.9725				108.7060	90.0470			
96	104.0080	131.4725				109.7395	95.2425			
97	118.1735	142.9055				111.6035	91.6615			
98	121.9840	146.0400				112.9200	91.3755			
99	131.6250	152.4350				112.2975	87.1310			
100	121.8705	146.8270				115.9595	94.1620			
101	122.5475	147.7500				117.8495	95.6670			
102	128.7005	151.6965				116.3140	91.7895			
103	137.7345	156.9550				115.4970	87.5430			
104	137.4310	157.2600				117.2170	89.2380			
105	135.6755	156.9380				120.1850	92.6100			
106	129.7930	153.8330				122.0190	96.6140			
107	130.0090	154.1445				122.9450	97.4025			
108	115.5970	143.7040				121.2525	101.3320			
109	113.4080	142.0595				121.2975	102.1815			
110	101.8975	133.7625				123.4510	108.4400			
111	105.1750	136.9525				125.9780	109.6050			
112	105.5490	137.2245				125.5015	108.8990			
113	111.1785	141.9980				127.1885	108.3520			
114	114.3810	143.8690				125.3335	105.4370			
115	103.5545	136.9105				130.8465	114.4640			
116	105.0785	139.1950				135.0235	117.6075			
117	99.8125	135.3160				136.3430	120.7935			
118	109.1030	142.9030				138.6525	119.4410			
119	113.7645	146.6645				140.2690	118.9920			
120	118.3555	148.3695				131.5515	109.3580			
121	130.3850	156.1480				131.0635	104.4805			
122	130.9770	156.7085				132.4910	105.5150			
123	126.2585	154.7920				138.4510	112.6565			
124	125.5490	154.6045				139.9520	114.2435			
125	120.9820	151.2175				137.1765	113.5225			
126	117.1160	149.4050				142.3785	119.7300			
127	105.4650	140.9040				142.1050	123.8130			
128	97.0055	133.7090				139.3935	124.5640			
129	95.8120	133.4490				142.8935	128.1480			
130	95.7640	134.1040				146.3245	131.2220			
131	102.7870	139.4455				145.2825	127.6785			
132	86.7150	127.6955				152.9115	140.4935			
133	97.8815	137.3690				158.6875	141.3475			
134	106.9350	143.6105				153.3180	133.2685			
135	117.1300	150.2710				149.4260	125.8980			
136	119.1930	151.6100				150.0610	125.7185			
137	106.3215	143.4325				163.2030	142.1900			
138	87.5160	129.3950				162.2765	148.4925			
139	93.0235	133.8880				165.9660	149.5340			
140	91.5575	132.7415				166.3285	150.4455			
141	91.1900	132.4390				167.4630	151.5940			
142	88.8925	130.6615				168.2230	153.0745			
143	95.1570	135.4280				169.5170	151.7335			
144	95.9590	135.8425				169.6350	151.5335			
145	82.2375	125.3160				166.1095	153.7145			
146	84.1320	126.7555				170.5140	156.7980			
147		71.0760	119.9375				18.4700	16.9395		
148		66.5190	113.4280				21.9025	19.3770		
149		63.6820	111.0595				21.7120	18.2500		
150		67.5505	113.2930				23.4230	21.3965		
151		56.6040	102.6555	128.9135			25.3190	19.9145	16.2265	
152		59.1280	102.8430	129.9770			28.4320	24.1390	18.9740	

Table 43

Original Measures of Positions of Selected Lunar Formations on Lick Observatory Photographs (See Table 23)

No.	Measured x (mm)					Measured y (mm)				
	1	2	3	4	5	1	2	3	4	5
153		61.8925	104.4475				30.0355	26.7425		
154		63.6220	106.7240				28.6220	25.8090		
155		64.7515	108.3605				27.5220	24.9830		
156		73.5950	116.8830				26.5665	26.7135		
157		77.8520	120.4185				27.4460	28.8880		
158		79.6140	122.9700				25.9030	27.7645		
159		81.1795	122.2935				29.7990	32.4670		
160		81.4780	121.9935				30.9130	33.7680		
161		87.6855	127.4940				32.0155	36.6810		
162		88.0750	127.5430				32.5725	37.3810		
163		62.5575	104.2300	132.1290			31.3750	28.4240	22.0190	
164		63.8995	104.5385				33.0565	30.6435		
165		59.5590	101.5640	129.2950			31.5250	27.6050	21.3720	
166		58.2330	100.0990	127.8275			32.0780	27.7700	21.4865	
167		54.4260	97.5510	124.7670			30.7790	25.1230	19.6360	
168		55.5585	97.9310	125.4320			31.8580	26.6270	20.6790	
169		56.7005	97.3805	125.4020			34.7120	30.0550	23.1200	
170		53.8780	94.8510	122.7420			34.9505	29.3780	22.6440	
171		60.6025	99.3905	128.0585			37.4505	34.2170	26.2205	
172		59.2895	97.2870	126.1235			39.4640	35.8970	27.4810	
173		67.2065	105.6380				36.8735	35.7125		
174		68.1275	107.0315				35.6800	34.7535		
175		74.8345	110.9490				40.6550	42.0055		
176		79.4850	115.3385				40.6125	43.3500		
177		84.5910	120.9960				38.9510	43.1310		
178		90.7465	128.0930				36.8865	42.6925		
179		49.4220	87.9235	116.3245			41.2545	34.5210	26.5525	
180		51.4355	88.1285	117.0355			44.4800	38.5665	29.6415	
181		55.3100	90.1940	119.5950			47.3845	42.8505	32.9455	
182		61.6350	96.0925	125.7740			46.7205	44.1770	34.0190	
183		62.1760	96.4375	126.1380			47.0660	44.7260	34.4600	
184		61.7510	95.5950	125.4170			48.1145	45.6630	35.2340	
185		62.9365	95.9770	125.9055			49.8105	47.7800	36.9570	
186		68.5025	103.6235	133.2445			43.7925	43.3045	33.4660	
187		68.4600	101.4320	131.4755			48.7975	48.4360	37.6285	
188		68.4530	101.1615	131.2130			49.4185	49.1255	38.1225	
189		70.1620	102.1760	132.3755			50.8155	51.0310	39.8190	
190		67.7575	99.5570	129.8025			51.7670	51.2750	39.9375	
191		72.2515	107.5495				42.8930	43.4645		
192		73.5745	107.3790				45.9960	47.0985		
193		75.6920	108.1990				48.7890	50.6330		
194		79.2505	113.0190				45.3810	48.1565		
195		81.9960	115.4870				45.7775	49.3935		
196		85.3680	115.7280				53.0610	57.8230		
197		87.4465	121.7750				43.4470	48.5785		
198		90.4915	123.6565				46.0065	52.0545		
199		95.3260	130.3815				41.7850	49.0540		
200		99.8400	133.3630				45.4805	54.0950		
201		102.7780	135.6250				47.1415	56.6135		
202		99.0345	130.3920				50.4580	59.0270		
203		56.3540	87.8825	117.9290			55.2270	51.1915	39.8030	
204		61.2750	91.3595	121.6970			57.6285	55.1960	43.1840	
205		64.3005	93.9370	124.4100			58.0650	56.5885	44.3860	
206		65.8535	96.9580	127.2340			53.9520	52.8985	41.2670	
207		67.6455	97.2260	127.7300			57.4820	57.0030	44.8005	
208		70.1640	99.4390	130.0335			57.6995	58.0000	45.7180	
209		70.8815	101.4830				54.3980	54.8745		
210		74.1240	104.8050				53.4840	54.9375		
211		73.7020	103.1330				56.8620	58.2110		
212		74.4780	103.8440				56.9245	58.5040		
213		81.1585	110.0950				57.1670	60.7390		
214		91.4965	121.2410				54.3855	60.9125		
215		96.2450	124.3960				58.5220	66.4460		
216		55.9590	86.1245	116.3025			58.8645	54.7580	42.8520	
217		57.8735	86.4585	116.8480			62.6310	59.1850	46.6455	
218		60.0670	87.8815	118.4505			64.0375	61.2540	48.4580	
219		61.2550	89.2725	119.8320			63.0305	60.6450	47.9000	
220		63.1050	91.0240	121.6345			62.9390	61.1110	48.3290	
221		61.9855	89.3240	119.9760			64.7500	62.5990	49.6100	
222		62.1795	91.1860	121.6640			60.2125	58.1260	45.6525	
223		74.5795	100.3315				66.7390	68.4510		
224		75.9755	102.1910				65.2475	67.3270		
225		78.7215	104.5990				65.7330	68.6595		
226		80.5605	105.5900				67.9905	71.4865		
227		81.1555	105.9475				68.5345	72.1945		
228		82.1895	109.5215				61.3700	65.3080		
229		85.9585	113.6930				59.9240	64.9560		
230		87.7160	114.6590				61.7200	67.2850		

Table 43

Original Measures of Positions of Selected Lunar Formations on Lick Observatory Photographs (See Table 23)

No.	Measured x (mm)					Measured y (mm)				
	1	2	3	4	5	1	2	3	4	5
231		87.0325	114.0195				61.7660	67.1115		
232		91.2290	119.0310				59.4235	65.9465		
233		105.4900	132.2580				62.7520	73.2565		
234		113.5120	140.2165				63.8705	76.4765		
235		60.1875	84.7750	115.4645			73.2865	70.5420	56.7315	
236		64.4210	89.7890	120.5685			69.8690	68.4240	54.8200	
237		68.2905	94.3070	125.1110			67.0190	66.8065	53.3495	
238		69.2635	95.3345	126.1715			66.7045	66.7820	53.3315	
239		71.5030	96.9990	127.8690			67.9220	68.6840	55.1105	
240		68.7220	92.2130	123.1035			74.5280	74.4125	60.2285	
241		69.4150	93.0790	124.0095			73.8395	73.9560	59.8165	
242		72.1660	95.4375	126.4025			74.3170	75.2570	61.0325	
243		72.7180	97.1890	128.0920			70.8010	71.9090	58.0260	
244		74.8305	99.3310				70.3340	72.0750		
245		76.6460	100.9775	131.9570			70.4665	72.7795	58.8865	
246		75.2965	99.0790				72.4240	74.3190		
247		74.0845	97.7040	128.6750			72.8335	74.3620	60.2665	
248		78.5590	102.1620	133.0920			72.3825	75.2700	61.2290	
249		84.1300	107.9060				71.2120	75.7285		
250		88.0350	111.0735				73.3110	78.9815		
251		86.9760	110.7220				71.1580	76.4260		
252		92.6705	116.9110				69.4140	76.3940		
253		97.4120	119.9680				74.2345	82.5250		
254		100.4525	123.7325				72.3270	81.4620		
255		104.1310	128.0880				70.6060	80.7210		
256		107.1035	130.1575				73.3855	84.3190		
257		56.6295	78.9580	109.4910			81.4725	77.5310	63.2460	
258		59.4965	83.1740	113.8610			76.2085	73.2410	59.2150	
259		67.0690	88.8220	119.6590			80.3420	79.6650	65.0660	
260		72.9310	93.9450	124.8340			81.3425	82.4670	67.7115	
261		73.1535	93.8660	124.7480			82.3180	83.4780	68.6545	
262		82.5710	102.8230				82.1780	86.1665		
263		83.7270	103.9175				82.3645	86.6830		
264		87.6080	108.2410				80.6820	86.1260		
265		95.7220	117.1020				78.0355	85.7995		
266		95.5790	116.4325				79.4950	87.2240		
267		94.2810	114.9385				80.3350	87.7010		
268		94.4820	114.9750				80.7350	88.1350		
269		94.3090	114.6700				81.2075	88.5380		
270		97.5700	117.9730				81.2250	89.4505		
271		100.4705	121.3395				79.6950	88.7490		
272		115.1750	137.4430				77.0120	90.0555		
273		65.9575	86.5970	117.3890			84.0690	82.9850	68.1865	
274		71.3535	91.1455	121.9250			85.7120	86.2315	71.2610	
275		74.7575	94.0205	124.7980			86.9395	88.4790	73.4080	
276		76.1910	96.2720	127.1695			83.7360	85.7720	70.8680	
277		77.7085	96.2445	126.9990			88.8245	91.2065	76.0460	
278		81.1275	99.7390	130.4790			88.1200	91.5190	76.4255	
279		84.8900	103.3750				87.9900	92.4800		
280		84.6550	103.6755	134.4350			86.1145	90.6130	75.6780	
281		92.3250	111.6785				84.5755	91.2810		
282		95.7830	114.6435				86.2375	93.8895		
283		98.9465	117.8890				86.0325	94.5700		
284		100.9910	120.2630				84.7985	93.9295		
285		100.3105	118.9320				87.0185	95.9180		
286		107.2545	126.4700				85.5650	96.3730		
287		108.2920	127.7870				84.8815	96.0050		
288		115.1940	134.4660				86.6945	99.5905		
289		66.1250	84.5630	115.1040			91.7435	90.5040	75.3415	
290		71.6685	90.7505	121.4670			88.1425	88.7000	73.5985	
291		78.8890	96.2350	126.8385			92.8905	95.5160	80.2180	
292		79.1380	96.8095	127.4650			91.6700	94.3830	79.1300	
293		84.9750	102.6250	133.2130			90.9430	95.4000	80.2840	
294		83.5190	99.4215	129.8115			97.2980	101.1630	85.8215	
295		88.4875	104.8915				95.0715	100.4445		
296		89.9880	106.4725				94.5615	100.3650		
297		107.0305	124.7910				90.6510	101.3260		
298		106.5180	124.0810				91.1240	101.6315		
299		107.2415	124.4195				92.6670	103.3240		
300		68.5625	84.9285	115.1840			98.7680	98.0960	82.6875	
301		70.6790	85.6960	115.7215			103.4170	103.2325	87.7665	
302		70.6600	85.4375	115.4090			104.3780	104.1490	88.6750	
303		71.0950	85.5380	115.4270			105.6695	105.5530	90.0610	
304		77.6680	93.0350	123.2625			100.4440	102.4570	86.9760	
305		78.9670	95.0135	125.4150			97.4880	99.9880	84.5670	
306		82.2345	97.6055	127.8760			99.6015	103.0345	87.6135	
307		83.1385	98.4035	128.6110			100.0655	103.7325	88.3170	
308		84.6840	98.8770	128.8495			103.8760	107.8375	92.4510	

Table 43

Original Measures of Positions of Selected Lunar Formations on Lick Observatory Photographs (See Table 23)

No.	Measured x (mm)					Measured y (mm)				
	1	2	3	4	5	1	2	3	4	5
309		84.3675	98.0920	127.8835			106.1015	109.9040	94.4895	
310		85.9865	99.8100	129.6425			105.3960	109.6960	94.3465	
311		86.6995	101.7510	131.9690			100.0660	104.7610	89.4400	
312		90.0225	105.4410	135.5425			98.7850	104.4855	89.3510	
313		94.3980	109.5790				99.6900	106.5820		
314		95.5610	110.5250				100.1380	107.3565		
315		95.9330	110.6870				101.1185	108.4510		
316		95.0630	108.8970				104.8050	111.7060		
317		97.5645	111.6735				103.6610	111.3155		
318		101.5540	116.0825				102.1220	110.9545		
319		65.5270	80.4495	110.2600			105.4270	103.5580	88.1985	
320		64.3720	80.3035	110.3380			101.6960	99.6185	84.2865	
321		62.6460	77.6680	107.3790			105.9185	103.1310	87.8720	
322		66.7100	79.2255	108.2970			115.2685	113.3340	98.1595	
323		68.7365	80.8885	109.9270			116.3160	114.9335	99.7555	
324		74.4750	88.2140	117.9885			107.7405	108.5190	93.0515	
325		76.3145	89.2090	118.7535			110.9675	112.1830	96.7400	
326		80.5530	92.6215	121.9100			113.5820	115.9215	100.6125	
327		81.5315	95.1850	124.9680			106.8080	109.7460	94.3090	
328		85.5370	97.4190	126.6045			113.8060	117.6070	102.4145	
329		88.4365	101.0725	130.4690			110.2970	115.0785	99.8875	
330		98.7960	110.1045				115.6245	123.0900		
331		98.1460	110.1740				112.5210	119.9315		
332		100.5600	112.7625				111.8590	119.9710		
333		101.9080	114.4855				110.1520	118.7260		
334		101.0310	114.1310				107.7945	116.1990		
335		107.4200	119.1870				114.6830	124.5420		
336		112.1300	124.9055				111.1900	122.4425		
337		107.9840	119.6850				115.0430	125.0040		
338		123.1985	137.4600				108.1085	122.2775		
339		123.5645	137.7070				108.7055	122.9485		
340		68.2210	79.5930	108.1940			120.4075	118.6670	103.6880	
341		66.9705	79.2245	108.2170			116.4665	114.5395	99.4055	
342		70.5200	82.0070	110.8125			118.9425	118.0380	102.8815	
343		70.6575	81.3245	109.7330			123.0125	121.8450	106.9780	
344		74.7625	85.5135	114.1240			121.3070	121.4845	106.4590	
345		78.5020	89.0685	117.6195			121.3910	122.7030	107.7030	
346		79.4220	89.7440	118.1905			122.4210	123.9620	109.0180	
347		76.8970	86.7795	114.9015			125.4540	126.0695	111.2845	
348		81.8155	91.7165	119.9665			124.0165	126.1705	111.3540	
349		84.4260	94.9035	123.4970			120.4510	123.5875	108.6130	
350		86.4280	97.3200	126.0470			118.4240	122.2245	107.2420	
351		83.1740	94.6095	123.6295			116.2470	119.2445	104.0185	
352		90.0405	100.9175	129.6230			117.8910	122.7330	107.8555	
353		91.3920	101.5415	129.9255			121.3325	126.4050	111.7130	
354		91.4010	100.9670	129.0220			124.5235	129.4195	114.9450	
355		92.5440	101.7545	129.5120			126.4345	131.5350	117.2280	
356		95.2050	105.0210	133.0560			122.9990	129.0440	114.6815	
357		117.2500	128.3600				120.7120	132.7965		
358		120.9885	132.1030				121.7645	134.7545		
359		71.0030	80.5190	108.1775			129.2530	127.8160	113.3485	
360		72.2725	80.9475	108.0165			133.8460	132.4660	118.3970	
361		78.7580	86.9960	114.0000			134.4250	134.9560	120.9460	
362		79.7215	89.1410	117.1030			126.9130	128.2610	113.6050	
363		87.6075	96.4820	124.0560			128.8820	132.4105	118.1125	
364		102.1885	110.2340				134.1100	141.3365		
365		107.2165	116.4115				128.0345	137.0450		
366		110.2160	119.9230				125.8500	135.7580		
367		110.6945	120.3910				126.1010	136.1100		
368		108.7865	117.1910				133.5140	142.5510		
369		110.7345	119.3050				133.1320	142.7140		
370		117.2145	126.8635				129.2135	140.7005		
371		118.6325	128.1130				130.8760	142.6445		
372		126.8735	137.5845				129.3135	143.1310		
373		73.9590	82.4360	109.4050			134.6300	133.7020	119.7010	
374		74.9375	83.0560	109.7615			136.5130	135.7580	121.9475	
375		74.7135	82.4190	108.6420			139.6590	138.5935	125.1465	
376		76.7415	84.1895	110.3870			140.1680	139.6200	126.3000	
377		76.8040	84.0795	110.0390			141.5160	140.8850	127.7550	
378		77.6655	85.7545	112.6090			135.6085	135.7320	121.8455	
379		78.8985	86.7150	113.3420			137.1895	135.5770	123.8645	
380		82.3100	90.0750	116.7510			136.5420	137.9720	124.2440	
381		83.2120	90.7165	117.1115			138.2865	139.8450	126.3385	
382		84.2555	91.4770	117.5170			140.2620	141.9755	128.7310	
383		87.2590	94.8075	121.3230			136.9510	139.8140	126.2885	
384		95.6395	102.9450	128.9155			138.5755	143.6500	130.7430	
385		103.3020	110.8810				137.6865	144.9575		
386		86.4510	93.1275	118.5445			143.9945	145.9785	133.3440	

Table 43

Original Measures of Positions of Selected Lunar Formations on Lick Observatory Photographs (See Table 23)

No.	Measured x (mm)					Measured y (mm)				
	1	2	3	4	5	1	2	3	4	5
465			77.0330	105.4795				40.1455	31.3960	
466			78.2825	106.4165				36.9865	28.9230	
467			81.5130	110.6610				43.9120	34.0465	
468			84.1310	112.9250				39.2445	30.2825	
469			58.8580	84.6895	46.8430			39.3295	33.5045	48.8340
470			59.4395	86.0240	48.8825			42.7030	35.6195	50.3845
471			63.3260	90.6880	53.2405			43.3910	35.3900	48.4890
472			66.3045	94.4790	57.5750			46.6430	37.4835	49.1250
473			69.6610	97.8555	60.0090			44.2660	35.2275	45.3130
474			72.1865	100.7295				44.9030	35.4985	
475			74.5955	103.4525				45.7820	35.9680	
476			76.5785	105.7190				47.2015	36.9715	
477			84.4285	113.9420				46.5640	36.0940	
478			85.0975	114.6875				46.6845	36.1500	
479			86.0875	115.6120				45.6730	35.3130	
480			58.6060	85.9275	50.6235			49.3455	40.8305	55.3395
481			58.4110	85.8900	50.9615			50.6855	41.8760	56.3450
482			61.8280	89.7245	54.1265			49.6540	40.4955	53.6650
483			65.3630	93.6570	57.6745			49.5950	39.9305	51.7025
484			66.4230	95.1420				53.4805	42.9900	
485			74.1660	103.4220				50.7070	39.9930	
486			76.2800	105.6730				50.4380	39.6410	
487			75.5155	105.1580				55.0230	43.5185	
488			77.7700	107.6730				56.4235	44.5770	
489			81.0630	111.1870				57.2165	45.0775	
490			82.2365	112.2575				54.9750	43.1425	
491			59.5830	87.6575	53.8270			55.1410	45.2110	58.8620
492			62.5230	91.1465	57.9045			58.0910	47.2680	59.5710
493			63.7865	92.6335	59.6965			59.5190	48.3210	60.0530
494			66.6250	95.8135				59.5435	48.0530	
495			66.7810	96.0320				60.4530	48.7920	
496			66.3455	95.6095				61.3180	49.5700	
497			70.7800	100.3620				60.4070	48.4265	
498			73.1630	102.8580				58.9050	46.9615	
499			75.1280	104.9435				58.3370	46.3720	
500			74.2890	104.2280				62.4160	49.8980	
501			75.7980	105.7580				60.6075	48.2755	
502			77.5420	107.5455				58.6780	46.5195	
503			49.4735	76.3780	46.0340			62.7500	53.4290	70.6950
504			50.9560	78.2355	47.9225			63.6690	53.8385	70.3860
505			51.2520	78.5490	48.4145			64.2430	54.3025	70.6655
506			58.3635	86.6585	54.6585			60.8460	50.1400	63.8865
507			71.2965	101.1020				65.6950	52.9670	
508			77.0035	107.1835				64.8445	51.8710	
509			77.4995	107.8110				67.0950	53.8460	
510			80.7140	111.2270				69.0275	55.4515	
511			49.8480	77.0780	48.3190			68.2505	58.0410	74.7880
512			54.9720	83.1745	54.2270			69.6545	58.2970	72.8555
513			56.4455	84.8360	55.8275			70.0335	58.3950	72.3780
514			57.0880	85.5950	57.0270			71.5760	59.7000	73.3040
515			62.5860	91.7820	62.4665			71.0295	58.4700	69.9630
516			63.4325	92.6305				70.1870	57.6585	
517			65.8380	95.2685				70.3685	57.6225	
518			66.2200	95.7115				69.6485	56.9425	
519			72.4085	102.5050				72.8160	59.2400	
520			71.5385	101.5805				74.8320	61.1540	
521			73.1720	103.3320				74.5495	60.8120	
522			79.4695	110.0100				72.5490	58.6810	
523			79.3565	109.9095				77.6740	63.3710	
524			83.7915	114.5400				76.7760	62.3710	
525			43.0940	68.8050	41.5495			69.6180	60.9710	80.4750
526			46.8470	73.6110	45.8910			70.2270	60.4310	78.2905
527			56.7765	85.3825	57.9755			75.2180	62.9835	76.5010
528			62.2770	91.4520				76.5765	63.5835	
529			63.2910	92.6305				80.0275	66.6440	
530			62.4155	91.6755				81.3880	67.9875	
531			38.4530	63.2165	38.5795			75.6840	67.8280	88.9730
532			44.8650	71.4365	47.9025			82.5555	71.9510	89.9530
533			50.3950	78.0330	52.5350			78.8355	67.3520	83.1960
534			55.3800	83.7990	59.2025			83.4625	70.7610	84.4175
535			56.5590	85.1510	59.4390			80.4160	67.7595	81.0995
536			57.6205	86.2655	62.6445			86.9930	73.7625	86.3685
537			60.9550	90.0175				82.3755	69.0715	
538			62.4900	91.7630				84.1840	70.5705	
539			77.7270	108.0410				88.5245	73.6415	
540			77.8820	108.1935				89.0055	74.0845	
541			38.1070	62.9220	41.6725			86.1600	77.4005	98.1355
542			43.1745	69.3065	47.6105			87.3700	76.9190	95.4025

Table 43

Original Measures of Positions of Selected Lunar Formations on Lick Observatory Photographs (See Table 23)

No.	Measured x (mm)					Measured y (mm)				
	1	2	3	4	5	1	2	3	4	5
543			45.3240	71.9845	49.3080			85.4080	74.5160	92.1575
544			48.5680	75.7410	54.4400			90.7920	78.8850	94.9745
545			49.3705	76.6860	55.4405			91.2495	79.1555	94.9015
546			50.4780		55.9840			89.2640		92.4190
547			55.6970	83.9900	61.6270			90.0755	76.9860	90.2490
548			55.5315	83.7495	61.8210			91.3500	78.1865	91.4990
549			57.4445	85.9400	63.6650			90.9535	77.5750	90.1165
550			57.4550	85.9255	64.1080			92.2550	78.8415	91.3260
551			59.1040	87.7265	66.4255			94.2300	80.5335	92.2950
552			60.8050	89.6900				91.7250	77.9280	
553			64.2010	93.3525				94.4965	80.2490	
554			67.0410	96.3770				93.7600	79.2925	
555			66.7880	96.2650				91.3945	77.0450	
556			66.3030	95.7740				90.9010	76.5970	
557			66.5750	96.0670				89.9385	75.6750	
558			71.4395	101.2980				90.5595	75.9095	
559			71.9390	101.8020				91.9375	77.1945	
560			71.8025	101.6075				92.9810	78.1910	
561			70.8985	100.6160				93.8380	79.0985	
562			77.0680	107.1480				94.1635	79.0920	
563			77.5565	107.7405				92.9400	77.8690	
564			78.5280	108.7690				92.2175	77.1440	
565			37.3245	61.7400	42.7800			92.3050	83.5625	104.4145
566			38.9390	63.8015	44.8755			93.3750	84.0370	104.0830
567			40.7595	66.1400	46.8050			93.0980	83.1350	102.3570
568			43.9405	69.9720	50.9705			95.5530	84.5880	102.4050
569			43.9705	69.9625	51.2750			96.4170	85.4190	103.2075
570			45.3700	71.6690	52.6715			96.1320	84.7750	102.0060
571			46.9230	73.6375	53.8640			94.5505	82.8680	99.4950
572			47.2940	74.1420	53.8175			93.1410	81.3885	97.9230
573			47.9585	74.8890	54.4925			93.0235	81.1570	97.4130
574			49.2485	76.3470	56.7100			95.9000	83.7235	99.3235
575			51.0075	78.4040	58.3280			95.1885	82.7050	97.6465
576			51.4090	78.8480	59.1650			96.4710	83.8525	98.5895
577			57.2550	85.6160	64.9170			95.4590	81.9075	94.3420
578			67.2370	96.3705				100.5120	85.8925	
579			70.0705	99.3730				101.1570	86.3040	
580			73.2010	102.9410				97.0760	82.0955	
581			78.1310	108.1350				97.6555	82.4215	
582			39.9760	64.7945	47.8935			99.5890	89.8255	109.2235
583			41.8935	67.2790	49.6125			98.3730	87.9740	106.6020
584			42.3955	67.9290	50.2720			98.6970	88.1395	106.5030
585			45.7300	72.0095	53.7400			98.3925	86.9720	103.9360
586			48.3190	74.9730	57.2290			100.8410	88.7855	104.6160
587			53.2610	80.7310	62.5510			101.3150	88.3465	102.1305
588			53.2405	80.6815	62.7100			101.9060	88.9270	102.7020
589			54.2735	81.8365	63.9185			102.3415	89.2355	102.5945
590			54.7585	82.2075	65.0725			104.6950	91.5070	104.6245
591			63.4310	91.9750				104.5295	90.2150	
592			66.4060	95.3680				102.2805	87.7350	
593			68.4780	97.3850				105.6765	90.9970	
594			69.0615	98.0235				105.6780	90.9350	
595			35.8290	58.9410	43.9720			102.3940	94.4735	115.6875
596			40.7440	65.5885	49.6130			102.5820	92.5745	111.5670
597			41.8295	66.9060	50.9485			103.1960	92.8155	111.3085
598			45.9230	71.7155	56.2475			106.1775	94.7915	111.4695
599			55.3320	82.3620	67.5505			111.1620	98.0495	110.7860
600			56.9455	84.2605	69.0820			110.6465	97.2795	109.3810
601			62.6055	90.5040				111.7290	97.6935	
602			62.5165	90.3545				112.1485	98.1215	
603			63.3910	91.6020				108.6350	94.4200	
604			64.6450	92.9175				110.2455	95.9350	
605			65.4805	93.7245				111.4205	97.0505	
606			71.5090	100.2645				111.2685	96.4285	
607			73.1890	101.8440				113.3175	98.4160	
608			73.9905	102.6835				113.6610	98.7485	
609			75.0310	103.8170				113.9000	98.9285	
610			75.6360	104.7185				110.3490	95.2570	
611			81.0920	110.5525				109.5150	94.1600	
612			38.4365	61.3050	49.6875			112.6600	104.3085	124.1890
613			39.0610	62.1175	50.4860			113.0930	104.4970	124.0750
614			39.9030	63.2390	51.5390			113.3405	104.4165	123.6030
615			38.8450	61.9580	49.8660			111.6800	103.0225	122.7540
616			39.9825	63.7510	50.7955			110.1155	100.8115	120.0205
617			40.7855	64.7720	51.7310			110.2955	100.7560	119.5660
618			40.0890	63.7405	51.2230			111.3365	102.1255	121.2055
619			41.9990	66.0640				113.0190	103.2730	
620			42.5155	66.8460	54.0845			111.9025	101.8630	119.8780

470

Table 43

Original Measures of Positions of Selected Lunar Formations on Lick Observatory Photographs (See 23)

No.	Measured x (mm)					Measured y (mm)				
	1	2	3	4	5	1	2	3	4	5
621			45.0315	69.7315	57.4515			114.2290	103.5990	120.5395
622			52.4525	78.9685	64.7840			111.9475	99.3575	113.2315
623			52.7550	78.9630	65.8765			114.9785	102.4840	116.1670
624			54.1260	80.5240	67.5380			115.6010	102.9530	116.0520
625			47.4935	72.3220	61.2205			118.2660	107.3090	123.1275
626			50.5900	75.9960	64.7180			119.0305	107.2845	121.7695
627			56.5585	82.6270	71.5725			121.4615	108.7565	120.7930
628			59.9160	87.0130	74.0730			117.5685	104.0935	114.8620
629			62.2210	89.5850				117.1160	103.3355	
630			65.9030	93.8220				115.5660	101.3220	
631			67.5250	95.2790				118.5730	104.3255	
632			68.9905	97.0150				117.0900	102.6550	
633			70.1665	98.1535				118.4200	103.9440	
634			69.7035	97.4720				120.0870	105.7470	
635			68.9640	96.6010				120.7540	106.5115	
636			75.3505	103.9745				116.0395	101.1440	
637			78.3795	107.0390				117.4325	102.4150	
638			39.2490	61.4190	51.9640			118.3380	110.4465	129.9895
639			46.7545	70.6280	61.4425			122.7330	112.4415	128.5740
640			57.9090	83.8050	73.7315			124.3875	111.7005	123.1805
641			62.1660	88.5440	78.3880			125.3600	112.1530	121.9350
642			69.7865	96.6205				128.0165	114.2385	
643			73.8630	100.9075				129.1240	115.1380	
644			72.6195	100.3255				122.3815	107.8430	
645			43.0380	65.2525	57.9490			125.3585	116.9925	134.8290
646			45.3245	68.1205	60.7250			126.2725	117.1480	133.9480
647			45.7945	68.6770	61.3030			126.5450	117.2615	133.8650
648			63.8215	89.9635	80.9750			128.7790	115.7270	124.8385
649			64.9540	91.1100	82.3880			129.6990	116.6410	125.3020
650			64.3545	90.3515	81.8215			130.0365	117.0820	125.9725
651			65.8605	91.6990				132.3435	119.4545	
652			71.4735	97.3130				135.7030	122.6420	
653			76.1510	102.7470				133.3315	119.6260	
654			43.1395	64.1730	58.7510			129.4510	122.2170	140.1510
655			52.5850	75.9460	70.2770			133.3180	123.1305	136.8250
656			53.8835	77.3175	71.8330			134.1835	123.7520	136.9475
657			56.2115	79.4050	75.1825			137.8710	127.7015	139.9220
658			66.5200	91.1170	86.4445			139.8690	128.0530	136.1365
659			74.6040	99.3820				143.5145	131.3945	
660			76.8675	102.1245				141.9160	129.3285	
661			80.1625	105.6015				142.5345	129.7835	
662			82.8605	107.7595				145.8190	133.4700	
663			64.0280	87.3970	84.6045			143.4745	132.8650	142.0055
664			68.7240	92.5725	89.5965			144.3085	133.0765	140.3855
665			69.6390	92.8865	91.1505			147.3310	136.6130	143.6330
666			71.1895	95.2965				146.6105	133.0585	
667			77.5845	101.4180				149.1185	137.8510	
668			80.0090	104.6405				146.7065	134.7590	
669			56.7080	75.2895	77.2950			150.1810	144.7805	157.6095
670			62.9515	83.4780	84.7425			151.4575	143.7680	153.7585
671			61.0880	81.0020	82.6855			151.6535	144.6980	155.4965
672			64.8390	86.5890	86.5135			149.5730	140.4415	149.4745
673			72.5165	94.3530	95.3285			153.7945	144.4840	150.5485
674			74.5845	96.8630	97.3790			153.3510	143.5760	148.8030
675			74.5500	97.1765	97.0990			152.0810	141.9255	147.1525
676			81.9880	105.6465				151.6555	140.5730	
677			83.2290	107.3520				149.7620	138.2085	
678			83.9845	107.9620				150.5595	139.0825	
679			85.0870	108.6600				152.7525	141.6855	
680			64.7985	85.1575	86.9515			153.1800	145.5630	154.8580
681			76.9505	98.5505	100.4700			156.7730	147.7855	152.2245
682			79.0635	101.1315				156.0950	146.5840	
683			80.4200	101.4325				159.5465	151.0945	
684			81.5720	103.4190				157.8110	148.5680	
685			87.8150	108.9990				161.4690	152.9025	
686			87.8540	108.8920				161.8915	153.4895	
687			88.5025	108.4550				164.6615	157.4130	
688			86.8825	106.4660				165.1195	158.3930	
689			91.3315	110.7990				166.3610	159.7170	
690			91.7320	111.2390				166.4940	159.8055	
691			90.2160	109.1615				167.1335	161.0825	
692			91.9425	110.6200				167.9065	162.1850	
693			83.6180	101.8215	108.4545			166.5855	161.5855	164.1400
694			89.0080	106.5470	113.9475			168.8080	164.4295	165.1665
695			95.3615	113.4600				169.5575	164.5555	
696			95.3250	113.0075				170.1215	165.6425	
697					60.2705					146.2035
698					97.6995					163.5880

Table 43

Original Measures of Positions of Selected Lunar Formations on Lick Observatory Photographs (See Table 23)

No.	Measured x (mm)					Measured y (mm)				
	1	2	3	4	5	1	2	3	4	5
387		94.7390	101.0445	125.5330			147.3425	151.3945	139.7380	
388		98.0995	104.6420	129.3175			145.3570	150.5000	138.7765	
389		98.5480	104.9815	129.3620			146.7200	151.8690	140.4510	
390		102.2410	109.1015				143.3945	149.8570		
391		103.1735	110.0450				143.6460	150.3280		
392		103.3135	109.9980				144.5985	151.2510		
393		105.8305	112.6775				145.1630	152.4035		
394		108.0305	115.3210				141.5635	149.7310		
395		112.8685	120.3190				144.2370	153.3920		
396		113.8290	121.3400				144.4305	153.8075		
397		114.9020	122.8945				140.0235	150.1080		
398		119.2650	127.5690				141.6715	152.6410		
399		117.3485	125.0775				146.5285	156.5220		
400		83.7635	89.6595	112.7715			155.7460	155.7920	145.2710	
401		89.4435	95.3965	119.2780			151.5625	153.7050	142.4820	
402		95.8785	101.7605	124.0675			156.8235	160.1375	150.6040	
403		98.3600	104.3450	126.9500			154.9595	159.1565	149.4030	
404		106.3440	112.8570				150.1420	157.0085		
405		107.6470	114.2200				151.9245	158.9285		
406		116.2035	123.6725				149.9465	159.2830		
407		116.4385	123.9110				151.5640	160.7750		
408		116.3750	123.8820				152.9135	161.9890		
409		116.9810	124.5215				152.8945	162.0270		
410		117.7380	125.3735				152.6240	161.9450		
411		119.3040	127.1495				152.2330	162.0080		
412		92.1620	97.9230	120.2730			157.5585	159.7545	150.1270	
413		94.7515	100.6200	122.6880			158.0550	160.8805	151.5990	
414		105.7120	112.1230				157.9505	163.6745		
415		107.9310	114.5050				156.8475	163.2700		
416		86.3475	92.4990	112.6815			165.6475	165.0295	157.6200	
417		90.6520	96.6220	117.2380			163.8500	164.7305	156.9030	
418		91.3235	97.4845	117.3635			165.9070	166.6185	159.6090	
419		112.6140	120.7770				164.3660	170.5630		
420		99.7635	106.6660	125.3010			167.1675	169.8370	164.4130	
421		107.1640	115.3875				168.2765	172.3430		
422		100.0730	106.8850	125.7150			166.6855	169.5210	163.9245	
423			98.0105	122.5120				14.1790	13.0310	
424			84.3000	108.5270	63.1390			16.9670	15.6410	23.4625
425			85.0270	109.6100	64.2670			17.7715	15.9070	23.3410
426			87.6295	112.7240				18.4490	15.9820	
427			87.0690	112.0155				18.1230	15.8675	
428			88.4120	114.0035				19.8140	16.6195	
429			87.8795	114.3280				23.3155	18.7825	
430			91.0725	117.2240				21.4465	17.4325	
431			90.8030	117.7425				25.0850	19.7945	
432			93.0625	119.8545				23.6590	18.7505	
433			93.7690	120.5185				23.3510	18.5215	
434			95.4720	121.2925				19.1520	15.8310	
435			96.2295	121.9680				18.5950	15.5035	
436			76.2895	100.5700	56.7380			20.8250	18.8270	29.3075
437			74.9000	100.3890	57.4665			25.2545	21.3720	31.7635
438			77.7845	102.8975	59.3400			22.7850	19.5215	29.0585
439			78.4710	102.8960	58.6680			20.0700	18.0425	27.7210
440			78.7415	105.5950	63.0740			29.0175	23.2255	31.6560
441			80.8720	107.9885				29.1875	23.1375	
442			82.4810	109.4815				27.8720	22.0975	
443			88.9235	116.6145				29.5980	22.9450	
444			69.5240	93.6695	51.2680			24.3240	22.0295	34.8705
445			65.1015	89.6450	48.6680			28.5540	25.3875	39.3980
446			66.1165	90.6450	49.4075			27.8800	24.7580	38.4195
447			67.8840	92.8275	51.2540			27.5190	24.0895	37.0420
448			71.7010	97.6955	55.7910			28.5705	23.8790	35.0885
449			70.9825	97.8680	57.3165			34.0620	27.6125	38.5540
450			74.6350	101.1300	59.2635			29.5555	24.0380	34.1855
451			77.6110	105.0125				32.5665	25.7720	
452			82.8675	111.2425				36.4130	28.2425	
453			85.2895	113.3740				33.2855	25.7560	
454			86.6940	114.8860				33.5160	25.8525	
455			87.9230	116.0760				32.8580	25.3255	
456			88.9970	117.4725				35.0015	26.8590	
457			89.9225	118.4770				35.3460	27.1055	
458			60.3105		43.7255			29.1570		43.4280
459			64.0130	90.5190	51.5530			37.5070	31.0995	44.4710
460			64.9265	91.2325	51.7150			35.7105	29.7115	42.9105
461			65.6040	91.6890	51.6950			34.0150	28.4225	41.5370
462			67.4845	93.9070	53.7350			34.1270	28.1270	40.4550
463			71.0310	99.0000	60.0340			40.5860	32.2550	42.5375
464			76.4040	104.9670				41.7885	32.6870	

The numbers in the column headings refer to the Lick Observatory photographic plates of the moon as described in Table 21.

APPENDIX 3

THE LUNAR TIDAL BULGE AS A FUNCTION OF THE MOON'S DISTANCE

The three axes, ζ, ξ, and η, of a homogeneous moon distorted by a tidal pull and centrifugal force of rotation are (1)

$$a\left(1+\frac{35}{12}\frac{M}{m}\frac{a^3}{r^3}\right), \quad a\left(1-\frac{10}{12}\frac{M}{m}\frac{a^3}{r^3}\right), \quad a\left(1-\frac{25}{12}\frac{M}{m}\frac{a^3}{r^3}\right),$$

where
- a = The radius of the mean equivalent lunar sphere,
- M = The mass of the earth,
- m = The mass of the moon,
- r = The distance of the moon in lunar radii.

The average of the ξ- and η-axes is

$$a\left(1-\frac{17.5}{12}\frac{M}{m}\frac{a^3}{r^3}\right).$$

Therefore, the measured excess of the radial, ζ-, axis would be Δa,

$$a\left(1+\frac{52.5}{12}\frac{M}{m}\frac{a^3}{r^3}\right) = a(1+\Delta a).$$

To allow for non-homogeneity, multiply the coefficient by 0.9; and, to allow for elastic tides, multiply again by $1 - k$, or 0.987. Therefore,

$$\Delta a = 0.00002943 \left(\frac{r_0}{r_1}\right)^3,$$

where

r_0 = The present distance of the moon,
r_1 = The distance of the moon when the tidal bulge of the moon equaled Δa.

References

1. JEFFREYS, H. *The Earth*. Cambridge: Cambridge University Press, 1924.

AUTHOR INDEX

[Page numbers for references are in italics.]

Aarons, J., 275, *285*
Abbot, C. G., 38, *50*
Abelson, P. H., *410*
Adams, L. H., *340*
Adler, I., *292*
Akabane, K., 271, *283*
Albritton, C. C., Jr., 8, *51*, 71, 72, *102*
Alderman, A. R., 26, 27, 28, *49*, 290, *292*, *360*
Aldrich, L. T., 398, *410*
Alexander, W. M., 299, *312*
Allan, D. W., *410*
Allen, W. A., 180, *185*
Allix, A., 62
Almond, M., *414*
Alperin, M., *384*
Alter, D., 114, *126*, 294, *312*, 316, 317, 323, 324, *332*, 351, 352, *359*, 368, 391, *393*, *394*, 416, *419*
Altshuler, S., 413, *414*
Amenitskii, N. A., 271, *284*
Andel, K., xix
Anderle, R. J., 245, *247*
Anders, E., 16, *48*, 287, *291*, *292*, 304, *313*, 397, 398, *410*
Arago, F., 249, *262*
Arthur, D. W. G., 141, *151*, 189, *196*, 234, 294, 296, *312*
Ashbrook, J., 248, *262*, 382, *384*, 416, *418*
Astapovich, I. S., 37, 38, 39, 41, 43, *50*

Baker, G., 288, *292*
Bakharev, A. M., 44, 45, *51*
Baldwin, R. B., 94, *104*, 112, *126*, 155, 160, 178, *185*, 230, *246*, *247*, 295, 309, *312*, *313*, 327, *332*, 353, 358, *359*, *383*, *389*, *394*, 395, *409*
Banachiewicz, T., 203
Barabashov, N. P., 250, 252, 253, 254, 256, 257, 259, 260, *263*, *264*, *265*, *266*, 275, 276, *285*, 353, *359*, 360
Barker, R., 391, 392
Barnes, V. E., 288, 289, 290, *292*
Barringer, D. M., 8, 14, 16, 120, *127*

Barringer, D. M., Jr., 8, 11, *47*
Batchlor, C. D., *247*
Bauer, C. A., *348*
Baussart, M., 62, *66*
Beals, C. S., viii, 58, 59, 60, 61, *66*, 76, 79, 80, 81, 83, 84, 85, 86, 87, *103*, *104*, 107, 148, *152*, *185*
Beard, D. P., 303, *313*
Bedford, 28
Beer, A., *126*
Beer, W., *246*
Belkowitsch, J., 203, *211*, 216, *246*
Bell, R., *66*
Bennett, A. L., 254, *265*, 353, *359*
Bentz, A., 78, *103*
Bergstrom, R. E., 100, *105*
Beringer, R., 271, *284*
Bessel, F. W., 220
Biermann, L., 278, *285*
Billerbeck-Gentz, F., 355, *360*
Birch, F., 398, *410*, *411*
Bjork, R. L., 171, 178, *186*
Blagg, M. (Miss), xix
Blanford, W. T., *66*
Blevis, B. C., *285*
Bonfanti, N., 64, *66*
Boon, J. D., 8, *51*, 71, 72, *102*
Born, K. E., 88, 89, *104*, *105*
Bornstein, *410*
Borough, H. C., 277, *285*
Borst, L. B., 293, *311*, 344, 345, *348*
Bowen, N. L., 392, *394*
Boyd, F. R., 18, *48*, 403, *411*
Bracewell, R. N., 271, *284*
Brady, L. F., *51*
Branca, W., *65*, 76, 78, *102*
Brandt, J. C., 345, 346, *348*, 349
Braun, W. von, 347
Brickwedde, F. G., 154, 184
Brock, M. R., 96
Brockhaus, K., *247*
Brown, H., 109, *112*, 349

475

Brown, R., 19, 49
Browne, I. C., *285*
Bruton, R. H., 281, *285*
Bucher, W. H., *65*, *73*, 76, 90, 96, 99, *102*, *105*
Buell, E. N., 353, *359*
Bülow, K. von, *383*, 385, *388*
Buettner, K., 192, *196*
Bullard, E. C., 336, *340*, *414*
Bunch, T. E., *51*, 82, 98, *105*, *360*
Burbidge, E. M., *410*
Burbidge, G. R., *410*
Burder, G. F., 260, *266*
Burns, G. J., *291*
Butcher, D., *47*

Cailleux, A., 52, *65*, 250, 251, *263*
Campbell, W. W., 367, *370*
Carey, S. W., 245, *247*
Carpenter, J., 351, *359*
Cassidy, W. A., 29, 30, *49*
Cassini, J. D., 201
Castelli, J. B., 275, *285*
Celis, R. de, 25
Chamberlain, J. W., *348*
Chamberlain, T. C., 303
Chandrasekhar, S., 253, *265*
Chant, C. A., *291*
Chao, E. C. T., 18, *48*, *50*, 79, 96, *102*, *103*, *292*
Chapman, D. R., 288, 289, *292*
Charters, A. C., 131, *151*, *313*
Chekirda, A. T., 252, 257, 259, 260, *264*, *266*, 276, *285*, *360*
Chubb, F. W., 22
Clark, H., *411*
Clark, J. F., 83, *104*
Clark, S. P., 398, *410*
Clegg, J. A., *414*
Clement, G. R., 345, *348*
Clopine, M. S. (Mrs.), vii
Coates, R. J., 270, *283*
Coes, L., Jr., *48*
Cohen, A. J., *51*, 82, 98, *105*, 245, *247*, 288, 289, *292*, *358*, *360*, *363*, *370*
Conder, 288, *291*
Condon, R., *414*
Conley, J. M., 277, *285*
Conway, R. G., 271, *283*, *284*
Cooke, H. B. S., vii, 56
Cooke, S. R. B., 392, *394*
Cooper, C., *414*
Copeland, J., 275, *285*
Costain, C. H., 343, 346, *348*
Craig, K. J., 281, *285*

Daly, J. W., *48*
Daly, R. A., 94, *104*, *384*
Darney, M., 325, *332*, 351, *359*
Darwin, G. H., 198, 199, 206, *210*, *211*
Davidson, M., *291*, 325, 326, *332*
Davis, E. G., 303, *313*
DeBoer, K., *313*
Debye, J. W., 403
DeCarli, P. S., 17, 18, *48*
Delporte, E., 67
Denisse, J. F., 271
Denning, W. F., *291*
DeSitter, L. U., 195, *196*

DeSitter, W., 197, *210*
Detre, L., *414*
Dicke, R. H., 271, *284*
Didion, I., 318, 326, 327, 328, *332*
Dietz, R. S., vii, viii, 8, *48*, *51*, *65*, 73, 74, 77, 89, 93, 94, 96, 97, 98, 99, *102*, *104*, *105*, 358, *359*
Diggelen, J. van, 253, 254, *265*, 353, *359*, 379, *384*
Dollfus, A., 250, 251, *263*, 343, 346, *348*
Donn, B., *267*, 397, *410*
Dubief, 46
Dubin, M., 299, *312*
Dubois, J., 261, *266*, *267*
DuFresne, E. R., 214, *246*, 335, *340*
Dugan, R. S., 254, *265*
Dunbar, C. O., *102*
Dwornik, E. J., *292*
Dzhapiashvili, V. P., 251, *263*

Eaton, J. P., 381, *384*
Eckels, A., 244, *247*
Edwards, W. F., 293, *311*, 344, 345, *348*
Elsasser, H., 346, *348*
Elsasser, W. M., 244, *247*, 413, *414*
Elsmore, B., 343, 346, *348*
Elvey, C. T., 63, *66*
Emrich, G. H., 100, *105*
Engel, K. H., *247*
England, J. L., 18, *48*, 403, *411*
Epstein, P. S., 272, *284*
Escher, B. G., 114, *126*
Evans, G., 19, *49*, 123, *127*, 148, *152*
Evans, J. V., *285*

Fahey, J. J., *50*
Fahrig, W. F., 59, *66*
Fauth, P., 268, *283*, 351, *359*, 415, 416, *418*
Fedorets, V. A., 253, 258, *264*, *266*
Fedoseyev, L. I., 271, *284*
Fedynski, V. V., *50*
Fenner, C., 288, *291*
Ferguson, G. M., *66*, *103*
Ferioli, C. P., 275, *285*
Fermi, E., 154, 171
Fesenkov, V. G., 38, 39, *50*, 249, 251, 253, *262*, *263*, *265*, 341, 342, *348*
Fielder, G., vii, xvii, 256, *265*, 306, 307, *313*, 317, 318, 325, 326, 331, *332*, *348*, 353, 355, *359*, *360*, 367, 368, *370*, 371, 372, 373, 374, 376, 377, 379, *383*, *384*, 385, 386, 387, *388*, *389*, 409, *411*
Figgins, J. D., *49*
Firsoff, V. A., 114, *126*, 257, 259, 260, *266*, *306*, *313*, 331, *332*, 345, 346, *348*, *383*, 385, 386, *388*
Fish, R. A., 304, *313*, 397, 398, *410*
Fisher, C., 11, *48*, *49*, *185*
Fisher, W., *291*
Flint, R. F., *340*
Foote, A. E., *48*
Ford, G. R., vii
Forsythe, W. E., 441
Fotheringham, J. K., 197, *210*
Fowler, W. A., *410*
Fraas, E., *65*, 76, *102*
Franz, J., 2, 204, *211*, 213, 214, 215, 216, 217, 220, 221, 224, 226, 229, 230, 232, 234, *246*, 450-63
Fremlin, J. H., 277, *285*
Frenkel, J. C., *414*
Fricker, S. J., *285*

Index

Friedman, I., *292*
Frye, R., *151*, 441

Gane, P. G., *340*
Garstang, R. H., 270, *283*
Gault, D. E., 75, *105*
Gaydon, A. G., 417, *419*
Gehrels, T., 249, 251, *262, 263,* 267
Geoffrion, A. R., *283*
Gèze, B. (Mrs.), 52, *65*
Giamboni, L. A., 357, 358, *360*
Gibson, J. E., 270, 271, 274, 275, 276, *283, 284, 285*
Gifford, A. C., 9, *47*
Gilbert, G. K., 303, 310, *312,* 325, *332,* 392, *394, 395, 409*
Gilvarry, J. J., vii, viii, *47, 50,* 154, 155, 160, 167, 178, 179, 180, 184, *184, 185,* 273, 274, *284,* 302, 303, *313,* 347, *349, 411*
Glasstone, S., 41, *50, 127*
Goeckermann, R. H., *127,* 148, *152*
Goel, P. S., 29, *48*
Götz, F. W. P., 252, 256, *264,* 265
Gold, T., *xix,* 123, *127,* 154, *184,* 273, 274, 281, *286, 287, 291,* 294, 295, 296, 297, 298, 299, 300, 301, 302, 309, *312,* 333, *340,* 424
Goles, G. G., 304, *313,* 397, 398, *410*
Goodacre, W., xix
Gottschick, F., 77, *103*
Graham, J. W., *414*
Grant, K., 26
Gratton, L. C., *340*
Grebenkemper, C. J., 271, *284*
Green, J., 114, *126*
Green, P. E., Jr., *285*
Gregory, H. F., *384*
Guppy, D. J., 30, *49*
Gutschick, R., 97, *105*

Habakov, A. V., *383,* 385, *388*
Hacker, S. G., *360*
Hackman, R. J., *104,* 109, 111, *112,* 296, 297, 309, *312*
Hager, D., *47*
Hales, A. L., *340*
Hall, A. L., 93, *104*
Hamaguchi, H., *410*
Hammer, W., 58, *66*
Hargraves, R., 94
Hargreaves, J. K., *285*
Harper, A. F. A., 274, 277, *284*
Harris, I., *348*
Harrison, J. M., 23, 24, *49*
Hart, H. B., viii, 95, 96
Hartwig, E., 203
Haskell, N. A., 195, *196,* 243, *247,* 423
Hastings, J. M., 293, *311*
Hawkins, G. S., 299, *312*
Hayn, F., 2, 203, *211,* 213, *246*
Heck, N. H., *313*
Hédervári, P., vii, 241, 242, *247, 411*
Heiskanen, W. A., 198, *210,* 244, *247*
Hendriks, H. E., 101, *105,* 370, *370*
Hendrix, W. C., *414*
Herring, A. K., *384,* 391, 392, *394*
Herring, J. R., 344, 345, *348*
Hess, H. H., 392, 394
Hewitt, A. W., 116

Hey, J. S., 45, *285,* 288, *292*
Heyl, A. V., Jr., 96
Hill, J. E., vii, *47, 50,* 155, 160, 167, 178, 179, 180, 184, *184,* 185
Hinds, H., 101, *105*
Hodgson, J. H., *340*
Hoffleit, D., *51*
Hollander, J. M., 398, *410*
Hooke, R., 392, *394*
Hopmann, J., 2, 202, *211,* 216, 219, 221, 230, *246*
Hospers, J., *414*
Hossfield, P. S., 64
Hoyle, F., *410*
Hughes, V. A., *285*
Hughes, V. H., 100, *105*
Huntley, H. E., *348*

Ilsley, R., 118, *127*
Ingalls, R. P., *285*
Innes, M. J. S., 58, 59, 61, *66,* 81, 83, 85, 86, 87, *104,* 148, *152,* 178, 179, *185*
Izsak, I. G., 245, *247*

Jackson, G. D., *66*
Jacobs, J. A., *410*
Jaeger, J. C., 269, 270, 271, 272, 274, 275, 277, *283, 284*
Jakosky, J. J., 14, *48*
Jakowkin, A., 203, *211,* 216, *246*
Jamieson, J. C., 17, 18, *48*
Janssen, C. L., 53, *65*
Jarrett, D. E., 440
Jastrow, R., *348*
Jeans, J. H., *349*
Jefferson, T., 8
Jeffreys, H., 197, 198, 199, 200, 201, 203, 207, 208, 209, *210, 211,* 239, 240, 245, 431, *432, 474*
Johnson, E. A., *414*
Johnson, G. W., 160, 163, 170, 179, 183, *185,* 440
Johnston, R., *360*
Joksch, H. C., 242, *247*
Jones, H. S., 197, *210*
Jones, J. E., *349*
Jordan, P. (Mrs.), viii
Junner, N. R., 57, *65*

Kajdanovsky, N. L., 275, *284*
Kalkun, J., 31
Kannuluik, W. G., 272, *284*
Karman, T. von, 329
Karpoff, R., 46, *51*
Katcoff, S., 293, *311*
Katz, A. H., *389*
Kaula, W. M., 245
Kawai, N., *414*
Keenan, P., 63
Kennedy, G. C., 17, *51*
Kepler, J., 21, 346
Khabakov, A. V., 259
Khaikin, S. E., 275, *284*
Khan, M. A. R., *348*
King, Capt. G. E., 441
King, P. B., 92, *104*
King-Hele, D. G., 245, *247*
Klavins, H. D., 44
Klein, H. J., 377, *383, 384*
Knetsch, G., 64, *66*

Knopf, A., *340*
Koenig, G. A., 16
Kohman, T. P., 29, *48*, 287, *291*
Komissarov, O. D., *312*
Korner, M., *283*
Kosyrev, N. A., 261, 262, *267*, 303, *313*, 416, 417, 418, *419*
Kozai, Y., 245, *247*
Koziel, K., 203, 204, *211*
Kranck, E. H., *66*
Kranz, W., 77, 78, *103*
Kraus, E., *49*, *151*
Krinov, E. L., 35, 37, 38, 39, 40, 43, *50*
Kuiper, G. P., vii, xvii, xix, 114, *126*, *127*, 204, 205, 206, 207, 209, *211*, 234, *266*, 295, 310, *311*, *312*, 323, 325, *332*, *340*, 344, *348*, 349, 354, *375*, 377, *384*, 397, *410*, *411*, *414*, 417, *419*, 424
Kulik, L. A., 35, 36, 37, 38, 40, *50*

LaGow, H. E., 299, *312*
Lagrange, J. L., 201
Lahiri, B. N., 402, *411*
Lampson, C. W., 116, 117, *127*, 130, 131, 133, 142, *151*, 163, 165, 183, *185*
Landau, A., *66*, *103*
Landerer, A. J., 249, *263*
Landolt, *410*
Lange, I., *414*
Langley, S. P., 268, *282*, *283*
LaPaz, J., *51*
LaPaz, L., 22, 24, 37, *49*, *50*, *51*, 63, 64, *66*, 288, *291*
Laplace, P. S., 201, 203
Laporte, L., *414*
Leadabrand, R. L., 285
Learner, R. C. M., 417, *419*
Lecomte, P., 403, *411*
Leland, Dean S., vii, 221
Lenham, A. P., 352, *359*
Leonard, F. C., 22, 24, *49*, *51*
LeRoux, E., 271
Lettau, H., 272, *284*
Levin, B. Yu., 39, *50*
Liberty, B. A., 83, *104*
Licht, A. L., 344, 345, *348*
Lindner, J., *414*
Link, F., 260, 261, *266*
Little, A. G., 346, *348*
Littler, J., *50*, *292*
Löffler, R., *103*
Lohrmann, W. G., 220, *246*, 415
Lombard, 11
Longwell, C. R., *340*
Lovell, A. C. B., 279, *286*
Low, A. P., *66*
Lowe, E. J., 260, *266*
Lowman, P. D., *292*
Luba, A., *49*, *151*, *186*
Lubimova, H. A., 399, 400, *410*
Luce, J. von, 31
Lugn, R. V., 75, *105*
Lyot, B., 249, 250, 251, *263*, 343, *348*

McCutcheon, T. E., *105*
MacDonald, G. J. F., viii, 198, 199, *210*, 278, *285*, 309, 313, 375, *383*, 399, 400, 401, 402, 403, 404, 405, 408, 409, *410*, *411*, 427

MacDonald, T. L., 352, *359*
McKnight, E. T., 100, *105*
MacLaren, M., 57, *65*
McMath, R. R., 145, *151*, 301, *313*
Macphail, M. S., 178
Madigan, C. T., *49*
Madsen, B. M., *48*, *102*
Mädler, J. H., 220, *246*, 415
Mainka, C., 216, *246*
Malott, C. A., *105*
Maree, B. D., 94, *104*
Markov, A. V., 250, 252, 255, *263*, *264*, *265*, 267
Marshall, R. K., 114, *126*, *360*, 392, *394*
Marshall, R. R., *411*
Martin, L. H., 272, *284*
Mason, W. C., *285*
Matheson, R. S., 30, *49*
Mebane, J. A., *291*
Medlicott, H. B., *66*
Meen, V. B., 22, 23, 24, *49*, 58
Meinesz, F. A. V., 244, *247*
Meithe, A., 256, 257, *266*
Melin, M., *414*
Merewether, Col. A., 58
Merrill, G. P., 11, 17, 36, *48*, *50*
Merson, R. H., 245, *247*
Meyer, R., *49*, *151*
Mezger, P. G., 275, *285*
Michailowski, 203
Middlehurst, B. M., *266*
Millman, P. M., 23, 24, *49*, 83, *104*, *151*
Milne, E. A., *349*
Minnaert, M., 253, 260, *266*
Minnett, H. C., 270, 271, 277, *283*
Mitchell, F. H., 275, *284*
Miyamoto, S., 114, *126*
Mohorovičić, A., 242, 334, 392
Molengraaff, G. A. F., 93, *104*
Monnig, O. E., 19, *49*
Monod, T., 45
Moore, P., xix, 114, *125*, 237, *247*, 314, 324, *332*, 361, 367, 369, *370*, 382, *383*, 390, 392, *393*, *394*, 418, 444, 445, 446
Moulton, F. R., 14, *48*, 178, *185*, 303, *312*
Muncey, R. W., 277, *285*
Munk, W., 198, 199, *210*
Murata, K. J., 381, *384*
Murphy, T., *414*
Murray, W. A. S., *285*

Nasmyth, J., 351, *359*
Naumann, H., 203, *211*
Nazarova, T. N., 299, *312*
Neison, E. (pseud.); see Neville, E. N.
Nel, L. T., *104*
Neugodov, L. N., *312*
Neville, E. N., *48*, 220, *246*
Newbold, *66*
Newton, 245
Nicholson, S. B., 255, *265*, 269, 270, 271, 272, 274, 277, *283*
Nininger, H. H., vii, 10, 12, 13, 14, 15, 16, 17, 19, 24, 29, 47, *48*, *49*, *152*, 170, 173, *185*, 287, *291*
Nordyke, M. D., *127*, 148, *152*
Noskova, R. I., 271, *284*

Index

Öpik, E. J., 9, 47, 123, 124, *127*, 158, 159, 160, 167, 169, 170, 171, 174, 176, 178, 179, *184*, 253, 265, 282, *283*, 294, *312*, 345, 346, *348*, *349*
O'Keefe, J. A., 244, 245, *247*, 287, 288, *291*, 353, *359*
Olsen, Lt. F., 118, 119, 120
Olte, A., 287, *291*

Parenago, P., 249, *262*, *265*
Parsons, W. (third Earl of Rosse), 249, *263*, 268, *282*
Partridge, W. S., *313*
Pawsey, J. L., 271, *284*
Peal, S. E., 268, *283*
Pease, F. G., 351
Peirels, R. E., 400, *411*
Perry, Adm. O. H., 7
Peters, H., *285*
Peterson, A. H., *332*
Petrie, R. M., 145, *151*, 301, *312*
Petrushevskiy, F. F., 249, *263*
Pettengill, G. H., 278, 279, *285*
Petterson, H., 298, 299, *312*
Pettit, E., 255, *265*, 269, 270, 271, 272, 273, 274, 277, *283*, *284*
Philby, H. St. J., 34, *50*, 290, *292*
Phillips, T. E. R., xix
Picard, E., 117
Pickering, W. H., 213, *246*, *291*, 347, *349*, 352, 353, 355, *359*, *360*, 371, 377, *383*, 390, 392, *393*, *394*
Piddington, J. H., 270, 271, 277, *283*
Pinson, W. H., Jr., 290, 291, *292*
Platt, J. R., 262, *267*
Plummer, H. C., 205, *210*, *211*
Poisson, S. P., 201
Poldervaart, A., *349*
Poloskov, S. M., *312*
Poulter, T. C., 304, *313*
Price, A. T., 402, *411*
Priddy, R. R., 100, *105*
Pritchard, 220
Pullen, M. W., 96, 97, *105*

Rambaut, A. A., *418*
Reck, H., 78, *103*
Reed, G. W., *410*
Reid, A. M., 98, *105*, 289, *292*, *360*
Reinmuth, K., 67
Reinvaldt, L., 31, *49*, 148, *151*, 183, *186*
Revelle, R., 337
Reynolds, J. H., 288, 293, *311*
Rich, J. L., 95
Rikitake, T., 402, *411*
Rinehart, J. S., 12, 14, 15, *48*, 63, *66*, 173, 178, 180, *185*
Rishbeth, H., 346, *348*
Ritter, H., 2, 216, 218, 233, 234, *246*
Rittmann, A., 78, *103*
Riyvés, B. G., 254, 255, *265*
Robinson, C. S., *127*
Roche, E., 198, 200
Rochester, M. G., 244, *247*
Rogers, A. F., 17, *48*
Rohleder, H. P. T., 55, 56, 57, *65*, 77, *103*
Roman, N. G., 281, *285*
Rose, E. R., *66*

Rosenberg, H., 252, 256, *264*, *265*
Rosse, Earl of; *see* Parsons
Rostoker, N., 178, *186*
Rotschi, H., 298, 299, *312*
Rottenberg, J. A., 58, 59, 61, *66*, 81, 83, 85, 87, *104*, 152
Rougier, G., *264*
Rubey, W. W., *349*
Rudnjer, D. D., *50*
Runcorn, S. K., 243, 244, *247*, 431
Rusakov, L. Z., *312*
Russell, H. N., 254, *264*, *265*
Russell, J., 415, *418*
Rymer, H., vii, 232

Saari, J. M., 276, *285*
Safford, J. M., 89, *104*
Salisbury, J. W., 393, *394*
Salomon, 329
Salomonovich, A. E., 271, 275, *284*, *285*
Sangster, R. L., 64, *66*
Saunder, S. A., 2, 213, 214, 215, 220, 221, 226, 229, 230, 234, *246*, *247*, 450–63
Savage, G. (Mrs.), vii
Sawyer, H. E., 145, *151*, 301, *312*
Schaeffer, O. A., 293, *311*
Scheiner, I., 256, 260, *265*
Schlüter, H., 203, 220
Schmidt, J. F. J., 220, *246*, 371, 382, *383*, 384, 415, *418*
Schmidt, R. A., 289, *292*
Schmidt, W., 58, *66*
Schoch, K., 197, *210*
Schoenberg, E., 353, *359*
Schönrock, A. M., 38, *50*
Schröter, J. H., 147, 148, *152*
Schrutka-Rechtenstamm, G., vii, 2, 216, 219, 221, 224, 225, 229, 230, 234, *246*, 450–63
Schuchert, C., *102*
Schweizer, F., *414*
Schwinner, R., 354, *360*
Scott, W. E., *414*
Seaborg, G. T., 398, *410*
Secchi, A., 249, *262*
See, T. J. J., 295, *312*, 333, *340*
Seeger, C. L., 271, *283*, *284*
Seegert, B., 256, 257, *266*
Sellards, E. H., 19, *49*, 123, *127*, 148, *152*
Senftle, F. E., *292*
Senior, T. B. A., 279, 281, *285*, 287, *291*
Shaler, N. S., 392, *394*
Shane, C. D., vii, 220
Sharanov, V. V., 252, 256, *264*, *265*
Shelton, A. V., *127*, 148, *152*
Shepard, E. M., *104*
Shoemaker, E. M., xvii, 10, 11, 12, 17, 18, 24, *47*, *48*, 55, *65*, 75, 79, 93, 100, *102*, *103*, *104*, *105*, 107, 109, 110, 111, *112*, 115, 119, 120, 124, 125, *127*, 146, 148, 150, *151*, 154, 160, 167, 170, 171, 172, 173, 174, 178, 179, *184*, *185*, 294, 296, 297, 309, *312*, 355, 356, *360*, *388*, 440
Shorthill, R. W., 276, 277, *285*
Shrock, R. R., 96, *105*
Siegel, K. M., 279, 280, 281, *285*, *286*, 287, *291*
Simon, F., 403
Simpson, E. S., 29, *49*

Singer, S. F., 346, *349*
Sinton, W. M., 250, *263*, 269, 271, 274, 276, 277, 278, *283, 284, 285*
Smith, C., 45, 288, *292*
Smith, C. F. O., *383*
Smoluchowski, M., 272, 273, *284*
Spencer, L. J., 25, 34, 45, *49*, 56, *65*, 288, *291, 292*
Spitzer, L., 346, 347, *349*
Spurr, J. E., 114, *126*, 353, *360*, 385, 387, *388*, 392, *394*
Squires, R. K., 244, *247*
Stair, R., *360*
Stanley, J., 328, *332*
Stanyukovich, K. P., *50*
Staude, N., 249, *262*
Stearns, R., 89
Steavenson, W. H., xix, 325, *332*
Sterne, T. E., 328
Stevens, R., 280, *286*
Stewart, J. Q., 254, *265*, 353, *359, 360*
Stishov, S. M., 18
Stone, M. L., *285*
Stoney, G. J., 346, *349*
Strassl, H., 275, *285*
Stratton, F. J. M., 203, *211*
Strominger, D., 398, *410*
Strong, H. M., 403, *411*
Struve, O., 341, *348*
Stubbs, P. H. S., *414*
Stutzer, H., 79, *103*
Suess, F. E., 58, *66*
Suess, H. E., *349, 410*
Sykes, J. B., *389*
Sytinskaya, N. N., 256, *264, 265*, 281, *286*

Tarr, W. A., 95, *104*
Tatel, H. E., *340*
Taylor, G. I., 198, *210*
Teyfel, V. G., 260, *266*
Thomas, L. H., 154, 171
Thorpe, A., *292*
Tilghman, B. C., 16
Tisserand, F., *210, 211*
Tomkins, H. G., 351, 353, *359*, 392, *394*
Toporets, A. S., 250, *263*
Torreson, O. W., *414*
Trexler, J. H., 278, 279, *285*
Troitsky, V. S., 270, 271, *283, 284*
Tschunko, H. F. A., 253, *265*
Turkevich, A., *410*
Turusbekov, M. T., 275, *284*
Tuttle, O. F., 392, *394*
Tuve, M. A., *340*
Tyler, W. C., 275, *285*

Umov, N. A., 250, *263*
Urey, H. C., vii, xvii, xviii, *xix*, 17, *48*, 114, *126*, 167, *185*, 206, *211*, 243, 244, *247, 267*, 281, *286*, 288, *292*, 294, 295, 303, 304, 309, 310, 311, *312, 313*, 322, 325, 327, 328, 330, *332*, 357, 358, *360*, 377, *383, 384*, 395, 396, 397, 405, 406, *409, 410, 411*, 417, 418, *419*, 424, 427

Vand, 358, *360*
Van Fleet, H. B., *313*
Varigny, H. de, *47*
Varsavsky, C. M., 287, *291*
Vaucouleurs, G. de, *348*
Very, F. W., 269, *283*
Vesnesenski, A. V. (*or* Vosnessenski), 36, *50*
Vestine, E. H., 344, 345, *348*, 413, *414*
Victor, W. K., 280, *286*
Vogt, J. H. L., 406, *411*

Wagner, P. A., 55, *65*
Wagner, R., *414*
Warner, B., 301, *312*, 379, *384, 388*
Watson, F. G., 25, *47, 49*, 68, *102, 262*
Watts, C. B., vii, 204, *211*, 219, 221, 234, 235, 236, 237, 238, 239, 240, 242, 304, *313*
Webb, H. D., *285*
Webb, T. W., *418*
Wegener, A., *49, 151*
Weimer, Th., 203, 204, *211*, 216, *246*
Werner, E., 77, 79, *103*
Wesselink, A. J., 269, 272, 273, *283, 284*
Westerhout, G., 271, *283, 284*
Wetherill, G., 398, *410*
Whipple, F. G., 37, 38, 43, *50*
Whipple, F. L., 281, 282, *286*, 299, *312*
Whitaker, E. A., 234
White, W. C., 180, 185
Whitehurst, R. N., 275, *284*
Whitfield, G. R., 343, 346, *348*
Whitney, H., *285*
Wiechert, E., 202, *210*
Wilkins, H. P., xix, 114, *125*, 237, *246*, 314, 324, *332*, 361, 369, *370, 383*, 390, *393, 418*, 444, 445, 446
Williams, H., 77, 79, *103*
Willmore, P. L., 83, *104, 340*
Wilsing, I., 256, 260, *265*
Wilson, C. H., *46*
Wilson, C. W., Jr., 79, 88, 89, 91, *103, 104, 105, 152*
Wilson, J. T., 335, *340*, 407, *411*
Winslow, A., *104*
Wirtz, C., 216, *246*, 252, 253, *264*
Wislizenus, W. F., 252, *264*
Witt, G., 67
Wright, F. E., 250, 256, *263, 360*
Wright, W. H., *266*
Wylie, C. C., 160, 178, *185*

Yaplee, B. S., 281, *285*
Yezerskiy, V. I., 258, *266*
Yoder, H. S., *384*, 403, *411*
Young, J., 141, 306, 307, *313*, 368, *370*

Zaharov, S. H., 44, 45, *51*
Zahringer, 288
Zelinskaya, M. R., 270, 271, *283, 284*
Zöllner, J. K. F., 248, *262*, 282

SUBJECT INDEX

Acceleration, secular
  moon, 197, 198
  sun, 197
Age
  of cryptovolcanic structures, 76, 79, 88, 89, 93, 95, 98–100, 102
  of earth, 293
  of earth's crust, 293
  of lunar craters, 300–309, 317, 321, 421, 422
  of maria, 111, 112, 295–97, 305–10, 316, 319–21, 331, 379
  of meteorites, 293
  of moon, 293–95
  of planetesimals, 294
  of rilles, 301
  of tektites, 288–91
  of terrestrial meteorite craters, 7, 18, 19, 24, 28–30, 35, 43, 46, 47, 52, 53, 60, 69, 82, 83
  of wrinkle ridges, 380, 381
Albedos, of terrestrial materials, 251, 255, 260, 262, 281
Angle of repose, 248
Asteroids, 44, 67, 68, 95, 111, 184, 249, 250, 296
Astroblemes; see Cryptovolcanic structures

Brightness, variations of, with phase angle
  of asteroids, 249, 250
  of moon; see Moon, albedo
  of planets, 249, 250

C_2 emission bands, 302
Cassini's laws, 201
Central peaks of impact craters, 4, 11, 31, 32, 59, 76, 77, 79, 88–91, 93, 95, 107, 108, 114, 115, 118–20, 122, 123, 165, 190, 192–94, 361–70, 418, 421, 425
Chain craters, 317, 364, 371, 376–79, 426
Chattanooga Sea, 88
Coesite, 8, 17, 18, 35, 52, 57, 74, 79, 82, 98, 99, 289
Contour maps of moon, 213, 215–19, 232–35
  Air Force, 234
  Army Map Service, 234, 235
  Baldwin, 232–44, 301, 335, 422

Franz, 217, 234
Ritter, 218, 233, 234
Schrutka-Rechtenstamm, Hopmann, 219, 234
Watts, 235–38, 240
Cosmic rays, 192, 287
Criteria for identification of meteoritic craters, 8, 74, 75
  anticline, ring, 6, 8, 19, 20, 45, 56, 57, 59, 71–73, 75, 89, 90, 92, 93, 95, 96, 98–101, 107–9, 120, 125, 317–20, 322, 331
  brecciated layers, 8, 11, 14, 17, 19, 27, 29, 31, 53–55, 57, 58, 67–69, 77–80, 82–89, 92–96, 98, 102, 107, 124, 148–52, 170, 171, 173, 178, 362, 418, 429
  coesite, 8, 17, 18, 35, 52, 57, 74, 79, 82, 98, 99, 289
  distortion of rock layers, 10, 12, 19, 20, 23, 29–31, 45, 55, 56, 59, 67, 71, 78, 86–88, 107, 108, 123–25, 140–51, 172, 173
  impactite, 16
  lechatelierite, 16, 17, 27
  meteoritic diamonds, 16, 17
  meteoritic metallic spheroids, 12–15, 34, 173, 290
  rebound dome, 6, 11, 55, 71–73, 75–77, 79, 88–96, 98–102, 107, 108, 118–20, 122, 140, 150, 151, 363, 364, 367, 418, 425, 426
  ridges around craters, 23, 26, 27, 56, 57, 62, 71, 75, 87, 89, 108, 109, 317–20, 322, 331, 332
  rock flour, 8, 15, 19, 20, 25, 27, 31, 69, 80, 89, 96, 98
  shatter cones, 18, 52, 57, 74, 75, 77, 79, 89, 91–94, 96–99, 101, 102, 108
  silica glass, 16, 18, 25, 27, 31, 34, 35, 58, 68, 290
  stishovite, 8, 18, 74
  suevite, 57, 78, 79
  syncline, ring, 6, 8, 57, 71–73, 75, 87–90, 92, 93, 95, 96, 98–102, 107–9, 120–22, 128–51, 315, 318–20, 322, 331
Cryptovolcanic structures, 59, 73–76, 88–102, 106, 107, 109, 111, 331
  Crooked Creek structure, 74, 96, 100–102, 107, 370
  Decaturville structure, 73, 95, 107
  Des Plaines structure, 100

Cryptovolcanic structures—*Continued*
  Flynn Creek structure, 88, 89, 91, 107
  frequency of, 109–11
  Howell structure, 98, 107
  Jeptha Knob, 98, 99, 107
  Kentland disturbance, 74, 75, 96, 97
    outlined by radio-wave ground transmission, 96, 97
  Kilmichael structure, 100, 107
  Rieskessel, 57, 75–79, 91, 107, 288, 354
  Serpent Mound, 99, 107, 362
  Sierra Madera Dome, 92–94, 107, 151
  Steinheim Basin, 57, 75–77, 79, 91, 99, 107
    Klosterberg, 76, 77
  Upheaval Dome, 100, 107
  Vredefort structure, 75, 93, 94, 111, 352, 375
  Wells Creek Basin, 89–93, 98, 107, 109, 150
    associated craters 89
      Austin Basin, 90, 91
      Cave Spring Hollow Basin, 90, 91
      Indian Mound Basin, 90, 91

Diamonds, meteoritic, 16, 17
Domes, 3, 382, 390–93, 415, 427
  central craters, 391
  color of, 392
  in Darwin, 391, 392
  locations, 391, 393
  on rille, 392
  Rümker, 391, 392

Earth
  angular momentum, rotation, 198–200, 209
  bodily tides, 198, 206–8
  density variations, 202
  dissipation of rotational energy, 198, 199
  friction in oceanic tides, 198–201, 206–10
  Himalaya Mountains, 195
  history of, 198–200, 203, 206, 207, 209, 210
  lava sheets on, 334, 335, 337, 338, 429–31
  length of day on, 197, 198
  meteoritic dust on, 298, 299
  seismic-wave velocities, 334
  shape of, 201, 205, 244, 245
Erosion, 93, 94, 107, 108, 114, 192–94, 297, 298, 333, 421
Explosions
  chemical, 75, 114–20, 128–51, 153–84, 436, 437, 438, 439, 440
  nuclear, 41–43, 75, 115, 124, 125, 133, 148, 149, 439, 440
Explosive craters
  Burton-on-Trent, 123, 357
  compression effects, 119–24, 140–51, 154
  effects of scaled depths of burst, 116–20, 125, 128–51, 153–84, 447–49
  mechanics of, 41, 42, 114–25, 128–51, 153–84, 314, 354
  scaling laws, 109, 110, 128–51, 153–84
Explosives, heat of, 441

Faults, 372–76, 382, 383, 385, 386, 426
  ring faults, 56–59, 77–79, 87, 94, 99, 101, 375, 376, 426
  Straight Wall, 3, 194, 248, 324, 382, 383
  types of, 372, 382, 383
Fluorescence on moon, 260–62, 423

Geologic periods
  Archeozoic, 67, 200
  Cambrian, 67, 95, 101, 200
  Carboniferous, 95
  Cretaceous, 20, 46, 67, 89, 93
  Devonian, 88–91, 97
  Eocene, 46
  Jurassic, 76, 77, 91, 100
  Mesozoic, 78
  Miocene, 76–79, 91
  Mississippian, 88, 90, 99
  Ordovician, 26, 45, 83, 88, 90, 95, 98, 101, 102
  Paleozoic, 73, 80, 82, 83, 96
  Pennsylvanian, 97, 101, 102
  Permian, 92, 93
  Pleistocene, 18, 30, 98, 99, 120, 195
  Pliocene, 46, 52
  pre-Cambrian, 30, 57, 82, 83, 95
  Quaternary, 139
  Silurian, 31, 90, 97, 99
  Tertiary, 73
  Triassic, 77
Gravity anomalies, 80, 81, 83, 85–88, 94
Grid system, 372, 385–88, 409, 426, 427

Height determinations, 213–35, 451–63
Hypsographic curve of moon, 241, 242
Hypsometric curve of moon, 241, 242

Impactite, 16
Internal origin of lunar craters, 376, 377
Isostatic adjustments, 3, 94, 193–96, 207, 210, 240, 245, 302, 304, 334, 335, 376, 381, 383, 421–23, 425, 426, 431
  equipotential surface, 201–4, 206, 207, 243, 423
  in Fenno-Scandia, 195
  hydrostatic equilibrium, 203, 204, 395, 423, 431
  viscosity of earth's asthenosphere, 195

Kepler's third law, 21, 22

Laccoliths, 57, 63, 78, 392, 393, 427
  Cratère de Semsiyât, 63
Lambert's law, 253, 254, 278, 280, 424
Lava flows
  on earth, 334, 335, 337, 338, 429–31
    Columbia River Plateau, 337, 338
    Deccan Trap, 338
    Oregon Plateau, 338, 381
    Snake River Plateau, 337, 338
  on moon, 115, 333–39, 428
    ages of, 111, 112, 295–97, 305–10, 316, 319–21, 339, 379, 386, 406
    origin of, 335–39, 406, 424, 426, 428–31
    withdrawal of, 381, 382
Lavas, viscosities of, 338, 430
Lechatelierite, 16, 17, 27
Lommel-Seeliger law, 253, 278
Lunar craters
  age effects on crater forms, 189–91, 421–23, 431
  ages of, 300–309, 317, 321, 421, 422
  ancient craters, 194, 195, 331
  central craterlets, 96, 101, 114, 367–70, 425, 426
  dimensions of, 443–46
  distributions of, 188, 189, 294, 296
    compared with shotgun patterns, 189
  frequency of, 109–12

Index 483

internal origin of, 339, 376, 377
names of
  Abulfeda, 378, 379
  Agatharchides, 307
  Agrippa, 320
  Albategnius, 378
  Aliacensis, 195
  Almanon, 194, 378
  Alpetragius, 194, 301, 361
  Alphonsus, 253, 303, 323, 325, 368, 416, 417, 418, 431
    gas emission, 416–18, 431
    ridge, 324
  Ancient Newton, 306
  Arago, 391
  Archimedes, 305, 307, 322
  Archimedes M, 305, 307
  Archytas, 323
  Aristarchus, 253, 255, 258, 261, 276, 277, 299, 321, 351
  Aristillus, 27, 109, 170, 177, 188, 305, 307, 331, 362
  Aristoteles, 255, 320, 369, 387
  Arzachel, 302, 387, 416
  Atlas, 320, 332, 369
  Bailly, 109, 237, 331, 390
  Beaumont, 306, 307
  Bessel, 170, 177
  Bode, 323
  Bohnenberger, 306, 307, 369
  Boscovich, 382
  Brenner, 316
  Bürg, 382
  Bullialdus, 109, 331, 355
  Bullialdus F, 170
  Cassini, 305, 307, 369
  Catherina, 314–17
  Cauchy, 383, 391
  Censorinus, 315
  Clavius, 170, 176, 177, 190, 298, 305, 332
  Cleomedes, 319
  Condorcet, 306
  Cook, 315
  Cooke, 306, 307
  Cooke B, 307
  Copernicus, 93, 111, 174, 255, 256, 258–61, 270, 276, 277, 299, 321, 332, 358, 369, 387, 390, 430
  Cyrillus, 316, 317, 387
  Daguerre, 307
  Daniell, 387
  Darwin, 391, 392
  Donati, 194
  Doppelmayer, 301, 306, 307
  Eimmart, 306
  Eratosthenes, 260, 277, 321, 369, 379, 381, 387, 390, 416
  Eudoxus, 320, 382
  Euler, 321
  Flammarion, 339
  Fra Mauro, 386
  Fracastorius, 301, 306, 307, 314
  Franklin, 320
  Furnerius, 351
  Gassendi, 306, 307
  Gemma Frisius, 195, 387
  Godin, 276
  Grimaldi, 250, 253, 260
  Helicon, 321
  Hellplain, 190, 380
  Hercules, 382
  Herodotus, 261
  Herschel, 324, 378
  Hippalus, 306, 307
  Hipparchus, 194, 320, 323, 421
  Hortensius, 391
  Hyginus, 375, 378
  Isidorus H, 307
  Janssen, 316, 387
  Julius Caesar, 386
  Kant, 315
  Kepler, 170, 177, 184, 258, 259, 299, 351–53, 357
  Kies, 391
  Lambert, 369
  Langrenus, 350, 430
  Le Monnier, 306, 307
  Le Monnier Z, 307
  Lee, 306, 307
  Letronne, 301
  Lick, 306, 307
  Linné, 415
  Littrow, 307
  Littrow A, 307
  Loewy, 307
  Macrobius, 256, 319, 320
  Mädler, 307
  Magelhaens, 315
  Maginus, 190
  Manilius, 323, 369, 375
  Mare Humorum C, 307
  Mare Humorum E, 307
  Mare Humorum H, 307
  Mason, 320
  Maurolycus, 369
  Maury, 320
  Menelaus, 392
  Mersenius, 307
  Mersenius D, 307
  Milichius, 391
  Moltke, 315
  Monge, 315
  Monge B, 315
  Moretus, 362, 365
  Neander, 315
  Olbers, 351–53
  Palisa, 302, 378
  Pallas, 323
  Piccolomini, 315
  Pitatus, 302
  Piton A, 170
  Plato, 257, 260, 261, 305, 306, 307, 321, 323, 379, 416, 418
  Plato D, 379
  Plinius, 370
  Polybius, 195, 315
  Pontanus, 387
  Posidonius, 301, 306, 307, 369, 387
  Proclus, 253, 256, 276, 319, 357
  Ptolemaeus, 93, 255, 302, 323–25, 339, 368, 378, 379, 386, 426
  Ptolemaeus G, 378
  Puiseux, 307
  Pythagoras, 170, 177

Lunar craters—*Continued*
  names of—*Continued*
    Regiomontanus, 261, 367
    Reichenbach A, 316
    Reichenbach B, 316
    Reinhold, 387
    Rhaeticus, 387
    Rheita, 316
    Riccius, 316
    Rosse, 358
    Sacrobosco, 387
    Schickard, 237, 250, 261, 298
    Seleucus, 351
    Snellius, 316
    Stadius, 378, 382, 390, 391, 430
    Stevinus, 352
    Tacitus, 378
    Thebit, 380, 382
    Theophilus, 109, 145, 146, 150, 151, 190, 192, 301, 314, 351, 362, 369, 387
      contour of, 190, 192, 301
    Timocharis, 367, 369, 370, 387
    Triesnecker, 339, 375, 376
    Tycho, 170, 177, 188, 253, 255, 257–59, 261, 276, 277, 299, 351, 354, 357, 358, 367
    Ukert, 323, 381
    Vitello, 306, 307
    Wallace, 305, 307
    Walter, 190
    Wargentin, 237, 302
    Werner, 195
    Yerkes, 306, 307
    Zagut, 369
  parameters for Classes 2–5 craters, 189–91, 421–23, 431
  secondary craters, 111
  volumes of, 193
  volumes of crater rims, 193
Lunar materials, suggested
  basalt, 242, 250, 256, 260, 428
  basic rock, 250, 429, 430
  C_2, 302
  diopside, 403
  diorite, 406
  dunite, 406
  fluorescent materials, 261, 262
  free radicals, 262, 423
  gabbro, 406
  granite, 250, 406, 407, 431
  gravel, 281
  iron quartzite, 259, 260
  meteoritic dust, 250, 262, 281, 282, 299, 302
  obsidian, 250
  olivine, 393, 407, 430
  peridotite, 392, 407
  porphyrite, 256
  pumice, 250
  quartz, 250, 251
  quartz porphyry, 259, 260
  sandstone, 250, 259
  serpentine, 393
  syenite, 406
  ultrabasic rock, 428, 430
  vitreous sands, 250, 281
  volcanic ash, 250, 251, 255, 256, 259, 260, 276
  water and water vapor, 303, 393

Magnetic field
  of earth, 401, 402, 412, 413
  of moon, 402, 412–14
Maria
  ages of, 111, 112, 295–97, 305–10, 316, 319–21, 339, 379, 386, 406
  dating of, 386, 406
  names of
    Lacus Mortis, 182
    Lacus Somniorum, 296
    Mare Australe, 233–35, 237
    Mare Crisium, 233, 257, 261, 296, 297, 306, 309, 315, 318–20, 380, 385, 425
    Mare Foecunditatis, 233–35, 257, 296, 315, 318
    Mare Frigoris, 236, 296, 322, 323
    Mare Humboldtianum, 236, 306, 315, 319, 380
    Mare Humorum, 233, 257, 258, 261, 296, 306, 309, 315, 316, 319, 375, 378, 380, 385, 425
    Mare Imbrium, 232–34, 257, 258, 261, 270, 295, 296, 305, 306, 309, 315, 318–25, 328–31, 339, 372–80, 382, 385–88, 425, 426
      origin of, 380
    Mare Marginis, 235
    Mare Nectaris, 27, 232, 233, 255, 296, 301, 306, 309, 314–19, 324, 330, 331, 375, 379, 380, 386, 387, 425
    Mare Nubium, 194, 233, 255, 257, 258, 261, 296, 324, 391
    Mare Orientalis, 237
    Mare Serenitatis, 232, 233, 255–58, 261, 296, 306, 309, 315, 319–22, 375, 376, 379, 380, 382, 387, 392, 415, 425
    Mare Smythii, 235
    Mare Tranquillitatis, 234, 256, 261, 296, 301, 320, 323, 376, 391
    Mare Vaporum, 272–74, 320, 322, 323, 339, 386
    Oceanus Procellarum, 233, 234, 236, 257, 258, 261, 322, 391
    Palus Epidemiarum, 296
    Palus Putredinis, 378
    Palus Somnii, 255
    Sinus Aestuum, 272, 322, 323, 339, 381
    Sinus Gay-Lussac, 305, 307
    Sinus Iridum, 305–7, 321, 331, 339, 380
      origin of, 380
    Sinus Medii, 255, 260, 261, 280, 322, 323
    Sinus Roris, 236, 322
  nature of, 3, 249, 250, 252, 253, 256, 301–3, 333, 338, 339, 407, 424
  origin of, 3, 301, 303, 305, 335–39, 406, 424, 426, 428–31
  origin of circular maria, 3, 305, 306, 314–32
  shelf areas at circular maria, 315, 318–22, 331
Materials found at meteoritic craters
  coesite, 8, 17, 18, 35, 52, 57, 74, 79, 82, 98, 99, 289
  enstatic granophyre, 94
  impactite, 16
  lechatelierite, 16, 17, 27
  meteoritic diamonds, 16, 17
  meteoritic metallic spheroids, 12–15, 34, 173, 290
  pseudo-tachylite, 94
  rock flour, 8, 15, 19, 20, 25, 27, 31, 69, 80, 89, 96, 98

Index

silica glass, 16, 18, 25, 27, 31, 34, 35, 58, 68, 290
stishovite, 8, 18, 74
suevite, 57, 78, 79
Meteorites
  age of, 293
  Ahnighito, 7
  chondrites, composition of, 406, 407
  Cyrillids, 288
  flattening on impact, 124, 166, 167, 172, 174, 175
  fragments, spalled off, 13, 14, 30, 46, 330, 331
  frequency of, 109–11
  Hoba West, 7
  iron oxide spherules, 15, 16, 29, 34
  kamacite, 28, 34
  mass required
    to produce specific craters, point-source model, 164–84
    to produce specific craters, surface-source model, 174–84, 314, 315
  meteoritic dust on earth, 298, 299
  meteors, 6, 7
  micrometeorites, 7, 192, 282, 300
  nickel-iron, 7, 8, 12–15, 18, 21, 24–31, 34, 37, 40, 44, 45, 57, 67–70, 164
  penetration into ground, 139, 140, 154, 166, 167, 170, 171, 174, 175, 179–83
  planetesimals, 294
  radioactivity in, 294, 296
  showers of, 288, 299
  size required
    to produce specific craters, point-source model, 164–84
    to produce specific craters, surface-source model, 174–84, 314, 315
  sluglets, 16
  stony, 22, 39, 40, 46
  striking velocities of, 6, 7, 9, 12, 15, 18, 25, 39, 154–84
  travel in groups, 21, 22, 25
  Widmanstätten figures, 28
Meteoritic composition, 15, 303
  C^{14} content, 19, 29
Meteoritic-impact craters
  ages of lunar craters, 300–309, 317, 321, 421, 422
  central peak craterlets, 96, 101, 114, 367–70, 425, 426
  Class 1, 115, 153, 157, 163, 175, 188–91, 193, 194, 297, 350, 362, 364, 365, 367, 369, 387, 420–22, 431
  Class 2, 115, 189–91, 193, 363, 365, 367, 420–22
  Class 3, 115, 189–91, 193, 363, 365, 367, 368, 420, 421
  Class 4, 115, 189–91, 193, 364, 365, 367, 368, 420, 421
  Class 5, 115, 189–91, 193, 364, 365, 367, 420
  compression effects, 119–24, 140–51, 154
  distribution of, on moon, 188, 189, 294, 296
    compared with shotgun patterns, 189
  double craters, 27, 31, 59, 79, 189
  effect of moon's curvature on crater form, 168, 169
  energies required, 28, 37, 39–44, 70, 109, 110, 124, 153–84, 314
  energy needed to produce Altai Scarp, 317
  mechanics of, 6, 8, 9, 12, 14–17, 20, 21, 23–26, 28, 34–36, 39, 40, 43–45, 67–71, 75, 76, 92, 108, 114–16, 123, 124, 131–51, 153–84, 304, 305, 314, 322, 331, 354
  penetration into ground, 139, 140, 154, 166, 167, 170, 171, 174, 175, 179–83
  secondary lunar craters, 111
  seismic soundings, 61, 80, 82, 83
Meteoritic origin of lunar features, 4, 8, 9, 113, 114, 115
Moon
  accretion of, 294, 395–97, 420, 427
  albedo
    of lunar features, 252, 253, 255
    of moon, 251–53, 262
    variations
      in brightness with phase angle, 248, 249, 252–55
      of rays, 255, 256
  angular momentum of revolution, 198–200, 209
  atmosphere of, 3, 207, 333, 341–48, 425
  bulge; see Moon, shape
  capture of, 294, 303
  changes on, 415–18
    parameters for Classes 2–5 craters, 189–91
  color on, 256–60, 423
  convection in, 243, 244, 423, 427, 431
  crust, thickness, 242
  degassing of, 302, 333
  density, variations of, 2, 201, 202, 204
  differentiation, chemical, 3, 393, 399, 400, 404, 406–9, 428
  distance, 2, 199–201, 203, 205–7, 209, 210, 239, 241, 422, 423
  eclipses of, 260, 261, 271–76
  erosion on, 297, 298, 333, 421
  expansion of, 377, 429
  grid system, 372, 385–88, 409, 426, 427
  history of, 198–200, 203, 204, 206–10
    age of moon, 293, 295
    Chamberlain-Moulton hypothesis, 303
    future of earth-moon system, 200
    models by
      Baldwin, 295, 305–10, 410–31
      Gilvarry, 302, 303
      Gold, 273, 274, 281, 294–97, 298–302, 309, 333, 424
      Kuiper, 295, 310, 311
      Urey, 294, 295, 303–10
        duration of intense bombardment, 303, 421
  inclination of axis, 201, 204–6, 209
  lava sheets, origin of, 2, 335–39, 409, 422, 424, 426, 428–31
  libration of, 208, 212, 221, 223–26, 228, 235, 237
  magnetic field, 345, 402, 412–14
  moments of inertia, 202–4
  month, length of, 198, 208, 209
  motion of node, 204, 205
  polarization
    compared with terrestrial materials, 250
    of craters, 250
    of maria, 249, 250
    negative, 250, 251
    of possible lunar materials, 250, 251, 256, 259, 260, 281, 303, 406, 407, 430, 431
  small clefts, 253
  uplands, 249, 250

Moon—*Continued*
  polarization—*Continued*
    variations
      with color, 250
      with phase angle, 249
  rigidity, 3, 194, 203, 210
  rotation, 201, 203, 207, 208
    Cassini's laws, 201
  seismic activity, 302, 303, 337, 426
  shape, 1, 2, 201–10, 212–14, 233, 238–41, 304, 331, 395, 409, 422, 423
    equipotential surface, 201–3, 206, 207, 243, 423
    on maria, 214, 215, 240, 241
    on uplands, 214, 215, 238–41
  slopes on, 248, 423, 424
  surface materials, nature of, 248–82, 336, 423
    bare rock, 250, 252, 423
    dust, 3, 249–53, 260, 262, 269, 270, 272–74, 276, 277, 281, 282, 297–302, 333, 382, 423, 424
    maria, 3, 249, 250, 252, 253, 256, 301–3, 333, 338, 339, 407, 424
    radar reflections from, 278–82, 424
    shattered rock, 248–50, 252, 276
    uplands, 2, 249, 253, 255, 256, 407
  temperatures, internal
    heat balance, 278, 294, 304, 309, 310, 395–409, 427
    history of, 206–9, 379, 393, 395–409, 418, 421–23, 426–29, 431
    sources of heat, 396–99, 427
  temperatures, surface
    affected by solar corpuscular radiation, 278, 281, 282
    black-body, 269
    measures of, 262, 268–82
      by infrared, 268–70, 274, 278
      by radio waves, 270–78
    variations of
      during eclipse, 271–76
      over lunation, 269–71, 275, 276
      over surface, 269, 270, 272, 276–78
      with wavelength, 270–78
  viscosity of, 423, 428
Mountain peaks, lunar
  Archimedes ζ, 322
  β, 322
  Caroline Herschel ζ, 322
  La Hire, 322
  La Hire a, 322
  Lambert γ, 322
  Mt. Argaeus, 320
  Mt. Huyghens, 322
  Pico, 322
  Piton, 306
  Pt. Acherusia, 320
  Spitzbergen, 322
Mountain ranges, lunar
  Alps, 321, 322
  Altai, 24, 195, 232, 315–18
    energy required to produce, 317
  Apennines, 239, 320–22, 377, 378, 424
  Carpathians, 321, 322
  Caucasus, 320–23
  D'Alembert, 237
  Doerfel, 237
  Haemus, 306, 319, 320, 323, 387
  Harbinger, 321
  Jura, 321, 331
  Pyrenees, 314
  Rook, 237
  Straight, 321
  Taurus, 320
  Teneriffe, 321

Origin of
  Alphonsus Ridge, 324
  Alpine Valley, 323, 325–30
  chain craters, 317, 324
  circular maria, 3, 305, 315
  dark areas, 3, 301, 303, 305, 424, 429–31
  domes, 392, 393
  lava sheets
    on earth, 334, 335, 337, 338, 429–31
    on moon, 335–39, 406, 424, 426, 428–31
  lunar craters, 114, 142, 143, 170, 378, 379
  lunar features
    meteoritic, 113, 114, 170, 177
    volcanic, 114, 378, 379
  Mare Imbrium, 321, 322, 380
  Mare Imbrium valleys, 322–31
  Mare Nectaris valleys, 317, 318, 386, 387
  moon
    accretion, 294
    capture, 294, 303
    Chamberlain-Moulton hypothesis, 303
  planetesimals, 294
  rays, 425
  Rheita Valley; *see* Origin of Mare Nectaris valleys
  rilles, 372–77, 379, 426
  Sinus Iridum, 380
  tektites
    lunar, 287–91
    terrestrial, 288–90
  wrinkle ridges, 380, 381

Planets
  earth, 249
  Mars, 249
  Mercury, 249, 250
  Saturn, 200
  Venus, 67, 249, 280
    radar reflection of, 280
Polygonal craters, 386, 427
Positions of lunar formations, 220–25, 450–63
Precession of equinox, 197

Radio-wave transmission by ground, 96, 97
Ray craters, 298–300
  diameters of, 354, 356
Ray systems, diameters of, 254, 356
Rays, 350–58, 425
  of Aristarchus, 261, 351
  circumlunar, 357
  of Copernicus, 258, 259, 298, 351, 353, 355
  craters on, 353, 354, 358
  effect on limb brightness, 357
  of Furnerius, 351
  of Kepler, 258, 259, 351–53
  of Langrenus, 350
  materials of, 253
  of Olbers, 351–53

Index

of Seleucus, 351
of Stevinus, 352
of Theophilus, 301, 351
theories of, 351–58, 425
of Tycho, 255, 257, 259, 292, 351–55, 357, 358
variations in brightness, 255, 256
Roche's limit, 198, 200, 209
Rock flour, 8, 15, 19, 20, 25, 27, 31, 69, 80, 89, 96, 98
Rilles, 3, 301, 324, 331, 336, 371–79, 382, 385, 387, 388, 426
of Ariadaeus, 320, 372–77
of Cauchy, 383
craters on; see Chain craters
dimensions of, 371
distribution of, 376, 426
faults, 372–76, 382, 383, 385, 386, 426
of Hippalus, 372
of Hyginus, 339, 372–75, 378
lunar grid system, 372, 385–88, 409, 426, 427
origin of, 372–77, 379, 426
of Sirsalis, 375

Salt dome, 100
Satellites, man-made, 244, 245, 299
Scaled depth of burst, 116–20, 125, 128–51, 153–84
Schröter's rule, 147, 148
Seismic waves, 198
Shatter cones, 18, 52, 57, 74, 75, 77, 79, 89, 91–94, 96–99, 101, 102, 108
Shelf areas at circular maria, 318
    Mare Crisium, 318, 319
    Mare Humorum, 319
    Mare Imbrium, 321
    Mare Nectaris, 315
    Mare Serenitatis, 320
Shields, 66
Shock waves, including ring anticlines, 6, 9, 12, 15, 16, 18, 25, 38, 41–43, 67, 69–71, 74, 75, 76, 94, 96, 107, 120, 122–25, 153–84, 317, 320–22, 331, 332, 356, 357, 375, 376, 379, 425
Silica glass, 16, 18, 25, 27, 31, 34, 35, 58, 68, 290
Spheroids, meteoritic metallic, 12–15, 34, 173, 290
Stishovite, 8, 18, 74
Suevite, 57, 78, 79

Tektites, 254, 287–91, 303
ages of, 288–91
composition of, 287–91
Cyrillids, 288
Darwin glass, 289, 290
Libyan glass, 290
origin of
    lunar, 287–91
    terrestrial, 288–90
radioactivity in, 287, 290, 291
Temperature
of earth, 399, 401, 402
of moon; see Moon, temperature
Terrestrial meteoritic craters
aerial photographs of, 79
cores, diamond-drill, 80, 83–86, 91, 92, 95, 96, 107
gravity anomalies, 80, 81, 83, 85–88, 94
heat effects, 36, 37, 42, 43, 45, 68–70, 86, 96, 173
seismic soundings, 61, 80, 82, 83

Terrestrial meteoritic craters, cryptovolcanic structures, and suspected meteoritic craters
Aouelloul Crater, 45, 46, 62, 63
Argentine craters, 25
Arizona Crater, viii, 5, 8, 10–18, 52, 96, 102, 108, 109, 119, 123, 125, 144, 146, 148, 149, 169, 170, 173, 178, 179, 331, 386, 442
Ashanti Crater, 56–59, 288
Austin Basin, 90, 91
Boxhole Crater, 29
Brent Crater, 76, 82–86, 88, 95, 106, 109, 137, 149, 179
age of, 83
Carswell Lake, 59, 106
Cave Spring Hollow Basin, 90, 91
Clearwater Lakes, 59, 106
Cratère de Talemzane, 46, 47, 52, 66
age of, 46, 47, 52
Cratère de Temimichat Gallaman, 62, 63
Cratère de Tennoumer, 62, 63
Crooked Creek structure, 74, 96, 100–102, 107, 370
age of, 102
Dalgaranga Crater, 29
Decaturville structure, 73, 95, 107
age of, 95
Deep Bay Crater, 86–88, 106, 108, 149, 179
age of, 88
Des Plaines structure, 100
Flynn Creek structure, 88, 89, 91, 107
age of, 88
Franktown Crater, 58, 61, 106
French craters
age of, 53
Faugères, 53, 139, 442
Le Clot de Cabrerolles, 53, 139, 442
Gulf of St. Lawrence arc, 61
Haviland Crater, 24, 106
Henbury craters, 25–29, 139, 170, 180, 183, 290, 358, 442
age of, 28, 29
Holleford Crater, 18, 79–83, 88, 106, 109, 137, 149, 179
age of, 82
Howell structure, 98, 107
Indian Mound Basin, 90, 91
Jeptha Knob, 98, 99, 107
age of, 99
Kaali Järv, 30–33, 76, 108, 146, 148, 149, 167, 170, 182, 183, 365, 442
Kaali Järv central peak, 31, 76, 108
Keeley Lake, 61
Kentland disturbance, 74, 75, 96–98
age of, 98
Kilmichael structure, 100, 107
Köfels "crater," 58
Labrador Crater, 58, 106
Lac Couture, 61
Lake Michikamau, 61
Lonar Lake, 63
Manicouagan Lake feature, 59
Mecatina Crater, 61, 106
Menihek lakes, 61
Murgab craters; see Pamir craters
Nastapoka Islands arc, 60, 61
age of, 60

Terrestrial meteoritic craters, cryptovolcanic structures, and suspected meteoritic craters—*Continued*
  New Quebec Crater, 22–24, 52, 82, 106, 108, 109, 146, 170, 178, 179, 442
    age of, 24
  Odessa craters, 18–22, 25, 56, 106, 108, 123–25, 148–51, 170, 171, 179, 181–83, 442
    age of, 18, 19
  Oesel craters; *see* Kaali Järv
  Pamir craters, 7, 44, 68
    Chaglgan Toushtou No. 1, 45, 170, 442
  Pretoria Salt Pan, 53–56, 57
  Rieskessel, 57, 75–79, 91, 107, 288, 354
    age of, 79
  Sault au Cochons, 61
  Serpent Mound, 99, 107, 362
    age of, 99
  Siberian Fall of 1908, 7, 35–43
    air burst of, 40
    air waves from, 36–40, 42, 43
    as comet, 43
    ground waves from, 36, 37, 42, 43
    heat effects, 37
    optical phenomena, 37, 38
    trees felled by, 37, 40
  Siberian Fall of 1947 (Sikhote Alin), 7, 25, 43, 44, 170, 182, 442
    as asteroid, 44
  Sierra Madera Dome, 92–94, 107, 151
    age of, 93
  Steinheim Basin, 57, 75–77, 79, 91, 99, 107
    age of, 76
    Klosterberg, 76, 77
  Ungava Bay, 61
  unnamed
    co-ordinates of, 63
    reported craters, 63, 64
  Upheaval Dome, 100, 107
    age of, 100
  Vredefort structure, 75, 93–95, 111, 352, 375
    age of, 95
  Wabar craters, 34, 68, 290, 305, 442
  Wells Creek Basin structure, 89–93, 98, 107, 109, 150
    age of, 89
  West Hawk Lake, 61
  Wolf Creek Crater, 29, 30, 52, 170 442
    age of, 30
Tides
  bodily, 198, 206–8
    phase lags, 198
  dissipation of rotational energy, 198, 199
  friction by, in oceans, 198–201, 206–10

Valley system near Janssen, 387
Valleys, radial
  origin of, 317, 318, 323, 325–30, 385, 387
  to Mare Crisium, 319
  to Mare Humorum, 319
  to Mare Imbrium, 322–25, 328–31, 339, 372–75, 385–88, 426
    Alpine Valley, 322, 323, 330, 386
  to Mare Nectaris, 316–19, 330, 331, 385–87
    Rheita Valley, 27, 316–19, 330, 331, 386, 387
    origin of, 317, 318, 386, 387
Viscosity
  of earth's asthenosphere, 195
  of lavas, 338, 430
  of moon, 423, 428
Volcanoes
  crater cones, 390
  internal-origin lunar craters, 339, 376, 377
  phreatic eruptions of, 78
  shield, 390–92

Wood's region, 257, 261
Wrinkle ridges, 336, 379–83, 385, 426
  dimensions of, 379, 380
  distribution of, 379, 380
  origin of, 380, 381, 392, 393

X-rays, 192

2
212-13